AF795621

Global Geostrategic
and
Politico-Military Perspectives
Through Millennia Past

Volume One

Global Geostrategic and Politico-Military Perspectives Through Millennia Past

Volume One

Major C.B. Verma (Retd.)

PENTAGON PRESS

GLOBAL GEOSTRATEGIC AND POLITICO-MILITARY PERSPECTIVES THROUGH MILLENNIA PAST

Major C.B. Verma (Retd.)

First Published in 2018

ISBN 978-93-86618-20-7 (Set)

Published by

PENTAGON PRESS
206, Peacock Lane, Shahpur Jat
New Delhi-110049
Phones: 011-64706243, 26491568
Telefax: 011-26490600
email: rajan@pentagonpress.in
website: www.pentagonpress.in

Printed at Aegean Offset Printers, Greater Noida, U.P.

In the Lotus Feat of His Divine Holiness
Shri Sai Baba of Shirdi (India)

Contents

VOLUME ONE

VOLUME TWO

Foreword

Study of the history of human civilisation, with all its explorative, investigative, and deductive challenges, cannot be but a rather complex exercise. It is a challenge that has enthused the interest of great scholars, explorers, scientists and philosophers down the ages for them to devote their lifetime to unearth, and more importantly, infer and record the story of mankind. The reward for that excursion comes by the way of the joys of discovery and spreading of knowledge among the naturally inquisitive humans. Indeed, the saga of evolution of the human society, politics, faiths and philosophies, with all its ups and downs, helps the contemporary generation in understanding the past and derive lessons from it to its future advantages. Obviously, therefore, while compilation of the story of civilisation is an intellectually as well as physically back-breaking endeavour, study of that compilation must be a very invigorating and satisfying experience for the readers.

Only a scholar of Major C B Verma's intellect, commitment and perseverance could plunge into a project so extensive, and complete it with the publication of his seminal work under the title of '**Global Geostrategic and Politico-Military Perspectives Through Millennia Past**' in two volumes of knowledge reservoir. It has taken the noted scholar, soldier, entrepreneur and socio-political observer over six decades to organise his time to do so. Admittedly, the facts discussed in these volumes are known and recorded, but the salience of this unique publication lies, firstly, in its analyses of events, and secondly, in presenting to the reader a storehouse of composite, multi-discipline and duly-compiled information in just two volumes. It is hoped that the hard-to-find features of this book would find the best utilisation from the academic as well as the knowledgeable world.

Volume One of the book starts with a projection of the ancient era, around 4500 BCE,and proceeds to cover the period up to the mediaeval age. It goes on to describe the phenomenon of 'birth of empires' wherein the Egyptian, Mesopotamian, Chinese, Indus Valley, Sematic and Aryan, Greek, Roman and European empires and civilisations, as the case be, are delved into. Notably, since the mankind's military history is closely intertwined with civilisational developments, at this stage the author has offered a rare insight into the military aspects that have shaped the world. The Volume, composed in six Chapters, ends with an exposition on the phenomenon of rise of the Western Empires and the advent of democracy.

Volume Two begins with the advent of what may be termed as the Modern period. It describes and analyses the courses and effects of the Industrial Revolution of the 17th and 18th Century before delving into the two landmark events of the 20th Century, namely the Great War or World War I (1914-18) and World War II (1939-45), which have shaped the world as it is seen today.

Appropriately, the geo-political, socio-cultural and politico-economic developments of that period is also discussed in continuation, thus building a holistic picture for the readers to look at. At this stage, a well-informed exposition on the role of modern facets of intelligence and politico-strategic application of Special Forces follow. Obviously, the author has highlighted these two modern-day attributes in the context of their salience as tools of geo-political power-play.

The later parts of Volume Two of the book are devoted to the examination of contemporary – described as 'super age' – developments in science and technology, and its assimilative effects on the society, particularly upon the Indian environment. Thus the assimilation of scientific inventions, computers and automated information processing, information technology, space sciences, etc. into the human societal construct has been delved into. Having described these landmark developments shaping up the modern era, the book shifts into an analytical role in interpretation and prognosis. In the true context of the modern times being conditioned by the overall effects of such multi-directional developments, this line of articulation is appropriate and timely. Accordingly, this crucial role is performed by touching upon the key nuances of nuclear, biological and chemical warfare, the scourge of terrorism and the 'great game' of international power-politics. Finally, the discussion is wound up with some thoughts over India's political, military and leadership attributes and challenges, with few hints for undertaking of reforms in line with the calls of the day.

A study so extensive and a work crafted over a long period of six decades requires many explanations and updates. The author has met that requirement through his 'Conclusions' and followed it up with a 'Post-script'. That indeed is an appropriate step to keep the readership in focus.

If it is a rewarding experience, the study of the history of human civilisation is also a serious business that requires quest and perseverance. One is sanguine that this painstaking contribution from Major C B Verma, and the missionary zeal with which the noted publisher Shri Rajan Arya has given shape to it, would raise readership interests, and its volumes would find a place of pride among the works of similar salience.

Jai Hind,

Lt Gen Gautam Banerjee, PVSM, AVSM, YSM (Retd)
Member Advisory Board and Distinguished Fellow
The Vivekananda International Foundation
New Delhi-110003

November 2017

Acknowledgement

I will always be gratefully indebted to Late Lt Gen. S K Sinha, PVSM, AVSM (Retd), former Governor of Jammu & Kashmir and Assam, for his most valuable guidance and inspiration, and for all the time he spared for me at his home, whether in Patna or New Delhi. He was indeed a 'living legend'. In fact, he had very kindly offered to write a 'Foreword' for this book and had taken the pains of meticulously going through the manuscript. Unfortunately, he was unable to do so because of his sudden demise.

I am also sincerely grateful to Lt Gen Gautam Banerjee, PVSM, AVSM, YSM (Retd) for his lifelong brotherly friendship, and for being a close family friend for many decades. I must say with heartfelt gratitude that this book has finally seen the light of the day due to his singular considerations and efforts. I do not have words to express my gratitude to him.

I remain indebted to Late Professor S Prasad, former Professor and Head of Department, Mechanical Engineering, College of Military Engineering, Pune, and Founder-Director of Kamla Nehru Institute of Technology, Sultanpur (UP). He had very kindly shared his insights about the fundamental aspects of modern technology with me.

Similarly, I must acknowledge the affection and invaluable guidance on many aspects that I was lucky to receive from Swami Madhusudan, Professor of Yoga at the Lal Bahadur Shastri National Academy of Administration, Mussoorie, (Dehradun), and the founder of the "Adhyatyam Yoga Mission" in Haridwar.

The United Services Institution (USI) Library, New Delhi, deserves a special mention for its valuable collections, which I could access being a life member for more than 50 years.

Last but not the least, I most sincerely wish to express my special thanks to Mr. Rajan Arya, Founder & CEO of Pentagon Press, New Delhi, for bearing with me and having my manuscript diligently edited.

Major C B Verma (Retd)

CHAPTER 1

Introduction

At the outset itself, even thinking about a desirable 'conceptual-framework' to commence on a preliminary critical 'journey' with strategic insight through humanity's evolutionary as well as revolutionary long history over an unending 'canvas of time'—specially keeping in mind the fascinating 'global geostrategic and politico-military perspectives', can indeed be a daunting undertaking. In any case, this 'journey' would perforce be a highly complex, multi-directional and multi-layered range of enriching studies and research, both in terms of its historic as well as futuristic parameters.

In this 'journey', failures would entirely be those of the author and at the same time, few worthwhile readable paragraphs/pages, if any, coming-up in this book, would totally be pleasantly "accidental fallouts", owing to my reading of the notably scholarly books, authored by various learned and famous Indian and foreign authors on allied subjects and also, many scholarly writings/ articles. In order to make this book, somewhat readable, the author has tried to liberally quote certain highly knowledgeable lines/paragraphs from the books and other such writings of the many scholars and intellectual giants.

History, in its core values, is tremendously valuable with enriching records in a methodically chronological order, for posterity, commanding multiple utilities. As time moves on unendingly, human civilisations too, keep on marching ahead as one of the mighty Mother Nature's almost countless and continuously altering phenomenon.

Now, coming back to our own adventurous 'exercise', since the appearance of Homo Sapiens on planet Earth and subsequently, gradual and somewhat slowly progressive dawn of various civilisations, mankind has been having a natural urge to move from relatively hostile environmental situations to more conducive and agreeable ones. In this respect, even in this 21st Century, it remains highly debatable as to the degree of legitimacy as well as authenticity of 'claim' that, how far and in what manner, mankind can truly be called even reasonably 'civilised.

It is already established that the mankind's insatiable urge to keep progressing in so many multiple ways in fact, has started with the 'dawn of civilisations' in different parts and in different geographical locations in our world and also, in different periods of time. Now, if one is permitted, then with the risk of abruptly 'jumping to a so-called meta-physical plane', one can say with reasonable legitimacy, that an inherent tendency in man to violently or otherwise 'clash' with fellow human-beings, both individually and/or in 'groups', has always been present since 'Homo Sapiens' appearance on this planet. Further, relevantly coming to a 'spiritual-level, it is well-known that since the earliest times, many saints belonging to many religions, have said that both the 'Demons' and the 'Angels' perpetually 'reside' inside every human-being and at the same time as per nature's 'dictates', demon is more aggressive and thereby constantly in the lookout for overpowering the angel. It decisively subjugates him/her to 'reign supreme' over that person(s), both individually or collectively, as per the various factorial-cum-emerging situations over a passage of time.

Well, these may be termed just as 'matters of conjecture' but, the 'points' being conveyed are definitely relevant. Logically, as the mankind emerged and became progressive and also 'dynamic', both in physical mobility as well as that of thought-processes', gradual appearance of a number of ancient civilisations started taking place in different and often far-away geographical locations, over a passage of many centuries. Consequently, as man's different 'innovative-skills' sharpened along with the urge to develop more and more with slowly enhancing mobility, different ancient civilisations started having 'contacts', resulting in both cooperative interactions as well as 'conflicts' too. It is such 'friction' or even conflicts of ideas and different interests that **Late Professor Samuel P. Huntington** of the Harvard University has perceptively termed as 'Clash of Civilisations'.

Professor Huntington says that, "Human history is the history of civilisations. It is impossible to think of the development of humanity in any other terms.... The idea of civilisation was developed by eighteenth century French thinkers as the opposite of the concept of 'barbarism'. Civilised society differed from primitive society because it was settled, urban and literate. To be civilised was good, to be uncivilised was bad... Civilisation and culture both refer to the overall way of life of a people, and a civilisation is a culture writ large. They both involve the 'values, norms, institutions, and mode of thinking to which successive generations in a given society have attached primary importance. A civilisation is, for **Braudel**, 'a space, a cultural area, a collection of cultural characteristics and phenomena.' **Wallerstein** defines it as 'a particular concatenation of worldview, customs, structures and culture (both material culture and high culture), which forms some kind of historical whole and which coexists (if not always simultaneously), with other varieties of this phenomenon.' A civilisation is, according to **Dawson**, the product of 'a particular original process of cultural creativity which is the work of a particular people', while for **Durkheim** and **Mauss**, it is 'a kind of moral milieu encompassing a certain number of nations, each national culture being only a particular form of the whole.' To **Spengler** a civilisation is 'the inevitable destiny of the Culture....the most external and artificial states of which a species of developed humanity is capable....a conclusion, the 'thing-become' succeeding the 'thing-becoming.' Culture is the common theme in virtually every definition of civilisation." (Huntington, 1977, pp. 40-41)

In India, and perhaps even in many other countries (largely excluding those from the Western world), there are several thinkers, authors and scholars, who are not fully agreeable to the perceptions of Huntington. This is quite natural and also logical, since they have weighty points towards their own perceptions. These views lead to analytically enriching reviews as well as critical studies of Huntington's book. In fact, these aspects are equally true as well as highly desirable of the academic and also of the professional spheres, in the areas of intellectual pursuits, in both inter-disciplinary and intra-disciplinary manners.

Coming back to gradual evolution of the mankind on this planet, it has been scientifically established that right at inception stage, of the 'Homo Sapiens', the mankind was gifted by an **"analytically thinking-cum-creative mind"**, though quite a primitive one in the beginning. And consequently, in the beginning itself, this singularly unique 'asset' gifted by the Nature, made all the differences and put the mankind in an unassailably advantageous position, with respect to the all other forms of life and species on this planet. But this is also true that this unique asset, available to no other species, was not able to make the mankind totally safe and insulated from any form of lethal or otherwise dangers and challenges. Threats of multiple challenges were always there, for the man on this earth and they are still there, though largely 'altering', in varieties, range and complexities. This continuing phenomenon has always been present, right from the 'beginning' of the process of Nature's own multi-dimensional and multi-layered unending creativity. It is also fascinating, both scientifically and "all inclusive spiritual-cum-religious" perceptions of the continuing processes of evolution.

This brings us to the fundamental meaning and the whole gamut of this process of evolution, under the absolute command of the all-powerful 'Creator', i.e. the 'Mother-Nature' on this planet. George Gaylord Simpson says, "The historian of life takes not only knowledge of fossils but also a tremendous array of pertinent facts from other fields of earth sciences and of life sciences and weaves them all into an integral interpretation of what the world of life is like and how it came to be so. Finally, he is bound to reflect still more deeply and face the riddles of the meaning and nature of life and of man as well as problems of human values and conduct. The history of life certainly bears directly on all these riddles and problems, and realisation of its own value demands investigation of this bearing". Further (page 9) he says that the meaning of human life and the destiny of man cannot be separable from the meaning and destiny of life in general. 'What is man?' is a special case of 'What is life?' Probably the human species is not intelligent enough to either question fully, but even such glimmerings as are within our powers must be precious to us. The extents to which we can hope to understand ourselves and to plan our future depends in some measure on our ability to read the riddles of the past. The present, for all its importance to dwell in it, is only a random point in the long flow of time. Terrestrial life is one and continuous in space and time. Any true comprehension of it requires the attempt to view it whole and not in artificial limits of any place or epoch. The processes of life can be adequately displayed only in the course of life throughout the long ages of its existence. The spatial scope of life as we know it is clearly defined as the planet Earth. On page 14, this author says that the **origin of life was necessarily the beginning of organic evolution and it is among the greatest of all evolutionary problems**. Then on page 281 of the same book, he says that **the first grand lesson learnt from evolution was that, of the unity of life.** One of the greatest achievements of early Christianity and some of the other religions was recognition, in principle, of the brotherhood of man. While refuting certain other intuitive conclusions of various Christian and other theologians, confirmation of the truth of evolution established this doctrine as a scientific fact." (Simpson, 1965, p. 3)

In fact, this entire concept pertains to the gradual evolution of all forms of life, including the Homo Sapiens since ages, on this planet Earth only, and these do have their own weighty scientific rationale and logic. But the larger scientific fact is that the Earth itself is not at all self-sustaining in 'isolation', and it sustains itself as part of the solar system. Further, with all its unique attributes, the gradual evolution of life in all its 'endless' forms could not have taken place in the first place, without the magnificent sunrays continuously falling on the Earth, along with so many unknown cosmic rays from distant outer space. Therefore, it can naturally be perceived that anything and everything in the Universe is inter-related, with their continuous 'birth' as well as 'destruction'.

A brief illustration of all these would be seen in the succeeding paragraphs of this chapter. In this respect, the so called "destruction", as understood by the people in conventional sense, is in fact notional only. Even looking at this phenomenon scientifically, the established reality is that in the whole limitless Universe and its entire unendingly continuous processes of both the so-called "creation" and "destruction" in fact, nothing gets ever created or destroyed. But in reality, there are unending phenomena of multi-dimensional, multi-layered and multi-directional simultaneous happenings of unconceivable as well as highly complex processes of "changing transformations" only. Logically now, these perceptions are inducing us towards scientifically 'conceptualizing' the highly delicate and yet assuredly stable and unalterable "Grand Balance" within the infinite Universe, without its beginning ever or any kind of impending end. And therefore, it can be said with a reasonably acceptable scientific basis that it is impossible to 'understand' and 'measure' the unfathomably complex mysteries of the infinite Universe, in all its magnificence and continuously altering manifestations.

The brief dwelling upon the theory of evolution of life on our planet in an all-inclusive manner, both in scientific as well as in multi-religious perspectives, the universal brotherhood of mankind gets irrefutably established. In addition to this, we also authentically find that indeed, the man is related to all forms of life on planet Earth.

Further, at this juncture, it is so humbling and also fascinating to remember modern scientific base of the great and magnificently spiritual thoughts of an internationally towering personality. **Swami Vivekananda**, and his unique concept of "Non-Duality", as initiated by his famous and benevolent master, **Shri Ramkrishna Paramhansa**. In fact, Swami Vivekananda, in his enlightening teachings, both scientifically and spiritually, took us even beyond the theory of evolution, universal brotherhood of man and his established relationship with all forms of life on not only our planet Earth but, with respect to those of the infinite Universe. That indeed is encompassing infinitely large dimensions in multiple scientific, religion-cum-spiritual and allied areas of this study.

While referring to some of the writings/teachings of Swami Vivekananda, we learn, "The Vedas teach us that creation is without beginning or end. Science is said to have proved that the sum total of cosmic energy is always the same. Then, if there was a time when nothing existed, where was all this manifested energy? Some say it was in potential form in God. In that case God is sometimes potential and sometimes kinetic which would make Him mutable. Everything mutable is a compound, and every compound must undergo that change which is called destruction. So God would die—which is absurd. Therefore there never was a time when there was no creation. If I may be allowed to use a simile, creation and Creator are two lines, without beginning and without end, running parallel to each other. God is ever Active Providence, by whose power systems after systems are being evolved out of chaos, made to run for a time, and again destroyed (pages 43 and 44).... The next idea that I want to bring to you that religion does not consist of doctrines or dogmas.... Books never make religions, but religions make books. We must not forget that. No book ever created God, but God inspired all the great books. And no book ever created a soul. We must never forget that. The end of all religions is the realisation of God in the soul. That is the one universal religion. If there is one universal truth in all religions, I place it here—in realizing God.... There is that beyond all books, beyond all creeds, beyond the vanity of this world, and it is the realisation of God within you."

In the context of this study-based 'critical journey' the logical point emerging here is that after the long-drawn process of the evolution of life on Earth, gradual fruition of appearance of the mankind here, again over millions of years, began to take place. Slowly from sub-human stage, mankind emerged to the primitive-man stage. It is also known that the 'primitive man' too, existed here for very many centuries and simultaneously continued to slowly come in sort of progressive 'shapes', within the grand framework of continuing and gradually evolutionary processes of the Nature. As mentioned earlier, as a distinct specie of life man has started coming to a particular form, enabling him to be known much later as, 'Homo Sapiens'. **Exclusively the mankind alone was gifted with a unique asset, i.e. an "analytically thinking, imaginative and innovative brain".** At the same time, this is also true that during the times of primitive man the human-brain too, was primitive and grossly underdeveloped.

In fact, what made the totally unmatched difference for the mankind has been the existence of this 'brain', which too started developing as he commenced his gradual and steady progressive march as per overall on the command of Nature over the infinite 'canvas of time'. Consequently as part of Nature's evolutionary processes, in course of centuries and with the gradual fruition of the birth of initial human civilisations in different geographical locations began unique developments for the mankind. Many kinds of developments were triggered by man, his thoughts and his multiple innovations, and as also driven by his pro-active as well as talented 'brain'.

Thus began the historically gradual advent of a number of human civilisations in different geographical areas/locations on our planet. It was a bumpy ride with violent periods spanning over many centuries. These aspects too have been substantially challenging as well as fascinating for many thinkers and scholars since ancient times, in their relentless pursuits of knowledge into the man's highly complex and checkered past spanning millenniums. A related and brief discussion will follow in the subsequent chapters. Now coming to a naturally allied aspect in relation to the birth of different human civilisations in different 'continents', scholars and eminent authors have indeed found that this appropriately so called allied 'aspect' is no less complex, challenging and also informative as well.

Logically this aspect has been that of the simultaneously evolving **"profession of Arms"**, with all its gradual manifestations. This naturally encompasses aspects of the arts, science and multiple crafts of soldiering, towards the assured defense of one's own civilisation's progress, achievements and both material and cultural assets, in their all multi-dimensional, multi-directional as well as multi-level connotations. An interesting and unique evolutionary consequence of all these has been subsequently a gradual development of the multiple military aspects, i.e. **the vast and highly complex "matters-military"**; *(interestingly, a terminology so scholarly often used by the Indian Army's one 'living-legend';* ***Lt. Gen. S.K. Sinha (Retd)****, and India's former Vice Chief of Army Staff)*, and also coupled with those of the military history as well. Obviously these too, though parallel and interrelated to the progressive advent of different human civilisations over many centuries, are by themselves vast subjects and no less challenging as well as knowledgeably fascinating and yet, they are, ironically a 'saga of violence' of the humanities' geostrategic and politico-military conflicts, battles and wars since manifested in man's grit, blood, pain, tears, and sacrifices.

A separate chapter gives brief glimpses and somewhat chronologically introductory presentation in some historic detail. It is further interesting as well as professionally challenging to note that the various weapon-systems, weapon-platforms and other many types of fast developing military technologies are getting more and more sophisticated and high-tech. All these gradual but consistently developing military technologies have naturally resulted in the multiple expansion of multi-dimensional complexities of the many aspects of military tactics and strategies, thereby tasks of formulating 'operational-plans' at different levels of military command and their mutual co-ordination, and professional tie-up among various 'arms and services'; consequently in fact, the entire gamut of warfare have also always seen a sea-change.

This magnificent and yet at times, even ruthlessly violent historic march of some ancient civilisations in different eras and in separate geographical areas in this world, did trigger the dawn of 'mediaeval-age' and building of empires. This era comparatively little more global in nature and reach, has lasted for a number of centuries. This prolonged period also saw the coming of little expanded sea-power as well. Initially in fact, the advent of primitive sea-power to start with, began during ancient times itself. There was certainly no question of even imagining then, any kind of ability for the man to be able to fly at any times during those such past era; though he was then also very captivatingly seeing the birds flying in the sky. In fact it is needless to say that the man, in reality, was able take to the sky only very recently that is, only during the beginning of the 20th Century.

In fact much earlier, the long era of medieval age as well as centuries of various empire-building began and these in turn, historically lasted for many centuries. Then much later around 300 to 350 years ago, a new kind of the highly revolutionary development in the lives of mankind began in the shape of **'Industrial Revolution'**. As a matter fact, this far-reaching beginning of a previously unforeseen development brought a new-age of the multiple changes in the areas of man's productivity, mobility, thinking-power, means of communications, and improvements in the overall quality of life and so on. There is again a separate chapter later, to deal in some brief details, with these aspects as well.

Much before beginning of the age of industrial revolution, relatively in the ancient times, an era of **'Agricultural Revolution'** had already started. This too, had brought multiple changes in the world as well as in the lives of mankind. Interestingly and quite knowledgeably, it has been said by a couple of scholars that the mankind since ancient times, can conveniently be compartmentalized in two relatively distinct 'worlds'. The first one can be known as 'primitive' or even as the 'aboriginals' and then, the one succeeding the first one as 'civilised. In modern context now, perhaps inadvertently or perhaps even remotely indicating that the aboriginals were in any sense 'uncivilised and thereby, conveying that they happen to be the same today would be considered somewhat relatively "offensive", but that in any case here, has never been any purpose. In fact, as already mentioned that all such elaborations being given here, happen to be entirely in a historical sense, as contextually necessitated by the central 'theme' of the main subject-matter under discussion. These aspects would repeatedly be seen in subsequent chapters and hence they too, would be entirely in an academically historic sense only.

Alvin Toffler, "During the last millennia when First Wave civilisation reigned supreme, the planet's population could have been divided into two categories—the 'primitive' and the 'civilised. The so-called primitive peoples, living in small bands and tribes and subsisting by gathering, hunting or fishing, were those who had been passed over by the agricultural revolution. The civilised world by contrast, was precisely that part of the planet on which most people worked on the soil. For wherever agriculture arose, civilisation took root. From China and India to Benin and Mexico, in Greece and Rome, civilisations rose and fell, fought and fused in endless, colorful admixture. However, beneath their differences lay fundamental similarities. In all of them, land was the basis of economy, life, culture, family structure, and politics. In all of them, life was organised around the village. In all of them, a simple division of labor prevailed and a few clearly defined castes and classes arose: nobility, a priesthood, warriors, helots, slaves or surfs. In all of them, power was rigidly authoritarian. In all of them, birth determined one's position in life. And in all of them, the economy was decentralized, so that each community produced most of its own necessities. During its dominance there were occasional hints of things to come. There were embryonic mass production factories in ancient Greece and Rome. Oil was drilled in one of the Greek islands in 400 B.C. and in Burma in A.D. 100. Vast bureaucracies flourished in Babylonia and Egypt. Great urban metropolises grew up in Asia and South America. There was money and exchange. Trade routes crisscrossed the deserts, oceans and mountains from Cathay to Calais. Corporations and incipient nations existed. There was even, in ancient Alexandria, a startling forerunner of the steam engine." (Toffler, 1981, pp. 37-38)

Further the modern-age including post-industrial-revolution era and its multiply far-reaching and multi-dimensional fallouts will be discussed. Further, these amazingly multiple advancements are continuously opening new frontiers of human-knowledge taking all these always to new heights, previously unforeseen.

Since the age of industrial-revolution systematic and fundamental scientific inventions, resulted in perfecting the design and development of **mechanical-power oriented engines**. A couple of their different types were in fact, based on **'steam-power'** and then later, **'oil-power'** based equipment and, in both, the required heat-energy was harnessed by burning, now so common 'fossil-fuels', i.e. coal and then 'petroleum' products respectively. Subsequently, through continuous processes of the multiply initial scientific and technological advancements, very far-reaching and revolutionary advent of the **electrical-energy** too took place. All these in turn, gradually resulted in the man's ability to simultaneously convert the heat energy into mechanical energy and then harnessing the same into the requisite electrical energy to be made available for further productive uses in various ways, in service of the man himself.

Later with the beginning of 20th century, a number of far-reaching events took place, some of them very violent and highly destructive too, which perforce brought-in major changes in the course of global history. In the beginning of this century, the era of different 'global' empires was at its zenith. Therefore, the principal European imperial powers, led by the British, were globally at their imperial glory; and then those major events started taking place one by one.

Far-reaching geostrategic interests of competing European powers started 'clashing' in the continental Europe and elsewhere, sparking violent clashes, resulting in the beginning of **First World War (1914-18)** and the same is also known as the Great War. Mankind had never seen such level and extent of deaths and destruction before with ruthless ferocity and indeed, so unprecedented. Further in continuation to this, the initial Aviation-technology started 'incubating' and this by itself, was indeed a revolutionary development with multiple fall-outs. These in turn, triggered rapid development of many scientific inventions, further resulting in various kinds of technological developments, with great potential of their uses in the service of the mankind, both for peaceful purposes or otherwise.

In fact 'dramatic' end of the First World War, as a consequence, contained the "seeds" of the Second World War (1939-45). In comparison to the First World War, the Second World War was extensively global and also much more violent as well as devastating. In fact, it was so destructive in both range and extent, that it did trigger permanently major and previously unforeseen turns in the overall course of history. Earlier generally considered the so called impregnable colonial empire, were delivered irreparable shock and crippling damages and consequently, they started gradually 'disintegrating' under their own weight. Then there was gradual grant of political freedom to a number of 'imperial' colonies, located in the vast continents of Asia, Africa and Latin America. In fact, this process began during the end of 1940s and continued to the beginning of 1970s or so. To the dismay of many people, this entire historic process did cause lot of turmoil and all sorts of socio-psychological, political as well as socio-economic upheavals in these highly complex centuries-old colonial regions in those continents.

One historically interesting point to observe here is that the then world's most dominant imperial power, the British-empire was forced to such bankruptcy mainly due to 'crushing' devastation of the Second World War. **The astonishing truth was that the Great Britain, for nearly latter half of the 20th century, did become a sort of "Basket-case" for the then already emerged mightiest superpower of the world, the USA**. This could definitely be considered as one of the biggest surprises of history and logically one may conclude that in any big and devastatingly violent conflict of such global reach, the end results could indeed be very tragic as well as historically far-reaching, and thereby bringing-in unprecedented changes in the subsequent global history of the mankind.

Coming to the post-second World-War era, we notice revolutionary developments, as well as unforeseen inventions/innovations in many areas of human endeavour and continuous thought-processes. These areas pertain to the scientific, technological, sociological, political, and also geostrategic spheres of the ongoing multi-dimensional progress of civilisations globally.

In this context, we see that efforts for a breakthrough in the sphere of nuclear-sciences were underway since pre-second World War times, to gain decisive global politico-military supremacy by the ruthlessly competing and the then imperial European powers as well as 'giant' USA with utmost secrecy and security. The Second World War perforce gave unprecedented boost to research and development in the unknown areas of atomic sciences research and development. Logically as a consequence then, the many-sided activities of espionage and counter-espionage too, followed with globally ruthless intensity and not at all refraining from all sorts of plans for even 'honey-trapping'; including criminal activities and violence. Well, it is historically interesting to notice that these multiple activities of espionage and counter-espionage have been continuously going-on in some forms or other, since the dawn of human civilisations. At the same time, we may remember that the natural and somewhat irresistible urge to perforce indulge in such activities have always been there in different forms since ancient times, as recorded by the continuing human history. It is also true that these areas of studies by themselves being so vast and multi-dimensional that they naturally constitute a distinctly different subject.

Now coming back to the subject of nuclear sciences, we find that during the last phases of Second World-War, USA had already perfected the making of initial but **highly devastating atomic bomb**. The US forces then dropped two of such bombs, one after another, over Japan to compel them to unconditionally surrender in the Pacific and thereby, bringing in the militarily 'successful' conclusion of the Second World-War for the Allied Powers in the year 1945. It has had such a devastating and socio-psychologically shattering experience for the entire humanity that even today, the global human societies have not been able to fully recover from the consequential shock, a sort of trauma and even revulsion.

At the same time, it is also an extremely sad fact that due to the unending global geo-politics and international power-games, relatively less 'sincere and dubious' scientific and technological research-work are continuously being carried-out in the areas of peaceful uses of this revolutionary nuclear-energy for the intended and multiple benefits of the mankind. Thereby, for the allied matters pertaining to conservation of overall ecology and protection of environment and 'regressively' these are also intended for more and more development of irrational as well as irretrievably self-destructive nuclear-weapons. Today mankind claims to be so civilised and going fast 'hi-tech' in an unprecedented manner, but ironically leading to their own downfall.

We further notice that revolutionary advancements have given amazing 'powers' in the control of mankind but have resulted in risks too, as a compulsive rider. It is a sad and gradual 'de-humanization' of the humanity as a whole and has resulted in ruthlessly multiple paradoxes. In this context, it can be affirmed that the highly enriching and immortally vibrant 'words of wisdom' as enunciated by a number of enlightened saints in almost every corner of this world ever since ancient times, have been almost 'wilfully' ignored by the mankind. Therefore logically a larger question arises as to where all these are leading us to?

*

CHAPTER 2

Gradual Human Evolution

Before coming to the fascinating subject of gradual human evolution, in the beginning, it would be more appropriate to briefly touch upon the timeless evolution of Mother Nature's phenomena of unfathomable continuing process of 'Creation' and thereby, slow and consistently steady evolution of life itself, particularly on our planet Earth, as known to us at present. It is already accepted that the Universe is infinite and at the same time, anything about its process of origin/birth is definitely not known in the sense that the entire process of its 'origin' can perhaps never be logically rationalized and scientifically interpreted.

Since ancient times, many thinkers, philosophers, spiritual visionaries as well as scientists belonging to different 'schools of thought', have been engaging their brilliant minds in extensive exploratory studies and research to satisfactorily 'crack' this **'Mystery of mysteries'**, but without any conclusively acceptable success so far. The relentless studies and researches nevertheless, are also continuing expansion as well as advancements of human knowledge, but at the same time these have 'ironically' or otherwise have also created many myths and misconceptions too.

In this context, it would further be interesting to quote few lines by **David Bergamini**, "Ironically, this age of human daring and resourcefulness is dawning at a time when man's concept of the larger cosmos beyond his own planet is more humbling than ever before. Whereas one's thought with utmost confidence that the entire universe—moon, sun, planets and myriad stars—revolved around him, he now knows that the earth is one smallish planet of a medium-sized star in the outskirts of one among uncounted billions of galaxies. Man has not yet begun to amass knowledge of even of the nearby moon at firsthand, but he has penetrated many of the greatest mysteries of space by observing celestial events at a distance and making careful deductions about what he has seen. Well-travelled astronauts of the future, reminiscing over the sunsets or moonscapes of other planets, will seem incomparably sophisticated. Yet not-so-dashing, earthbound astronomers of here and now have already glimpsed the overall vistas and overwhelming perspectives of the universe. Tracing its outlines has been an enthralling adventure for the bold minds which have carried out the task." (Bergamini, 1962–69, pp. 9-10)

Well, these areas of extensive and continuing scientific studies and research are by themselves very vast as well as entirely different and highly specialized multi-disciplinary subjects and therefore, much beyond the main 'subject-matters' and planned 'parameters' as intended for this book. In fact the intension has been just to make some preliminary efforts for giving some introductory glimpses of them by quoting few lines from some internationally well-known books written by highly acclaimed specialists and also in simple language meant for general readers, and in a perspective of logical sequence.

Further in this context, we need to bear in mind that today, howsoever man may, on any pretext, consider himself 'important', but the plain scientific fact remains well established that the man, for the planet Earth on which he exists with all his innovative dynamism, earth's solar-system and also the galaxy of which it is a part; he is absolutely unimportant and of no consequences at all. It is also naturally interesting to note that with respect to this unfathomably complex and multiple phenomenon, eminent scientists of our world have advanced a number of scientifically well-argued 'theories' to explain them, based on enlightening as well as continuously further developing multi-disciplinary knowledge of the sciences. Here, we find that in addition to others, two well-known theories are the "**Black hole theory**" and the "**Big bang theory**". The complex and multi-disciplinary scientific logic, well-formulated interpretations and also the related 'debates and conclusions' etc. in this regard, are exclusively for the concerned experts and hence, outside the purview of largely any layman like this author.

In the editorial column of *The Times of India*, New Delhi/Ranchi(2005), we find a very enlightening statement as 'Thought for today' by **Henri Bergson** as, "To exist is to change, to change is to mature, to mature is to go on creating oneself endlessly." Further, the editorial column itself of the same newspaper says:

> "The universe probably makes a lot of sense—albeit on grand scale and in vast cyclic forms. On the union-skin ecosphere of one insignificant planet revolving around an average sized star located at the edge of an inconsequential galaxy—one among trillions such in the cosmos—an unprecedented tsunami wrecks cataclysmic havoc. As a result, more than 150,000 lives are lost, entire landscapes are devastated, and the livings are left with the onus of limping back to normality. Meanwhile elsewhere, NASA scientists monitoring the X-ray observatory named after Indian born astrophysicist **Subramanyan Chandrasekhar** chance upon another, far more powerful, natural cataclysm. In fact they reckon it's the most powerful eruption ever witnessed in the universe to date. The event which has lasted more than 100 million years and generated energy equivalent to hundreds of millions of gamma-ray bursts, has apparently been caused by gravitational release, an enormous amounts of matter has been falling into a super-massive black-hole. Since astronomers are not sure where such a colossal amount of matter—equal in mass to 300 million suns—could have come from, they theorise it may be stellar gas from the black hole's host galaxy itself which, for reasons not understood, cooled down catastrophically and was wholly cannibalised thereafter. But to paraphrase an old saying, what goes in must come out. ...**Scientists think the interior of certain black holes could be connected to a corresponding 'white' hole in such a way that matter which has fallen into the black end could pop out of the white one through a link called a wormhole**. Where it pops out is though anyone's guess, it could be some place billions of light-years away from where it entered, in a totally different region of the space-time continuum, or in a past dimension. It could even be in another universe altogether. Wherever it might be though, it initiates yet again another cycle of **continued eternal existence**. Because in its end is always a new beginning."

Indeed, all these, scientific perceptions are based on well-formulated deductive logic and also on so far 'latest' scientific-knowledge guided 'conclusions'. Here, at this level of man's relentless and magnificent scientific pursuits it is enlightening that, ***"science so coincidently is endeavouring to 'meet' the humanity's seeking of again Infinite Spiritualism."***

Now, coming back to the 'title' of this chapter, in this overall highly complex multiple processes of the beginning of life itself on earth, various kinds of micro-organisms started evolving gradually. Insects, fishes, reptiles also started existing and then happened the evolution of initial mammals. Further many kinds of mammals started evolving and many perished too, as per the 'laws' and processes of evolution and also those of the mother Nature.

In this long-drawn process of the highly complex and multiple evolution of life, spanning over billions of years **mother Nature's most fundamental and fascinating 'law of the survival of the fittest'** in all categories of life on this planet Earth has been simultaneously in action as well. These aspects have been equally true in case of all the species of mammals. It will be interesting to recall that man or 'Homo Sapiens' to be more appropriate here, is a mammal too. But man as a mammal is distinctly quite different from all other mammals.

Therefore, it is a scientific fact that evolution of man on this planet-Earth has been in fact a much later development. A relevant question that comes to our mind here is that since, the 'study' undertaken pertains to the 'core-meaning' of this book's title, then why at all take the trouble to explain certain well-known facts regarding the whole infinite Universe, evolution of life and subsequently, include the evolution of man himself on our planet too? This 'question' calls for an attempt to provide an explanation.

In this context, it would certainly be befitting to quote few lines by **Alvin Toffler**, "It is also intended to demonstrate a method. Previously men studied the past to shed light on the present. I have turned the time-mirror around, convinced that a coherent image of the future can also shower us with valuable insights into today. We shall find it increasingly difficult to understand our personal and public problems without making use of the future as an intellectual tool." (Toffler, 1975)

Here, it appears that the deep meaning of the above quoted lines has some inherent dimensions in the sense that since, for meaningfully understanding various problems confronting the man today. It helps to understand the future and then make the same as an intellectual tool for proceeding further with a desired perspective.

The point is that for making any such study comprehensive, the past the present and the future are perhaps inseparable as composite and unending continuous processes and therefore, should not be attempted to be compartmentalized. Though for any understandably in these regards, opinions may naturally vary.

It is scientifically well established that man's gradual evolution even in a highly primitive state to begin with, has been a much later phenomenon on earth. **George Gaylord Simpson** says, "In considering the record of life, we sought progressive changes that involve life as a whole, and not only the evolution of particular groups within the total process. Only one was found the tendency to expand, to fill in all the available spaces in the livable environments, including those created by the process of that expansion itself. This is one possible sort of progress (page 243).... Man is part of nature and he is kin to all life. Yet the metaphoric 'brotherhood', while a valid expression of the fact of essential unity, is evidently too strong a term if we consider it more literally and in greater detail. Man is related to every other organism that has existed, or that does or will exist on the earth, but it is perfectly obvious that he is more nearly related to some than to others. It is obvious that man is an animal and not a plant, which is a taxonomic way of saying that he is more closely related, not only in structural and functional resemblances but also in material origin, to all animals than to any plant. Among the animals it is equally obvious that, in the same sense, man is a vertebrate and not an invertebrate. Among vertebrates, he is a mammal and not one of the various sorts of fishes, an amphibian, a reptile, or a bird. Among mammals he is a primate, a member of the Order Primates and of no other of the numerous mammalian orders... Man's attempts to understand the history of life have had an evolution of their own, an intellectual evolution in which progress toward truth and comprehension has been, if not constant, at least continual. The history of this progress casts light on the meaning of evolution to man. Consideration of successive contributions and of rejected alternatives shows not only how the current understanding of evolution and of man's place in it was achieved but also why and with what degree of confidence the present views on these subjects are to be received. Such consideration, necessarily summary, is the remaining necessity before turning to the final phase of this enquiry, examination of the philosophical and ethical implications of all that has gone before." *(These lines have somewhat the same in-depth meaning, as those of from the famous writings/preaching of Swami Vivekananda, on Vedanta, as quoted in the earlier chapter of 'Introduction')*

Needless to say here again that certain contextually specific and also narrating certain historic facts, the above quoted paragraph would not have materialized if man was not 'gifted' with that exclusively unique asset, i.e. 'Brain' to start with. At this juncture, it would be interesting as well as relevant to look at the fascinating anthropological processes of the gradual evolution and ascent of modern man, since more than hundred million years ago. In this respect, it is definitely useful to reproduce the precisely appropriate 'data' from the book *A Brief Survey of Human History* below:

1.	Fishes and Reptiles	More than 100 million years B.C.
2.	Birds and Mammals	
3.	Monkeys and Apes	Type of life
4.	Sub-Man	50,000 years B.C.: Glacial culture.
5.	Neanderthal	Earlier Paleolithic.
6.	Cro-Magnon	25,000 years B.C.: Later Paleolithic.
7.	Sapiens	12,000 years B.C.: Neolithic.
8.	Homo	6,000 years B.C.: Copper.
9.	MAN	3,000 years B.C.: Bronze
10.	Modern Man	1,500 years B.C.: Iron.

The above table gives the chronological order of the gradual evolution of man on earth. Here, we notice that the Neanderthal 'men' vanished many thousands of years ago, with gradual coming of man at a much later stage. Traditionally, there is a general opinion that the Neanderthals were totally devoid of any kind of 'brain' and also they were stupid, ugly and 'self-destructive' creatures. Well, scientists by means of further research are now coming to fascinatingly new logical conclusions.

In this regard, if we refer to a newspaper, *The Statesman*, featuring an interesting news "You bl**di Neanderthal...what a beauty." where the Press Trust of India brings out that, "But new findings have now suggested that they were probably the original metrosexual as they did their best to make themselves look quite presentable by wearing make-up 50,000 years ago. British archaeologists claim to have found evidence that Neanderthal men experimented with cosmetics

50,000 years ago, a discovery which proves the human subspecies weren't the 'half-wits' people assume but they were very much capable of symbolic thinking, the *Daily Mail* reported. Professor Joao Zilhao, who led a team at Bristol University, said the find shot down 'the view of Neanderthals as half-wits. To me, it's the smoking gun that kills the argument once and for all. The association of these findings with Neanderthals is rock-solid and people have to draw the associations and bury this view of Neanderthals as half-wits. This is the first secure evidence that, some 50,000 years ago—ten millennia before modern humans are first recorded in Europe—the behaviour of Neanderthals was symbolically organised. **Prof. Zilhao** was quoted as saying.... According to the archaeologists, the brightly coloured shells may have been worn by Neanderthals as jewellery or neck pendants for their makeup. 'This is the first secure evidence of their use of cosmetics. The use of these complex recipes is new. It's more that body painting', Prof. Zilhao said. During the Upper Palaeolithic period—or late Stone Age, around 40,000 years ago—Neanderthals and humans are believed to have co-existed. But Prof. Zilhao explained that the findings dated around 10,000 years before this contact."

It is also a fact that ironically, the commonly held view that the Neanderthals were 'dim-witted', still persists and refuses to wither away, in spite of the above quoted and proven scientific findings, as recorded by reputed experts. In any case, that is a well-known nature and socio-psychological fixations of 'modern man', largely with respect to anything or any thought-process in general, among human societies in this world. But, the pursuits of science go on for multiple enrichment of human-knowledge and thereby, new and new technological innovations as well as their imaginative applications for overall human-progress and related improvement in the man's 'quality of life'.

Further again coming back to **George Gaylord Simpson**: "In the first place, it must already be evident that there is no such thing as the rate of evolution. The record has demonstrated that evolution is not some over-all cosmic influence that has been changing all living things in a regular way throughout the periods of the earth's history. Some groups have been changed rapidly while others were remaining practically unchanged. The same group is commonly seen to have changed rapidly at some times in its history and slowly or not at all in others. Within a given group some parts may change while others are static... Opportunities for evolutionary change are thus ultimately both created or limited by what can result from determinants already present in a population of organisms and from mutations among those determinants, which in fact, are themselves the results of earlier mutations... Among large numbers of organisms and over long periods of time such as involved in the history of life, mutations are exceedingly abundant, but they are also exceedingly erratic. In experimentation, some ways of speeding up their appearance and slowing it down have been discovered, but no general principle as to their nature or the orientation to their effects has yet been found... Evolution works on the materials at hand: the groups of organisms as they exist at any given time and the mutations that happen to arise in them. The materials are the results of earlier adaptations plus random additions and the orienting factor in change is adaptation to new opportunities."

Now these lines substantiate that the entire process of fascinating evolution on this Earth has been a very long-drawn as well as unending process.... In addition, we learn from his this valuable book that this certainly not completely knowable process of evolution is quite erratic, even disorderly and not necessarily 'interrelated or interconnected' or for that matter, appearing as a 'composite-whole'.

At this juncture, it may be also of some interest to refer to a well-known scientist, thinker and **Professor Stephens Hawking**: "Whether we like it or not, the world we live in has changed a great deal in the last hundred years, and it is likely to change even more in the next hundred. Some people would like to stop these changes and go back to what they see as a purer and simpler age. But as history shows, the past was not that wonderful.... Anyway, even if one wanted to, one couldn't put the clock back to an earlier age. Knowledge and techniques can't just be forgotten. Nor can one prevent further advances in the future.... Moreover, on cannot stop enquiring minds from thinking about basic sciences, whether or not they are paid for it... If we accept that we cannot prevent science and technology from changing our world, we can at least try to ensure that the changes they make are in the right directions.... At this moment, the public has a rather ambivalent attitude towards science. It has come to expect the steady increase in the standard of living that new developments in science and technology have brought to continue, but it also distrusts science because it doesn't understands it... The science people learn in school can provide the basic framework. But the rate of scientific progress is now so rapid that there are always new developments that have occurred since one was at school or university." (Hawking, 1994)

Further Prof. Hawking's certain perceptions may also appear to be somewhat relevant to the prevailing concepts of the overall process of the evolution of man on this earth as he writes, "One has to keep the investigation of the fundamental laws of science and study of human behaviour in separate compartments. One cannot use the fundamental laws to deduce human behaviour... But one might hope that we could employ both the intelligence and the powers of logical thought that we have developed through natural selection...if one believes that everything in the universe is determined. It doesn't make much difference whether this determinism is due to an omnipotent God or to the laws of science. Indeed, one could always say that the laws of science are the expression of the will of God...how can the complexity of the universe and all its trivial details be determined by a simple set of equations?...the trivial details of the universe arise because the fundamental laws incorporate quantum mechanics with its elements of uncertainty or randomness...if everything is determined by some fundamental theory, then what we say about the theory is also determined by the theory—and why should it be determined to be correct, or rather than just plain wrong or irrelevant? My answer to this was appeal to the Darwin's theory of natural selection: only those individuals who drew the appropriate conclusions about the world around them would be likely to survive and reproduce." (Hawking, 1994)

Therefore, Prof. Stephen Hawking's scientific perceptions too, substantiate that the entire unending process of evolution of Nature's Creation on this planet, since thousands of billions of years, are unfathomably complex, multi-layered and multi-directional phenomena which in turn, are not 'regulated or orderly' but definitely unpredictably random and erratic. As also enunciated by Prof. George Gaylord Simpson. And it is also interesting to note that within that much larger and highly complex process of 'total' evolution on earth, gradual and complex phenomenon of the evolution of man here has got initiated, at a much later stage, as elaborated in earlier paragraphs.

Though the advent of man being a much recent phenomenon on earth, his features, complexion and his many other physical attributes have become highly diversified in different geographical locations. This is a fascinating and scientifically interesting developmental process of evolution and accordingly, it may appear appropriate to quote from **Shri G.S. Mudur**, "...Scientists believe that these genetic changes are primarily driven by population expansion and exposure to new environments and infections. 'Genes primarily involved in adaptation to different environments are favourably selected', said **Partha Majumder**, head of the human genetics unit at the Indian

Statistical Institute, Calcutta. 'Because of the adaptation and natural selection, human genome pools exposed to different conditions may have diverged more rapidly than what is expected by chance alone', Majumder told 'Knowhow'. Mutations that favour light skin, appeared among Europeans. Researchers believe light skin is needed in cold, dark climates to allow sunlight to synthesise vitamin D. When populations were exposed to new infections, genetic variations that protect against the diseases became favourable.... Most mammals lose the enzyme to break down milk after they are weaned away, but Europeans and east Africans acquired it about 7,000 years ago. This was a new adaptation—directly linked to the domestication of animals and use of milk as food. **Harpending** is now studying whether this mutation might have spurred the spread of the Indo-Europeans from Central Asia to Europe and India 5,000 years ago. He believes the ability to digest milk gave lactose-tolerant Indo-European speakers more energy, allowing them to conquer more area. While the study identified about 3,000 genes in each of the population studied, the role of many genes is still unclear. 'For many of these genes, we don't know what the new version does better, but we can tell that they must have done something'. **Hawks** said. Chinese and Japanese populations, for instance, have evolved genes that produce dry ear wax instead of the wet and sticky ear wax found in other populations. The best guess about what these genes are doing is the suppression of odour,' said Harpending." (Mudur, The Telegraph, 14 December 2007)

Even at this stage, it appears to be of some interest in the modern era context and in India's colonial era perspective that, keeping in mind their then 'over-riding' and as perceived geostrategic imperative, the British identified certain sizeable sections as comparatively more 'martial' than other sections of the populations, settled in vast expanse of Indian rural areas. This is debatable though. Such perceptions in this regard were principally based on the then available human knowledge of science concerning 'human-genes', and their highly complex pro-active multi-directional development, coupled with the 'consequent' development of sociocultural, psychological and physical attributes of the various sections of India's population-mass, spread over the vast Indian territory under the colonial rule.

The British proceeded to gradually conceive and organise the various battalions and regiments of the British Indian Army, put in service under their command. Then after independence, the Indian Army adopted such 'concept' as tried and tested 'systemic-tradition' but nevertheless, with a totally apolitical, nationalist outlook and unquestionably abiding by the principle of complete loyalty to the Constitution of India, as well as to the Indian-Nation and thereby, settling-down well in their 'profession of arms'. It is naturally well-known to all concerned that traditions and fundamental principles of the Armed Forces and the profession of arms the world over are crucially valuable in many ways, pertaining to the militaries and the matters-military.

But at the same time, it is necessary to note a very pertinent point of far-reaching national importance that now, after the year 1947, the independent India's Indian Army is unquestionably nationalist in professional ethos and loyal to the Constitution of India. These aspects have definitely been able to completely maintain the apolitical professional ethos of the Army along with its age-old and also well-tried military 'customs and traditions'; with suitable modifications as per the changing times and circumstances, in a healthy manner, without any contradictions, with desired flexibility and soldierly professional dynamisms. These are certainly unique, perspectives, particularly in South Asia, with an appropriately valuable message along with substantially creating a highly inspirational 'role-model' for all concerned in the region. These aspects would be dealt with in some detail in subsequent chapters.

Further, in spite of man being fundamentally a sort of 'animal' and at the same time, being comparatively very 'soft' and vulnerable, in the overall phenomenon of evolution and also being exposed to all sorts of threats and challenges, he is decidedly the highest specimen of animal on this earth. The unique and most favoured gift of "Brain" bestowed on man gives him the superiority.

But to begin with, primitive man's 'primitive-brain' was able to make revolutionary differences for him in comparison to all other animals on this earth. With Nature's evolutionary processes, many thousands of centuries passed, before man developed gradually and thereby, was able to 'enter' into his "pre-historic era", as recorded by many well-known thinkers and scholars of history.

*

CHAPTER 3

Pre-Historic Period

At the outset, it may be mentioned that in, the 'pre-historic period' refers to that era of which methodically compiled and reasonably authentic records or even 'primitive' anthology in some scientific form or other do not exist. In fact, it may be said with a reasonable certainty that during the pre-historic era it was too early and 'unreasonable' to expect from the then man to be even remotely capable of having done the same in any such form or manner for posterity. But at the same time, modern era man has been able to find many different kinds of ways, like ancient cave-drawings and interesting symbols, etc., that have been left by them around the world for 'posterity'.

These subject-matters are by themselves very vast and hence, requiring detailed specialised studies by anthropologists and allied experts. Through multi-disciplinarily continuing research and studies; the concerned knowledgeable experts have been able to create well-founded records of the pre-historic period of man on this earth as an endeavour towards expansion of human knowledge.

Prof. S.R. Sharma writes, "Anthropologists have marked the successive stages of man's ascent to manhood by observing his physical development, the evidence of which has been discovered in ancient graves. According to the places where such discoveries are made, the successive types have been named the Java, Heidelberg, Piltdown, Neanderthal, Aurignacian, and Cro-Magnon (i.e. skeletons or parts thereof, representing the various stages in the ascent of man). The last named appears to have been over six feet high and had, all the features of a modern European, though he must have lived between 50,000 and 10,000 years ago. **Before, the close of this period man had distinguished himself by his superior tool making capacity which gave him a distinct advantage over all other animals**. (*Rightly so but at the same time all these could only exclusively be possible, due to in fact, his inherent **'analytically thinking and creative mind'**.*). He could even fix it on a wooden handle or stick and use it as a spear or axe, and perhaps also send it flying at an enemy by means of a sling or bow. More than anything else, he had discovered 'the red miracle of fire' which he could light or extinguish at will. And what is still more wonderful, he could communicate with his fellows through intelligible speech. What a catalogue of discoveries and inventions. Nothing like it had existed on earth before. It was as if the Creator had suddenly turned alchemist or come upon an Eldorado. From the first chipping of wood and stone, and the first lighting of the

Promethean spark, and the first intelligible glibber, the modern progress of man in all directions has followed as the culmination of a logical series. **The fruits and flowers and thorns of modern civilisation are borne aloft on the stem of history whose roots are embedded in the pre-historic soil of dim antiquity**." (Sharma, 1948)

We naturally are aware that among all the animals, barring some exceptions, man cannot claim to be very strong, tough and even capable of defending himself on his own or in certain circumstances even in groups, against all types of dangers or any kind of lethal threats to his life from any other dangerous and violent animals. As the saying goes that **'necessity is the mother of invention'**. Therefore if viewed in this context, we notice that the process of evolution was exclusively special to man. But the question comes up that why the mighty Nature was so considerate to bestow special kindness exclusively on man? The point is that this was certainly no matter of chance and there has to be a scientific reason for this exclusive kindness to man, somewhere in Nature's grand, multiple and ongoing evolutionary processes on our Earth. This in turn made all the difference for man in this world. Interestingly, this also establishes another moot point that man's strength does not lie in the body muscles alone but, the real strength lies hidden in the 'brain', which has multiple faculties and huge potential.

Extensive studies, research and highly complex experimentation have most fascinatingly established that even highly gifted and remarkably brilliant persons, relentlessly engaged in studies, research and the many kinds of challenging and creative-work in different areas of human endeavour have not been able to utilise the full 'potential of the brain'. Well, another very interesting point here is that such is the continuous progressive dynamism of the 'nature and the indeed vast range of science' since the advent of man on this Earth.

Further, it may be relevant to note that all these analytical perceptions certainly do not mean a lack of total absence of any kind of 'brain' at all in animals. They too have some type of brain in themselves. But in their case, the brain is mostly 'static' and also not having any kind of innovative, thinking and imaginative capabilities and definitely not similar to that of the human beings. The mammals of different kinds have comparatively more 'developed' brains, having relatively more versatile 'capabilities' than other animal species. Among the mammals themselves, certain mammals have comparatively more developed as well as more capable 'brains' than other mammals.

Now coming back to the gradual beginning of the evolution of primitive man in the distant past, we find that his brain too was primitive. As primitive man slowly started developing and progressing over a 'canvas of time', his brain too proportionately started developing. In fact, this entire evolutionary process keeps unendingly taking place, as per the overall command of creator—Nature. Logically therefore, the pre-historic man too had a 'proportionately underdeveloped' brain as his natural gift and asset. But to reinforce, no other animal on this planet, to start with, has ever been gifted the kind of brain as possessed by man.

In our modern world today, we find so many fascinating and enduring, and hence preserved, mountain and rocky 'caves-paintings' located in the many far-off and different geographical locations in the world. These cave paintings have indeed very interesting and fine drawings depicting clear figures of various animals, birds and other objects. Further these cave-paintings don't belong to any one particular year of time or so but, they have been drawn in many different centuries and in various locations that are in no way connected or 'inter-linked' with each other.

In this context, the obviously pertinent point that comes in mind is, that for the man to be able to draw these figures and drawings then, he must have been able to innovate and 'develop' certain kinds of tools. In fact, these as well as other related developments slowly and steadily enhanced and multiple capabilities of the man were caused by gradual activation and development of his unique 'brain' and thereby, further progress of his mental faculties, though to begin with then in relatively primitive manners only. These are natural, as per Nature's own evolutionarily continuing 'laws of the Creations, Transformations and the so called destructions'.

Again referring to **Prof. S.R. Sharma** we learn that, "Scholars have divided the earliest history of man into two periods: Palaeolithic or Old Stone Age, and Neolithic or New Stone Age. Though this seems arbitrary it is not without reason. No other animal is known to have possessed the tool-making faculty which has given man such great mastery over Nature. Hence it is most natural to denote his early progress by the quality and materials of his tools.... Since even axe and spear heads were made out of hard stone or flint, it is obvious that metals had not been discovered. *(Metals were in fact discovered during much later periods).* Wood was cut with a flint piece sharpened at the edge, and fire was also kindled with its help. No new discoveries were made for long ages, but during the 50,000 or so preceding Neolithic Age (about 12,000–8,000 B.C.), considerable improvements were made even in the stone implements. On this account the later Paleolithic period is often distinguished separately as the Middle Stone Age. During these fifty thousand years well-shaped stone implements with wooden handles were made, instead of the crude flint instruments. Even the bones and horns of animals killed in the hunt were now utilised for various purposes. For instance, needles and whistles of such material, made by primitive man, have been found by archeologists. Though the Middle Stone Age man had not discovered the arts of weaving and pottery, he wore garments of skin and used skin bags for storing things. His most remarkable contribution to civilisation was, however, in the field of art.... The fundamental discoveries of the Palaeolithic Age were fire, tool-making and speech. Those of the Neolithic Age were weaving, pottery, cottage building, domestication of animals (sheep, goats, pigs), agriculture, wheeled transport, boat construction, metals, and some sort of thought-recording which later developed into regular writing. A few comments may be made regarding each of these. Weaving must obviously have been learnt from the observation of spiders' webs, birds' nests, and the natural plaiting and interlacing of creepers and branches of trees. Basket and mat plaiting might have preceded the spinning of yarn. Similarly, cave-dwelling and groves of trees suggested artificial houses. At first constructed with branches and twigs, cottages were soon improvised with matted screens and thatch. Dobbing of walls with wet mud and allowing them to dry in the sun, led to abode and brick construction; while the baking of clay for bricks, either in the sun or fire, naturally suggested pottery. The oldest artificial dwellings still surviving are, perhaps the pile constructions in the midst of lakes are of Wangen in Switzerland and other places. For purposes of greater security Neolithic men sometimes built their cottages on wooden support driven into the beds of lakes, connecting when with the shore by a removable bridge. Some two hundred of such dwellings were discovered in Switzerland in 1854. But the most revolutionising change that came over human life in Neolithic times was due to agriculture. All the other innovations of the period are associated with agriculture." (Sharma, 1948, pp. 7-9)

In the above lines the imperative of security aspects as perceived by the then Neolithic men have been emphasised. Logically therefore, aspects of '**defence**' with all their perceivable dimensions then, had already germinated. In fact, it is interesting to note that **natural instincts for self-defence and preservation of their own specie** have been present much earlier, right at the inception stage, among all insects, reptiles and birds, as they got evolved. And naturally therefore, this has held good also for the mammals too.

Further it is also well-known that the same aspect has been true for the man as well. Since the man has been so specifically gifted with an exclusive kind of "**analytically thinking-cum-creative mind**" during the beginning of his long-drawn process of evolution on our planet and hence, this particular attribute has made all the differences for him and also making so continuously and thereby, has been at the roots of the consequentially multi-dimensional and also the on-going progressive, as well as otherwise gradual march of civilisations, as time keeps rolling-by. Interestingly, the same is also the no less evolutionarily true—for his multiple security aspects and needs—with all their strategically inferred or 'imaginative-cum-motivating' manifestations, as per the any prevailing environments around him in the world.

Now coming back to the fundamental question of 'reason(s) as to why there have mostly been 'differences'/conflicts and then struggles, battles and extremely violent and ruthlessly devastating wars among man's groups, tribes, communities and then States; since ancient as well as very distant past times? This question is not only very interesting and fundamental in nature but in addition it is linked with the infinitely micro-cum-macro levels grand processes of evolution of the Creator-Nature within the infinite Universe and subsequently much later more specifically; with the slowly gradual advent of man on our planet Earth and also these aspects are still continuing as well as would most likely go on 'perpetually', till the man survives. Again the likely causative factors for this phenomenon as related to man, have been briefly touched upon in the beginning paragraphs of the first chapter of 'Introduction' of this book itself, as to the inherent presence of both the 'Angels' as well as the 'Demons' inside men and then these since getting 'manifested' in so many highly complex manners, as time rolls-by along with the unendingly arising different and complex situations in this world. These aspects as they exist are so governed again by the 'inherent nature/laws' of the Nature's perpetual processes of contextual 'Creations-cum-Destructions' on our planet Earth, and also within the infinitely vast and indeed also the unfathomable Universe.

It appears relevant to mention here, specifically with respect to the man in this world, the above referred 'questions and the probable answers' do inseparably have the aspects of all sorts of conflicts/wars, constituents of the profession of arms, the military and thereby the politico-military complex situations and dimensions, as per dictates of the changing times as well as the continuing history in the making. It has been very appropriately said that conflicts and wars are basically the 'children' of the continuous rise and fall of different civilisations, as the man's 'living-history in the making' keeps-on marching ahead. Therefore it can logically be inferred, that both the complex 'causes and effects' of conflicts/wars are totally inseparable from the inherent nature of man with apparently no control over them. In fact, facts are facts but at the same time as already emphatically mentioned earlier that since man has been exclusively gifted, he with the best of his power/abilities, both individually as well as collectively, endeavours to 'prevent' or at least 'minimise', as far as possible, any type of conflict/war in any given or emerging situations anywhere in this world.

Coming back to the subject of the 'genesis' of all sorts of conflicts/wars among men, their different historically emerging groups, tribes, settlements-based neighbouring communities, states/ nations in different geographical locations and also among various differing civilisations, since the advent of man on this planet Earth, we find that initially violent clashes/conflicts to begin with, had not got 'germinated' at a relatively later stage with any beginning of civilisations, but in fact much earlier. At the same time, it is also equally true that these initial conflicts, were definitely small and localized and also, in the beginning not so devastating.

The 'birth' of the 'profession of arms' as a distinct one, took place principally through the

advent of ancient civilisations in this world. Logically the much larger and more organised 'wars' in different range and dimensions are definitely 'children' of civilisations, as historically to begin with. These ancient civilisations began emerging, progressing and also clashing since their beginning, in different geographical locations of this world; as a relatively smaller part of the 'overall unending grand-processes of evolution', within the mightiest domain of mother-Nature. **Hence, it has been very appropriately said, that one of the fundamental natures of the all civilisations is 'war' itself, besides that of the inherent urges to make progresses' in every field.**

Therefore historically, it substantially gets established that in the beginning during the very distant era of primitive man, as centuries rolled by; small clashes and conflicts between men began and also, these were routinely taking place as naturally triggered by their fundamentally inherent nature and also, as per the man's psychological and socio-psychological attributes and 'make-up'. Here we come to a point that now we confront a fundamentally unalterable fact so 'staring in our eyes'; as many civilisations keep marching onto the 'infinite canvas of time', that conflicts and wars, in their all kinds of conceivable or otherwise manifestations; are inherently a natural and inseparable 'part' of the basic as well as evolutionary 'traits' of the man and therefore, the same just cannot be wished away. And consequently therefore, it appears that he perforce is destined to live with this extremely turbulent and devastating fact of his life in this world. This is indeed a most unsettling and devastating 'irony' for man in spite of the magnificent advancements, continuing progress in the spheres of micro-cum-macro economies and also, that of socio-spiritually multiple 'enlightenments' in his life.

Inescapably therefore, these continuing developments consequently throw-up menacingly multiple-challenges for man and his well-being. Also, threateningly appear to be so 'unsettling' of his many advancing and enriching civilisations again in today's world. While scanning through these irrefutable historic facts civilisations also caused the 'births' of all sorts of conflicts/wars. These very negative fallouts, i.e. the same 'conflicts/wars' have now been so menacingly threatening the very 'survival' of him, as well as his 'civilisations'.' **And here the "the theory of the 'clash' of civilisations", as earlier referred to, comes into one's mind and in addition thereby his this well-known, as well as 'controversial' theory both historically and presently, is found to be prominently in a 'pro-active mode'.** Indeed it is precisely perhaps, the greatest 'irony' today for the man in this world.

Now, again coming back to the aspects of 'pre-historic period' we may mention in a somewhat general term here that, this 'period' could little more precisely be identified as a long 'era', preceding the beginning of the learning-cum-intellectual culture of the formal as well as in a methodically chronological order, the systematic 'recording' of the mankind's 'history', in the then as comprehensively as possible; as the same started 'unfolding' with the on-going passages of the centuries in this world, over so vast a 'canvas of time'. Logically therefore, we may also note here that the well-known historic 'dawn' of the various human-civilisations in many geographical areas in our world and also, in different periods of time, began to take place consequentially, as a part of the slow and gradually on-going evolutionary progress of man.

In addition, we may also bear in our minds that, it cannot be historically possible to draw any kind of 'distinct lines' between different periods of history of the mankind as they perforce overlap too, quite inseparably. Hence, the so-called 'boundary-lines' historically themselves consist of some 'periods of time' of one or more centuries; as the man's 'living-history' itself does keep-on gradually unfolding in this world. So it clearly brings us to a fact that, the entire exercise of 'history-writing' itself is a continuing process in the best interests of posterity. Therefore, this challenging and continuing 'exercise', as undertaken by many scholars anywhere in the world also ensures the desired 'self-corrections' of the multiple 'records' of the living-history, covering almost all possible areas of man's relentless endeavour as well as, consequentially gradual progress of various human civilisations. As a composite 'whole', they themselves are perpetually ongoing processes.

If we may care to recall that earlier, in the chapters of 'Introduction' and also that of the 'Gradual Human Evolution'; the entire grand phenomena of the multi-directional and multi-layered mighty Nature's evolutions within the limitless Universe itself, the 'laws' of evolution dictates everywhere and in every context,'. The mighty Nature's all-encompassing perpetual 'command' of the 'survival of the fittest, is equally applicable to all forms of life and also to all species of life, including the mankind, without any differentiation or distinction whatsoever.

Man has been given many special responsibilities and also the numerously mandatory duties of his own, of our world and thereby that of this planet Earth as well. It is obvious that since the earth's ecology as well as overall tranquility of its environment, supports the continued survival of the all forms of life and they in turn, ensure survival of the mankind, it is imperative that they too are allowed to survive properly.

According to **Professor S.R. Sharma**, "Man's life is a moment between two eternities. It is the hyphen that joins the past with the future. Most people are content with the animal satisfaction that is derived from the present. But this is not a quality that makes for progress. If we want to live intelligently, we must 'look before and after'. In this faculty, which does not belong to the sub-human world, lays the supreme privilege of man. However keen the struggle for existence, however engrossing the living present, we shall be the better equipped for a happy and successful life, if do not ignore the past. While the future for which we seek to provide is uncertain, scientific investigations have so enriched our knowledge of the past, that the vital dependence of our present upon the 'living past' is now considered quite axiomatic. Indeed, one imaginative writer has compared the present to a pygmy riding on the shoulder of the giant past, and pointed out how much we are benefited by this position of vantage. **The present may be truly described as the child of the past and parent of the future. Life is a continuous whole. History is, therefore, record of only one stage in human life: it deals with man's past on earth. Whatever may be the future of our species, man to the best of our knowledge, is the last word in animate creation**. It was **Charles Darwin (1881-42)**, a nineteenth century biologist, who revealed to the modern world the continuity, not of human life only, but of all life, in his 'The Origin of Species and the descent of man'. If we take into account the long process of evolution by which the human species grew up on this planet, our history appears so short that we cannot but marvel at the rapid progress man has made since he evolved out of the ape. While it took millions of years to bring man into existence, the dynamic intelligence of man (Homo Sapiens) has transformed life on earth so tremendously in the course of a few millennia which constitute the sum of human history. There is hardly a story more enthralling in its character than this truth about our human adventure which is stranger than fiction." (Sharma, 1948) (*And this has indeed been continuously triggering man's multi-dimensional civilisational advancements, since his beginning. We would naturally be discussing these aspects with all their complex manifestations, in our specifically laid down perspectives in more detail later*).

The above paragraph authenticates the Nature's unending evolutionary reality of man's so far unique exclusivity in this world. Hence just to recall the cause for all this is the Creator-Nature's gift to man, i.e. his highly imaginative, thinking and innovative "brain", which no other animal of any kind has so far been gifted at any point of the history. This unique privilege endowed to man did make all kinds of differences and, is still making in so many countless ways. This precisely is the point, which further enriches the arguments and discussions on this specific topic here, as being a relevant 'part' of the main subject-matters of this book.

Earlier in this chapter we have since also talked about certain aspects; pointing towards the modern era bringing-in grave dangers to the continued survival of man and his various civilisations, as well as various multi-directional and multi-disciplinary complex progress and achievements so far in this world. In fact, these are mainly due to the mindless and also the most reckless disregard towards the mandatory and continued protection and preservation of the mightiest Nature's ecology and environment in all their probable manifestations in this world. These are definitely the modern era developments; tragically triggered by the man's irrational desire for unquestioned 'supremacy' over his fellow-beings and in addition, towards his insatiable greed for blind materialism with sub-human 'regressive-fixations'. But the 'seeds' for these 'traits' of man can be traced back to his pre-historic period, visible in certain apparent manners.

In this respect, we also find that, in fact the most ancient initial human-civilisations had begun 'sprouting' in different centuries during this pre-historic period itself. It is further interesting to note a point here, that during the times of prolonged 'primitive-age', man was slowly and gradually getting evolved from sub-human stage to the 'primitive-human' stage only. At this stage of the interesting historic saga of man's slow and gradual evolution in this world, we notice that the 'sub-human/primitive-man' was primarily engaged in fighting, individually or in small groups, mainly for self-protection, during hunting for food, for a 'female for mating' or at the most for securing a comparatively safe night-shelter etc.; or could also be for any other similar reason. Therefore beyond all these at that stage of man's evolutionary beginning itself, any question of man thinking or even imagining any kind of even little 'enlarged' conflict among themselves, was totally beyond his primitive mind's range and capacity. So there was definitely no question of any kind of even highly limited 'wars' among them, crossing his primitive 'brain' then, in any manner whatsoever.

We find that also exclusively due to that special kind of 'brain' so bestowed onto man, he did have the unchallenged capacity to continuously 'trigger' many kinds of multi-directional developments as well as achievements gradually benefiting him in many seen/unseen directions. Initially at the man's inception stage to start with, this overall process was relatively much slower but as some centuries passed, again principally due to the beginning of some such benefits getting gradually accrued to him his capacity to keep achieving and developing more and more got enhanced as well. We may both logically and scientifically perceive that this phenomenon too, has indeed been part of the Nature's own 'laws' of man's 'special' evolution in this world. And at the same time, this Phenomenon is also continuing even in the presently ongoing modern high-tech era and in addition also getting increasingly complex and multi-directional too, both in range and speed in this regard. Well this is an accepted fact of paramount consequential importance in our present-day world.

The fundamental reasons for all these, have again been naturally man's attributes to relentlessly keep progressing multi-directionally, in all sorts of manners, with totally mindless disregard for a number of vital supreme 'laws' of the all-powerful Nature in this world. Needless to say that now most ironically we find this particular extremely tragic consequence in this regard, has also been the biggest 'undoing' in this world. At this point, we may justifiably emphasise that since man has been so bestowed with a 'brain', it is mandatory on him and expected of him by the mother-Nature to exercise his inescapably essential need to visualise the Nature's evolutionary processes, laws, and connected 'constraints' in the requisite larger contexts in this world. In addition accordingly, man is always required to 'discipline' himself and suitably control and regulate his urges to keep achieving all sorts of 'material-progresses', in his own much larger and long-term interests.

We may further mention that the aspects regarding 'self-discipline' of man and also enabling him to suitably control and regulate his above mentioned 'urges' in our present context of discussions here, do perforce assume much larger dimensions as well as deeper meanings. The point is that these aspects, in fact, are not at all linear as 'straight-line' subject-matters, having obviously noticeable some 'simplistic-meanings'. Therefore, few paragraphs from a learned article on, 'My Experiences of Spiritual and Mystic Phenomenon', written by **Mr. L.N. Jhunjhunwala** are significantly relevant here; and these are thus quoted as: "Spirituality on the other hand is the science of developing higher consciousness in men and making one transcend from mere body consciousness to cosmic consciousness and further on to super-consciousness so as to enable one to understand the working of the divine plan.... *(As the lives of many high spiritual saints and divine personalities, the world over, have so enlighteningly shown to the mankind since ages)*.... With each one of them the evolution moves a step forward. Various levels of human beings are clearly discernible. The lowest level is eat, drink and be merry type of individuals. Indian philosopher Charvak has written a big treatise even on this type. The second higher level is the worshippers of energy, force, strength and power. The third level is he who communicates his aesthetic perceptions through words, sounds, figures, colours, shapes. These followers believe that 'truth is beauty and beauty is truth'. The fourth level is people who extol ideas who think and understand, conceptualize, analyze and synthesise. The highest level is people who devote their whole life to contemplation and communion with God. That could be called the super consciousness.... The motivation behind science and spirituality are similar—curiosity about the world around us. The significant difference is that science is more about the physical and mechanical. Spiritual is more about the unknown. As things become known and established, they pass into the realm of science. But when they remain mysterious, they approach the realm of spirituality. The tree is not visible when it is in the form of a seed of the size of a grain of sand. When it starts sprouting and becomes a large tree, it becomes a subject matter of botany. The Universe is too large for us for human comprehension. Thousands of Galaxies moving around continue to be mysterious phenomena. Similarly when we break the particles, we find electrons and neutrons and then the quarks as small as a billionth of a millimeter. Nano technology has come up as science only in the last few years of study.... Science is concerned only with that phase of creation which can be conceptualized. Spirituality is more widespread phenomena. Man is always eager to maximize his own happiness.... It is arguable whether there can be, or is, a scientific approach to spirituality. It is conceivable that in such an approach, one could not assume that any proposition is true simply out of faith, no matter how appealing that proposition is to one emotionally or culturally. One cannot believe any proposition that one is asked to assume to be true and not question. One must not believe or disbelieve anything that one cannot prove to be true or false." *(Here again, we so captivatingly see the connection of both the aspects of science as well as those of spirituality, in enlightening perspectives.)*

Interestingly the well-known saying that "Every coin has two sides", is equally applicable in our deliberations too, regarding man's 'evolutionary' nature as well as allied phenomena, as continuously 'caused' by the mother-Nature in this world. Perhaps, this aspect too, is a part of the all-powerful Nature's evolutionary 'laws'. Well, whatever be the 'perceptions and conjectures' in these regards, one fact remains established as already explained in some detail earlier that since the advent of man on Earth, he has only been exclusively gifted with unique "brain" and so, that has made all the differences for him and continuously making so in this world. But keeping in mind the above quoted well-known saying, we clearly notice that as per the Nature's absolutely sovereign and unending processes of evolution within the infinite Universe, especially in this world within the 'parameters' of man's continued survival, the "onus" lies only with the man alone. This is totally an irrefutable fact and hence, the same should be always accepted as such without any 'reservation' in any manner and thereby particularly with respect to man's every wishful thought, planning or endeavour, with whatever motivations therein, at anytime and anywhere in this world.

Over thousands of centuries, man too had suitably developed as a pre-historic man with a relatively evolved and developed 'brain' as well. Therefore, logically, during this pre-historic period, spanning over many thousands of years, a number of very ancient pre-historic civilisations slowly began to sprout in many different locations and during different centuries in this world. That in fact precisely is a very significant part of the continuous march of history. Consequently, it is equally true that since this 'pre-historic period' and onward, man's unique as well as relatively capable 'brain' too has been developing.

Therefore in this context too, as mentioned earlier the "onus", specifically with respect to all his omissions/commissions, gradually began to rest exclusively on man's shoulders. Now we may come to an interestingly relevant point here that what does the above mentioned historic-fact reveal to us with respect to the basically inherent 'constitution' of man, since the advent of prehistoric period? These aspects further lead us to the following two facts:

1. Man's natural instinctive tendency is to defend himself and keep him protected from all possible likely threats/dangers to his survival, as far as possible. This particular 'instinct' has always been present in all other species, including animals, since the beginning of evolution. So naturally the same has equally been true since the beginning for the man and accordingly so, for the primitive-man as well.
2. Man's another exclusive natural instinctive tendency is to keep making multi-dimensional **'progress'** as much as possible, and also in all sorts of feasibly conceivable manners in this world. This tendency has never been present in any other specie or animal at any given time/period in the history of the processes of evolution in this world. What has specifically triggered this special phenomenon, so exclusive for the man, since his evolution into a 'complete' man in this world? The simple answer is the presence of an exclusively unique "brain" that has helped him in his evolutionary progress into a 'complete' man from the stages of 'sub-human', to the 'primitive-man'. Logically, here 'lies the rub' and thereby, slowly emerged all sorts of problems for man and consequently, in differing patterns for other already surviving species and forms of life, at various stages in this world. Unfortunately the hard tragic fact is that man himself with his recklessly greedy behaviour and indiscrete initiatives, has been responsible for his numerous problems.

We find that the above mentioned ironical truth paints quite a mixed scenario with respect to the mankind since relatively ancient of times. Well, these may be considered matters of 'conjecture' only and their further in-depth exploration could befittingly be areas of analytical studies and research for the concerned learned experts as well as academicians. But at the same time, even such perceptions do also have a reasonably acceptable logic and accordingly in the fitness of our venture we may accept them so, in the furtherance of our deliberations. In the above paragraph we have identified that since the pre-historic period man has been 'driven' in highly complex and multi-dimensional manners, by his two natural and exclusive instincts, i.e. those of 'survival' and 'progress'. This aspect is special to the man only, and not to other animal or species. Logically here, the point arises as to what has been at the 'root' of this interesting phenomenon? We may proceed to find some answer to this by looking into certain relevantly specific thoughts, as enunciated in an article, 'Mysterious Tune Played by an Unknown Piper' as follows: "What is that we seek most in life? **Advaita guru Ramesh Balsekar** says that whether we acknowledge or not, the one thing that all of us want most is to be at peace, both with ourselves and with others—including one's family, friends, colleagues—with whom we need to interact during one's life. Most of dissatisfactions stem from an analysis of 'what is' to a projection or expectation of 'what should be'. The 'what is' is what life presents us with from moment to moment. The 'what should be' is the mischievous play of **mind,** conditioned over years of dissecting issues and events of the past and projecting these into possible scenarios in the future. Unwittingly, we fall into a trap which hampers peace. Expecting others to respond or behave in a manner we are comfortable with is the fundamental problem-creator. Thereupon, inevitably, the **mind** gets more and more entangled, with growing expectations that include behavioural, emotional, financial and societal needs, spanning individuals, groups, nations and the world. **Gurus like Ramakrishna and Ramana Maharishi have said that in the phenomenal cycle of evolution, change, decay and regeneration, there never was or will be a time when the world or society will be totally bereft of concerns or calamities. Life necessarily involves individuals who are programmed distinctly and differently and who enact and play various roles**. Some are leaders, some champion a cause, some act as disrupting influences and many are bystanders, participating and contributing to the flow of events, sometimes with success and other time without, despite the best of efforts.....**we all dance to a mysterious tune intoned in the distance by an invisible Piper**."

What all we find in the above quoted 'paragraph', appears in our context at this point of our deliberations as truly enlightening. Moreover, leads us further to a higher 'plain' that overlaps both scientific and spiritual enriching of certain 'thoughts'; specifically with respect to the unendingly complex phenomena of the evolution as well as multi-dimensional development of the man.

Interestingly we may also say here that man's "brain" may quite befittingly be considered as 'hardware' and then, the "mind" therein as 'software'. This unique and unparalleled 'combination' is perpetually active. And therefore, this highly fascinating 'combination' has been continuously triggering and driving all kinds of the consequential aspects, 'sprouting' as per all sorts relentless activities of man's; they are both so fascinating and at times very disturbing and unsettling too, in this world. In fact, these precisely are the aspects which have been somewhat deliberated upon, in certain basic 'scientific' as well as 'spiritual' perspectives, in the above paragraphs in this chapter on 'Pre-Historic Period'.

Now coming back to man's relentless activities with apparently precise 'driving forces' behind them, as brought out in the above paragraph, we may logically highlight the specific 'reasons' behind the 'triggering-mechanisms'. Those particularly specific two 'reasons' are

1. Man's natural evolutionary instinct for **'survival'**.

2. Then as he very slowly and gradually began developing and so transforming as a 'complete-man', with the dawn of 'pre-historic period' in this world and thereby simultaneously as per Nature's command, he did acquire the irrepressible urges to continuously keep making multiple '**progress**'.

Therefore, now we see man's natural instinct for 'survival' and also, his late acquired urges for 'progress', as 'core-reasons' for all these phenomena; pertaining to man's multi-dimensional changes as well as development, along with their multiple 'fallouts', as gradually initiated by the man himself in this world. In fact, these too, are part of the highly complex phenomena of mankind's multi-dimensional developmental phenomena. Interestingly this specific 'scientific-perception' here, appears to be in fascinating harmony with the spiritually illuminating 'article' by **Vijay Vancheswar**, titled as, "Mysterious Tune Played By An Unknown Piper", a significant part of which has already been quoted above in this chapter itself.

As centuries passed, man did achieve multi-dimensional developments and these in turn, enabled him to organise his kinds into 'primitive tribes', build primitive shelters and keep making efforts to hunt and gather for a living. During the beginning of this pre-historic period a number of different independent tribes sprang-up' in unconnected geographical locations in the world. In turn thus, one particularly great as well as very special and far reaching "idea/ideas" began to 'germinate' and take initial shape, in his exclusive gift of 'brain'. This aspect too was in turn a 'consequence' of since his unique 'mind'—as built-in as earliest and most amazing 'software'—and already having been placed inside his that brain; got gradually more and more activated in surprisingly complex and multi-directional manners. Interestingly, this was, in fact, a beginning of a primitive form of **"organised-defence"** and thereby as a significant consequence, initial fructification of the **"matters-military"**.

The initial conceptions of the 'organised-defence' and 'matters-military', were indeed a singularly great turning-point in the 'human-domain'. Further another significant point is that these were gradually born during the pre-historic period itself. At this stage of our "Strategically Critical Journey", we may keep one important point in mind that in fact, the beginning of the essentiality of ensuring requisite 'self-protection', was already present in the most primitive 'mind' of the man, i.e. even much before the dawn of 'pre-historic period' in this world. This natural instinct of self-protection has always been present in all animals as well as in all forms of life, since the very beginning.

The 'birth' of different 'human-tribes' in different geographical locations and also during different centuries in the prolonged pre-historic period in this world, along with the above mentioned two far-reaching significant 'ideas' encompassing the man's effective protection aspects,] did ensure as a consequence perhaps the 'greatest' turning-point in the remarkable history of man's evolutionary process. These fascinating aspects; right from their beginning had multiple and highly complex 'fallouts', that persist even today.

Today, these socio-cultural 'negative' fallouts, especially in the present modern-era, have also been ironically instrumental in pushing the mankind into typical "catch-22" situations. Here again especially in the present modern-era paradoxically, the future 'survival' of the human species itself is now facing a shattering 'question-mark'. Hence there are greatly 'tragic' challenges facing the modern man. These aspects with their allied and highly 'menacing' new fallouts would be discussed in some detail, appropriately in the later chapters.

While studying some 'historic' narrations of this prolonged era by certain noted scholars, we learn certain very interesting points specifically with respect to the above explained 'protection aspects' of man and his various tribes. Since a very long time it is already an established fact that, all these on-going multi-dimensional human-advancements, during hundreds of millennia, have been possible as principally triggered by man's "analytically thinking, imaginative and pro-actively innovative mind". **Fr. Vellamthadathil** has written in his article on 'Civic Education' that, "*The two distinctive characteristics that distinguish man from animals are rationality or intelligence and moral or ethical sensitivity. The prospects and progress of human development depend entirely on the development of these two dimensions of human personality.* One who does not know his rights and duties is a slave needing a master. Infants and children whose intelligence and moral sense are yet to develop need and are naturally under the parental tutelage until they reach the age of maturity."

As per above, it is also true that after man gradually developed into the stage of 'complete-man' owing to his initial exploratory 'activities' and 'initiatives'; both at individual as well as at the tribal levels in different geographical locations over some centuries. As many centuries man became more 'focused' and intense.

Naturally as a number of pre-historic human-civilisations were born there had to be gradual fallouts in civilisations, again over a few centuries, at then socio-psychological, cultural and socio-political levels. Moreover, since the gradual 'births' of the civilisations were seldom smooth and peaceful these in turn created many even fiercely conflicting tribal-cum-social groups within any particular civilisation in any specific geographical territory. This development was also due to the gradual increase of populations as well as consequential expansion of the various civilisations based on the then 'socio-political' territories. As a further logical consequence, these developments significantly altered many factors in man's 'security-considerations' and thereby affecting and reworking on the existing aspects of the initial 'organised defence', as well as also that of the allied and already 'sprouting' preliminary 'matters-military' too.

These in turn again as a part of the overall evolution of man in this world further aided and abated the 'birth' of the distinct and highly significant "profession of arms", having different orientations-cum-value-systems as well. This was another turning point. Further, many kinds of 'conflicts' began, as well as their nature and range too expanded and altered. In this perspective, **Field Marshal Viscount Montgomery of Alamein** in his book *A Concise History of Warfare* (pages 21 and 22) writes that, "War has been constant in human affairs since the earliest societies of which there is a record. In 7000 B.C. Jericho was strongly fortified by a wall 21 feet high; out of 2,500 citizens 600 would have been fighting men armed with bow and arrow. But the art of war had clearly been evolving for long before those days. How then did war first arise? *War is a basic part of history because it is concerned with the essentials of life. Food and a secure place in which to live were the two basic necessities for primeval man—just as they are for us. But these things which man needs, as well as many other things which he desires, such as mates, wealth, power and prestige, are often available only in short supply. The basic reason why men and societies have almost incessantly fought each other lies in this economic fact—they have always had to compete for the minimum conditions for existence. Animals fight each other for the same reason. The urges of rage, aggression and fear are natural only because in a competitive situation each creature represents a frustration or a danger to his neighbour. Human beings of course, have been much less successful than animals in avoiding the resort to violence. What traces we can discover of the primitive peoples of ancient times show that from the first, men have had to organise themselves for defence.... Few generalizations can be made about the bellicosity of primitive peoples. Motives of war vary, and the degree of ferocity depends on circumstances.* Some tribes slaughter all their prisoners, while other treat them leniently.... Tribes ruled by kings are usually more prone to war than those with more democratic form of government. The most common and infallible cause of war among primitive peoples is overcrowding. Population has to be kept down, and this has generally been brought about by war or by disease. The history of Red Indian tribes of North America illustrates well how territorial possessiveness can lead to war.... War can appeal because it is exciting, it can be amusing; it can also offer easy profit, colour and romance, discipline and ritual, self-sacrifice and comradeship, status, and the admiration of women.... But a more fundamental explanation of the frequent occurrence of war may be the deep rooted desire of men to belong to groups. If loyalties and a sense of group identity and patriotism are to develop in a society, it must perhaps be positively antagonistic to its neighbours.... The factors which underplay all military and political developments in the history of the Near East during the four thousand years before Christ was overcrowding of population. The first civilisations arose in great river valleys, Babylonia and Sumeria on the lower reaches of the Euphrates and the Tigris, and Egypt astride the Nile."

So in other words we can also say that since hundreds of centuries of man's 'living-history' though often extremely ruthless and violent, ironically the 'war' and 'conflicts' have also been singularly instrumental in promotion, enrichment and territorial expansion of the various human-civilisations. In this context in fact, many prominent examples can be cited such as those of Egyptian around Nile, Greek, Chinese, Indian, Jews in Palestinian region (i.e. the land of Bible.), Mesopotamia region (i.e. modern day Iraq), Babylonia and Sumeria in the lower areas of the Tigris and Euphrates, the Arabia, the ancient Romans, the regions of the Mediterranean Sea, vast Asian and African regions; and also the European Region and so on. It has also been emphasised earlier that in fact, war and conflicts are as old as the history of man itself. Logically therefore, we may say that both the man's own evolutionary process of development and progressive advancements of various human-civilisations have always been quite inseparable from all possible 'manifestations' of the conflict and war.

In his scholarly book *A Military History of Ancient India* (pages 48, 49 and 64), **Major General Gurcharan Singh Sandhu** writes, ***"A constant struggle for survival against the elements, the environment, wild animals and fellow human beings is a major part of the heritage of man.*** The fashioning and use of tools distinguishes Homo Sapiens from other higher animals, the discovery of tools could double up as weapons against animals and for self-protection must have followed not long afterwards. Conflict between individuals is thus as old as mankind although more organised warfare between groups probably followed the establishment of semi-permanent settlements on reaching the food gathering stage.... *(In Indian Sub-Continent)* Microlithic tools found in the sand dunes of southeast coast in conjunction with pottery can be dated back to around 4000 B.C. These are beautifully faceted pieces of chalcedony set in handles of bone, wood or horn. Weapons and tools found there are javelins, arrows, harpoons, knives and sickles. Other late stone-age tools and weapons like arrowheads and sickles are found all over the country. Such small implements which belong to a period, 5000 to 10,000 years back, also found in Kurnool district, *(In India)* signal a transition from a food gathering to a grain harvesting stage.... The economic prosperity of the Indus makes it probable that a military class of some kind could have come into being. The main accomplishment of such a class, if it existed, was the vast extent and uniformity of the Indus culture; it could not have spread over such a large area without the use of some form of coercion."

Like other prominent and very distant ancient human-civilisations in other continents, dating back to even pre-historic times; the Indus-Valley civilisation too, has a very fascinating saga to narrate for posterity. **Military tactics, strategy, military technologies and 'Revolution in Military Affairs (RMA)'** too equally advanced in a 'parallel' and historically fascinating mode. In fact it is equally true that since the pre-historic era necessities have always been triggering very important breakthroughs, through the accelerated and surprising cycles of the multiple human innovations and inventions. These aspects would be taken-up later, in the fifth chapter on the 'Military Aspects and Their Parallel Progress'.

Again specifically in the Indian context, the first Prime Minister of independent India, **Late Jawaharlal Nehru**, in his book *The Discovery of India* (pages 49 and 50), has asked certain weighty questions pertaining to India: "What is this India, apart from her physical and geographical aspects? What did she represent in the past? What gave strength to her then? How did she lose that old strength? And has she lost it completely? Does she represent anything vital now, apart from being the home of a vast number of human beings? How does she fit into the modern world? *(And then on page 50)*....but surely India could not have been what she undoubtedly was, and surely could not have continued a cultured existence for thousands of years, if she had not possessed something very vital and enduring, something that was worthwhile? What was that something? I stood on a mound of Mohenjo-Daro in the Indus Valley in North West India, and all around me lay the houses and streets of this ancient city that is said to have existed over five thousand years ago; and even then it was and old and well-developed civilisation. 'The Indus Valley civilisation', writes Professor Childe 'represents a very perfect adjustment of human life to a specific environment that can only have resulted from years of patient effort. And it has endured; it is already specifically Indian and forms the basis of modern Indian culture.' Astonishing thought: that any culture or civilisation should have this continuity for five or six thousand years or more; and not in a static, unchanging sense, for India was changing and progressing all the time. She was coming into intimate contact with the Persians, the Egyptians, the Greeks, the Chinese, the Arabs, the Central Asians, and the peoples of the Mediterranean. But though she influenced them and was influenced by them, her cultural basis was strong enough to endure."

Well, the above quoted magnificent historic facts concerning India are no doubt true and, internationally too are well recognized. The paradoxical reality is that since the advent of this great as well as one of the oldest civilisations, as way back to the many millenniums, its dawn in the now so-called the 'Indian Sub-Continent' and thereby since ages, its inhabitants now logically known as the 'Indians' ironically; have been so regressively lacking the **"multiple-aspects of essential Strategic-Culture and Thoughts"**. The extremely sad fact is that since the beginning, even historically, culturally, as well as socio-psychologically it is not that simple to 'rationalise' or even somewhat remotely 'justify' this so 'embarrassing' a lacuna—among both the Indian masses as well as the noteworthy 'torch-bearers'—of indeed the great, so glorious, and also the very ancient Indian-civilisation.

Perhaps the sadly lingering 'passiveness of Hindu religion', as well as its ironically misguided and also totally misplaced 'spiritual sense of resignation'—with respect to the geo-politically strategic 'pro-activeness'—in the domain of the then so-called 'India's socio-political life', could be the principle reasons behind this lacuna. Further, another unfortunate lacuna has been that since the beginning, Indian civilisation was not able to cultivate any 'vital-cum-fruitful sense of territorial geography', as well as even the then prevailing initial scientific techniques of the geographical 'map-making' in any manner whatsoever. Nor they at any time, even cared to gradually evolve and suitably imbibe the same at any given timeline of the Indian 'living-history as being part of India's own suitably visualised strategic-imperatives. Unfortunately, this 'regressive' phenomenon, in fact, continued till the arrival of the British and also that of the other competing and seafaring European colonial-powers in India, during around 1600 A.D. when she was under the rule of the powerful but crumbling 'Mughal dynasty'. Well, these are certain plain facts of India's long history and hence, the same now may be taken only with a desired sense of historic objectivity.

The point here is that there is no 'intention' of any kind to even remotely criticise or to 'deride' the Indian or any other civilisation, culture, faith, value-systems or politico-strategic social 'dynamism' in any manner whatsoever; but pragmatically always to bear in mind the larger 'focused-aim' to keep proceeding with the "Strategically Critical Journey: Through Humanity's Past, Present And Possibly Future: In Global Geostrategic And Politico-Military Perspective" with necessary historic-objectivity in these regards. But at the same time, any 'slip-up' or any kind of literary or perceptive 'fault-line', ever having taken place during the course of carrying-out this task in hand, howsoever unintentional it may be, the responsibility/guilt for the same, would be entirely that of the author. And the author therefore, well in advance, apologises for the same.

At this juncture now if we may care to have a fleeting glance at the ancient Jewish peoples with their unparalleled geostrategic unbroken sense of proud tenacity, sense of unshakeable confidence, faith in their civilisation and then, in this historic perspective, we find the unique and enduring magnificence of their continuing saga since last thousands of years. In this respect it may be sufficient to refer to just the 'Introduction' part of **Leon Uris**'s well-known book, 'EXODUS', *where he talks about the magnificent birth of modern Israel as a superbly self-confident and dynamically strong Nation-State, on the eastern coast of the Mediterranean Sea in the Middle-East, says that, "(this) is the story of the greatest miracle of our times, an event unparalleled in the history of mankind: the rebirth of a nation that had been dispersed two thousand years before. It tells the story of the Jews coming back after centuries of abuse, indignities, torture and murder to carve an oasis in the sand with guts and blood.* All the cliché Jewish characters, which have cluttered up our American fiction, have been left where they rightfully belong, on the cutting-room floor (*yes indeed.*). I have shown the other side of the coin and written about my people who, against a lethargic world and with little else than courage, conquered unconquerable odds. Exodus is about fighting people, people who do not apologize either for being born Jews or the right to live in human dignity. Their story was a revelation to me as I discovered it in the farms and cities of Israel" (*how true and also historically irrefutable too.*).

As we proceeded with a brief and yet so much fascinating discussion about the very prolonged-era of the 'Pre-Historic Period' in this chapter in hand; it has been considered relevant as well as necessary too, to write the immediately preceding few paragraphs with respect to the next chapter on 'Advent of Human Civilisations and Ancient Era', as per the demands of imperative continuity here; mentioning certain specially interesting facts regarding few so ancient and well-known human-civilisations. As we have seen that the magnificent 'seeds' of ancient human-civilisations, with 'in-built' potential for achieving continuous and marvelous multi-dimensional 'expansive' progress and advancements, were gradually 'sown' during this period itself. Further in this respect, it may be interesting to note that again centuries had past, before those 'seeds' even began to 'visibly' sprout in their concerned geographical areas in this world.

Then again the inbuilt factor of complex 'opportunism' in the Nature's own unfathomable methodology is perpetually and pro-actively present; and where somewhat even marginally apparent 'phenomenal game of the free for all' always taking place, as it may sometimes or otherwise naively appear to be the case. But in fact, that is never the scientifically correct 'picture'. One important scientific reality always required to be borne in mind is that, in spite of the presence of above mentioned so-called complex 'opportunism' there is always mother-Nature's own evolved and at the same time continuously evolving "system" of amazing 'checks and balances'. And thereby these in turn are unendingly in 'action' and hence, pro-actively keep initiating the requisite 'balancing acts' too, about the factor of 'opportunism', as per Nature's own unseen patterns of perpetual 'command and control'.

Interestingly in our context of the on-going discussion on this 'sub-topic' at this stage here, we may also find the presence of certain 'visibly' consequential differences, with specifically 'selective' reference to the aspects of certain perhaps evolutionary features among various noted ancient human-civilisations in this world. We notice that though the 'progress' is always incorporated into the unending processes of Nature's evolution; with reference to man here, this phenomenon of progress is not always 'positive or for the good. It may be prudently relevant to look into certain specifically important paragraphs particularly relating to man by **Professor George Gaylord Simpson.**

"Man has certain diagnostic features which set him off most sharply from any other animal and which have involved other developments not only increasing this sharp distinction but also making it an absolute difference in kind and only a relative difference of degree. In the basic diagnosis of *Homo sapiens* the most important features are probably interrelated factors of intelligence, flexibility, individualization, and socialization.... In man all four are carried to a degree incomparably greater than in any other sort of animal. All have as their requisite and basis the still more fundamental evolutionary progress 'in the direction of increase in the range and variety of adjustments of the organism to its environment,' which involves increased and improved means of perception of the environment and, particularly of integrating, coordinating, and reacting flexibly to these perceptions.... *(On page 294, he writes that)* Man is a moral animal. With the exception of a few peculiar beings who are felt to be as surely crippled as if the deformity were physical, all men make judgments of good or bad in ethics and morals. All feel some degree of compulsion to value and promote the good, to condemn and eliminate the bad.... Man almost inevitably acquires an ethic and this responds to a deep need in any normal member of the species.... *(On page 308, he writes that)* The meaning that we are seeking in evolution is the meaning to us, to man. The ethics of evolution must be human ethics. It is one of the many unique qualities of man, the new sort of animal, that he is only ethical animal. The ethical need and its fulfillment are also products of evolution, but they have been produced in man alone....*(On page 313 and 314, he writes that)* Human responsibility requires, in each individual as well as in society as a whole, that the search for knowledge be search for truth, as unbiased as is possible to human beings; that probable truths as discovered be tested by every means that can be devised, that these truths be communicated in such a way as is most likely to ensure their right utilization and incorporation into the general body of human knowledge, and that those who should receive this knowledge seek it, share in its communication, and in their turn examine and test with as little prejudice as possible whatever is submitted as truth. This is large order, indeed, but a necessary one. It involves responsibilities for every living person, and responsibilities that cannot be ethically evaded; that is their evasion is morally wrong.... (*Then on page 325, he writes that)* Evolution is clearly going on around us today. There is no obvious reason for its not continuing for many millions of years to come, as regards both man and the rest of the organic realm. It would be futile, in the present state of knowledge, to try to predict just what this future evolution will produce."

The above quoted lines from an already earlier referred book, do also convey the noted author's learned perceptions with respect to the Nature's grand phenomena of man's evolution in this world to our great benefit here. **First point is Nature's exclusively unique gift of "brain" to the**

Homo Sapiens, **in the beginning itself. And the other is an all-encompassing aspect of nature's 'law' of "survival of the fittest", which is equally applicable to all other forms lives everywhere.**

Here we may logically say that man's 'attributes' and so on, must have been applicable to him, only after man had already evolved into a 'complete man', through a passage of thousands of centuries. Further during this prolonged process of slow and consistently evolutionary development of man in this world, he must have gradually evolved from 'ape' to 'sub-human' and then subsequently to the earlier mentioned 'complete man' stage.

In this regard, a particularly interesting point to bear in mind is that the man at his inception stage itself, had already been bestowed with a "brain", with many special qualities. Now as this continuing evolutionary development of man progressed, correspondingly his unique 'brain' too, gradually began to develop in an appropriately befitting manner. Another point of very special interest at this stage is that by the time man's 'brain' had developed into an 'analytically thinking, imaginative, and innovative one', a very long period had already passed over the unending 'canvas of time' and also thereby then, the 'dawn' of this 'Pre-Historic Period', under discussion here, had gradually taken place in this world.

Now therefore, his most important features of the interrelated factors of intelligence, flexibility, individualisation and that of socialisation were decisively responsible for making exclusively far-reaching and powerfully dominating differences in favour of only the man, in comparison to all other animals and species, in this world.

In clear scientific perspective, we find that the unending 'processes of evolution' is Creator-Nature's most important natural, inherent and unstoppable 'attribute', in fact since the basically hypothetical 'birth' of the Universe itself. Here factually as well as perceptively and irrefutably established logical point is that the 'Universe' is infinite and the fundamental 'nature' of Nature's process of evolution is again the unending phenomena of 'creation, growth, multi-dimensional development, advancement (again, multi-directional and multi-layered.), perpetually competitive 'survival of the fittest' and finally selective 'destruction' under the overall built-in and 'continuous command and control' of the all-powerful mother-Nature. This unfathomably complex and unending phenomena is always mind-boggling and also fascinatingly continuing in the entire infinite Universe itself. Logically therefore, these are equally as well as appropriately applicable to our world too.

Now at this point of our interesting discussion, as already brought out earlier too, that the phenomenon of 'destruction' is notional only and therefore, certainly not in commonly understood sense. In fact, in the domain of Nature's unending as well as always unfathomably complex 'processes of evolution', this so-called phenomenon of 'destruction' is a misnomer and nothing else, and the same has been clearly explained in the chapter of 'Introduction' itself. Logically now, these perceptions are inducing us towards scientifically 'conceptualising' the highly delicate and yet assuredly stable and unalterably 'Grand Balance' within the infinite Universe, yet without its beginning ever or any kind of impending end. And therefore, it may be said with a reasonably acceptable scientific basis that these appear to be in fact, the unfathomably complex mysteries of the infinite Universe, in all its magnificence and continuously altering manifestations.

Nature has its own certain special 'features' specifically applicable to this planet Earth, since the slow and gradual beginning of life here. We also know that such special 'features' are not there in other planets in our solar-system. At the same time to the best of latest scientific knowledge and exploratory discoveries of the man today, no trace of any known 'form of life' has been 'found' anywhere in the outer space so far. Now coming back to our own world over here, in the course of our present specific deliberation we have noticed certain very fascinating, complex, multiple and many innumerable outcomes due to such special 'features', since the beginning of life on this Earth. **The most surprisingly fascinating outcome in this regard, is certainly the "advent of man" on this earth**.

Indeed the exceptionally long history of this unending process and nature of evolution in this world is unfathomably complex and fascinating to us. Therefore, naturally the subject-matter of this multi-dimensional history has been very attractive and intellectually very challenging to so many scholars as well as research scientists all over the world, since a very long time. During the pre-historic period, **man as always being activated by his so exclusively unique 'brain' was naturally able to trigger multi-dimensional revolutionary changes and advancements in his own patterns of life as well as around his surroundings in multiply far-reaching manners. Thereby the most fascinating development, as a highly special outcome has been the long process of the advent of various 'human-civilisations', during different periods of time and also at different geographical locations in this world**.

Accordingly in fact, these processes of the 'births' of the various ancient 'human-civilisations' did in fact, take place at different times, during prolonged pre-historic period. This entire phenomena of the fascinatingly complex 'sprouting' of these civilisations, have also been great areas of highly engaging studies and research; since ages for the scholarly world defying all sorts of political-cum-national 'boundaries' and, the same even now is continuing.

In this context we wish to again turn to the 'Future Shock' by **Alvin Toffler (**pages 28 and 29) where he says, *"All things—from tiniest virus to the greatest galaxy—are, in reality, not things at all but processes. There is no static point, no nirvana-like un-change, against which to measure change. Change is, therefore, necessarily relative.... We know, for example that compared with the biological evolution of the species, cultural and social evolution is extremely rapid. We know that some societies transform themselves technologically or economically more rapidly than others. We also know that different sectors within the same society exhibit different rates of change—the disparity that* ***William Ogburn*** *labelled 'cultural lag'. It is precisely the unevenness of change that makes it measurable. We need, however, a yardstick that makes it possible to compare highly diverse processes, and this yardstick is time. Without time, change has no meaning. And without change, time would stop. Time can be conceived as the interval during which events occur.* Just as money permits us to place a value on both apples and oranges, time permits us to compare unlike processes.... Time is the currency of exchange that makes it possible to compare the rates at which very different processes play themselves out."

Here the learned author has so scholarly made efforts to enable us to conceptually visualise the magnificent interplay of cosmos-based dynamic and evolutionary inter-dependent 'relationship' among the processes, time and space, as unendingly playing themselves out, under Nature's universal 'laws of continuing evolution', at both the 'micro' as well as 'macro' levels. All the above brought out the interplay of those principal 'factors', such as the processes, time and space, and as visualised in the infinite universal perspectives; these entire phenomena are, in fact, both unfathomably vast and complex. In addition, they are unknowably mysterious too, and also they remain capable of baffling the otherwise magnificent reaches of fast advancing now the modern-era frontiers of the sciences and technologies as well. Further in this context, it is very interesting to learn that even for the highly accomplished saints with great spiritual-vision, the Infinite Nature's these again infinite and unending evolutionary phenomena, both in their immeasurable depths and dimensions remain insolvable mysteries. Here we are able to somewhat visualise in a perspective, the infinite magnificence of the mother-Nature's unending grand processes of evolution.

Therefore in our present specific context of deliberation, it appears to be highly relevant to quote **Professor George Gaylord Simpson**, *"This is not to say that the whole mystery has been plumbed to its core or even that it ever will be. This ultimate mystery is beyond the reach of scientific investigation, and probably of the human mind. There is neither need nor excuse for postulation of nonmaterial intervention in the origin of life, the rise of man, or any other part of the long history of the material cosmos. Yet the origin of that cosmos and the causal principles of its history remain unexplained and inaccessible to science. Here is hidden the First Cause sought by theology and philosophy. The First Cause is not known and I suspect that it never will be known by the living man.* We may, if we are so inclined, worship it in our own ways, but we certainly do not comprehend it." *(Here surprisingly and yet, intellectually in a highly enriching perspective, we find again a confluence of the excellence of man's perpetually on-going pursuits of both science as well as 'spiritualism'.*

As earlier already brought out that, in this macro-level processes of competitive evolution, man as an 'animal' to start with and since especially 'blessed' so exclusively by the mother-Nature; does command an unchallenged competitive advantage as well as 'supremacy', with respect to other animals or any other form of life on our planet-earth. Therefore, as a consequence such unrivalled 'assets' of man, have been perpetually making a 'world of difference' for him. Logically man's these gifted aspects have so interestingly enabled him to gradually transform in the developmental mode, from 'ape' to the 'sub-human' and then to the stage of 'complete-man' in the distant past, over a period many thousands of years and even before the advent of the pre-historic period. And 'complete-man' as a distinct 'specie', is in fact never static or even irrespective of the resultant 'good' or 'bad' for him or even to his environment or for that matter, for the overall well-being of the world at large, in which he himself continues to live in.

Subsequently, as a number of millenniums went by on the 'infinite canvas of time', man got transformed from masquerading crowds/herds to that of the primitively settled tribes. And at the same time, man's livelihood gradually changed from principally hunting to that of different forms of primitive agriculture; to start with, in the different and far-off geographical locations. We may prefer to add here that man's multiple developmental process is also a continuous phenomenon, as a part of the overall completely unknowable and unending phenomena of the Nature's grand processes of evolution.

Coupled with them, **the initial challenges of the security-aspects too, as per the then emerging various kinds of threats to such settled and slowly getting 'civilised' tribes; began to get noticed by the men belonging to other tribes and thereby, leading to some kinds of primitive actions/ precautions then, for organizing 'defence' as well as ensuring some security.** These phenomenal aspects and many other allied matters would be discussed in some detail in the next chapter.

*

CHAPTER 4

Advent of Human Civilisations and Ancient Era

In the previous chapters at a number of places, we have had some glimpses of the causal phenomena leading to the advent of various human civilisations during different times and at different geographical locations in this world. In fact these evolutionary developments with respect to the man, during distant ancient times have indeed been historic and magnificently phenomenal as well. In addition, again the above noted 'evolutionary developments' had taken place very slowly as well as gradually for some thousands of years. But at the same time, these observations as per above, with respect to the time-period in quantitative terms, are just 'relative' and therefore in absolute-terms any such narration of a so-called any specified 'time-period' as per well-accepted scientific rationale in fact, means nothing. These perceptions regarding 'time' as a commodity, asset or even 'currency' have been very convincingly illustrated by the well-known scholar and writer Alvin Toffler, as so appropriately and gratefully quoted in the end pages of the last chapter on 'Pre-Historic Period'. Indeed this well-known author in his usual scholarly manner has been able to put this far-reaching conceptual framework concerning 'Time' as such, in a clear perspective and also so understandable even by any average 'student' of these subjects.

Further as already mentioned and earlier quoted above, **Alvin Toffler** has also scientifically clarified that in comparison to the process of biological evolution of species, cultural and social evolution is very fast, as per the above elaborated and conceivable 'template', as needed for quantitative expression of 'Time' as a distinct and consumable/spendable 'commodity' in any given situation. Now coming to the specific 'topic' in hand at present for our discussion, it is interesting to note that the above mentioned process of relatively very fast cultural and social evolution, in fact specifically pertains to the man, as distinctly and rapidly evolved along with the Nature's exclusive gift of his so 'unique brain', since already contextually talked about a number of times in earlier chapters.

We may befittingly add here that without this brain no such rapid cultural and social evolution would have been possible and at the same time, with such evolution, no 'triggering' of the advent of various ancient human-civilisations, would in any case have taken place.

Another interesting point to appropriately notice in this regard is that though the 'birth' of a number of diverse ancient 'civilisation' was indeed taking place very early at different far-off places and in different centuries, man was not so much knowledgably aware about these far-reaching evolutionary developments specially concerning him and also, he never then 'named' them any 'civilisation' as such, as we now understand the substantial 'meaning' of this term, 'civilisation'. In this context, late **Professor Samuel P. Huntington** has indeed very informatively written, **"The idea of civilisation was developed by eighteenth century French thinkers as the opposite of the concept of 'barbarism'. Civilised society differed from primitive society because it was settled, urban and literate**. To be civilised was good, to be uncivilised was bad.... Civilisation and culture both refer to overall way of life of a people, and a civilisation is a writ large. They both involve the values, norms, institutions, and mode of thinking to which successive generations in a given society have attached primary importance." This conception with respect to all thinkable dimensions so far-reaching, as well as depth and range of overwhelming influences of any major historic civilisation in the world, are indeed perceptively very enriching.

Therefore, we may now be able to say that the fascinating advent of various diverse human-civilisation since ancient times, with all their highly exciting and unprecedented manifestations, is indeed the biggest as well as 'turning-point' in the entire evolutionary process of man, since his inception in this world. Each different civilisation, as gradually sprouted in different centuries in the distant past and also in different far-off geographical locations in this world, does have its own distinct features and multi-dimensional attributes. Further we also find that all of them do have one generally common specific feature, and so it happens to be that a civilisation is always dynamically vibrant, living and progressive as its fundamental nature of being itself. Hence always driven by this so special and common feature, a civilisation keeps throwing-up new kinds of 'needs' for itself. And since it has been correctly said that 'necessity is the mother of invention', capable members of any particular civilisation, do keep relentlessly striving to usefully unfold new inventions for their own advancement as well as betterment of the overall qualities of their lives in various spheres.

As a consequence, due to these interesting initiative and activities on the part of man in any particular civilisation anywhere in the world, did gift various types of 'connectivity', though definitely quite 'primitive' to begin with, ensuring further multiple progress for the mankind. In fact, right from the distant ancient times, such connectivity was not only meant for trade, commerce or even for advantageously conducting conflict against one's rivals/adversaries, but in addition also for fruitful exchange of thought-processes and valuable ideas. In this perspective, it has been rightly enunciated by the noted scholars that no civilisation has been able to develop in any kind of any thinkable 'splendid isolation'. Therefore, now we are able to see more clearly, the magnificently decisive role of the above mentioned highly creative 'gift' of this unique idea of 'connectivity' to the man by himself, even if quite 'primitive' in its nature and also in its 'infra-structural incarnation' to start with, in the beginning; it did become a great boon to him since the distant ancient era itself.

Interestingly and so informatively, **Professor S.R. Sharma** writes that, "We have also noted how the civilisation of the Indus Valley was not developed in isolation, but had connections with the neighbouring tracts, particularly Mesopotamia. In the hinterland between India, China and Mesopotamia, is the roof of the world or the Pamir Plateau with Afghanistan and Persia below it and the head of the Arabian Sea. Beyond these towards the west are Mesopotamia, Arabia and Asia Minor.... It is vain to discuss the origins of civilisation with respect to any definite country or any precise period of chronology. (From: 'When and Where Old Civilisations Begin?' by **M.C. Burkitt** in *Wonders of the Past*, I, 429-34). The progress of primitive man from dark discomfort to enlightened comfort depended upon two factors: First, his racial stock and native intelligence which enabled him to learn from experience and improve either rapidly or slowly. *(The cranial capacity of man has varied from his stock. That of the Neanderthal man was 1600 cubic centimeters. That of the Cro-Magnon man ranged from 1590 to 1715 c.c.)* Second, his geographical environment, including climatic conditions and natural resources like water supply, facilities for food-production, transport and quality of minerals and building materials available in the vicinity. Egypt was happy in the possession of all these pre-requisites of civilisation. Hence her marvellous achievements *have been* during antiquity. By 4000 B.C. she had emerged from nebulous pre-history and entered the more definite region of history. Of this she has left wonderful testimony in her pyramids, sphinxes, obelisks and mummies etc." (Sharma, 1948, pp. 17 & 73)

Since his inception of course, man is known to be a social-animal. But so are many other animals too, then what all have ensured man's overall unchallenged superiority over all animals and other different species? His multiple capabilities and multi-dimensional faculties, that have been singularly responsible for this height and unassailable commanding position in this world, at which he so securely stands, in comparison to all other animals, species as well as different forms of life. This is indeed an irrefutable statement of fact. Naturally, his exclusive asset of 'analytically thinking-cum-creative mind' and allied special attributes have been causal for his astonishing all-round progress over a very short span of some millenniums and at the same time as mentioned earlier, man's many pro-active as well as his amazing incentive-oriented multiple activities have been instrumental in the birth of many historically fascinating civilisations in various locations and also, during different 'time-frame' in the distant past.

Now before we engage ourselves in the brief exercise of narrating few important features of the leading ancient civilisations, it would be informatively interesting to recollect different highly scholarly and perceptive 'view-points', as available to us from well-known learned personalities and authors on these subject-matters, which are very authoritatively relevant to us for our present discussion at this stage. Interestingly we may recollect in our own ways that such highly attractive subject-matters, as pertaining to those of the civilisations, culture, values, norms, social-peculiarities, institutions and even that of thoughts and specific historic 'baggage' on 'imperialism' and so on, have always been engagingly 'fertile' areas since centuries for enriching discussion as well as knowledgeable debates by noted scholars, authors and thinkers.

Professor Samuel P. Huntington writes that, "In the summer of 1993 the journal *Foreign Affairs* published an article of mine titled 'The Clash of Civilisations?' That article, according to the Foreign Affairs editors, stirred up more discussions in three years that any other article they had published since the 1940s. It certainly stirred up more debate in three years than anything else I have written. The responses and comments on it have come from every continent and scores of countries. People were variously impressed, intrigued, outraged, frightened and perplexed by my argument that the central and most dangerous dimension of the emerging global politics would be conflict between groups from differing civilisations. Whatever else it did, the article struck a nerve in people of every civilisation."

Well, that's what specifically has been the conceptual 'foundation-pillars' of Professor Huntington's definitely weighty, thought-provoking and even sensationally highly controversial 'argumentative thesis' in this regard. It may also be said with reasonable certainty that this 'thesis' naturally had to stirringly do, what all it actually did the world over among the scholar and writer fraternity, belonging to the fascinating concerned areas of intellectual endeavour. In this context one thing needs to be fairly granted that in any case Professor Huntington too, had been entitled to his views and perceptions.

Interestingly in this regard **Professor Amartya Sen** writes that, "A remarkable use of imagined singularity can be found in the basic classificatory idea that serves as the intellectual background to the much-discussed thesis of 'the clash of civilisations', which has been championed recently, particularly following the publication of Samuel Huntington's influential book, *The Clash of Civilisations and the Remaking of the World Order.* The difficulty with this approach begins with unique categorization, well before the issue of a clash—or not—is even raised. Indeed the thesis of a civilisational clash is conceptually parasitic on the commanding power of a unique categorization along so-called civilisational lines, which as it happens closely follows religious divisions to which singular attention is paid. Huntington contrasts Western civilisation with 'Islamic civilisation', 'Hindu civilisation', 'Buddhist civilisation' and so on. The alleged confrontations of religious differences are incorporated into a sharply carpentered vision of on dominant and hardened divisiveness.... The 'clash of civilisations' was already a popular topic well before the horrifying events of September 11 sharply added to the conflicts and distrust in the world. But these terrible happenings have had the effect of vastly magnifying the ongoing interest in the so-called clash of civilisations.... Let me illustrate the issue by considering the way my own country, India, is treated in the classificatory system. In describing India as a 'Hindu civilisation', Huntington's exposition of the alleged 'clash of civilisations' has to down play the fact that India has many more Muslims than any other country in the world with exception of Indonesia and very marginally Pakistan. India may not be placed within the arbitrary definition of the 'Muslim world', but it is still the case that India (with its 145 million Muslims—more that the whole British population and the entire French population put together) has a great many more Muslims than nearly every country in Huntington's definitions of the 'Muslim world'. Also it is impossible to think of the civilisation of contemporary India without taking note of the major roles of Muslims in the history of the country." (*Interestingly as well as very knowledgeably we indeed find here very powerful 'counter-arguments', as enunciated by Professor Amartya Sen in this regard.*). (Sen, 2007, pp. 10, 40, 46 & 47)

In addition, we also interestingly come across some highly knowledgeable writings in the scholarly and well-received books of famous learned thinker and writer, late **Professor Edward W. Said.** Professor Edward Said writes that, "Although these categorizations are not exhaustive, they do, I think accurately describe a general policy with considerable social authority, which created what D.C.M. Platt called in the British context a 'departmental view.' The leading academic figures analysed by Gendzier—Huntington, Pye, Verba, Lerner, Lasswell—determined the intellectual agenda and the perspectives of influential sectors of the government and academy. Subversion, radical nationalism, native arguments for independence; all these phenomena of decolonization and the aftermath of classical imperialism were seen within the guidance provided by the cold war. They had to be subverted or co-opted; in the case of Korea, China and Vietnam they required a renewed commitment to expensive military campaign." (*We feel that certain interesting points in these contexts may be that in the domain of internationally reputed scholars and authors, belonging to different 'continental/national' backgrounds, it is perhaps logically natural that their respectively differing 'view-points' would be based on highly illuminating as well as intellectually penetrating and scholarly 'argumentative perceptions'; while debating such 'controversial ideas/issues' in inter-disciplinarily elevating manner. These well-known aspects are indeed very much desirable and highly valuable too, in the best* *interests of the multiple advancements of human-knowledge in this regard. N.B.: Italic is this 'author's' here.)* .... Further, Professor Edward Said very interestingly writes that, "In a quite constant way, Orientalism depends on its strategy on this flexible *positional* superiority, which puts the Westerner in a whole series of possible relationships with the Orient without ever losing him the relative upper hand. And why should it have been otherwise, especially during the period of extraordinary ascendancy from the late Renaissance to the present? The scientist, the scholar, the missionary, the trader or the soldier was in, or thought about, the Orient because he *could be there*, or could think about it, with very little resistance on the Orient's part." (Said, 1994)

But the substantially important as well as historically far-reaching point very much 'irrefutably' stands at this stage of our deliberations too, that resorting to any type of 'clashes, conflicts or even may be war' remains, if any given situation 'perforce' warrants so, as one of the last options to 'advantageously' settle any 'pressing' dispute(s)/problem(s) among different 'human-groups' in this world. The point is that this 'perception', otherwise 'factual' too, remains corroborated, both as per the authenticated recordings of long human-history as well as the inherently fundamental 'laws' of unending evolution under the unchallengeable command of mother-Nature, in this world. In this context, we may wish to again refer to the highly perceptive lines from the scholarly book, "A Concise History of Warfare", by **Field Marshal Viscount Montgomery** of Alamein, as quoted in the last chapter itself. The Field Marshal correctly and very appropriately has stated that the war has been constant in human history since the earliest era and at the same time contextually, war is very much a basic part of the history as it is commonly perforce triggered by the essentials of life. All these may sound quite ironical but in any case, 'facts are facts'. Hence in any war, which side is 'right' and which is 'wrong', may naturally depend on one's perceptive stand/views and thereby, his/her 'line of debate' in this regard. Therefore in any manner, 'war' very much remains an irrefutable 'facts of life' even now, and historically that of human-society at large.

Contextually in the earlier chapters only, we have already briefly touched upon the historically basic parameters identifying the man's inherent urge to continuously engage himself, both individually as well as 'socio-strategically' in 'groups or political-entities', in his multiple pro-active thought-processes and initiative-oriented activities for his overall betterment/enrichment in this world. In addition here, we may also interestingly recollect certain basic facts concerning the process of evolution in this world and the inherently many types of threats/dangers too, start coming-up and taking shape with all sorts of complexities, menacing its very survival and its growth. Logically therefore, all sorts of multiple as well as mutually competing 'struggles' relentlessly begin in the concerned 'arena' for applicability-based survival and growth. These 'fascinating' but totally ruthless processes, inherent in the much larger and unfathomably complex overall phenomena of Nature's competing evolution too, may again naturally be termed as one of the segments of the infinitely all encompassing 'laws' of the same evolution. Therefore, these aspects of mother-Nature's unending multiple processes of evolution are equally and comprehensively true for the mankind as well since its inception in this world.

Thereby, to make the sum-total of these aspects comprehensively complete, when man's multi-dimensional thought-processes and initiative-oriented activities, triggering the far-reaching complex phenomena of the **advent of various human-civilisations** in surprisingly such fascinating manner too, are suitably integrated with them, naturally then the entire complex 'picture' becomes amazingly exclusive in this world.

These in turn, make the man a truly unique 'social-animal', unrivalled by any other animal or form of life, at least specifically on our planet 'Earth'. At this stage, it may be important to bear in mind that the above mentioned highly 'complex picture' is not static and inert or lifeless but, the same is in reality, quite dynamic and lively in multiple evolutionary-ways. 'Life' as such means continuous **'struggle for survival'** and ensuring the on-going 'developmental growth', especially for the man in this world, these in turn, create the compulsive ideas for complex **'threat-perceptions'**. The natural phenomenon of fascinating "cause and effect", is supremely being directed by the Nature itself.

As then once again interestingly we find here that the relentless 'struggle for survival' leads to the complex 'threat-perceptions' and further compulsively drives the man into the mode of tirelessly working hard for ensuring suitably reassuring **'organised-defence'**. When we proceed further we again very fascinatingly find, that man is unstoppably marching ahead, since the very distant past, coupled with also the continuous fructification of the **'matters-military'**.

As already discussed earlier, the 'pivotal' driving-force behind these unique and truly magnificent processes of continuing advancements of man's multiple development since ages, has indeed been his 'relentlessly active and analytically thinking-cum-creative mind'. **Dr A.P.J. Abdul Kalam**, a well-known scientist, teacher, popular intellectual-leader and former President of India, as well, he writes that, "A man is said to pass through different stages in his lifetime. **Dr Wayne W. Dyer**, in his book *Manifest Your Destiny*, makes an interesting categorization of them as athlete stage, warrior stage, states person stage and spirit stage. It occurred to me that nations, too, make a similar transition and in extending this analogy to them I have termed the last two stages big brother and self-realisation stages respectively. The stages do not follow in sequence necessarily; they can be coexistent with one aspect dominant. In the first athlete stage, a nation fresh from an independence struggle, or some other transition, embarks on an energetic pursuit of performance and achievement. This has happened in Japan, Singapore and Malaysia. When a nation leaves this stage behind, it generally enters the warrior stage. Proud of its achievements, it finds ways to demonstrate its superiority over others, perhaps through conquests. Ego is the driving force. During this stage people are busy with goals and achievements in competition with others and this, as Dyer points out for the individual, generates anxiety. Convincing others of its superiority becomes the theme." (Kalam, 2002, pp. 11 & 12)

Similar 'thoughts' had been first put-across by **Dr Wayne W. Dyer** with respect to the man as distinct individuals. Subsequently **Dr Kalam** has elaborated and made it more relevant to the nations as well. And therefore logically, at our this present point of deliberations over here, we may now very fascinatingly find that these very 'ideas' in their still enlarged and bigger context, are also very precisely relevant and even more applicable both in historic as well as futuristic perspectives, into the vast domain of the 'chequered' and still continuing multi-dimensional 'developmental-march' of the various ancient human-civilisations since ages in this world.

Though **Professor Samuel Huntington** in his already earlier referred book, has tried to identify different civilisations and accordingly named them, based on their particular dominant 'religion' or even religion-based specific 'culture'; but needless to say that this 'methodology' of identifying them has naturally been controversial. But understandably, Professor Huntington too, has his own logic and concerned justifying arguments in this regard. In any case for our purposes here, we may somewhat more befittingly proceed to identify the various human-civilisations, based on their identifiable geographical locations or regions, or by addressing them by the name(s) of their appropriate nations. Even if perforce any need arises then, we may try to mention any of them through the concerned communities as well. In these larger perspectives, we may also need to bear in mind that quite a few numbers of then the leading ancient civilisations have since long become extinct and thereby, they have already become 'subject-matters' of historic-past themselves. Well in any case, it is also known that such interestingly complex phenomena are very much part of the much larger 'dictates' and 'laws' of the mighty mother-Nature's unending process of evolution in this world.

Now coming back to **Dr Wayne W. Dyer's concepts** and also so contextually enlarged by Dr Kalam in his above cited well-received book; if we closely study the substantially apparent multiple aspects of the four different 'stages' of 'athlete', 'warrior', 'big brother' and then that of 'self-realisation', then logically as well as with great relevance, we may visualise that different human-civilisations at their different stages of existence and development do pass through these four 'stages' in their own 'self-governing' pattern. Further as Dr Kalam has so illustratively mentioned in the context of nations and here too, with respect to the human-civilisations, these 'stages' may not follow in sequence necessarily, and in addition any two or even more of those four 'stages' may co-exist at the same time, with a particular one remaining dominant for some period of time. But as we proceed further, we may notice that with the appropriate application of these 'four' interestingly precise evolutionary 'stages' to the aspects of human-civilisations, hidden multiple complexities do so fascinatingly get thrown-up into 'light' for further exploratory studies.

Here while referring to **Dr Kalam's** above mentioned book, as in the case of nations and so also logically as well as appropriately to that of various civilisations, comprehensive application of those four stages of 'athlete', 'warrior', big brother' and then 'self-realisation' would be taking us to many vast and fascinating areas of studies in altogether in different manners, in the case of different ancient civilisations. Then we would also be noticing certain historically factual relevance of even Professor Huntington's perceptions in this regard, howsoever these may be otherwise 'controversial' too, in high scholarly circles. It may suffice to say here that the above mentioned second 'warrior' stage would perforce be bringing us 'face to face' with certain factual aspects of the *"clash of civilisations"*, both in historic as well as present modern-era perspectives. Further, such 'civilisational-clashes' during different periods of human-history and in different far-off geographical areas in this world, may be seen both as 'inter-civilisational' and/or even 'intra-civilisational' as well

In any case, these are 'irrefutable' facts of history and hence, these precisely happen to be the reasons that we have earlier already called such vast evolutionary phenomena as chequered, since they happen to be always ruthlessly competitive. And further, the multi-dimensional 'developmental-march' of various ancient human-civilisations, since the very distant past in this world, is also equally 'chequered'. The above mentioned 'developmental-march' of various ancient human civilisations have not always been peacefully smooth and hence they have had 'mixed fortunes' at many a times. Further, there have indeed been such 'clashes' sometimes, as brought out above and therefore, these have also been termed 'chequered' as well. Now, we may wish to list some prominent ancient human-civilisations below:

1. Egyptian Civilisation: 4000 to 3400 B.C. onward.
2. Ancient Mesopotamian Region (Present-day Iraq and adjacent Areas): 4500 to 3500 B.C. onward.
3. Semi-historical China and the Dawn of history: 2852 to 2697 B.C. onward.
4. Indus-Valley Civilisation: 3500 B.C. onward.

5. Ancient India: 3500 to 2000 B.C. onward.
6. Semitic and Aryan: 2800 to 1925 B.C. onward.
7. Ancient Greece: 3600 to 1200 B.C. onward.
8. Ancient Rome: 1000 to 753 B.C. onward.
9. Europe: 476 A.D. onward.
10. Rise and also 'negation' of Islam: 570 A.D. onward and beginning of medieval as well as then, modern history in the making.

The data is based on the book *A Brief Survey of Human History* authored by **Professor S.R. Sharma of Fergusson College, Pune (India).** We notice that the civilisations belonging to the very old 'Indus-Valley' and that of 'Ancient India', have been listed separately for more focused attention in this regard as they are generally known but in fact, they are largely one in continuity and at the same time, both of them belong to the same ancient larger territory of the Indian Sub-Continent. Since ages, human civilisations with their gamut and multiple dimensions have indeed had many facets.

(a) Egyptian Civilisation: 4000 to 3400 B.C. onward

With respect to the Egyptian civilisation and all other civilisations we must first pay attention to the extraordinarily unique as well as allied assets belonging to this civilisation.

Even in today's modern world very befittingly, Egypt is so commonly known as the country of extra-ordinarily ancient Pyramids, Sphinx, Mummies and other historic relics and sites. And naturally these in turn do act as highly valuable assets for further exploratory studies. It is well known that most of the leading ancient civilisations had developed along the course of a perennial river-system and also where other important environmental factors have been supportive for their all-round growth. The Egyptian civilisation as too had come-up along the course of the well-known river Nile. In fact, the river Nile, in this context, has historically been unique and important in many ways, as we can naturally say so in case of the river Ganges, with respect to the many legends as well as highly fascinating historic details that are associated with it..

Hence very befittingly with respect to the Nile in Egypt, **J.H. Breasted** has said that, "The Nile has become for us a great historical disclosing stage by stage the first advance of man from bestial savagery to civilisation." No doubt such unique multiple contribution of the Nile towards the prolonged and truly magnificent as well as highly enriching growth of the Egyptian civilisation has indeed qualified it to lead the process of exploratory studies, in any context, of the leading ancient human-civilisations of the world.

The point here is that among the noted ancient civilisations in our world, the Egyptian civilisation is one that has been able to preserve its many magnificently enriching heritage sites as well as amazingly massive and beautiful monuments, constructed during different centuries of their prolonged civilisational glory, in a unique enduring manner for posterity.

Interestingly, the known history of this Egyptian civilisation dates back to much earlier in the distant ancient past in this context and even before the beginning of the long era of Egyptian pyramids, dating back to 4000 B.C. and even earlier. In fact, the long era of Egyptian pyramids lasted even little more than 500 years, from around 3000 B.C. onward. And before the advent of this era, the old Kingdom was established founding the very first ruling dynasty belonging naturally to the first Egyptian royal families, with the beginning of 3500 B.C. During the age of the old

Kingdom, the King occupying the throne was known as Pharaohs and accordingly, King Menes who merged the northern and southern Kingdoms and majestically ruled over the whole ancient Egyptian territory. Befittingly, he as ruler of the then Egypt, was known as first of the Pharaohs.

It is naturally logical that the founding of any kingdom at any given time in the long history of man in this world could not have been possible without constituting the possible and befitting military-force, suitably led by the concerned king/ruler wielding the sword. Therefore as already mentioned it is a historical fact that even before the advent of ancient human-civilisations during the distant past, man had already gradually developed the very primitive fighting-skills and accordingly, he occasionally engaged in fighting for his self-defense and also for looking after his various 'interests' as situations and challenges confronted him.

Now coming back to the so fascinating aspects of the ancient Egyptian civilisation, we notice in the history as available to us for our learning, that the unique age of the Pharaohs lasting for nearly 500 to 600 years or so, gradually began during around 3000 B.C. In fact, this was indeed a unique historic era of the glorious constructions of the timeless enduring monuments built with tireless human efforts and mind-boggling skill, which are again historically unparalleled in the world history. We learn that **King Khufu** (Cheops) during his 77 years' reign from 2898–75 B.C., constructed the great Pyramid of Gizeh in ancient Egypt, which even today, stands majestically at its site in the desert there. Further, **King Khapre** (Chephren) during his regime from 2867 to 11 B.C. in his kingdom in ancient Egypt built the Great Sphinx and the same is even now imposingly available to the modern era man for his unique viewing as well as for multi-dimensional studies and research by the concerned scholars/academicians from around the world. The most fascinating aspect in this regard, even in the present high-tech scientific age of the super-computers as well as multiple space-sciences, these monuments dominatingly stand in the unending 'canvas of time' before man with their unique historic splendor and glorious majesty.

As we proceed further, in our brief 'bird's eye-view' of magnificent Egyptian civilisation, we find the era of Middle-Kingdom (nearly 2000–1580 B.C.), and then dawn of the age of New-Kingdom (1580–1150 B.C.) and so on. We learn that during the last phases of magnificent 'Age of Pyramids', 'ground-work' for the establishment of the Old-kingdom was already on then and subsequently during the era of this kingdom, unchallenged union of both the northern as well as southern kingdoms in the ancient Egypt was established. In fact, historically the age of Pyramids dawned during the long regime of old-kingdom itself. Therefore it won't be any exaggeration to say that beginning of old-kingdom era did ensure firmly laying of enduring foundation-stone of truly magnificent Egyptian-civilisation.

Kingdoms flourished along the course of magnificent river Nile in Egypt. Rise, expansion and multiple advancement of any such famous ancient civilisation in the world in fact, have many facets. And important among them are known to be construction of large and beautiful monuments as well as other civilisational construction activities progressively; cultivation and enrichment of art, music, culture and literary-work; and also no less important, the multi-dimensional development of art and science of warfare along with improvement of weapon-systems/platforms ensuring decisively the increase in the military-prowess of the concerned kingdom/empire.

These historic facts also substantiate that the profession of arms and matters-military are indeed very much integral to the multi-dimensional development of man along with his many noted civilisations in the world. In fact since ages, it has been indeed inconceivable for the man anywhere in this world to be able to enduringly establish any kingdom/empire, without organizing appropriately adequate and suitable military means. These may even sound ironical to some people but in any case, these remain irrefutable facts of man's history, since his initial advent in this world.

Then subsequently coming to the aspects of middle-kingdom in ancient Egypt, we interestingly find that this era became instrumental in the gradual sprouting of the culture of 'feudalism'. We may even mention that historically this has been perhaps one of the oldest recorded germination as well as flourishing of man's now well-known life-style based culture of 'feudalism'. Egypt has definitely been many 'firsts' in the fascinating domain of the gradual advancement of man's a number of well-known ancient human-civilisations in our world. But at the same time, this also true that the long era of this middle-kingdom, spanning over a period of nearly seven to eight centuries, was indeed a golden age of the magnificent Egyptian civilisation. Further, the recorded history shows us that this golden age of the ancient Egyptian civilisation also naturally did have many kinds of 'fallouts'; some recorded as noted multiple civilisational advancements and then at the same time, some rather ruthlessly 'written' in the pages of history, in human-blood and grit as well.

The moot point is that civilisations generally don't achieve multiple advancements and enrichment, without resorting to the expansion of territories for socio-political as well as socio-cultural enrichment, for politico-military coupled with strategic dominance visible or otherwise, and so on. Therefore all these and much more, strategically even 'compels' man to resort to wars. Naturally then, it is obvious that wars cannot be resorted to in any such given situation, without the availability of a suitably organised, trained, equipped and competently led 'military-machine'. And therefore we need to bear in mind that the entire processes of 'creating' any such efficient as well as result-oriented military-machine for any kingdom/empire anywhere since ages, have perforce always been highly complex, time-consuming and also certainly not the least, very expensive multi-dimensional tough activities.

The historically as the Sun was setting over the middle-kingdom in ancient Egypt, the gradual dawn of era of New-kingdom over there, started taking place, during 2000 to 2100 B.C. approximately. Again this era of new-kingdom too, endured for nearly five or six centuries. Interestingly then within the Egyptian ancient civilisational perspective, we notice in the pages of the recorded history, that this era had also been known for the military conquests as well as building of empires. New-kingdom in ancient Egypt thereby, is also noted for producing great conquerors as well as kings and queens. Further during this ancient Egyptian age of magnificent imperial majesty, the grandeur as well as royal politico-military-cum-strategic domination of the New-kingdom did encompass Babylon, Assyria and other adjoining territories.

During that period covering 1479-47 B.C., the imperial ruler of the new-kingdom was **Thuthmose-III**, then the Pharaoh of Egypt and history also calls him the world's First Man. Then subsequently King Amenhotep-III ruled over the new-kingdom from 1410–75 B.C. as ancient world's magnificently dominating imperial monarch. Then another great king, **Amenhotep-IV,** ruled there for the next nearly twenty years and interestingly, he is also known in history as ancient Egypt's Akbar. He is also noted for his initiative to destroy idolatry and in that place, initiated the 'cult' of the worship of Sun during the age new-kingdom. He was then followed by the king Tutenkhamen, as the subsequent ruler. He is specially recognized for providing magnificent 'grave-furniture' in the history of ancient Egypt.

Well the more important point in this respect is that these bygone eras of ancient Egypt are not just noted for great kings, queens as well as imperial grandeur but in addition, in the vast fields of ancient world's art, culture and science too, they had magnificently been able to make great contributions to the larger human-civilisations. In fact, these were in addition to the then these ancient Egyptian kingdoms' aspects of great innovative contributions in the multiple fields of matters-military.

Prof. S.R. Sharma mentions that, "The tomb, temples and palaces of Memphis, Thebes and Amarna, constitute a veritable museum of antiquities, an invaluable 'Book of the Dead' illustrated with live sculptures and paintings, the most remarkable of their kind found anywhere in the world. Here we have a wonderful record of human struggles and achievements, dreams and fulfillments, as well as failures and disappointments, all depicted in such vivid line and colour, that he who runs may read. A mere inventory of all that the Egyptians did or attempted to do would in itself constitute the most remarkable commentary on their civilisation and culture. Apart from what has been described already, they built the first Suez Canal between the Mediterranean and the Red Sea; they invented the first calendar with 365 days; they devised the first shadow-clock for measuring time, and the first instrument for observing the movements of the heavenly bodies; they used the decimal system; they knew elementary algebra, and could calculate the cubical contents of cylinders, hemispheres, etc. ***(Refer: Chapter on Egyptian mathematics in 'The Endless Quest' by F.W. Westaway.)***; they knew physiology and anatomy fairly accurately as is borne out by their medical treatise contained in the Edwin Smith papyrus; they knew that the heart is in the center of vital power in the body, and that the brain is in the center of the nervous system; they carried out surgical operations and had among them dentists, eye, stomach and other specialists; their furniture as preserved in Tutankhamen's tomb, has a modern appearance; they used finger-bowls, cups, spoons, razors, wigs, hats face paints, cosmetics, kilts, sandals and wooden head-rests; their children played with mechanical toys, their sculptors produced the most life-like images, and were called 'vivifiers'; indeed, to carve was for them 'to give life'; their jewellery would excite the cupidity of a modern woman of fashion; they were the first people to have census and a postal system, to use sail-boats, and copper-pipes as conduits; to use glass, paper, ink and to devise a system of education." (Sharma, 1948, pp. 31-33)

The magnificently enormous contributions made by the prominent ancient Egyptian civilisation for their own as well as towards the larger humanity's overall betterment as described above, are unique. These are indeed historically praiseworthy. Looking them objectively in a historic perspective, one won't hesitate to emphatically confirm that, in addition to such imperial grandeur and multiple civilisational advancements, during the long glorious era of theirs, the ancient Egyptians were also able to make befitting progress in the areas of profession of arms as well as all related spheres of the **matters-military**. We would naturally discuss ancient Egyptians' these subject-matters in the next chapter.

At this juncture, it would definitely be of interest to also note as to what famous scholar-cum-author Edward W. Said has to say in this regard. **No wonder that when Napoleon the Great entered Egypt during his military campaign in the northern Africa, he too was overwhelmingly mesmerized by the magnificent relics, monuments and highly acclaimed treatises belonging to their glorious civilisational past.**

Professor Edward Said writes that, "Egypt was the focal point of the relationships between Africa and Asia, between Europe and the East, between memory and the actuality. **Placed between Africa and Asia, and communicating easily with Europe, Egypt occupies the center of the**

continent (*Hence, always multiply commanding great strategic importance.*) This country presents only great memories; it is the homeland of the arts and conserves innumerable monuments; its principal temples and palaces inhabited by the kings, still exist, even though its least ancient edifices had already been built by the time of the Trojan War. Homer. Lycurgus, Solon, Pythagoras, and Plato all went to Egypt to study the sciences, religion and the laws. Alexander founded an opulent city there, which for a long time enjoyed commercial supremacy and which witnessed Pompey, Caesar, Mark Anthony and Augustus deciding between them the fate of Rome and that of the entire world. It is therefore proper for this country to attract the attention of illustrious princes who rule the destiny of nations. No considerable power was ever amassed by any nation, whether in West or in Asia, that did not also turn that nation towards Egypt, which was regarded in some measure as its natural lot. **Because Egypt was saturated with meaning of arts, sciences, and government, its role was to be the stage on which actions of a world-historical importance would take place. By taking Egypt then a modern power would naturally demonstrate its strength and justify history....**" (Said, 1979)

So far the aspects of 'Orientals and Orientalism' are concerned, Egypt, whether ancient or modern, is certainly not the whole story since, it is a part of their long history as well as their vast geographical areas encompassing more than one continent. And the same facts with their many fascinating historic facets, have also been amply substantiated in a historic-cum-sociopolitical manner, by Professor Edward Said in his book.

(b) Ancient Mesopotamian Region (Present-day Iraq and adjacent Areas): 4500 to 3500 B.C. onwards

This is indeed another historic example as a glorious ancient human-civilisation in the long and fascinating saga of man in this world. In fact, recent excavations during the last couple of centuries have shown concluding archeological findings as well as remains to the experts and scholars that this civilisation to start with, was even independently much older and also quite multiply developed in comparison to the magnificent and highly publicized ancient Egyptian civilisation. That is altogether a different matter that these unique facts of human history are definitely not so well known. To be specific in this context, we learn from the pages of history that the scientifically methodical excavations in much wider areas in the valleys of Tigris and Euphrates rivers as well as in and around Isthmus of Suez have authentically confirmed the prolonged and highly flourishing existence of the ancient Mesopotamian civilisation.

'Mesopotamian-civilisation' too had taken root and flourished for many centuries along the valleys of the twin river-systems of Tigris and Euphrates. Hence logically scholars and historians the world over, have very befittingly called the general areas between these two rivers as the 'fertile crescent'. These general areas encompass both sides of these rivers and then extend towards the Arabian Peninsula.

The interesting point to be noted here is that before the advent of the era of some form of 'written/engraved' historical-records intended for further learning as well as leaving them preserved for posterity, enduring graves and well-built monuments of various kinds were the only tangible and reliable sources of knowing and re-structuring history of the distant past. These aspects have been equally true for both the ancient Egyptian as well as Mesopotamian civilisations. It is another matter that such ancient Mesopotamian monuments and graves have not been as magnificent and spectacularly visible as those Egyptian ones.

At the same time it is also historically true that as applicable to various situations and cases, both the above mentioned 'historical-records' as well as such graves/monuments have been simultaneously helpful as unique and highly valuable sources for the requisite exploratory studies during much later periods. It is natural that civilisations don't advance and flourish on a 'single-track' and in a 'linear' manner alone over any given passage of time. The entire process in this regard, is perforce highly complex and naturally multi-dimensional as well.

In much later periods during the present modern-era many highly surprising sites in Mesopotamia have been carefully dug-open and in the process, fascinating ruins of magnificent ancient cities were exhumed, belonging to another this ancient great civilisation in its own right. That is altogether a different matter that for many centuries since ancient times, larger humanity ironically remained quite ignorant about the historic existence of this magnificent civilisation with all its remarkable glory, advancement and multiple achievements (*Hence interestingly, it is similar to the case of Indus-Valley civilisation*).

In fact, it was only during last nearly three centuries or so that people from the world over have started making consistent scientific efforts to systematically dig into the past and thereby, learn more about the Mesopotamian region's glorious history and also its very old and fascinating civilisation with all its remarkable manifestations. Then at the same time another unique historic fact remains that in certain specific locations and also in certain interesting manners, this region's highly glorious civilisation has been even much older than the famously magnificent ancient Egyptian civilisation as well.

In this context, in **Professor S.R. Sharma** writes, "....Yet **Herodotus, the Father of History**, knew very little about them (with the exception of Babylon) in the fifth century B.C., while Berosus, a Babylonian historian writing about 250 B.C., vaguely refers to race of monsters coming out of the Persian Gulf and introducing the arts of agriculture, metal-working and writing, in the legendary age, and since that time no further inventions have been made.... Though excavations have been carried on in the Mesopotamian region since Layard unearthed, in 1842, the ruins of ancient buildings filled with sculptures and inscriptions, the most startling discoveries have been made more recently by the Anglo-American expeditions under Woolley and Langdon.... Corresponding to Memphis, Thebes, Karnak and Amarna, we have very ancient cities like Kish, Susa, Ur, Babylon and Nineveh. Some of them have disclosed a civilisation even older than that of Egypt. The relics of Kish and Susa begin with Palaeolithic and Neolithic ages. In some places there are as many as ten layers superimposed one above another, representing different strata of civilisation.... Susa (modern Sushan) had a history not unlike that of Kish. It was the capital of Elam to the east of Sumer. Here have been found human relics 20,000 years old and remains of an advanced culture dating from about 4500 B.C. They contain copper weapons and tools, hieroglyphic writings, mirrors and jewellery, finished vases 'elegantly rounded and delicately painted with geometric designs, or with picturesque representations of animals and plants. **The Elamites seems to have discovered the use of wheel, both for vehicular and potter's purposes** *(Putting in 'bold' is by this author)*, earlier than the Babylonians or Egyptians. Susa flourished during six thousand years and witnessed the rise and fall of more than half-a-dozen empires: those of Sumer, Babylon, Egypt, Assyria Persia, Greece, and Rome. But by far the most important and interesting of the cities of Mesopotamia were Ur, Babylon and Nineveh...it is necessary to acquaint ourselves with the earliest makers of this civilisation. They were the Sumerians, a people like the Hyksos of Egypt, of very uncertain origin.... Formerly the Euphrates and the Tigris entered the sea by separate estuaries. This region is referred to in the Bible as 'Shinar'. It is not strange, says one writer, that tradition should have located here Paradise, that primeval garden 'out of the ground of which God made to grow every tree that is pleasant to the sight and good for food.' **Hence, obviously the struggle for its possession which throughout characterises Mesopotamian history. It was in its racial aspects a struggle between the Semites and non-Semitic people which, in a sense, has not yet come to an end. The inhabitants of a rich oasis in the midst of an arid desert are compelled to be fighters**. It is not surprising therefore that even the comparatively peaceful Sumerians fought both among themselves and with the surrounding nomads of the desert by whom they were finally overthrown.... The real founder of the greatness of Babylon was **King Hammurabi**, who ruled for forty-three years, 2123-2083 B.C. His fame has been revived in this world, after four thousand years, on account of the discovery of his remarkable code of laws inscribed on a diorite pillar. *(It was excavated near Susa by French archeologists in 1897 and is now preserved in the Louvre Museum, Paris. There is a very good reproduction of it in the British Museum, London.)* It will now endure as long as our language and civilisation last. 'In its wisdom, justice, and tolerance the Hammurabi code forms one of most remarkable of early landmarks of civilisation.' **The Hebrew code of Moses was directly derived from that of Hammurabi, as Hammurabi was derived from the more ancient Sumerian. It is very comprehensive and contains over 280 sections**." (Sharma, 1948, pp. 39-44)

At the same time, advent of man on earth has definitely been a much later and also in this respect, quite a recent development under the overall all-encompassing evolutionary process of the mighty mother-Nature. Then we may say that further evolutionary development of few or more 'sub-species' within a particular specie again is logically understandable but, surprisingly a unique fascinating point with respect to man has been historically perforce putting his kind into different socio-cultural 'compartmentalized' racial representation.

In fact as per the supreme evolutionary-laws of the mother-Nature, any kind of racially differentiating representation within the mankind in this world has no anthropologically scientific-basis of any kind whatsoever. Factually these have happened due to various kinds of socio-psychological and cultural 'fixations' of man, as getting 'sprouted' in different far-off areas and in different given periods in this world. Naturally it appears to be quite logical that this being a 'fallout' of the gradual advent and steadily multiple advancements of various ancient human-civilisations within the on-going march of the human-history over the Nature's unending 'canvas of time'. At the same time, as different human-civilisations advanced as a fascinating historic-process, the various aspects of man's regressive 'racial-fixations' got more and more pronounced and finally man is found to be a helpless 'prisoner' of his own extremely regressive attitudinal 'trap'. It is indeed a great ironical 'negation' of man's own civilisational as well as other multi-dimensional advancements.

Now the ancient Mesopotamian civilisation being very old and also remarkable in certain respects, perhaps one of the oldest ones, naturally this too have had its big share of the highly ironical civilisation-based unsettling 'negation' as well. In fact as already mentioned earlier, ancient history of Mesopotamia is also known for the 'mindless' racial-based violent conflicts between the Semites and non-Semitic segments of the then that region's population. **Therefore it has been quite appropriately said, concerning all human-civilisations, that any kind of conflict or war in all their aspects, have always been their inseparable parts in any particular given situation or 'point of time' as applicable, in the continuously evolving human-history**.

We also find in the pages of its history that the Mesopotamian region, has had different kinds of people/communities living and at the same time interestingly, there had also been somewhat distinctly different 'sub-civilisations' with a variety of socio-cultural and socio-psychological moorings and there-by, **different kinds of 'matters-military' as well**. Further these facts of history are not exclusively unique to this particular civilisation alone but, they have definitely been so with all other noted ancient human civilisations as well.

Just mentioning one point here during ancient times, if the existence of any **'feudalism'** was there in Egypt, then similar was the case with Mesopotamian region as well. But the forms and socio-cultural as well as political methodologies for practicing this system of 'feudalism' in each of them was quite distinctly different with different fixations. Well these are certain historic facts as applicable to various ancient civilisations but at the same time, one specifically noticeable 'factor' largely common in this regard has been, the ironically cruel socio-psychological-cum-cultural fixation of the thoughtless application of extremely sadistic "ruthlessness", with hardly any escape ever available to the concerned victims.

Hence, these paradoxically undeniable facts of human-history are there to stay in its pages, as a sad saga of always an unfortunately sizeable section of the concerned populace, perpetually suffering and 'bleeding' in many ways. Therefore it has been rightly observed by the learned historians that along with the many magnificently shining records with respect to any particular noted ancient civilisation, there are also few gloomy records of extreme human cruelties too co-existing together. Here we may interestingly notice as an analogy that in the panorama of long human-history in this world, along with the periods of relative peace and tranquility in various human societies, there have always been periodic occurrences of wars or some kinds of conflicts as well.

We find that probably the beginning of the well-known ancient Mesopotamian civilisation took place in the distant past, sometime during 5000–4500 B.C. in an around a place over there, where subsequently along with the advancement of this particular civilisation, ancient city of 'Kish' was founded. We have already mentioned earlier in this chapter that in fact, Kish was one of the oldest cities anywhere in any ancient human civilisation with many enchanting civilisation-based facets and heritage, and therefore, it has consistently been a center of special attraction for now modern archaeologists as well.

More importantly, the city of Kish then had also been the capital of ancient Sumerians and hence, their rulers were naturally well versed in the prevailing systems of matters-military. No wonder they innovatively laid the well-planned defence of their capital-city and in addition, remained prepared for the protection of their kingdom. History also informs us that in addition to Sumerians' gifts of both their laws and systematic script for writing and maintaining useful records, they along with ancient Babylonians too, had made world's first beginnings in the attractive areas of mathematics and astronomy. For the larger humanity these aspects are indeed remarkable, as history shows us that, in fact, they had already begun their serious interests in these scientific areas since even around 3000 B.C.

Further within Mesopotamia, Sumerians had even established distinctly their own 'sub' civilisation as early as 4000–3500 B.C., and then, gradually established the flourishing ancient cities of Larsa, Lagash, Ur, Nippur and Uruk, with their many fascinating facets as well as great archeological treasures. History also informs us that the Sumerians were not only brave warriors but in addition, they were great scientific innovators as well as imaginative builders too. The Sumerians too became known for building famous Ziggurats, which are in fact, their historically unique and ancient 'temple-observatories'.

Ironically history shows us that in the multi-faceted work of art and craft, Mesopotamians were relatively poor and certainly, no match to the magnificent Egyptians in this regard. Interestingly, it is also a fact that since they were fearless warriors, rightly or wrongly, they not only fought with other desert-inhabiting neighbouring tribes, but fought among themselves as well. In any case, these historically fascinating aspects of fighting, etc., were largely common features of any noted ancient civilisation anywhere in this world.

In ancient Mesopotamia, there has been another well-known desert-tribe, called the Semites. Though they were different from the Sumerians and others but they too, were no less 'advanced' and innovative in their own way. To start with, they were able to establish their own 'sub' civilisation' nearly around 3000 B.C. and further they became particularly famous for founding the magnificently flourishing ancient city of Babylon. Interestingly, the literary meaning of the word 'Babylon', as per their ancient script, means "Gate of The God". Their Sargon King was also known as the kings of Akkad and Sumer. It is historically true that, Babylon was decidedly the most magnificent and famous city in the entire ancient Mesopotamia.

In fact, the reach of the glory and advancement of Babylon was so extensive and captivating that, almost the whole southern segments of the valley of twin-rivers, the Euphrates and Tigris, were befittingly known as Babylonia. This Semitic desert-tribe, who founded the ancient magnificent city of Babylon, earlier belonged to the northern areas in the region known as Akkad and they fought with the Sumerians, who were their arch rivals, and subsequently subjugated them, establishing their own kingdom. History also shows that the entire magnificent valley between those twin-rivers, also called as the "fertile crescent" in the ancient era, repeatedly changed hands between different kings and their kingdoms and that kept this area always in ferment.

The interesting point to be noted here is that, this entire 'crescent' in the whole of Mesopotamian region and along with Egypt too, have been having a special geostrategic as well as politico-military significance even now in the present times. These historically important aspects, with all their significant dimensions would be discussed in more detail in the subsequent chapters. In this respect interestingly, one obvious point clearly comes out that during the initial beginning of the 'Industrial-revolution' itself; as for the then leading and competing European colonial-Powers, the geostrategically crucial location of this so called "fertile-crescent" did assume almost central-region of great attraction. Hence naturally attracted by these aspects, they with their own different plans as well as politico-diplomatic strategies; went whole hog to achieve decisive control over this region craftily, with a clear larger imperial-aim to expand and consolidate their colonial-empires.

Edward Said says in this context that, "Britain was in Egypt and Mesopotamia; through a series of quasi-fictional treaties with local (and powerless) chiefs it controlled the Red Sea, the Persian Gulf, and the Suez Canal, as well as most of the intervening land mass between the Mediterranean and India. France, on the other hand, seemed fated to hover over the Orient, descending once in a while to carry out schemes that repeated de Lesseps's success with the canal; for the most part these schemes were railroad projects, such as the one planned across more or less British territory, the Syrian-Mesopotamian line. In addition France saw itself as the protector of Christian minorities—Maronites Chaldeans, Nestorians. Yet together, Britain and France were agreed in principle on the necessity, when the time came, for the partition of Asiatic Turkey. Both before and during World War I secret diplomacy was bent on carving up the Near Orient first into spheres of influence, then into mandated (or occupied) territories. In France, much of the expansionist sentiment formed during the heyday of the geographical movement focused itself on plans to partition Asiatic Turkey, so much so that in Paris in 1914 'a spectacular campaign was launched' to this end. In England numerous committees were empowered to study and recommend policy on the best ways of dividing up the Orient. Out of such commissions, as the Bunsen Committee would come the joint Anglo-French teams of which the most famous was the one headed by Mark Sykes and Georges Picot. Equitable division of geographical space was the rule of these plans, which were deliberate attempts also at calming Anglo-French rivalry. For as Sykes put it in a memorandum, 'it was clear...that an Arab rising was sooner or later to take place, and that the French and ourselves ought to be on better terms if the rising was not to be a curse instead of blessing.' ...The animosities remained and to them was added the irritant provided by the Wilsonian program for national self-determination, which as Sykes himself was to note, seemed to invalidate the whole skeleton of colonial and partitionary schemes arrived at jointly between the Powers." (Said, 1979, pp. 220 & 221)

We may say that the lines quoted here, gives a glimpse of the periods before and during the First World-War times but that all the more, highlights the continuing geostrategic as well as politico-military importance of these areas in the Middle-East and south of the Mediterranean. And of course in the on-going modern era, geostrategic perspectives as well as politico-economic and other important parameters have gathered major dramatic alterations, as many centuries have already rolled-by into the pages of history, bringing-in revolutionary far-reaching multiple developments in the lives and times of the larger humanity in this world. Further in this context since more that last one century or so, **the paramount strategic factor of "OIL" with its multiple dimensions and 'fallouts' have perforce created many mind-boggling complexities in the entire game of the internationally multi-faceted power-politics in today's world**.

Now again coming back to the aspects of ancient Mesopotamian civilisation, we find that, if Babylonians had earlier militarily subjugated the Sumerians and then similarly later, those Babylonians too were overpowered and conquered by another desert-tribe, then thriving in the highland areas in the north-side of the ancient Mesopotamia. They were Kassites, another warrior-class of a tribe then thriving in the area and who are credited to bring-in horses there and needless to say that, this animal with many attractive properties was not only decisively useful in battles then but also proved to be strong, robust and highly dependable 'beast of burden' for man, and at the same time, who as master of this wonderful animal was able to train him for many valuable tasks.

Then appeared the Assyrians there in and around 1300 B.C. in the historic framework of ancient Mesopotamia. History informs that they were settled on the banks of river Tigris and originally the ancient city of Ashur was the capital of their kingdom. They have derived this name from their earlier capital in Ashur itself. Later on they founded the city of Nineveh and then, shifted their capital over there. Their kings, noted in history, are Sargon-I and then much later Sargon-II, during the sizeable period of 1300 to 700 B.C. These Assyrians too, were a warrior tribe in their own right in the ancient Mesopotamian region with the then expansionist designs. In fact, their king Sargon-II is known to have fought many battles, notably against Hittites, Armeans of Syria and also, against Babylonians of Mesopotamia as well. Further their Kings noted in the pages of history, were Senncherib and Assurbanipal and they ruled one after another respectively, during a period ranging from nearly 700 to 625 B.C. They were also great warriors in those times in Mesopotamia.

Then subsequently came Chaldeans, who have their own chequered history in the ancient Mesopotamia during nearly 620 B.C. Interestingly earlier, the socio-economic well-being of the ancient city of Nineveh under Assyrians coupled with their luxurious cultural and military practices then, were heavily burdensome on the Assyrian kingdom and also, their military had gradually become quite oppressive over the people too. These factors together resulted in provinces adopting rebellion path against the king and as a result, the Chaldeans as their powerful enemy attacked the ancient Assyrian kingdom in around 605 B.C. and during their decisive battle, the Chaldeans destroyed the capital city of Nineveh. Here, we find that the history for the Assyrians had turned 'full-circle', i.e. as earlier the same Assyrians had destroyed the city of Babylon, during their war with the Sumerians. Well then, the Chaldean king named, Nebuchadnezzar, during his long rule in 625 to 526 B.C., rebuilt the city of Babylon. He was also a warrior and a noted military commander then, who attacked Judah during his reign and brought Jews captives as prisoners to his city of Babylon. In this then on-going turbulent saga of ancient Mesopotamian civilisation, finally Medes and Persians came there during around 540 B.C. and lot many conflicts between the rival forces resulted, culminating in the eclipse of Chaldean empire in the region.

Since then hundreds of centuries have passed and ancient rare relics of this region have got destroyed by both the ravages of time, 'ruthlessness' of weather-cycles, as well as mindless destructive actions of man. In fact, Mesopotamia has undergone far-reaching geographical, geo-political and socio-cultural as well as religious changes and now, it is almost unrecognizably different from its so ancient and chequered past. Its history spanning thousands of years have already got buried in the bottomless 'dust-bin' of time. In any case, that's the way history keeps evolving having perhaps closest of 'kinship' with the unending time itself and also, history is always living as well.

Unfortunately the latest mindless destruction of invaluable relics was savagely caused at the end of 1991 'Iraq-war' preserved in Museum of the capital city of Baghdad in today's Iraq, mainly due to callous attitudes of the occupying U.S. military-forces. 'Military-campaigns' in certain specific situations are understandable or may even be considered justified, but irrational destructions of any such irreplaceably invaluable heritage legitimately belonging to the entire humanity are mostly avoidable if the military commanders concerned are reasonably knowledgeable and imaginative enough to be able to prevent such vandalism.

Well in any case, man's relentless quest for 'structuring' the human-history with maximum possible authenticity, inclusive of all its multiple facets goes on and that's precisely the way, history keeps getting recorded and continuously being studied by the larger humanity the world over.

(c) Semi-historical China and the Dawn of History: 2852 to 2697 B.C. onward

The ancient Chinese civilisation is one of the most prominent human-civilisations in this perspective, **Professor S.R. Sharma,** in the same book says that, "The last three countries, constituting South-West Asia, from a block of territory inhabited mostly by the various branches of the Semitic race, like the Hebrew or Israelites, the Phoenicians and the Arabs. The Fertile Crescent (Euphrates-Tigris valley) was the melting pot or confluence of three main streams of peoples belonging to the Mongolian, Aryan and Semitic races. The civilisations developed by each of these ethnic groups had distinctive features.... They are all children of Asia whose contributions to human history have been both versatile and rich."

Hence it becomes obvious that China and her remarkable civilisation is definitely an important segment of Asia's highly colourful ancient history.

Regarding China, **Jawaharlal Nehru** mentions that, "In China and in the other countries of eastern Asia, like Japan, Korea, Indo-China, Siam and Burma, we have not to deal with the Aryan people. We have here the Mongolian races. About 5000 years or more there was an invasion of China from the west. These invading tribes also came from Central Asia and were fairly advanced in their civilisation. They knew agriculture and kept large flocks and herds of cattle. They built good houses and a well-developed society. They settled down near Hoang Ho, which is also called the Yellow River, and organised their State. For many hundreds of years they continued spreading over China and improving their arts and crafts. The Chinese people were largely farmers, and their chiefs were really patriarchs....more than 4000 years ago from now—we find a man named Yao calling himself emperor. But in spite of this title he was more of a patriarch than an emperor of the kind Egypt or Mesopotamia had. The Chinese people continued to live as farmers, and there was not much of a central government."

No wonder that the Chinese civilisation has had a remarkably unbroken continuity since thousands of years and also, been capable of outshining in many ways a number of other ancient civilisations in the world. Their long history too, substantially confirms the tenacious toughness of the Chinese people, with their quite uncommon and visionary sense of history. Earlier we have already noted that as per fascinating history of the advent of various human-civilisations in the world's different geographical locations and also during different centuries, almost all prominent ancient civilisations have taken initial roots and thrived by the side of noted perennial river-systems in the concerned area-locations. Ancient Chinese-civilisation too advanced by the sides of already well-known Chinese major river-systems namely, Yangtze-Kiang as well as Whang-Ho and other tributary-rivers and so on. Therefore, it has been quite befittingly said that natural river-systems anywhere in the world have perpetually been fundamental and perennial life-giving sources for the man.

Geographically, China has been a very vast country and since ancient times, river-systems have made gigantically vast tracks of land available to the numerous Chinese population, as highly fertile valuable gifts of Nature. Earlier we have also noted that when both evolutionary as well as revolutionary "civilizing-process" of man began in a sustained manner in this world since the very distant past, the first 'revolution' in this regard, which gradually started and gathered momentum, was indeed the advent of the art, craft and 'science' of the agriculture with its surprisingly multiple facets. At this point, it appears somewhat relevant to refer to a well-known Scottish author and historian **John Keay**'s book *China: A History*, which is a substantially in-depth with a historic account of the glorious history of China and its magnificent civilisation; beginning from its very ancient past to the presently on-going modern age. *India: A History* by John Keay also comprehensively covers unique multiple facets of China.

In the paperback edition of the year 2009 of the above mentioned book, on pages 21 and 22, the author very informatively mentions that, "Certainly China merits a grand treatment. A vast country with an interminable pedigree, an idiosyncratic culture, a traumatic recent past and an exciting future can hardly be taken at canter. But it should not be supposed from all the groaning shelves that China's history is therefore altogether unlike to that of other nations. It is not. In China, too, empires rise and fall, personalities shine, progress is fitful, peace ephemeral, social justice elusive. The difference is one of pedigree, not kind, of scale, not character. Forewarned of the difficulties, the reader will find China's history just as instructive and rewarding as any other—only more so."

These comments are indeed very precise, thought-provoking and historically enlightening as well. No wonder that besides its vastness on a grand geographical-scale China has been engaging so extraordinary attention the world over. At the same time, China has remained a greatly fascinating mystery too; in many aspects such as its multi-faceted turbulent history since a very distant past, its civilisations, its complex socio-psychological cultural-dimensions; it magnificent legacy of art, literature, philosophies, religious-sides, and many other enchantingly multiple creativities and last but not the least, its multi-dimensional aspects of **matters-military**, coupled with some traditions for **"essential strategic-culture and thoughts"** as well.

No wonder that such legend-based Chinese historical narrations 'date' back to thousands of centuries. These legends and their dramatically narrative stories are indeed multiply colourful as well as always captivating since, they never go 'stale'. Various kinds of 'debates', for establishing their 'authenticity' have routinely been taking place globally.

Zhong Jingwen has been a well-known Professor in the 1920s-30s and head of the Chinese Department of Beijing Normal University. In addition, he was also then Vice-Chairman of the Chinese Folk Literature and Art Research Society and a National Committee Member of the China Federation of Literary and Art Circles. **Professor Zhong Jingwen** himself says that, "Today there still exist many tribes or peoples who have lagged behind in their social evolution and are still leading a rather primitive life. *(There are even now many such distinct tribes in India's jungles and mountain as well.)* Yet they all possess a rich store of folk songs, myths, legends, folktales, painting, sculpture and dance. Many are of a high standard, and we are often astonished and delighted to discover their true beauty. These artistic achievements which form a fundamental part of the civilisation of these tribes or peoples make an important contribution to the fund of culture so treasured by the whole of human society. The existence of this great diversity of folk cultures is today a well-known fact. The realisation of its significance has, however, been a slow historical process. It has taken a long time for us to recognize that folk culture is a phenomenon that has been widespread since the dawn of history. *(So interesting indeed, and these aspects have been so magnificently true for any society/nation, with a colourfully fascinating prolonged civilisation spanning over many centuries.)*.... In the history of China, the development of folk literature and art as part of the life of the people can be traced back to Neolithic times, and its progress can be charted alongside the advance of society. Its character underwent an important transformation. While the appearances of a class structure in society led to the stratification of culture, Chinese folk literature and art continued to grow and develop in its own course.... During the last years of the Qing dynasty (1644–1911), the character of Chinese society underwent a tremendous change. These years witnessed the invasion of China by foreign capitalist countries, the economic stimulus provided by the new capital mode of production, development of national capitalism and the collapse of feudal dynastic rule. These developments, coupled with the contacts of many progressive intellectuals with modern cultures in other parts of the world, resulted in the spread of many new and nationalistic ideologies and in the launching of many related practical activities.... Another example is the story of **'Cinderella'**. Everyone has heard this story which exposes the evils of a cruel step-mother. This is perhaps one of world's most famous fairytales and it has become an important research topic for scholars of folklores and folktales in every country. As early as end of the last century *(i.e. 19th century, in fact.)*, it drew the attention of experts in this field. Such work has continued to this day. The earliest relatively complete record of this story dates from Tang dynasty documents of the 9th century. A somewhat modified version of this story is popular today among the Han and Zhuang peoples in the Guangxi Zhuang Autonomous Region and Guangdong province. This cultural treasure house of China's various nationalities is an important contribution to the spiritual nourishment of all peoples of the world.... **Although the creators of these folklores emphasize the importance of collective wisdom and strength, they in no way negate the role of the individual in society. As a matter of fact, they highly extol the wisdom and strength of the individual.... Folktales throughout the world share many common characteristics in artistic expression. This is due either to the spread of ideas or to the similarity of the life experiences and creative imagination of the creators.** Character types, personality contrasts, repetition of plot, problems presented and the happy ending... Everyone is familiar with these universal traits. But this does not mean that the folktales of various peoples can be reduced to a simple formula like in physics or mathematics. Apart from their similarities in structure, plot, set expressions and so on, the folktales of various peoples all have their own unique images to interest and captivate the reader." (Jingwen, 1983, pp. 5-7, 9, 11, 15)

The above quoted lines from a beautiful Chinese book, do give us an enchanting glimpse of the multi-faceted and colourfully charming wonders of various ancient folklores and folktales, belonging to the various primitive Chinese as well as many other societies of the distant past, and reassuringly surviving even at present, in different far-off geographical locations in this world.

With understandable fascination, history knowledgeably informs us that **Fu Hsi**, who ruled in China, approximately during 2850 to 2738 B.C., was indeed a genius and also instrumental in first-time shaping-up the beautifully figurative script-oriented expression of Chinese letters and words. His designs of the Chinese-script did also display his remarkable knowledge of the basic symmetry and other related aspects of the fundamentals of geometry, as well. In addition, he was also instrumental in inventing the Chinese music, musical-instruments and also their ancient calendar. Then his son too was no less creative and multiply talented, who ruled over there as his father's immediate successor, during the period spanning from 2737 to 2705 B.C.

History records his name as **Shen Nung** and further, he is also himself known as the then 'father' of agriculture, commercial activities in China and the medicine. Then we come across another prominent personality of ancient China by the name of Su-ma Ch'ien, who is recognized in the Chinese history as the earliest historian, a visionary besides being highly learned and therefore has left a valuable treatise on the various aspects of colourful Chinese history. His historic record begins the period of Yellow Emperor of ancient China. But we also find that, there are even more prominent eras and emperors in this regard, in Chinese history. Prolonged and highly complex ancient-history of imperial-China, spanning many centuries are indeed unique as well as subject-matters of great attraction for China-experts and scholars.

Another very important noted personality in more distant past in China has been **Emperor Yu**, who magnificently ruled there during the period 2205–2197 B.C. and at the same time, he is known as the founder of a remarkable dynasty by the name of 'Hsia dynasty'. Interestingly in Chinese language, the word Hsia means 'civilisation' itself. During the long stable period of Hsia dynasty, seventeen emperors came in succession to occupy the royal-throne and ruled over that country. This dynasty endured for 439 years from 2205 B.C. onward. Unique and long ancient history of China further shows us that after the Hsia dynasty, immediately in succession followed the magnificent advent of another famous ancient Chinese royal dynasty by the name of Yin or say, Shang dynasty, whose colourful and highly contributory rule lasted for another long period of nearly 650 years naturally, from 1766 B.C. onward. This dynasty is well-known for producing brilliantly crafted great works of art and crafts.

In fact scientific excavations have been carefully carried out at the selected sites, like Honan, pertaining to the period of this noted Shang dynasty. Many astonishingly beautiful objects like, items of jade, bronze, deer-horn, tortoise-shells, pieces of seasoned wood, bamboo and so on, with artistic engravings of attractive designs, figures as well as letters in Shang-period script on them, have been found for further study and research by the concerned archeological experts. Further, during the prolonged rule of the magnificent Shang-dynasty in ancient China, development of arms, weapons and other items as well as strategic and tactical concepts of significant military-value such as fortification and building of then imaginative field-defence network, also advanced in a far-reaching manner. During this era, people there also became very creative in literary-work and composing poems as well. Interestingly some of them have now been recovered for further enriching the Chinese history of that bygone era.

One well-accepted fact of history so globally common everywhere and also in every ancient human-civilisation is that, wherever any famous and so enduring royal-dynasty comes-in to rule then, there remains sprouting of the related reasons/activities as well, for triggering that dynasty's downfall, since already perhaps destined to come sometime in future. Similarly, in ancient China too when the time came, an ironically similar fate visited over the Shang dynasty as well. During the coming end, the royal-court of the Shang-dynasty sank deeper and deeper in mindless luxury and merry-making, causing an 'uprising' in the empire, and then a revolution against the Emperor himself.

One personality by the name **Fa** then, assumed the leadership in this turmoil and pulled-down the Shang dynasty in China. Subsequently, Fa established another dynasty by the name, Chou-dynasty. Another long and truly magnificent era of this new royal-dynasty began in 1122 B.C. and then endured for 867 years. The most noteworthy aspect of the long reign of this Chou-dynasty has been that history records it as most glorious and thereby, as a golden era the unique Chinese-culture, in ancient China. In these aspects the exceptionally long and unbroken history of magnificent Chinese civilisation has been unique in many ways. No wonder that the rule of this Chou-dynasty too, did last in China for nearly 870 years, remarkably bringing in 37 emperors belonging to this dynasty in succession and also the emperors were regarded as having been celestially appointed.

During this glorious era, the 'profession of arms' and the matters-military were given strategically very high importance. Emperors were highly careful in selecting their officials in various required capacities and hence, rigorous 'test-systems' were always adopted in their screening and selection, including adhering to good standard of proficiency in handling of swords, daggers, horsemanship, archery, leadership-qualities, reading, writing, drawings, music and fair expertise in royal-court ceremonies and so on. During the rule of thus Chou-dynasty, a number of schools and other Institutions of learning were also established and successfully maintained. No wonder that this golden era of Chinese history, produced a number of talented and learned scholars, saints, thinkers, philosophers and experts in many areas of creative human-endeavour in ancient China. Among them, **Confucius, Mencius and Lao-tse** are specially known the world over.

We have been able to touch upon quite a number of ancient Chinese dynasties ruling over China in succession and then, fading-away accordingly; as centuries kept rolling-by and, the unending human-history moved-on. But interestingly in our specific context, this just doesn't complete the whole story. We may in fact, be able to take into account of only, some of the historically noted major Chinese dynasties in the process; as a part of our on-going 'Strategically Critical Journey' here. And at the same time, surprisingly another historic reality is brought before us that in between them, nearly a hundred totally individualistic dynasts kept sprouting at a sort of irregular intervals.

We may also wish to mention that interestingly, many of such individualistic-dynasts devised peculiarly innovative-methods for being able to 'bask in the reflected glories'; by even unhesitatingly propagating as a type of 'fake-clone', that they intended to revive the magnificent glories of some of earlier prominent dynasties. And amusingly for all these they did not hesitate to adopt even the names and royal-titles of those earlier prominent-dynasties for recognition as parts of their blue-blooded lineages. Since ages, amazingly man the world-over, has been throwing-up perhaps countless ways of fructifying his insatiable desires, greed and wishful thinking; stranger are the facts than fiction.

But at the same time, the historically multiple-glories and lustre of many prominent Chinese-dynasties cannot be wished-away; if the desired objectivity is to be maintained in any such analytically perceptive studies. In this context, **John Keay (*China: A History*)** mentions that, "Mercifully some dynasties acquired a semi-permanency and soldiered on for centuries, winning a reputation for administrative integration, military endeavour, political stability, cultural distinction and personal magnificence. The five imperial dynasties that lasted longest—each for three to four centuries *(as since recorded in the earlier paragraphs)*—constitute the great plateaux of Chinese history and are well worth memorising. Cross reference to contemporary empires may help. They are:

HAN (Former and Later): 202 B.C–AD 220, coeval with the Roman Republic and their early Empire.

TANG: 618–907 AD, coeval with the expansion of Arab empire.

SONG: (Northern and Southern), 960–1279, coeval with the Crusades.

MING: 1368–1644, coeval with early Ottoman and Mughal empires.

QING (or Manchu), 1644–1912, coeval with Europe's Global Expansion.

Many other dynasties of note will be encountered. Ironically the one that most nearly approached the Chinese imperial boast of ruling 'All under Heaven' was not Chinese at all but Mongol. This was the Yuan dynasty (1279–1368), under one of whose emperors the Venetian Marco Polo supposedly found employ."

It has been appropriately observed by many historians, that the well-known ancient Chinese-civilisation happens to be no different substantially from other similarly prominent ancient human-civilisations but at the same time; the unique difference lies in its range and multiple-vastness with respect to the others. Historic turmoil apart, in fact, history has recorded that the Sun never set over the glorious and turbulent ancient Chinese-civilisation, and thereby a remarkably dynamic civilisational-continuity has been maintained. In this context, we may also say that peoples' own civilisational-dynamism, relentless culture of multiple-vision, their multi-faced creative-genius, their grip over the matters-military with all their otherwise inseparable geostrategic as well as politico-military dimensions and so on have played their own decisive roles.

Professor S.R. Sharma writes that, "Unlike **Lao-tze**, who would repay evil with good, **Confucius** as a practical philosopher said: 'Repay kindness with kindness; but repay evil with justice.' According to him, for everyone called to the government of nations and empires, there are nine 'cardinal directions to be attended to:

1. Cultivating his personal conduct.
2. Honouring worthy men.
3. Cherishing affection for, and doing his duty towards his kindred.
4. Showing respect to the high ministers of the state.
5. Identifying him with the interests and welfare of the whole body of public officers.
6. Showing himself as father to the common people.
7. Encouraging the introduction of all useful arts.
8. Showing tenderness to strangers from far off countries.
9. Taking interest in the welfare of the princes of the Empire.'

For more than 2000 years the philosophy of Confucius moulded and dominated the Chinese mind. It is tempting to go on with the Chinese philosophers, but we shall conclude with only one more: viz. **Mencius**. He was a contemporary of **Plato** and **Aristotle** in Greece, having lived between 372–289 B.C. He was to **Confucius** what **Plato** was to **Socrates**." (Sharma, 1948)

No wonder that the brilliant democratic thoughts of **Mencius** in China even during those ancient times are so magnificently relevant to the present modern age as well. Such great philosophers and thinkers of very ancient times, belonging to Chinese and other prominently old civilisations, including those from India too, were indeed the visionary forerunners to today's now powerfully and so positively entrenched democratic thoughts as well as concepts of statecraft and systems of political governance in the modern world.

In fact, the multi-disciplinary pro-active thoughts and brilliant concepts of Confucius are valued so high that even in the present-times, the well-reasoned critics of 'Communist' China, emphatically maintains that the real and far-reaching modern-era salvation of China can only take place, if the people and regime over there, both substantially and sincerely, again go-in for adopting the unique 'treatises' of Confucius, with perhaps some due modifications as thoughtfully needed, not only for China but in addition, for the larger humanity of the world as well. Another pertinently interesting point here is that as during the periods of various other dynasties, the society in this Chou-dynasty China too, was rigidly feudalistic in both socio-cultural ethics and behavioural pattern.

Well, we may also maintain that so was the case then with ancient India too and also, in many other ancient civilisations in this world. In fact, the history of all noted ancient civilisations, since their inceptions during the very distant past in this world clearly shows that, as they slowly but consistently began sprouting during different centuries, from among the various primitive human-societies in far-off different geographical locations to start with, gradual germination of the hidden 'seeds' of the self-actuating human-nature based '**feudalism**' too, started as an accompanying natural-process. Another interesting aspect comes to light that in such extraordinarily ancient societies and their otherwise overwhelming civilisational-accomplishments; ironically also witnessed evolution of many kinds of socio-psychological as well as socio-religious myths; and also typically accompanied by their many even illogical and rigid mental-fixations. The socio-culturally powerful negativities of these developments were also instrumental in regressive stratifications in these ancient societies, such as those of **India** and **China** and thereby, as a socio-political consequence the ruthless paradoxical multiple-exploitation of weaker-sections of human-societies by the stronger and dominating segments, mainly consisting of the ruling and rich classes.

In this interesting context, **John Keay** writes that, "India's mythology matches this *(i.e. with the Chinese legendary stories regarding the origin of man on the planet-Earth.)*, with a dismemberment myth. Out of the corpse of a sacrificial victim the Vedic gods supposedly hacked a hierarchy of caste, with the priestly Brahmin being born of the victim's mouth, the martial *ksatriya* of his arms, the house-proud *Vaisya* of his thighs, and the wretched *Sudra* of his feet. The **Brahminical imagination** responsible for this conceit overlooked the possibility of a section of the human race being derived from an intestinal infestation. Perhaps only elite, as sublimely superior as China's could have assigned to their raven-haired countrymen an origin so abject. When in later times foreigners came to resent the arrogance of Chinese officialdom, their grounds of complaint were as nothing compared to those of China's unregarded masses. From both the above examples an early insistence on social stratification—on a superior 'us' and an inferior 'them'—is inferred; and it is thought to be corroborated in China by the numerous other myths emphasizing that heaven and earth had to be physically separated." (Keay, 2008, p. 27)

This remains one of the peculiar socio-psychological evolutionary traits, exclusively specific to the human-beings in this world. Interestingly since man is a thinking social-animal, as always 'activated' by his exclusive kind of brain, perhaps these aspects of **'feudalism'** too, may have also been triggered in the same on-going process, in different forms and content, as per historically different emerging situations in different places and during different times. Well in any case, since this so far remains one of the many evolutionary mysteries pertaining to human-beings, we may call the above 'elaboration' as just a matter of conjecture only. But in any case the fact remains that the lingering 'feudalistic-fixations', with different degrees of intensity as well as socio-psychological ruthlessness, have always historically remained with man, since the very beginning of all the human-civilisations.

Interestingly, the various aspects of ancient human-civilisations and the gradual historic process of the primitive human-societies getting 'civilised', have ironically multiple facets. Hence, when important 'judgmental-factors' in this regard are critically considered along with their accompanying finer nuances, in that case even the modern age, various highly developed human-societies claiming themselves to be again highly 'civilised' too, can be very substantially questioned and thereby, causing great embarrassment to those societies even very legitimately. **Well, human-beings and their so-called modern societies, paradoxically and then persistently, do have many 'dark-corners', still so glaringly visible and these just can't be wished away**. In this context, we may also perceive that any kind of display of racialism, apartheid, lingering caste-system, 'class-based' differentiation, 'birth/family-based' differentiation and so on, are also so many differing types or manifestations of the same age-old 'feudalistic-fixations'. In any case, these are irrefutable realities of human-societies since ancient times and, these are surviving in many ways in almost all the modern societies as well in our world, in varying degrees.

Then subsequently we find a historic rise of the rule of Ch'in dynasty in China, surviving for a short duration only from 255 to 210 B.C. This dynasty had just two emperors occupying the throne namely, **Chen and Shih Whuang-II**, in that order. This brief dynastic era was known for two great things in its ancient history, one for building a positive as well as truly unparalleled structural 'monument' along its long southern borders, and the second one for mindlessly destructive work. *This construction-work then, resulted in the tireless building of 'Great Wall' of China and the second one as 'destruction' of the highly important unique and heritage literature left by one of the greatest philosophers of all times, named Confucius.*

Interestingly in this specific context, it appears quite relevant to again quote **John Keay**, "That Confucius was a formidable scholar and an inspirational mentor with a well-defined mission is more relevant. In a thumbnail autobiography, his professional inspirations receive not a mention. *(He says that)* At fifteen heart was set upon learning at forty I was no longer perplexed; at fifty I understood Heaven's Decree; at sixty I was attuned to wisdom; at seventy I could follow my heart's desires without overstepping the mark."

The above quoted lines indeed enable us to have a wonderful look into the magnificence of both heart and mind of perhaps the greatest Chinese ever born in China. Well, naturally it is no wonder that even during the present high-tech age Confucius is still considered probably China's greatest, thinker, philosopher, scholar, visionary as well as mentor of the awe-inspiring ancient Chinese-civilisation. His great thoughts and inspiring vision remain so valuable, not only to the Chinese peoples but to peoples the world over. It remains indeed a great paradox that the Communist Party of China presently ruling-over this great ancient nation since last sixty-two years or so have not perhaps cared to give importance to the great Confucius.

After the demise of the Ch'in dynasty, Han dynasty was established in China. This dynasty lasted from 205 to 823 A.D. This dynasty is recognized for a number of historic achievements in China. **First and foremost in this regard is the revival of the magnificent Confucian literature.** They effectively initiated far-reaching administrative-reforms then in China, with a substantial 'dose of socialistic-thoughts. In fact, they displayed a considerable concern for the welfare of the people of their kingdom. No wonder that they are particularly known for visionary nationalisation of land for the peasants and then carried out equitable distribution among them. This ancient dynasty's specially noted emperors have been **Wu-Ti and Wang Mang**. During Han dynasty's regime, aspects of the 'profession of arms' and 'matters-military', in accordance with those ancient times had also substantially progressed, in the desired long-term interests of the Chinese kingdom.

Now, in our specific context, looking into the basic historical aspects of **Japan Richard Overy** writes that, "Japan is first mentioned in Chinese records of 3rd century AD. These state that the archipelago was divided into a series of small kingdoms, a number of them tributary to the Chinese emperor and most acknowledged the spiritual leadership of the Empress Himiko. Over the next three centuries, a line of emperors, possibly Himiko's descendants, unified most of what is today Japan either by absorbing or destroying other leading families.... Japan was consistently overshadowed by its much larger and more powerful neighbor—so much so that Japan consciously model itself on China. Embassies were dispatched, imperial records were on the Chinese model and **Buddhism introduced** *(Those were the times to start with, sowing the seeds of the later period prominent Japanese Nation-State)*. The changes culminated in the Taika revolution of 645.... Further impetus to reform came in 668 when T'ang Chinese and Sillan armies unified the Korean peninsula.... Fear of invasion proved a powerful spur to reform. Early in the following century, Japan accordingly sought to model itself more closely still on the highly centralized Chinese state, continuing the 'Taika' process begun in 645." (Overy, 2008, pp. 114 & 115)

Here we do find, not only the interesting phenomenon of the initial birth of Japan but in addition, a short glimpse of the ancient history of Japan as well. Further in our context, the facts remain that in the beginning Japan had gradually come to shape as a State, under the overwhelming influence in multiple ways of its very large, much more civilised, developed and powerful neighbour, i.e. China. In fact in the very beginning Japan too had its own distinct dark-age for some centuries and, it also remained populated with violent, uncivilised and very primitive tribes such as the northern Emishi tribesmen. While the mighty China as its huge neighbour was already fairly civilised and, not only socio-culturally but politico-militarily a highly organised and well-governed State.

From reclaiming equitable distribution of land for the agricultural-purposes *(mainly for cultivation of rice)* among the ancient Japanese populace and then its ultimate control by the State organising the society on hierarchy-based as farmer's shiki and samurai's shiki, etc.; various kinds of revenue-generation and collection; health-services; recruitment, training and organisation of the military forces and so on had all got gradually evolved, both geostrategically as well as politico-militarily on the Chinese model, and also with active assistance of China. It is altogether a different matter that today the relationships between both China and Japan haves taken different turns and, in any case those are separate aspects of their history.

Interestingly, we find in today's China that more than ninety-percent of the population there, are in fact the Han Chinese and consequently, this is indeed a totally sweeping majority. Therefore, other Chinese ethnic minorities have hardly any say in the affairs of the State as well as those of the society at large. These realities, and their unavoidable consequences, have perforce caused many socio-political 'fault lines' in the present-era Chinese society. Even the twenty-first century China appears to be considerably far-away from achieving a truly acceptable transformation, to the stage of liberally open democratic society as a modern Nation-State. Perhaps this will still take considerable time but it is also a fact that some enduring reforms too are being pushed by the modern educated and futuristic Chinese-youth population, belonging to the present fast progressing I.T. and Space-age world. In fact historically, socio-culturally as well as socio-psychologically; Chinese do possess certain unique attributes. Their focused determination for hard-work, zeal for progress and ruthless dedication for competitiveness in the modern world (perhaps without much caring for imperatively larger aspects of 'moral') are today known all over the world. These aspects would be dealt with in more detail, in the Modern-Era chapter later.

Now coming back to the ancient-era China again, further we find from the recorded history, the gradual establishment of two noted dynasties, who ruled over then the feudalistic China, namely, Tang and Sung dynasties respectively. Their combined periods of dynasties, lasted from 615 to 1020 A.D. Tang dynasty is known for their effective 'modernisation' of the Army and also expansion of the empire. In addition, they had also brought considerable improvement in the functioning of civil-service and thereby, ensured better functioning of the Government machinery. The Sung dynasty is noted in history for bringing considerable reforms and improvement in the then system of management of the people's commercial and trading activities as well as running of the industries. Naturally, as a result, these achievements were in fact, instrumental in considerable improvement of the overall economy of both the people as well as the State.

(d) Indus-Valley Civilisation: 3500 B.C. onward

Whenever one is required to have a glimpse of the ancient 'Indus-Valley civilisation', which took 'root' in the north western part of the highly colourful Indian sub-continent around 3000 B.C. and, remarkably flourished for more than 1200 years or so, it perforce becomes an uphill task. In fact the main reason being that no ancient-era record pertaining to this civilisation has ever been traced, for even preliminarily facilitating any such venture. What is indeed even more surprising is that for more than 4500 years or so, world was totally in dark about the existence of any such magnificently ancient human-civilisation in this part of our world.

The many 'layers of curtains' got lifted during the early part of the 20th century, when India was interestingly in the last phase of the 'British-rule'. The very initial marvellously multi-dimensional glimpse of the extensive range of unparalleled 'Indus-valley civilisation', overwhelmingly captivated both the Western and Indian experts/scholars and left them totally dumfounded. Even now, multiple scholarly research-projects are continuing to unearth remaining mysteries in this regard and to learn more and more, so that peoples the world over are able to enjoy truly panoramic views of this glorious ancient civilisation.

John Keay, in context of Indus-valley civilisation, writes that, "Not until nearly four thousand years later, in fact in the early 1920s, was its existence even suspected. It was pure chance that Indian and British archeologists, while investigating later more visible ruins at Mohenjo-Daro in Sind and at Harappa in the Punjab, made the prehistoric discovery of the twentieth century. They called their find the 'Indus valley civilisation', and drew the obvious comparisons with those of Egypt and Sumerian. Indeed they thought that it might be an offshoot of the latter. Later, as its sophisticated and surprisingly uniform culture became more apparent, the Indus Valley civilisation was accorded distinct status. And when the extent of its cultural reach was found to embrace a host of other sites, many of them well beyond the valley of the Indus; it was renamed after one of these sites as the Harappa civilisation. Suddenly India's history had acquired a rich prehistoric pedigree of archeologically verifiable antiquity. Here, it seemed, was a worthy comparison to that Sanskrit literary heritage of equally impressive, though maddeningly uncertain, antiquity as comprised by the Vedas and associated texts—the Brahmans and Purana's as well as epics such as the Mahabharata. Perhaps these two very different sources, the one purely archeological and the other purely literary, would complement one another. An ancient and immensely distinguished civilisation would thus be revealed in multi-dimensional detail. The Harappan finds included buildings, tools, artefacts, jewellery and some sculpture. Intimate details about Harappan housing, diet and waste disposal came to light. Maritime trade with Sumeria was attested and led to some cross-dating. The Carbon 14 process produced comparative dated accurate to plus or minus a century or so.... Clearly Harappan settlements were not just India's first cities and townships but its first, indeed the world's first, planned cities and townships. Town-planning not being conspicuous in the subcontinent's subsequent urban development, they have been hailed by the only such examples until, in the eighteenth century AD, Maharajah Jai Singh decided to layout his 'pink city' of Jaipur in Rajasthan.... Harappan tool, utensils and materials confirm this impression of obsessive uniformity. Unfamiliar with iron—which was nowhere known in the third millennium BC—the Harappan sliced, scraped, beveled and bored with 'effortless competence' using a standardised kit of tools made from chert, a kind of quartz, or from copper or bronze. These last, along with gold and silver, were the only metals available.... What made all this consistency even more remarkable was the area throughout which the Harappans sustained it. With Mohenjo-Daro and Harappa nearly six hundred kilometres apart, it was immediately obvious that the 'Indus valley' civilisation was more extensive than its contemporaries—Egypt's Old Kingdom and Mesopotamia's Sumeria." (Keay, 2008)

The Indus-valley peoples' sense of sophistication and also their immense awareness of the 'geometry' of architectural planning and design for comfortable community-living during that ancient period of human-history are indeed surprising. In fact, when the world did come to know about all these, India was a British-colony and hence, naturally due typically lingering 'smallness of minds' of the largely arrogant and callous 'imperialism'; nothing much was scientifically done to explore and further study these invaluably unique assets of man's unique history. Tragically therefore, some valuable sites and relics of the Harappan civilisation have been lost forever. Ironically indeed, this is permanently a great loss for the entire humanity.

This is also very surprisingly true that how the peoples of this Indian subcontinent, situated in South Asia, having a magnificent history and multiple cultural enrichment spanning over many thousands of years, so 'callously/regressively' would have forgotten such ancient high civilisational acumen continuously for thousands of years. If this and many similarly other paradoxes of India's age-old history and peoples, as well as their 'saga of the modern-age', are comprehensively explored and objectively studied with requisite strategic-criticality then, many more extremely 'embarrassing-surprises' would start glaringly appearing before the same peoples, since already identified as Indians. What is the root-cause and also, what are so identifiably 'ailing' them even today?

The first and foremost cause has been that since the very ancient times, in spite of many marvellously civilisational as well as magnificently multiple socio-cultural enrichment, peoples of the Indian subcontinent have been unexplainably lacking the **"essential strategic-culture and thoughts".** This in turn ironically and also very tragically, has ensured on the part of those very same Indian peoples, as persistently lacking again the strategically **"essential sense of history".** Further, as an extremely damaging consequence since thousands of years of amazingly unbroken continuity, again the same very Indian peoples have perforce been found wanting in their typical socio-psychological as well as socio-cultural abilities in this regard, to be able to enduringly cultivate the vital strategic acumen to keep **"fruitfully learning from the history".** These undeniably important aspects, are also saddled with the above explained multiple 'irony of ironies.' with their many dimensions and fallouts; and these would be discussed more appropriately in detail in later chapters.

Now again coming back to certain prominent aspects of the ancient Indus-valley civilisation, we further notice that this one was quite distinctly different in a number of ways, from almost equally ancient Chinese, Egyptian and Babylonian, Sumerian and similar other Mesopotamian civilisations. In fact since the advent of various ancient human-civilisations in this world's different far-off areas and also, during different centuries in the very distant past, certain such exclusive features with proven authenticity, have only been found with the Indus-valley civilisation. These distinctive exclusivities pertaining to the Indus-valley civilisation are further able to establish that its socio-economic, socio-cultural and socio-political features had hardly any usual civilisational commonalities with the world's other ancient civilisations.

In this context of the Indus-valley civilisation, again **Prof. Sharma** writes that, "Here, for the first time, we come to a chapter of history where nothing may be said about kings and conquerors, empires and armies, but everything about the people, their habits and customs, their arts and religion etc. It is the history of peace rather than of war, unlike that of Assyria and Babylonia or even Egypt. We have here no Dynasties or dates to commit to memory, and no strange names to remember. It is one even picture of man as man, and his doings undisturbed by the inhumanities of war and imperialism. If there were kings in the Indus valley, we know nothing about them. It is just possible that the people of Mohenjo-daro and Harappa governed themselves without the expensive luxury of monarchs.... Unlike Sumer and Babylon these two cities were built entirely of burnt red bricks and have consequently better endured the ravages of time. The spade of the excavator has here unearthed layer after layer of buildings superimposed one over another like the nine cities of Troy.... The people of Mohenjo-daro, as well as those of Harappa, were used to cosmetics, fine jewellery and a high standard of living as indicated by their excellent habitations and furnishings.... Every house had its own well in the center of the courtyard as may be witnessed even today *(Interestingly as even in the present 'modern-age', we usually find in almost every traditional house in whole of the Indian subcontinent.)*.... Each house had a well-constructed sink from which the water was drained off into the underground sewers in the streets.... That the Indus people were punctilious about cleanliness is evidenced not only by their perfect drainage system, which has no parallel anywhere in the world until we come to our own times but also, by their provision for private as well as public bathing.... Their weapons were simple, and axes, maces, spears, bows and arrows practically exhausted their memory. Only two copper swords, or dirks, not more than $18^1/_2$ inches long, have so far been discovered. 'Swords of so early a date are extremely rare at any site' writes Dr. Mackay.... In the words of Sir John Marshall, by the discovery of the Indus civilisation, "we have established the fact that in the third millennium before Christ, and even before that, the peoples of the Punjab and Sind were living in well-built cities and were in possession of a relatively mature culture with a high standard of art and craftsmanship and a developed system of pictographic writing." (Sharma, 1948, pp. 59, 61, 62, 65 & 69)

At no stage during our presently ongoing "Strategically Critical Journey", there is unintentional displaying of any kind of 'ignorable' bias or any 'hidden' programme to get engaged in an 'uncalled-for' criticism as well as in any manner, even remotely/indirectly minimising the truly extraordinary magnificence and well-acknowledged historic marvels of the Indus-valley civilisation.

The historic fact remains that since the gradual dawn of the very ancient Indus-valley as well as Indian civilisations perhaps paradoxically, the peoples of Indian sub-continent have been always found cultivating a primitively requisite 'feel' of the geography of their vast land-mass, so exquisitely gifted with highly valuable natural resources of invaluably multiple kinds. And in addition when gradually the appropriate time came, the same peoples lacked capacity to imbibe any desired sense of history and consequently these again perhaps 'crippling' limitations on their part, perforce then failed them 'hopelessly' on the path of slowly and historically evolving the inescapably "essential strategic-culture and thoughts", singularly most crucial for the growth, multiple advancement and also healthy survival of any human-civilisation.

Since ages the fact remains that the vast geographical mass of the Indian sub-continent has never remained socio-politically as well as also geostrategically one vast geographical entity. The principal reasons perhaps being that to begin with, peoples inhabiting here or even those who during very ancient times came here and settled down were not of the same racial stock, colour of the skin, body and facial features and then, and over the years the different race/groups of peoples facing different climatic conditions and situations, cultivated their own distinctly varied cultures, food-habits, ways of life and subsequently, different religions during differing periods of history.

No wonder that since the dawn of history in this regard, the desired socio-political unity had never enduringly taken root and survived in the Indian sub-continent and hence causing many 'fallouts' and associated complexities. Interestingly we find in the varied history of mankind's few other relatively ancient civilisations in different far-off geographical areas in our world, such as Mesopotamia, Egypt, Middle-East and the European continent, fairly similar historic phenomena have taken place. Hence, in those areas too, various civilisations and empires flourished quite independently and at the same time, without much coming into conflict as a routine.

But certain peculiarities with respect to these very ancient civilisations in the Indian or in South Asian sub-continent are that, even for the experts it is not easy to formulate a chronological history with details in any orderly sequence. Whatever they may be so, it is logical that the history and allied archeology of the chain of Harappan cities/settlements during a fairly long period of Indus-valley civilisation, takes precedence over other Indian or Aryan civilisational-literature since, the former commands greater historic antiquity here. We also notice that in the numerous Harappan civilisation-era cities and marvellously laid-out and built human-settlements, so extensively spread over hundreds of miles of geographical areas in the north-western region of the Indian sub-continent, terracotta-work of art and crafts as well as bronze sculptures were of certainly very high standard and sophistication. One particularly world-famous and truly magnificent bronze-sculpture is, that of the 'dancing girl'. Noted connoisseurs of 'sculpture-art' from the world over, have indeed poetically tried to observe and explain many finer artistic sides of this very ancient and captivatingly beautiful little figurine. No doubt, it is a treat not only to see and admire this pleasing piece of 'delicate' and beautifully crafted figurine but also to read the accompanying explanatory narrations.

Further, **John Keay** writes that, "Skeletons unearthed in the Indus valley, however, attest that the Harappan people were of several different racial types, amongst them that, related to Australia's native people and still represented in parts of India, of proto-Australoid cast. Furthermore, although most of the surviving Harappan stone sculptures were found at Harappa itself, whence contacts with Namazga seems to have been closest, and the 'dancing girl' *(since already mentioned in some detail above.)* was found at Mohenjo-daro, whose external was more orientated to the Persian Gulf and Mesopotamia.... Trade, both within the sprawling Harappan world and without, was essential to the development of its culture. Bronze or tin (for making bronze), silver and certain precious stones like lapis lazuli and soapstone are not found within easy reach of the Indus valley, and must therefore have imported from elsewhere. Likewise it is clear that the Mesopotamian cultures obtained numerous commodities from the Harappans, including copper, gold, timber, ivory and probably cotton textiles. Harappan sealings and seals have been found in Sumerian sites, and Sumerian documentation makes frequent reference to the relations with the distant lands of 'Dilmun', 'Magan' and 'Meluhha'. The first seems to have been in the Persian Gulf, possibly Bahrain, and to have been something of an entrepot. 'Magan' is usually identified with the coastal regions of Iran and Baluchistan, the modern Makran coast. And 'Meluhha', by a process of deduction from the trade items associated with it, looks to have been the Harappan civilisation." (Keay, 2008, pp. 15 & 16)

The above quoted lines have again authenticated that there did in fact exist a thriving commercial system of trade, based on exports/imports of goods, between the ancient Mesopotamian civilisation and the Indus-valley civilisation peoples. This fact of man's ancient history of these regions in addition, highlights another interesting fact that peoples of these two and other contemporary very ancient civilisation, had already remarkably evolved then a viably engaging 'sea-fairing' culture as well. This naturally indicates that they already had acquired then the necessary skill for making suitable primitive boats too, which could sail through the sea or at least all along the concerned sea-coasts for the purpose. Therefore these peoples must have been one of oldest men in history to be able to plan, design and construct such boats. A well-known Indian weekly news magazine *India Today* (September 12, 2005), had published a colourfully pictorial and very informatively illustrated news in this regard (pages 90, 91, 92 and 93), "Four thousand five hundred years later after this hypothetical scenario, the names may have changed—Magan (pronounced Majaan) is now Oman, and Meluhha, the ancient name of the spectacular Indus Valley Civilisation, is now India. **The reed boats plying the World's busiest shipping straits may have since given to lumbering steel leviathans with belly loads of oil for refineries in Gujarat**. But the spirit of scientific enquiry persists. What did these boats which navigated the seas of India over 1000 years before Mahabharata look like? Did ancient mariners really sail flimsy vessels made of reed across the Arabian Sea? *(In* *this specific perspective in fact, here lies the 'catch', as these issues would very soon be reasonably explained little later.)* The answers could well be found in the picturesque coastal town of Sur 350 km Oman's capital Muscat where the newest where the newest tourist attraction is a black sailing vessel. It looks like a giant wicker basket and, thanks to the water-proofing smells like a freshly tarred road.... The Indus Valley Civilisation carried on a brisk seaborne trade with Mesopotamia, a fact established by hundreds of Harappan potshards uncovered there. It was, however, almost entirely dependent on copper mined from the mountains of Oman, which fell midway to Mesopotamia.... By dashing across the Arabian Sea, the expedition hopes to disprove the theory that ancient trade between these two countries was actually coastal. The boat breezed through its first sea trials last month, clocking a healthy 3 knot speed.... But the Magan expedition faces many challenges. A flat bottomed reed sailing vessel is inherently fragile. It lacks a keel which stabilises a boat and prevents it from drifting sideways, and is vulnerable to the vagaries of nature. It could spring leaks while at sea, be blown off course or be toppled by a whale."

First of all the question concerning the main material more likely having been used by then the Mesopotamian and/or Harappan peoples since, plenty of timber was available as a natural resource in the Indus-valley and also the same was being exported to Mesopotamia. So logically they would have conveniently constructed more robust and durable boats of timber for the purpose. Another point also comes in that with all probabilities during those so remotely ancient era, the Harappan as well as Mesopotamian peoples would have perhaps, definitely selected coastal-water route to conduct trade by means of such timber-boats, rather than avoidably undertaking more risky-route across the high-seas in their regions. And, as perhaps was likely then, and also further reported by the newspapers later, this scientific experiment couldn't succeed and that model-boat soon capsized in the sea.

So far as the aspects of trade as well as other forms of relatively regular contacts of the Indus-valley civilisation peoples with those of other civilisations are concerned, Jawaharlal Nehru has mentioned that, "'The Indus civilisation,' writes **Professor Childe**, 'represents a very perfect adjustment of human life to a specific environment that can only have resulted from years of patient effort. And it has endured; it is already specifically Indian and forms the basis of modern Indian culture.' Astonishing thought: that any culture or civilisation should have this continuity for five or six thousand years or more; and not in a static, unchanging sense, for India was changing and progressing all the time. She was coming into intimate contact with the Persians, the Egyptians, the Greeks, the Chinese, the Arabs, the central Asians, and the peoples of the Mediterranean. But though she influenced them and she was influenced by them, her cultural basis was strong enough to endure." (Nehru, 2002)

In the absence of credible anthropological evidences as well as proper historic records, we do not know about the system of political control and the type of administration the Harappans had practiced for more than a millennium in their regions. Further how such a thriving age-old human-civilisation, practicing the world's oldest and advanced systems of well-planned and highly organised urban lifestyles, could have got 'buried' in the sands of time and also, without any authentically traceable reasons in this regard so far.

But it is a fact that in the numerous archeological sites in the north-western regions of the Indian subcontinent, the Harappan ruins have been found to be covered by layer after layer of sand and this phenomenon in turn, has logically induced many scholars and experts to infer that some catastrophic devastation might have caused the sudden 'wiping-out' of the world's

most probably oldest as well as unique civilisation. This scholarly perception naturally appears to command reasonable scientific credibility, as mentioned by well-known writers, like **Late Jawaharlal Nehru**, **John Keay** and others, in this regard.

We have also reasoned that major river-systems in the Indian subcontinent have been singularly instrumental in triggering the gradual dawn of the magnificent ancient civilisations and their subsequent flourishing. But as already mentioned earlier it is also true that these rivers are flood-prone too, and therefore, extremely devastating floods in the river Indus and its other noted tributaries during sometime in that very ancient-era, might have caused perhaps the sudden demise of the unique Indus-valley civilisation.

Some scholars have been of the opinion that when **Aryans** came from north-western side of the Indian subcontinent, since they were warriors-like, they subjugated the Harappans in the Indus-valley areas and thereby, their even older civilisation got 'wiped-out' as a consequence. While interestingly, the other speculative 'theory' put forward says that during those remote historic times, the Aryans had preceded the Harappans and therefore, the peoples of Indus-valley civilisation were in fact descendants of the Aryans only.

In any case there is no archeological or historic evidence, which even remotely points towards this perception. Further, to begin with, the Aryans did not have any legacy for urban-living, but otherwise, they had a magnificent and very ancient heritage of creating their own epics, treatises as well as literature and logically therefore no such Aryan document anywhere historically substantiates any of such conflicting opinions. At the same time, it is also credibly accepted that Aryans were one of oldest horse-riders in the world, with fleets of remarkable chariots and, rearing very large herds of cattle and other animals, for their multiple uses. They were perpetually organising, training and maintaining marvellously big and imposing armies, consisting of a number of arms/branches for their strategic purposes. In fact in this context, these were the first phenomena then taking place in the Indian sub-continent. Hence there are magnificent records of their unparalleled military-exploits in many ways. But as events keep on occurring, in many ways and in multiple perspectives and thereby; histories are also their on-going reflections on the same unending 'canvas of time'.

That's the methodology by which the same history too, keeps evolving, taking shape and simultaneously, getting recorded in multiple ways by the man himself for posterity. And accordingly, in the much wider perspectives of the evolutionary laws of supreme-Nature in this world, again the same history never spares or forgives any person, community or even civilisation whatsoever. Logically therefore for any civilisation or nation, as the multiply consequential many realities and issues in this regard, with all their possibilities as well as manifestations, since continuously and proactively appearing so far, and also most likely to be the same in future as well; the singularly crucial importance of the already earlier highlighted **"essential strategic culture and thoughts"** just cannot be over-emphasised. These and so many other allied aspects would further be discussed more appropriately in the subsequent chapters.

(e) Ancient India: 3500 to 2000 B.C. onward

Since centuries scholars the world over have almost unanimously accepted that the ancient India, her multiple manifestations with wonderfully fascinating colours and varieties as well as her unique history encompassing more than three millenniums, have indeed been unparalleled.

No wonder that a highly perceptive scholar, C.E.M. Joad has so thought-provokingly said that, *"Whatever the reason, it is a fact that India's special gift to mankind has been the ability and willingness of Indians to effect a synthesis of many different elements both of thoughts and peoples, to create, in fact, unity out of diversity."*

This is precisely a remarkable observation, with multiple meanings and depth, on India and her peoples and also her unique as well as continuing civilisational-cum-cultural antiquity. **The history of ancient India is as old as of Egypt, Mesopotamia and China. In fact similar to them, the fascinating history of ancient India goes back further in time by a couple of millenniums**; when already discussed the Indus-valley civilisation is also befittingly included, as the pre-Aryan era, with its many unrivalled dimensions and socio-cultural innovations. At the same time it is also true that, mysteriously the post-Harappan era rather had begun abruptly, with the arrival of Aryans in the Indian subcontinent during 2000 B.C. with its many civilisational impact and creativity.

Concerning the Aryans contextually here; it may perhaps be of interest to refer an old book *Heroes of European History* written by **Mrs. Louise Creighton** (pages 1 and 2) has written that, "When we first begin to know anything about Europe, we do not find it divided up into nations as it is now. The history Egypt and India and China goes back to many hundreds of years before the birth of Christ, when Europe was still peopled only by wild, wandering tribes.... The nations that have made Europe great, as well as the chief nations of Asia; to this people the name of Aryan has been given. History tells us nothing about the Aryans in their early home. It is because all the Aryan nations have the same words for the common things of life, that we know that they were once one people. They passed out from their first home south-wards into **India** and westwards into Europe. There the first who made themselves a great people were the Greeks, or the Hellenes, as they called themselves."

The Marquess of Zetland in his highly scholarly 'Introduction' to a book *The Legacy of India* says that, "**The legacy of India; how rich a heritage, drawing contributions, as it does, from diverse races and many epochs both preceding and following the great Aryan incursion from the lands lying beyond the snow-capped ranges of the Hindu Kush**. From the prehistoric civilisation of the Indus valley, brought to light by excavations at Harappa and Mohenjo-daro, to British India in the twentieth century is a span in all probability of from five to six thousand years. During all these centuries the peoples of India have lived their lives and evolved their civilisation in comparative isolation." (*This is indeed a highly reflective perception of India's remarkably great living ancient civilisation.*). (Marquess, 1934-35)

Now as already noted earlier that relatively sudden 'eclipse' of the Indus-valley civilisation is one of the great mysteries of man prolonged history with no certain answers to them emerging so far. And naturally in this regard, interestingly, quite a number of hypothetical explanations from scholarly circle the world over have been coming from time to time.

Now coming back to the post Indus-valley era of the ancient Indian civilisation beginning with the advent of Aryans during 2000 B.C. and their far-reaching multiple historical impact, we notice that this ancient age does have a number of civilisational periods within, in continuity and in a sequence, starting from the Vedic-age onward. Then we find the magnificent dawn of Epic-age from 1000 to 500 B.C. Following the Epic-age then, the Buddhist-age begins heralding altogether a new spiritual philosophy and thoughts, spanning a short period from nearly 560 to 485 B.C. Each of these different periods in ancient India beginning from 2000 B.C. onward, is magnificently unique in their own right not only in historic sense but in addition, they were multiply creative too, in all their marvels in the perspectives of spiritual, intellectual, architectural, art, sculpture and socio-cultural aspects as well.

Subsequently then in ancient India, a couple of foreign invaders entered the Indian subcontinent, crossing the rugged mountain-barriers and through the north-western frontier passes and then entering the plains of now called Punjab, astonishingly even during the pre-Christ era itself. First foreign invader to arrive during the period of 520 to 480 B.C. was **Darius** from ancient Persia. After this during the ancient period of 350 to 325 B.C., another invader came broadly through the same route, who was a Greek named **Alexander**, and who perforce got engaged in a closely contested as well as ferociously fought battle with **King Porus** of that region, which even now remain a specially noted historic battle of ancient India. Later on interestingly, history did acknowledge this famous Greek warrior, charismatic leader, great military-strategist, innovative tacticians in battles and campaigner, as **'Alexander the Great'**.

In this context, interestingly one exclusive point emerges that none of these two ancient invaders from relatively far-off places, were able to enduringly establish their footholds historically on the Indian soil, for some compelling historic reasons or might be according to their own strategic-plans. Then the colourful history of ancient India records the dawn of a number of dynamic as well as colourfully fascinating periods/historic-dynasties in a sequence encompassing a long and continuous age of many centuries from approximately 320 B.C. to nearly 620 A.D. and thereby, beginning the **Magadha empire, also known as the famous 'Maurya-empire'**, with the ancient city of Pataliputra located in the Gangetic-plains in Eastern-India as their capital and subsequently; culminating into the last phase of 'Hindu-era' in the Indian subcontinent, with end of the rule of King Harshavardhana.

In this perspective we have already noted that the historic turning point regarding gradual dawn of the ancient Indian-civilisation, has been arrival of the Aryans into Indian subcontinent in around 2000 B.C., through its north-western frontiers. Regarding the method and system of their gradual entry and then all over the vast land-mass of that distant past of this entire subcontinent, for settling down all over the expansive north, eastern and towards the central regions; there have been different perceptions among the concerned experts and perceptive scholars. **Besides these aspects, another interesting question also remains as to who were precisely these Aryans, we are talking about here?** These are highly complex specialised subject-matters for them exclusively for their further research and explorative studies. For a layman, like this author, few quotations from a couple of well-known books of scholars of international fame may be of great value here, in this ongoing study.

Jawaharlal Nehru mentions that, "The Aryan migrations are supposed to have taken place about a thousand years after the Indus Valley period; and yet it is possible that there was no considerable gap and tribes and peoples came to India from the north-west from time to time, as they did in later ages, and became absorbed in India, We might say that the first great cultural synthesis and fusion took place between the incoming Aryans and Dravidians, who were probably the representatives of the Indus Valley civilisation. Out of this synthesis and fusion grew the Indian races and the basic Indian culture, which had distinctive elements of both. In the ages that followed there came many other races: Iranians, Greeks, Parthians, Bactrians, Scythians, Huns, Turks (before Islam), early Christians, Jews, and Zoroastrians; they came, made a difference, and were absorbed. It is odd to think of India, with her caste system and exclusiveness, having this astonishing inclusive capacity to absorb foreign races and cultures. Perhaps it was due to this that she retained vitality and rejuvenated herself from time-to-time." (Nehru, 2002, pp. 73-74)

The above quoted lines do give a glimpse of Nehru's historically perceptive thoughts and yet, they are poetically emotional too. Hence the question remains to start with that, whether the gradual arrival as well as remarkably historic processes of settling-down in this vast Indian subcontinent, were relatively peaceful or violent or even both, in different regions and areas? As we are able to notice so far perhaps that, there are no broadly accepted answers in this regard among the scholars.

Further appropriately in this context, Professor S.R. Sharma writes that, "The first of these impacts on a large and ascertainable scale was between the Aryans and the Dravidians. The latter name is here applied to all the peoples of India who already existed in this land and came into hostile or peaceful contacts with the Aryans. If we may link up the Aryan immigrations into India with Mohenjo-daro on the one hand and the Iranian and Hittite evidence on the other, we might place the Vedic period of our history approximately in the second millennium B.C., i.e. 2000-1000 B.C." (Sharma, 1948, p. 90)

No wonder as to why the ancient Indian civilisation is so unique, colourful and remarkably mysterious too, and at the same time compositely encompassing many further colourful sub-civilisations. These in turn, have given marvellously integrating socio-cultural enduring unity in diversity as well. Another related aspect, is that though these astonishingly civilisational phenomena are both inclusive and integrating but simultaneously those diversities too, do continue to exist keep on thriving with multiple harmony. Naturally and consequentially, one exclusive unique fact of history remains well-accepted that, in spite of so many turbulent trials and ups and downs, this wonderful mosaic of the unique 'unity in diversity', regarding India's age-old civilisation, historically continues to survive with an ongoing process of dynamic growth and adaptation.

These perceptions are well-taken in a perspective, but as commonly accepted that every 'coin has two sides'; Late **Nirad C. Chaudhuri** has perhaps highly controversially called the Indian subcontinent as, **"the continent of 'Circe'."** He might have had his even own reasons for such an 'extreme' opinion but, this may not be acceptable by everyone concerned. Interestingly he has also authored a book, by this specific 'title' itself, as a collection of his engaging as well as highly reflective essays on 'the peoples of India', which too, as his a number of other books, has been internationally well-received. More so that the many controversies concerning his multiple range of highly engaging opinions apart, he could never be ignored along with his powerful writings as well.

The relevant point here is that one is compelled to reflect on his highly argumentative 'analytical-reasons' for calling India as "The Continent of Circe" and then also followed by interestingly calling the Indians as "the victims of Circe". Though in his writings he has not spelled out clearly that along with many relatively glaring paradoxes regarding India and the Indians; if one cares to read in 'between the lines', one may perceive that the regressively lingering almost total lack of earlier already mentioned, **"essential strategic-culture and thoughts"** does appear to be as the 'fulcrum' of India's age-old and at the same time still persisting, many paradoxical 'ironies'.

When the Aryans came to the ancient Indian subcontinent during as early as 2000 B.C., if one goes by the well-accepted history in this regard, then we find that during that bygone ancient era they were not only singularly dynamic, progressive and knowledgeable with many remarkable attributes, and simultaneously then to begin with in the Indian subcontinent, but were most probably not suffering from the above mentioned perhaps 'crippling' paradoxical 'ironies'. Then in the process of their permanently settling-down here, with remarkably comprehensive integration with the lands, peoples as well as rich natural multiple resources, forests and wildlife what went 'wrong with them and also how many kinds of complacencies gradually set-in? *And then the earlier historic question, concerning precisely who were the Aryans and where from they came-in into the Indian subcontinent too, remains.* A noted Scottish author has very precisely attempted to come to an intellectual grip with these interesting questions in his already referred book on 'India a History', and hence these need to be relevantly to be reflected upon.

In these contexts, **John Keay** writes that, "Numerous writers continued to warn against the assumption that a shared language necessarily meant a shared ethnicity. Yet the idea of a single race sowing the seeds of civilisation from Bengal to Donegal proved intensely exciting, and ultimately irresistible. To **Friedrich Max Muller**, the distinguished German Professor of Sanskrit at Oxford in the mid-nineteenth century, it seemed that the Aryans had a 'mission to link all parts of the world together by chains of civilisation, commerce and religion'....Muller, too, warned against drawing any simplistic conclusions about race, but already Aryan descent was popularly seen as the mark, if not yet as a master race, at least of ethnic distinction.... *(Here, this is historically a far-reaching point with greatly harmful consequences; at times as much later we find that how, Hitler as irrational prisoner of this highly erroneous perception, when coming to power in Germany during 1930s, could get mindlessly transformed into a singularly most dangerous menace to the whole world.)* Gratified by the discovery of their proud historical pedigree, India's aspiring nationalists embraced the Aryans as readily as did Europe's cultural supremacists, Given the vast spread of the Indo-Aryan languages, an Aryan homeland was soon being sought somewhere in the middle of Eurasian landmass. Most scholars favoured the steppes of southern Russia and the Ukraine, or the shores of the Caspian. Nomadic pastoralists, the Aryans needed plenty of room. Thence, in a series of sweeping migrations spread over many centuries, they supposedly took their language, plus their gods, their horses and their herds, to Iran and Syria, Anatolia and Greece, Eastern Europe and northern India. Indian Aryans were therefore originally immigrants, and to judge by their exploits as recorded in the Vedas, highly combative ones. Aided and encouraged by deities like the fire-breathing Agni and thunderbolt-throwing Indra, the Aryan conquistadors were seen as having hurtled down the passes from Afghanistan to career across the plains of Punjab. Dealing death and destruction from fleets of horse-drawn chariots, they subdued the Indigenous peoples and appropriated their herds. As *dasa* (slaves) or *dasyu* (demons), these indigenes or aborigines were characterised as dark, flat-nosed, uncouth, incomprehensible and generally inferior. The Aryans, on the other hand, finer-featured, fairer, taller, favoured over others in the excellence of their gods, their horses and their ritual magic and together a very superior people.... The Aryans, spreading their superior culture right down the valley of the Ganga and then deep into the peninsula, had conferred on India an unprecedented cultural integrity an enviably high degree of civilisation. In time, however, the purity of the Aryan race had become hopelessly diluted; manliness, creativity and drive had succumbed to the enervating effects of an intolerable climate and an insidious social system. Hence no serious resistance had been offered either to the thrust of Islam or to the advent of the colonial powers. India had slumped into seemingly irredeemable decadence and degeneracy. Then, in the nick of time, out of the west came the British. No less fair, no less manly and no less confident of their superiority, they were the neo-Aryans, galvanising a naturally lax people into endeavour and industry, showering them with the incomparable benefits of a superior civilisation and a humane religion, and ushering in a new and golden age. Or so, some liked to think." (Keay, 2000, pp. 21-22) *(Now coming to the last few lines as quoted above: well, shall we call these as astonishingly 'arrogant' thoughts, or perhaps may even be, not that of John Keay; but on the part of some of the other fellow country-men an indiscreet display of probably the lingering fixation of 'the white man's burden',* *more so during beginning of the 21st century, since lots of water has already flown down the Thames in London.)*

Well, certainly not wholly but, then definitely contextually there are, considerably some factual perceptions here, in John Keay's arguments. Interestingly, these may even possibly be seen in the above given 'reflections' on **Nirad Chaudhuri's** somewhat 'controversial' book. In the same perspective, one may also like to refer to **V.S. Naipaul's** book on *India: A Million Mutinies Now*. If one is objectively inclined to consider these perceptions, precisely within fascinatingly explorative debating parameters then, one may even possibly call, these 'controversial' ideas, as "India-bashing" and so on. In any case these may be so but, many other internationally well-known scholars and thinkers, like Jawaharlal Nehru, Edward Said and others, have given equally weighty and perceptive counter arguments in this regard and therefore, these too, need to be delved into befittingly. These aspects and also relevantly more so, would be appropriately discussed later in the sixth chapter on the, 'Medieval Ages and Birth of Empires'.

As since already noted earlier, with the gradual immigration of Aryans into the Indian Subcontinent during 2000 B.C. some kind of contact with peoples of the Indus-valley civilisation must have taken place; specially if those Harappans' and their otherwise thriving urban and other habitats as well as their other living assets won't have possibly been 'wiped-out' earlier, by some kind of natural calamities or so. Whether these contacts were largely peaceful and mutually perhaps 'complementary' or violently conflicting, remain even now debatable and shrouded in mystery. Substantiating as per the archeological and other corroborative findings so far, and also discussed earlier here, some noted scholars have perceived that the Harappans were largely peace-loving and hence, hadn't evolved any culture of even perhaps marginal weaponry for their effective self-defence. During that distant era, they appear to be far-away from any kind of then perceivable, "essential strategic-culture and thoughts" and thereby, most likely they might have been 'practicing' kind of now unexplainable 'strategy of peace'. If that were so then, their reasons and details in this regard too, are again shrouded in mystery.

Now so far as the reasonably and well-accepted details regarding the migrating Aryans are concerned, they were much sturdier, more war-like, better organised, equipped and well-trained in various types of weapon-systems as per that ancient age, having numerous impressive horses and fleets of again horse-drawn chariots, herds of multiple live-stocks, their rich language and many other sociopolitical-cum-socioeconomic assets and heritage.

Vedic-age, as we understand, has been in fact then a historic-gift to the Indian subcontinent, by the Aryans, with their multiple material, cultural as well as socio-political and strategic assets and in addition then, along with their indeed unique multi-dimensional attributes. At the same time some prolonged contact and then gradual multiply socio-cultural-cum-political as well as 'genetic' fusion between Harappans and the Aryans; could never be completely ruled-out. But the perhaps some persisting opinions of violent contacts leading to the massacre of Harappans are not at all borne by either any archeological findings or Vedic descriptive evidences. Even the Rig Veda, possibly the world's first 'library-book' too, doesn't give narration or any indirect hint towards any such war-like conflict and massacres. In fact again in present-day Punjab and other spread Indus-valley townships/habitats, where the Aryans initially perhaps came and settled, some fairly sophisticated ceramic and other potteries have been found, which indicates some kind of interesting and fruitful fusion of both Harappan as well as Aryan such skills; and also these most likely could have flourished under the Aryan patronage.

Further, some scholars do opine that relatively the Indus-valley peoples were somewhat 'illiterate' and also not aware of some agrarian skills, but they did use transport pulled by ox as well as they had become skillfully tuned to making some useful uses of other domesticated animals for then whatever indeed, remarkable furtherance of their individual community based quality of life. But at the same time, it appears that they too, acquired many things from the Aryans like, relentless passion for dairy-farming, horse-riding, learning highly enriching Aryan languages, cultivating multiple Vedic-values, religious-rituals and dedicated spiritual-pursuits evolving higher consciousness towards 'meta-physical' dimensions and so on. In many ways again, these multi-directionally socio-cultural interactions were sometimes also mutually supporting as well. **For example, it is also opined by some scholars that originally Lord Shiva was not a 'Aryan/Hindu-God', but in fact a Dravidian 'Deity'; and if that be so then interestingly, even in religious spheres too, there were highly fascinating and so much mutually enriching interactions as well.**

At this point contextually it appears relevant to critically and briefly delve upon fundamentally ancient-era humanity's religious-cum-spiritual aspects. First and foremost, Man's 'Idea of God as Supreme Lord' and 'His Multiply Divine Manifestations' Have Never Created any religion. Therefore, religions are the creations of man himself, during different centuries and again in different far-off geographical continents as per multi-dimensional continuing march of various flourishing human-civilisations again in those very continents/areas. Further trying to peep into the most fascinating as well as fully inconceivable dimensions of Spiritualism, one may perceive that Lord (Or God) Is Himself/Herself Infinite. At this point, little enlarging this most enriching and enchanting perception, one may be tempted to even say that this truly infinite Universe, with its infinite energy-cum-matter, does in fact appear to be again an Infinitely Grand Manifestation of the Same Supreme Lord, without any kind of any Beginning/End.

Hence in this somewhat even initial, whether precisely clear or hazy perception of this supremely grand and truly infinite 'scheme of things', under the supreme command of mother-Nature, perpetually evolving-cum-altering, where does the tiny man, with his so much earlier discussed unique 'brain'? All these are mind-boggling indeed and hence, never fully comprehensible. We may say at this point of our discussion that with the gradual arrival of Aryans, and along with the advent and magnificent flourishing of the Vedic-age, during 2000 to 1000 B.C., the above delved supremely enchanting "**Concept of God**" is in fact, part of the vast and captivating Vedic-values; since both got perhaps simultaneously evolved and enriched during the same Vedic-age. We again find the same wonderful glory of the infinite spiritualism, harmoniously embracing the dynamically advancing multiple knowledge of science, displaying all the grand magnificence of their again mutually complimentary meeting.

Here contextually, while 'entering' into the much larger (perhaps infinitely large.) enlightening domain of 'Spiritualism'; the fascinating depth and dimensions of Einstein's 'Theory of Relativity' acquires a much wider and enthralling perspective as well as greater meaning. In the same perspective, perceptively of course, in the highest all-encompassing visualization of the blissful 'synergy'; so superbly manifested by this 'grand-synthesis' of both the supremely enlightening spiritual as well as objectively analytical scientific perceptions then, enable a person to experience their indescribable and exclusively phenomenal 'rainbow' like magnificence. If this whole perpetually continuing phenomena is truly imbibed by a person then the mind and thoughts get transcendentally elevated to a 'state of bliss'; i.e. achieving a state of total 'thoughtlessness' and hence, in other words, that is the highest state of 'sainthood'. Well these may sound naively simplistic but, all these are easier said than done. In plain reality of this world, a true saint may even spend entire life in relentlessly meditating and striving, to be able to achieve that highest state of 'sainthood'.

Now few words about religions here could also be perhaps mentioned, since it is needless to say that spiritualism and religion are different in many ways, though the ultimate purpose of both remains the same. Interestingly, we also say that both the religion and the spiritualism are 'two sides of the same coin'. Spiritualism is more abstract in nature and naturally, its pursuits too, are abstract both in form and content and thereby for any person, it perforce perhaps has to be mostly powerfully 'obsessive' and at the same time, it is always all through an uncharted 'meta-physical path' for that person; and then finally leading to the state of 'self-realisation' and 'enlightenment'.

The teachings and writings of various Saints so knowledgeably informs us that the ultimate purpose of all these pursuits is reaching that most sought-after state of the supreme 'enlightenment' and then remaining there at that person's own will. Hence we may also say that, such true pursuits of spiritualism with all its form and advancing manifestation, are never or perhaps cannot be so, any religion-specific or confined to any particular religion in any manner, at any stage of 'evolution' with respect to that person, undertaking that 'spiritual-journey'. Whereas religion gives the man, specifically a well-defined defined 'path of worship' to the same 'Supreme Lord'; Whose Divine Manifestation may again be only at the most, superficially different in so-called different religions. In other words, externally those different paths, as shown in various religions, may appear to be different but, the ultimate goal/destination for all of them remains the same.

So all religions in their true sense as well as their gradually and precisely unfolding 'forms and content' are humane, in their meaningful depths and dimensions. In fact, it is only the various kinds of their 'distortions' that may or perhaps even do make them otherwise. Now in this perspective, we need to say that, such distortions consequentially may turn somewhat 'polluting' and so, may even 'rob' perhaps some 'humaneness' from any religion. In addition, those 'distortions' in this respect, (as dictated at times, by some unscrupulously narrow-minded/vested interests.) are mostly deliberately or inadvertently man-made and nothing else. Therefore, we may conclusively say that man (as mostly always.) are at fault here too, and certainly not any religion in any manner. Again, the briefly described perceptions as above too, are part of the same vast Vedic-values, as so magnificently illustrated in the enormously enriching Vedas themselves. All these, in fact, are remarkably enchanting revelations of the ancient Vedas with all their glory, to man.

At this stage, **another equally interesting as well as far-reaching question now arises, as to what are these still surviving Vedas and what is their origin?** Vedas are perhaps the oldest literature, or definitely one of the oldest ever created by man. There are in total four Vedas and the oldest one is the Rig-Veda. Vedas are not only brilliantly conceived, gradually evolved, having remarkably multiple depth and range and also written with precision as magnificently comprehensive treatises.

In this historic perspective of great civilisational value, **Jawaharlal Nehru** has written on page 77, in his earlier referred book, 'The Discovery of India' that, "Whatever the exact date may be, it is probable that this literature is earlier than that of either Greece or Israel, that in fact it represents some of the earliest documents of the human mind that we possess. Max Muller has called it: 'The first word spoken by the Aryan man.' **The Vedas were the outpourings of the Aryans as they streamed into the rich land of India. They brought these ideas with them from that common stock out of which grew the Avesta in Iran, and elaborated them in the soil of India. Even the language of the Vedas bears a striking resemblance to that of the Avesta, and it**

has been remarked that the Avesta is nearer to Veda than the Veda is to its own epic Sanskrit. How are we to consider the scripture of various religions, much of it believed by its votaries to be revealed scripture? To analyse it and to criticise it and to look upon it as a human document is often to offend the true believers. Yet there is no other way to consider it."

In other words perhaps particularly in this context, there appears to be a need to be relevantly flexible and also, appropriately keeping in mind the well-accepted historic antiquity of the Vedas. **John Keay** in his earlier referred book, has understandably in some detail, commented upon certain aspects of the desirability on the part of man; with specific reference to certain religiously 'sacrificial-rituals'. Keeping in mind his modern Western-European socio-cultural as well as socio-psychological background and ethos, his such perceptions and comments are again contextually understandable; but he has done so perhaps, by little rigidly adhering to the modern-era concepts and value-system, while probably 'ignoring' the Vedas' antiquity. Certain ideas/thoughts, as available in any ancient epic or book anywhere, naturally may appear to be antique and even obsolete, and therefore unacceptable. These may desirably be taken with some flexibility and contextual relevance; by going into little deeper to visualise perhaps the hidden ideas therein and hence, not always paying any particular attention to the apparently literal meanings on the surface itself.

Professor S.R. Sharma has also written that, "The Rig-Veda is not only one of the oldest books in the library of the world, but it is the first piece of scientific composition produced by man. In grammatical literature there is nothing older or more perfect than what the Indians produced in the centuries preceding the Christian era. Indian mathematics began in the Vedic age and registered great advances in succeeding times. **Aryabhatta, Bhaskaracharya** and **Lilavati** are still honoured by astronomers and mathematicians. The Cipher and the decimal notation were Indian inventions. Algebra was introduced into the world by India through the Arabs. In medicine and surgery India produced the works of **Charaka** and **Sushruta**, respectively; they bear testimony to the remarkable achievements made through experimental research in those two fields of curative science. But the highest achievements of the Indians, by universal acknowledgement, were made in philosophy. **Schopenhauer** acclaimed the *Upanishads* as having provided him with the 'solace of my life and the solace of my death,' **Max Muller** proclaimed the glories of the culture embodied in our literature enthusiastically to the West in glowing terms. The *Ramayana*, the *Mahabharata*, the Bhagavad-Gita and the dramatic and poetic works of **Kalidasa** ranks with the best in the world's literature in all languages." (Sharma, 1948)

The above quoted scholarly and contextually gives a very brief glimpse of the superb magnificence of Vedas and some of the principal Epics of Ancient India, belonging to a distant and bygone period of history which is yet to be scientifically 'authenticated';. More about those epics would be discussed little later. Indeed it is well known that in fact, there are total four Vedas. Out of them Rig-Veda is the oldest one and the remaining three are the Samaveda, Yajurveda and the Atharvaveda. These three are mainly supplementary to the oldest and generally known as most important and commanding, the Rig-Veda.

It may appear to be quite befitting to submit at this precise point that, it definitely takes much more in intellectual-cum-spiritual perceptions, with vast depth and dimensions and also much higher-levels of pious perseverance as well as imaginative range of human minds to be able to reflectively come face to face with the highest levels of true essence of the Vedas. As it is always seen that the highly accomplished scholars and historians, both Western as well as the Eastern the world over have undertaken many kinds of intellectual exercises to fully decipher and explain the superb and rather, enchantingly difficult to fully fathom the many depths and dimensions of the Vedas.

These in away contextually, some discerning perceptions, would appear to be fully reinforced; if we may refer to a very captivating book titled, *The Holy Vedas: A Golden Treasury* and, beautifully shaped-up by **Pandit Satyakam Vidyalankar**. Though at the same time, it is also known there are many highly scholarly books on these timeless Holy Vedas the world over. The original texts of the Vedas are in classical Sanskrit; the World's most ancient language and also appearing to be perhaps a very difficult one that is not comprehensible by most people.

The timeless and universally relevant Holy Hymns, so enchantingly incorporated in fact, in all the four Vedas are principally in six parts namely: Hymns on Creation, Hymns on Devotion, Hymns on Revelation, Hymns on Action, Hymns on Splendour and then, Hymns on Positive Sciences. In the just above recommended book, we notice that, "The Rig Samhita has 10,589 verses, the Yajur 1,975, the Samaveda 1,875, and the Atharva 5,977 verses, and thus the total come to 20,416 verses.... **The Vedas constitute the back-bone of our entire culture and development through the millennia not only in India but also abroad.** For most of us they constitute the first literature that dawn on us at the earliest time of man's appearance on this globe. In India, we regard them as revealed knowledge. What the effulgent sun is to animate and inanimate activity on the terrestrial earth, the Vedic enlightenment is to the prestigious life of man on this planet for the majority of humanity." (Vidyalankar, 1983, Foreword)

Ironically, in a kind of perhaps intellectually rushing through historically and also, in a routine manner; some scholars are seen to be estimating some datelines, regarding the origin of the Vedas; which may appear to be logical. First we may be required to bear in mind that the Holy Vedas are certainly no classical treatises belonging to any specific religion, as perhaps routinely misunderstood to be, by many people and even some scholars. In fact concerning the Vedas, it has been rightly said that the fundamental Vedic concepts of God, are superbly ethical and even scientific with respect to the Mother-Nature Herself. When we talk of the mother-Nature here, we don't refer her with respect to our Earth or the Solar-system alone but, with all-encompassing and integrative reference to the entire Infinite Universe. If one may pause and care to reflect on the truly enchanting verses of Vedas, one may find that they uphold the high moral values and then show that, God is always and Unchangingly, Supreme Truth Personified, all Activities Personified and also, the Purity Personified. Hence, it would be a great injustice done to the Holy Vedas if one may makes efforts to perhaps 'chain' Vedas into any religion-specific erroneous dogmas, which may not logically be possible as well in any perhaps convincing manner.

No wonder that concerning the Vedas, **Svami Satya Prakash Sarasvati** has mentioned that "We are told that the divine revelation came to man at a time when the world was in its infancy. I shall not take you to primitive man and his group as conceived by an evolutionist of the modern age—a society which was least conducive for the type of revelation we received from the divine source. Undoubtedly, the primitive and mentally unevolved man could have been least receptive to the highest type of enlightenment. I shall not refer you to the history or the geography of the event of revelation, for the time-space reckoning must have started very long ago in our history. I am talking of days when man had no language, though he had a complete set of vocal and hearing organs. Think of the days when man has existed without a vocabulary.... In the midst of such a state of affairs now inconceivable, the divine knowledge was revealed to him through exceptional personalities with high receptivity, stupendous memory and superb understanding. The earliest contribution of men of this group was to assign *names* to the objects of surroundings in the most general terms. The language of the divine Rigveda itself has an astonishing stock of 35,000 words in 10,000 verses with immense intrinsic potentiality for coining new terms.... Without having an instinct to communicate to posterity, man could not have made any history, culture, philosophy, science and technology. The communication could have been possible only through a *language*, as divine as Creation.... Thousands of years have passed since the divine knowledge was first revealed to a small group of seers (the four Samhitas to the first four—Agni, Vayu, Aditya and Angiras, so named traditionally). There was another group of seers with stupendous memory, who passed on this knowledge to the successive generations. The art of script and writings was invented and developed at a much later stage." (Vidyalankar, 1983, pp. 10-12)

Over the centuries Prophets too, came for their high spiritual preaching to the mankind. And in these phenomena, gradually different religions took their births. All these preaching by different Prophets and during different times of the man's living-history too, are by themselves revealing and humanly enriching through the same Divine Grace. It is a different matter that these Divine revelations through those spiritually evolved and saintly Prophets happened during different periods of the man's chronologically recorded living-history and thereby, also indicating their firm datelines for posterity.

Interestingly and quite relevantly with respect to the above, in the earlier referred book *The Holy Vedas: A Golden Treasury* (pages 11 and 12) it is mentioned that, "In the British Museum, one may see a written Bible of the third and fourth centuries A.D.; the Holy Quran of thirteen hundred years ago beautifully scribed, but one would rarely find a script of the Veda of such an ancient date (*related to the origin of Vedas.*). Such an amazing phenomenon of preserving the most ancient texts of 20,000 verses through all the years of history could not have occurred in any other land. The credit goes to the traditional Brahmanas of India who against all hazards of human history could keep the texts so well preserved with the right phonetic accents and accuracy to this day." (*Hence in these very engaging contexts, it gets perhaps reasonably established that it would not be so easy to conclusively come to the grips with the historicity, regarding the origin of Vedas in the modern-era mankind's generally accepted and scholarly conventional sense.*).

Again coming back to the multiple-aspects of the Vedas in the earlier referred book and, on the page 13, it is further beautifully mentioned that, "**The Veda takes you even a little beyond this creation. While the Vedic texts present to you a little picture of the mysteries of this creation, they lift you up a little beyond the physical or metaphysical reality. They raise you from creation to the Creator**. They take from the Sun, to the Sun that shines behind the Sun, to the Fire that glows behind the mundane fire; they take you to Light that enlightens all the lights familiar to us. They take to Beauty and Pleasure behind the so called beauty and pleasure that exist in our everyday life.... In the lower stages, all the disciplines of knowledge are distinct and separate.... The highest knowledge is merely one, the integrated knowledge, and this is the knowledge, not gained through our sense-organs, vital organs or through our mental behaviour. This is the final knowledge that we aspire for. This is then the establishment of a personal link between an aspirant and the Supreme One. The Vedic verses enable you to enjoy the glory of God in His creation."

Now at this precise point, coming back critically to the historicity of our continuing discussions in our larger perspectives, we also find that, when the Vedas were getting conceptualised by the Aryans and also, their then slowly on-going advances in northern India and gradually that being extended towards east, all along the vast plains of the rivers Ganga and Yamuna, as centuries passed two unique and far-reaching historic multiple processes too, were taking shape and advancing. **These two again unique and exclusive processes in fact, over many centuries were also simultaneously in progress, i.e. 'Aryanisation' of India as well as 'Indianisation' of the Aryans, in a multi-dimensional as well as multi-layered manner.** In other words, these aspects in fact, were definitely the most decisive 'core-component' of perhaps singularly unique and evolutionary as well as multiply gradual and highly complex advancement of the ancient Indian-civilisation, over many centuries and covering almost the whole of the vast Indian landmass. These ancient civilisational processes were all-integrative as well as inclusive that they did remarkably encompass the whole range of socio-cultural, socio-psychological, socio-political and socio-economic dimensions of the then 'newly-transforming' Indian populace; during India's post-Harappan, and historically most fascinating era.

Further the span of this historic era has been nearly four millenniums and yet enchantingly, maintaining unbroken civilisational continuity. In addition, during the post-ancient era, since the beginning of the early middle-age and then right through the modern-age; socio-cultural aspects of the ancient Indian civilisation have also been vibrating, living and healthily continuing. There have been, no doubt, periodical changes in the systems of governance as well as far-reaching multiply socio-political historic transformations and then there have been wars; but even during very modern times there has never been any 'revolution' as we now understand the widely accepted meaning of this terminology. Besides all these, perhaps the singularly most remarkable historic fact remains that there is no other contemporary and magnificently parallel ancient civilisation, which has been able to maintain any such multiply socio-cultural, socio-psychological and civilisational fragrance as well as unbroken living continuity. With all the historic turmoil, heat and dust along the geostrategic as well as politico-military ravages of so many centuries, ancient Indian-civilisation marvellously keeps breathing, fascinatingly keeps accepting all streams of multiple thoughts and inputs from others and keeps marching. This is perhaps again the central theme of so many messages of the Rig-Veda since thousands of years of unhampered continuity, both in lively thoughts and practices, even in contemporary India.

Then as further centuries passed, sometime during 1000 B.C. in ancient India, there was historically a magnificent dawn of the unique **Epic-Age**, heralding indeed the earliest golden-era of super as well as immortal literary creations. In this otherwise very enthralling context, many scholars both from India and abroad, have been rather very strictly 'critical' regarding the basic historicity, sequence of their creation and accuracy of various datelines of these well-known epics from ancient India. In the absence of now scientifically acceptable records/evidences, substantiating any answers to these obvious questions, such scholarly criticality is understandable. But in the existing circumstances, when we are focusing on so many truly enchanting aspects of very ancient as well as remarkably pioneering human-civilisations, and at the same time when the concerned civilisation happens to be the ancient Indian-civilisation spanning over an exceptionally long period of more than five millenniums and still lively continuing with all its multiple manifestations and varieties, are these questions really of importance? **Well opinions may naturally vary but, as already noted earlier, these epics' amazing messages of highly enriching spiritualism, coupled with their antiquity as well as unique literary qualities and lyrical-poetry, are indeed mesmerizing.**

Another no less important aspect is that, it is not always of any dominating magnitude that whether these epics took birth in the Indian subcontinent, or in China or in the then domain of any other ancient civilisation but in fact; the singularly most enthralling reality remains that these gems of the immortal and multiply fragrant literary-marvels did sprout in the 'intellectual-womb' of the humanity of that distantly bygone age of such antiquity. It perhaps cannot be any exaggeration to say that, these unique and remarkably ancient literary-pearls truly and immortally belong to the whole humanity and thereby, always undeniably remain matters of great pride as well as perpetual sources of enriching spiritual fulfillment. In addition, it would always be a great cruel-injustice as well as most uncalled for; to ever identify these most ancient and immortal epics, as perforce belonging to any narrow and naturally very suffocating confines of any particular religion-specific forced parameters.

In fact there are 'mind-boggling' range and versatilities of these unique epics. Aspects of their critical historicity of apart, the unique geniuses of their definitely gradual creation over many centuries, as well as their captivating literary range and poetic fragrance; have always kept many scholars the world over spell-bound. We all know that there are a wide variety of these enchanting epics. There are four Vedas and the Rig-Veda being the oldest one, there are twelve Puranas, and then there are *Ramayana*, the *Mahabharata*, for the Hindus' perhaps the most sacred epic the Geeta, the Upanishads and so on. Further many ancient 'Shastras' too, also belong to the multiple varieties of the same Epics.

Now coming back to the principal subject-matter of our on-going "Strategically Critical Journey" too, many of these epics splendidly and universally encompass the always challenging multi-dimensional aspects of the **'essential strategic-culture and thoughts'** as well. In this unfortunately the most crippling process of all-pervasive extreme negativity, in the Late **Nirad Chaudhuri's so-called 'The Continent of Circe'**; even geostrategic and politico-militarily most valued "essential strategic-culture and thoughts" with all their potential for multiple creativities too, were neglected to vanish into the 'thin-air' probably forever. In this context paradoxically, what could be a bigger tragedy for this ancient civilisation and in addition lately, for the 'unfortunate' India's Nation-State? We shall touch upon these aspects of great import more appropriately in subsequent chapters.

Now again coming back to certain aspects of the enduring rich varieties of these timeless epics, **Late Jawaharlal Nehru** says that "...some of the famous books which the Aryans produced in those days, the *Vedas* and *Upanishads*, and the epics, the *Ramayana* and the *Mahabharata.* And we have followed them spreading out over northern India, and even penetrating to the south and, and in contact with the old Dravidians, building up a new civilisation and culture, which had something of the Dravidian in it and a great deal of the Aryan. Especially have we seen how their village communities grew up on a democratic basis and developed into towns and cities, and forest *ashrams* became Universities." (Nehru, 1983)

Further remarkable aspects of the some of these epics are also noteworthy that their magnificent influence as well as far-reaching and enriching socio-cultural and spiritual impacts, did not remain within the confines of only the vast Indian subcontinent but, in many areas of south-east Asia and into the Far-East. So far as Japan and China are concerned, subsequently the splendid advent and spread of Lord Buddha' immortal preaching had then played a much greater, enduring and far-reaching role in this role in this regard. We shall also talk more about Buddhism, little later in this chapter itself.

Now coming to many legends-based and historically over many ages in India; it is also true that specially **both Ramayana and Mahabharata** have undergone interestingly, many 'incarnations' and also stand repeatedly composed and translated in many versions as well as in many languages, again both Indian and foreign as well. Therefore contextually though, historicity is definitely important and which here in any case remain very difficult to conclusively authenticate; but no less significant are the superb civilisational brilliance as well as spiritual and poetic-cum-literary richness of these timeless epics. Hence we may also pragmatically affirm that these enduringly living epics; do perforce traverse much beyond the relatively narrow and dogmatic confines of any particular religion. Some of these indicative and truly universal teachings of these wonderful epics for the larger humanity are relatively more obvious and apparently clear but at the same time, far more superbly illuminating and timeless preaching of these ancient epics; are very subtle, indirectly indicative as well as at times hidden too, and logically thereby, it remains imperative on the part of man to dig multi-dimensionally much deeper, to be able to have a highly enriching glimpse of them. Such are indeed the vast universal range and depth of the storehouse of unique treasure of wisdom, contained in these magnificently timeless ancient Epics.

At times it is also true that since the texts of these epics, such as Vedas, the Puranas, the *Ramayana*, the *Mahabharata* and so on, are very ancient and at the same time in different ages, some of them have been reworked, originally in Sanskrit and subsequently later in different languages, there are some contradictions in datelines and even in some details of events and episodes. Contextually these matters again pertain to only their historicity and then in any case, the above elaborated wisdom-content of these epics do remain timelessly lively as well as universally relevant. Socio-economically as well as politico-strategically as per the man's evolutionary aspects and the ancient historic era to begin with, human-societies, in every ancient civilisation, were initially organised on tribal-lines and sustaining principally on hunting as well as in addition perhaps, on wild fruits and other allied 'edible' natural forest-produces. Then as centuries passed, since already elaborated in earlier chapters; gradually man in groups started migrating into more congenial location along the extended riverbeds for settling down, simultaneously along with the gradual beginning of the far-reaching and multiply both evolutionary and revolutionary socio-economic food-generating methods of agriculture.

This was indeed a great turning-point in the long and on-going aspects of remarkable human-history on this planet Earth. Fascinatingly indeed, this whole process of man's gradual multiple progress triggered the great advent of various human-civilisations in different geographical locations/regions. As in ancient India so in other equally ancient and unique civilisations thereby; in different centuries with the multi-dimensional fascinating expansion of man's creative-abilities too, began, resulting in the dawn of epic-ages. In this context it may also be emphasised that this entire process, was socio-culturally, intellectually as well as spiritually on meta-physical plain, was indeed the zenith of man's multiple development during prolonged ancient ages. With respect to the ancient Indian civilisation, in the above paragraphs, we have tried to cover some of the remarkably fascinating aspects of the ancient India's various timeless Epics. It also remains well-accepted that these marvels of man's great literary creations indeed, command again antiquity as well. Simultaneously, as the mental horizons of man's unique intellectual abilities for literary creativities got gradually enhanced; in course of centuries many fascinating 'myths' too, got poetically intertwined and wonderfully integrated with the concerned historic facts as well. Considering the various aspects of those ages of the past and also the remarkable antiquity of these wonderful epics of ancient India, they are in fact, understandable and though, opinions may again naturally vary in this regard.

Subsequently in ancient India then, magnificent dawn of unique 'reformist' **Buddhist-Age** took place, nearly around 560 B.C. Further, India's ancient-history indicates that the period of

Buddhist-age in the Indian subcontinent, may be taken approximately from 560 to 485 B.C. Then in fact, it is equally interesting to note here that, during that age in ancient India, another remarkable and highly enriching dawn of Jainism too, had taken place. Interestingly further, like Christianity, Buddhism too, did not expand much and also did not endure much but they remarkably expanded in many foreign lands. Jainism did spread in India, with dedicated belief in uncompromising non-violence and peace and, the same is still healthily surviving in India. **In Jainism there are twenty four 'prophet-like' Tirthankars, from perhaps Adinathjee to lord Mahavira and then subsequently, they are having two systems of the 'Jain-faith', i.e. Digambara and Shwetambara as well.**

Now coming back to the Buddhist-age, by the time **Lord Buddha** appeared on the ancient Indian scene, era of 'monarchies and empire-building' had already begun and was thriving too. In fact the enduring historic foundation of magnificent monarchies as well as great ancient empires in the Indian subcontinent were already gradually laid during the unique Vedic-age itself. No wonder the timeless Epics of the bygone ancient Indian civilisational era, commanding great antiquity do beautifully contain magnificent and unparalleled poetic saga of great monarchs and fabulously extensive empires. During those times the extra-ordinary appearance of Lord Buddha on the then Indian scene, socio-psychologically as well as spiritually, was indeed subsequently a great turning-point.

Though after his amazingly captivating enlightenment to begin with, his principal mission happened to initiate far-reaching multiple reforms in the Indian societies as existing then; with maximum possible visionary comprehensiveness. The reasons being that, he could see many depressing 'decay and distortions' in the socio-religious, socio-cultural and spiritual spheres, as already prevailing then, both among the Indian rulers and aristocratic classes in various kingdoms, as well as among the toiling-masses. Since the Vedic and great Epic ages, many centuries have already passed and therefore, by Lord Buddha's time the ancient Indian societies at macro-level, were already in the process of overall unfortunate degeneration and thereby, the naive peoples had been gradually taken 'for a ride' by the regressively entrenched powerful vested interests, so harmfully having sway over them. This was the overall social scenario that **Lord Buddha's** superbly sensitive and great visionary mind as well as divinely elevated-soul, began to 'confront' in the Indian subcontinent.

In a broader framework Jawaharlal Nehru writes that, "In India, in that sixth century before Christ, we had the Buddha and Mahavira; in China, Confucius and Lao-Tse; in Persia, Zarathushtra or Zoroaster; in the Greek island of Samos, Pythagoras. You may have heard these names before, though perhaps in different connections. The average school boy or girl thinks of **Pythagoras** as a school boy who proved a theorem in geometry ...this theorem deals with the squares on the side of a right-angled triangle and is to be found in Euclid or any other geometry. But, apart from his discoveries in geometry, **Pythagoras** is supposed to have been a great thinker, (then on page 43) Buddhism today is the religion of the greatest number of people in the world. Other religions which have the largest number of followers are Christianity, Islam and Hinduism. There are besides, the religions of the Hebrews, of the Sikhs, of the Parsis, and others. Religions and their founders have played a great part in the history of the world, and we cannot ignore them in any survey of history." (Nehru, 1983)

Further, **Nehru** in his other well-known book*The Discovery of India* says that, "**Buddha's method was one of psychological analysis and again, it is surprising to find how deep his insightwas**

into this latest of modern sciences. Man's life was considered and examined, without any reference to a permanent self, for even if such a self exists, it is beyond our comprehension. There is an emphasis on the pain and suffering of life, and the 'Four Noble Truths' which **Buddha** enunciated deal with this suffering, its cause, the possibility of ending it, and the way to do it. Speaking to his disciples, he is reported to have said: 'and while ye experienced this (sorrow) through long ages, more tears have flowed from you and have been shed by you, while ye strayed and wandered on this pilgrimage (of life), and sorrowed and wept, because that was your portion that ye abhorred, and that which ye loved was not your portion, that all the water which is in four great oceans. (Then on page 130) **Buddha's way was the middle path, between the extremes of self-indulgence and self-mortification. From his own experience of mortification of his body, he said that a person who has lost his strength cannot progress along the right path. This middle path was the Aryan eightfold path: right beliefs, right aspirations, right speech, right conduct, and right mode of livelihood, right mindedness, and right rapture. It is all a question of self-development, not grace**." (Nehru, 1983)

Historically it remains reasonably well-established that both lords Buddha as well as Mahavira, independently in their rights, socio-religiously and psychologically, during more than five centuries before Christ were refreshingly revolutionary and great thinkers and visionary philosophers. During those ancient times, both were born in highly distinguished clans, entitled to royal privileges in their respective kingdoms or perhaps non-monarchial states, the vast plains of the northern part of the Indian subcontinent. Their teachings were so original and spiritually inspiring that in the process, these were destined to come in ideological-clash with the ruthlessly powerful and socially exploitative 'vested-interests' of the deeply entrenched Brahminical priesthood in the ancient Indian societies.

In this context interestingly, **John Keay** mentions that, "For the lives and teachings of the great founding fathers of Buddhism and Jainism quickly inspired a host of didactic and narrative compositions which supplement and sometimes contradict orthodox sources like *Puranas*.... They also incorporate significant information on places other than Ayodhya and on state systems other than monarchies. These alternative state systems have been variously interpreted as oligarchical, republican or even democratic."

The appearance and illuminating preaching of these two charismatic visionaries, who were capable of thinking brilliantly 'out of the box' during that ancient era, were indeed great turning-points, in the then existing overall social-scenario. These naturally had to create many socio-religious complexities as well as controversies and thereby, the entrenched powerful and regressively caste-based priesthood felt seriously threatened and hence, they were up in arms against the two enlightened reformists, since intended to trigger multiple reforms through their inspiring thoughtful vision.

So far as certain well-known initially evolutionary concepts regarding socio-politically organising the larger human-societies and putting a chosen systems of governance in place, whether **Monarchy and building of indigenous 'ancient-empires' within the Indian subcontinent** or any other system, are concerned; they in turn were causative to many kinds of clashes, not only of ideas but, of multifarious political ambitions and thirst for power. These whole greedy processes of urge and activities were mostly violently confrontationist. These were indeed paradoxical ironies, since created by the historic aspects of many gradually developing as well as advancing, and also otherwise, fascinating ancient human-civilisations, themselves anywhere in the world; with varying degrees and inclusive of different socio-cultural and geo-political textures and then later on, including a variety of even rigid 'religious-fixations'. Here in this context, it may also be emphatically asserted that, no monarch or ruler with his/her empire, whether small or large, ever comes to power by any means, and subsequently establishes again his/her dynasty; not again, for any kind of benevolent and/or philanthropic reasons or even intentions.

In the beginning of the first chapter on 'Introduction itself, it has been mentioned that thousands of years have passed over the mighty Nature's unending 'canvas of time', before the mankind could come to a stage of even remotely be able to claim themselves to be progressively 'civilised'. In addition then it has also been emphasised that, 'even in this 21st century, it remains highly debatable as to the degree of legitimacy as well as authenticity of 'claim' that, how far and in what manner, mankind can truly be called even reasonably civilised'. Considered objectively in a perspective, this is certainly not a casual statement and in addition, factually for the present ongoing exceptionally high-tech era, it has many again 'embarrassing' aspects for the modern man, as well. Paradoxically, these civilisational 'negations' are not limited to only the monarchies in the world but perhaps equally so in case of the republican or democratic systems of governance as well. Astonishingly, man appears to be pathetically 'trapped' by his own greed, vested-interests and deeds/misdeeds in the, otherwise very vast and yet highly complex as well as constantly altering 'civilisational-arena'.

In this context a brief 'article', on **"The Urge for Democracy is Universal"**, by **Gwyyne Dyer**, says that, "The default mode for human beings is equality. Every pre-civilised society we know about operated on the assumption that its members were equals. Nobody had the right to give order to anybody else. What drove this was not idealism but pragmatism. In hunting-and-gathering groups, nobody can own more than they can carry, so there is no way to accumulate wealth. If you want meat, then you'll have to co-operate in the hunt. These were the societies where nobody could control anybody else, and so they had to make their decisions democratically. They were all very little societies: Rarely more than 50 adults (who had all known one another all their lives). On rare occasions, when they had to make a major decision, they would actually around and debate it until they reached a consensus. Direct democracy, if you so like. People have been running their affairs that way ever since we developed language, which was almost certainly before we were even anatomically modern human beings. So 99.9 per cent of our history, say. That is who we are and how we prefer to behave unless some enormous obstacle gets in our way. The enormous obstacle was civilisation. All hunting-and-gathering societies were essentially egalitarian. The mass societies that we call civilisations arose less than ten thousand years ago, thanks to the invention of agriculture. Until very recently all of them, without exception; were tyrannies, pyramids of power and privilege in which the few decided and the many obeyed. What happened?"

Here **Gwynne Dyer** has appropriately asked as to 'what happened'. Well since long, man's long and continuing living-history definitely shows that many things have happened and are also happening, regardless many of them fabulously positive and multi-dimensionally indeed, exclusively enriching, but simultaneously in addition, some of them have been 'nauseatingly' great negations of the same perpetually advancing human-civilisations as well.

Factually, these entire highly complex phenomena are indeed part as well as under command of the supremely powerful Nature's again unstoppably continuing and infinitely multiple evolutionary-process, over the same unending 'canvas of time'. Now looking with pragmatic criticality, at this perhaps most fascinating as well as at the same time, certainly fully incomprehensible multiply complex, this entire process of Nature's continuing evolution; everything everywhere in a context, are definitely not in man's hands or within his control. But at the same time, the same 'living-history' shows that, many things do crop-up principally due to the man's deliberate deeds/misdeeds and logically therefore, the same supreme-Nature, very substantially allows man to be able to rationally maintain decisively regulatory control on many far-reaching aspects in this regard, of great consequences for the same man.

Hence, here in a specific context, in the above quoted lines from an article of the British Journalist **Gwynne Dyer**; we find that he calling the civilisation itself, an 'enormous obstacle', does have an understandable logic and hidden 'rationale' as well. Now interestingly at this point of our on-going discussion, a specially pressing question very relevantly arises as to, *"what are the central reasons for all these otherwise extremely violent as well as unjustifiably exploitative historic happenings for the fellow human-beings, since the advent of human-civilisation itself and then at the same time, perforce triggered by many continuing and deliberate deeds/misdeeds of the men themselves; and in turn thereby, forcing the man's 'living-history' rather ruthlessly being 'written' by human-blood and massive human-sufferings, over the same unending 'canvas of time'?"*

Now quoting few lines from the Introduction, we recollect that, 'Further relevantly coming also to a spiritual-level, it is well-known that since the early times, many saints belonging to many religions, has said that both **Demons** and the **Angels** perpetually reside inside every human-being and at the same time as per Nature's dictates, demon is more aggressive and there by constantly in the lookout for overpowering the angel present by the side and decisively subjugate him/her to reign supreme over that person(s), both individually or collectively, as per the various factorial emerging situations over a passage of time.' In addition, we also interestingly notice duly mentioned that the same at times especially 'benevolent' supreme-Nature, has indeed exclusively gifted the mankind since the beginning, an **'analytically thinking-cum-creative mind'**. Therefore in turn we may even astonishingly as well as justifiably perhaps 'conclude' that, both the 'Demon' and the 'Angel' residing as 'neighbours' inside the man and simultaneously, the man already having been gifted with the above described exclusive 'mind'; are surprisingly indeed, an extremely **"lethal" combination".**

Interestingly, with respect to the above mentioned that, "The Vedic verses take delight in referring to the eternal conflict between Indra and the Sun and the shrouder, the clouds, which obstructs light and warmth of the sun. Ultimately, it is sun that becomes victorious. But again the story is repeated everywhere. The Divine Poet of the Veda is never tired of narrating this parable; and He takes us to another conflict of the same nature which exists with the interior of all of us—a constant struggle between our divine tendencies and our devilish ones. The incessant conflict remains between Truth and Non-Truth, between Good and Evil, or Enlightenment and Nescience, Knowledge and Ignorance. While the Vedas narrate this parable, their reference to the sun and clouds is merely symbolic. **The real conflict which they intend to stress is between the self and the dark forces within our own personal make-up. On one side we have truth, light and immortality and on the other evil, darkness and death**." (Vidyalankar, 1983, p. 14 & 15)

Well in the specific perspectives the above explained interesting 'combination' of the 'Angel' and 'Demon' are definitely the overriding aspects and that too, universally as well. Now, coming back to the 'recipe-based' story, i.e. concerning the highly visible and always the inherent urge to exploit the fellow humans, since the pre-historic times we find the same historically present in the Indian subcontinent as well. And another enormously far-reaching gradual **civilisational-created**

'factor' during that ancient Vedic-age has sprouted, since belonging to even now, the extraordinarily continuing and healthily 'self-altering' Aryan-era. This enormously dangerous 'combination' further logically leads us to visualise that besides so many other things/complexities, civilisation too, is definitely none else's but the same combination's own phenomenally highly complex product. Further, that "factorial-icing" on again the same "combination-based cake" was in fact, added much later.

Now quite relevantly, the clear identification as well as accompanying details, concerning the above stated civilisation-created "factor" is obviously called-for. Well, that far-reaching 'factor' is nothing but India's age-old and still regressively lingering **"caste-system"**. Since many centuries, leading both social and political scientists have very scholarly written volumes on this otherwise engaging subject, but in any case, this precisely lingering phenomenon has defied any reasonable rationalisation. This is amazing, isn't it so? Well, indeed, it is so. This has indeed been paradoxically, a highly depressing reality of otherwise, unique, humane and very colourful societal landscape of India, since ages. Multiply comprehensive global-histories of the mankind show that every ancient human society, with its own specific civilisational-orientation and evolved 'fixations' unexplainably does have socio-psychologically, some peculiarly lingering 'kinks' again, in its own socio-cultural making. Hence, howsoever, one may even nauseatingly dislike Indian societies perhaps most unfortunately since ancient ages; have remained 'diseased' with this highly harmful 'caste-system'. This has indeed been a great social-negation. But to India's good luck, after its independence in the year 1947, thanks to the certain visionary steps then taken by the founding-fathers of the Independent India's liberally democratic "Constitution" the vice-like grips over the Indians' social-psyche, are fast loosening. And therefore in course of time, as further some years roll-by, this should wishfully wither-away.

In this context, **R.P. Masani** has said that, "The division of mankind into groups based on fundamental differences in disposition, capacity and character is a common feature of society all over the world. Such groups have tendency to isolate themselves into separate classes holding intercourse chiefly among themselves; but the evolution of a caste-system postulating hereditary orders functioning within rigidly circumscribed spheres of social-intercourse and yet sharing the large life of the community is a phenomenon peculiar to the organization of Hindu society; including as it does an elaborate code of ceremonial purity and defilement, unapproachability and untouchability, commensal restrictions and connubial prohibitions, penance and excommunications." (Masani, 1937) This is indeed historically an irrefutable 'social-diagnosis' with high penetrating precision, of very regressively conservative segments of the larger Hindu-societies in India, by a noted Indian scholar. All these ironically do paint clearly an unlikable 'social-picture' of substantially the majority community, in even the contemporary India, but facts do remain facts.

Then in this context **Jawaharlal Nehru** says that, "Almost everyone who knows anything at all about India has heard of the caste system; almost every outsider and many people in India condemn it or criticize as a whole. Perhaps there is hardly any one left in India who approves of it in all its present ramifications and developments, though there are undoubtedly many still who accept its basic theory and large number of Hindus adhere to it in their lives. Some confusion arises in the use of the word caste for different people attach different meaning to it." (Masani, 1937) Well again ironically, there is still some truth persisting in the present early 21st century Indian societies in this regard, though Nehru had written these historic words since; before otherwise the 'Incredible' India's independence-year of 1947.

It then appears that since these being so, where does man realistically stand? Well, ironically, the man perforce stands both ways, i.e. as at times the situational 'victim' and then 'victimiser' too. In other words, man is, in fact, fellow-man's own 'victim' and then as per the altering situations, man is also man's own 'victimiser'; all these phenomena are continuously taking place within the larger processes of various advancing human-civilisations, again since remotely ancient times as well. This is indeed a phenomenal paradox of perhaps all the paradoxes in the already duly 'claimed' man's own 'civilised-world'. Well in any case, that's the way they really are and then, the supreme-Nature's unending process of multiple creations move-on regardless, 'unquestioningly' adhering to the supreme-laws of evolution of the same Nature. These aspects have been delved-upon more appropriately in some detail, in the first three Chapters themselves.

Referring to Nobel Laureate Late **Rabindranath Tagore** knowledgably says that, "Man's history is being shaped according to the difficulties it encounters.... These difficulties have been different in different peoples of the earth, and in the manner of our overcoming them lies our distinction.... But man, when confronted with difficulties, has to acknowledge that he is man, that he has responsibilities to the higher faculties of his nature, by ignoring which he may achieve success that is immediate, perhaps, but that will become a death-trap for him. For what are obstacles for the lower creatures are opportunities to the higher life of man.... But all the same moral law is the law of humanity, and the exclusive civilisation which thrives upon others who are barred from its benefit carries its own death-sentence in moral limitations. The slavery, that it gives rise to unconsciously drain its own love of freedom dry. The helplessness with which it weighs down its world of victims exerts its force of gravitation every moment upon the power that creates it. And the greater part of world which is being denuded by its self-sustaining life by the Nation will one day become most terrible of all its burdens, ready to drag it down into the bottom of destruction. Whenever power removes all checks from its path to make to its career easy, it triumphantly rides into its ultimate crash of death. Its moral brake becomes slacker every day without its knowing it, and its slippery path of ease becomes its path of doom." (Tagore, 1937, pp. 3, 4, 22)

It is important to clarify here that the above quoted scholarly thoughts of both Gwynne Dyer and Tagore have been originally written concerning two different historic happenings of great significance of their own distinctly different times. Gwynne Dyer's article pertains to the beginning of 21st century's surprising awakening and revolutionary agitation for freedom and democracy in probably the entire Arab world. These are, indeed, going to have far-reaching long-term impact on the entire global geo-politics as well as modern-era fast-changing leading nations' international power-game oriented strategic-alignments. All these are important segments of history in the making. Then, Tagore's visionary perceptions on certain important and broader aspects of the 'civilisation-cum-nationalism', was delved upon then, principally concerning the early 20th century colonial-era India's relentless struggles for freedom from the British-rule, which are now part of history. Hence we are able to clearly visualise that they are universal and then always highly relevant, irrespective of the changing times and situations and precisely therefore, in our context too, and particularly at this point of our ongoing discussion, they are significantly relevant as well.

Now, while glancing over the above quoted perceptions as well as related explanations, it may appear rather ironical, as to why, these indeed universally relevant aspects of certain contextual scholarly thoughts are perhaps deliberately being included in this 'sub-chapter', specially dealing with a brief 'review' of the ancient India's civilisation? The messages, as contained in the above quoted lines are unquestionably wholly universal in their highly knowledgeable and thought-provokingly vast range of intellectual discoveries. That is precisely the intended purpose of this exercise, for considering them in a perspective here.

We may appropriately bring-in a reasonably befitting 'example' in this regard, perhaps for little more substantiation. In this respect, we may look back into a brief study of the ancient Chinese-civilisation, since given above in the earlier 'sub-chapter' on the "Semi-historical China and the Dawn of History: 2852 to 2697 B.C. onward". There as may interestingly be seen that, some glance has been pre-arranged on the fascinating aspects of ancient-era folk-stories of the Chinese societies belonging to that very distant past, by quoting some paragraphs from Professor Zhong Jingwen's book on, "Favourite Folktales of China", and translated in English by Professor John Minford. There are similarly very fascinating ancient folk-stories and folklores, belonging to the ancient India's fascinating civilisations as well; such as 'Panchtantra' and 'Jataka' stories. They are living creations of great ancient minds, belonging to various equally ancient human-civilisations, since remarkably flourishing in different centuries of time as well as in different areas of the world. In other words, they are immortal products of great ancient human-geniuses, of many ancient ages and hence, most appropriately always belong to the whole world.

Further in this respect, it may also be said that certain historically striking aspects of any noted ancient civilisation in the world, are by themselves not rigidly exclusive to any particular ancient civilisation since its inception'. History shows that, in keeping with the ongoing passages of time, as various ancient civilisations took roots in different geographical locations and kept advancing, they in any case, couldn't remain isolated, even during those bygone ages. Then, certain issues-based 'flow of thoughts' and in addition, various 'inventions/innovations', having universally much wider connotations too couldn't remain exclusively isolated and there-by, always fascinatingly appear to be integrative as well as collectively 'inter-faced' for many of those marvelously ancient civilisations belonging to again the same long bygone ages. We shall touch upon these fascinating aspects appropriately in the later chapter on 'Modern-Era'.

Another interesting point in the similar context here is the aspect of multiple-relationships between the so-called **'tribal' and 'non-tribal' populations**. Well at the outset itself, it may be kept in mind that these social factors are not exclusively 'India-specific' only; though since many centuries, at both the socio-cultural as well as socio-religious levels, the multiple Indian societies have definitely been having certain different kinds of 'allied-factors' in this regard. In addition in the same Indian societies, these allied-factors have also been both, 'positive' and 'negative' ones too. And then, all of them living and interacting with socio-culturally integrative as well as harmonious spirits, within a democratic framework, enjoying equal rights and privileges as well; as always guaranteed by the Constitution of India. They have been no small achievements. Another exclusively unique feature only in India in this regard, has been; the remarkably so many varieties of distinctively different tribal communities, observing different cultures, traditions, customs languages and ways of religious practices; settled throughout the vast Indian landmass, since their inceptions and then yet now, they are socio-politically harmonious part of the same large democratic India. Well, the huge African continent can also claim to have many different aboriginal African communities, they are seldom living peacefully together and at the same time, they now belong to different political entities and thereby, paradoxically indeed, claiming to have different nationalities.

No wonder therefore, it is even internationally well-accepted that socio-politically, democracy has already taken firm-roots into the vast Indian landscape having remarkable varieties of multicoloured socio-cultural as well as socio-psychological aspects and connotations, so fascinatingly in this regard. India has, indeed, come a long way since the fruition of her freedom in the year 1947; but at the same time, it is also equally undeniable that India as a modern and socio-economically fast developing democratic Nation-State, has still miles to go before, she can legitimately earn a 'claim' to be called a truly and multiply developed country in the modern comity of nations. The point is that it is one thing to keep rather naively indulging in otherwise, avoidable habit of 'self-patting of own back' and then; quite another to view things with critically pragmatic-realism, having strategically desired comprehensiveness. Discerning prudence as well as again strategically critical sense of history too in this regard, consistently expect for the same. Now coming back to specifically certain important aspects of the relationships between tribal and non-tribal segments of the various human societies, it would appear contextually relevant here to quote some paragraphs from the books of some learned authors.

A well-known professor of sociology from India, **Andre Beteille** says that, "Early anthropological studies established certain conventions, and in using those conventions—as with words such as 'tribe', 'aborigine', 'indigenous people'—we must keep in mind their original context. Anthropologists in the past studied race, language and culture together, since it appeared natural for them to assume that the differences between themselves the people they studied were not merely of language, religion and culture but also of race. The assumption of a close relationship between language, religion and culture on the one hand and race on the other, often led to a serious misrepresentation of the relations between social groups.... The relationship between tribal and non-tribal populations have been of a very different kind in the ancient civilisations of India, China and the Islamic world. This is true to such an extent that in India, for an example, it is difficult to draw a sharp line of distinction between tribal and non-tribal communities on the basis of race, religion or language. In Australia, the term aborigine has a very definite meaning; it is difficult, if not impossible, to give such a term a definite meaning in India.... The great paradox of Indian society is that, despite its rigid rules for the maintenance of social distance between communities, there has been pervasive interpenetration between tribal non-tribal populations for well over 2000 years. This can be seen if we compare tribal and non-tribal ways of life in most regions.... The co-existence of tribal and non-tribal communities, population movements of both tribals and non-tribals, and the displacement of one type of community by another were no doubt accompanied by a great deal of exploitation and expropriation, but they were slow and gradual rather than sudden and cataclysmic as in the new world *(i.e. for example, American continents and Australia.)*. Throughout inner India...there have been movements of both tribal and non-tribal populations from time immemorial. Some tribal populations have displaced other tribal populations and have in turn been displaced by non-tribal populations." (Beteille, 2000, pp. 181, 182 & 183)

Interestingly in the same continuing context regarding some weighty questions concerning the 'fate' of the Australian and American aborigines, since last three to four centuries in these two large 'continents', we may very relevantly refer to the 'lifetime's original reflections' of a professor in history at La Trobe University, Australia, wherein **Professor John Hirst**, writes that, "... But even if **Keith Windschuttle** were right about the number of those killed by direct violence, he cannot deny that forty years after the European settlement, all the Tasmanian Aborigines had either perished or had been removed to offshore islands. **Empires even good empires believe in conquest and by any standards this was a complete and rapid expropriation**.... I am surprised at Windschuttle's lack of sympathy for the plight of the dispossessed Aborigines.... This was a plunder and murder merely.... It is a great conceit on the part of these two judges to think that a difference could have tamed the force of European colonialism and given us a history with no dark aspects. From all that we know of what the land meant to the Aborigines, they could not have yielded it without a fight. The judges insult them by thinking that it could have all happened peacefully.

From 1833 the law in the United States on Indian land *(i.e. Red Indian tribals.)*.... Indian land could officially only be yielded up by treaty. But the Indians were not free to make or not make treaties. They were pressured into treaty-making after they had been defeated in battle or in an attempt to save some of their lands from the on rush of the settlers, who did not wait for official sanction before pressing into Indian country.... If our policy is to be more realistic and effective, we have to sweep away the woolly thinking that supports the present policies. I identify five fallacies in the current official thinking about Aborigines:

(i) All Aborigines possess a rich culture.

(ii) Land rights are key to Aboriginal regeneration.

(iii) Aborigines are naturally co-operative.

(iv) Aborigines should determine their own future.

(v) White prejudice is the chief barrier to Aboriginal advancement." (Beteille, 2000, pp. 181, 182 & 183)

So far as the nature of multiple 'chequered-history' of the centuries old interesting relationship between the so-called 'tribal' and 'non-tribal' populations in a few leading nations as applicable, are concerned; relevantly informative as well as historic-facts' based perceptions of a couple of internationally well-known scholars, as quoted above, are indeed very revealing. Here astonishingly we also notice, the embarrassingly glaring, some socio-cultural 'chinks' in the 'civilisational-armour' of the prominent nations belonging to Samuel Huntington's that multiply well-advanced, scientifically visionary and also, the entire humanity's unique path-finder-cum-beacon-light type: the Western-civilisation. But at the same time, since the relatively fast advent of highly fascinating "Industrial-revolution" in our world during the presently ongoing modern-era around three centuries ago, and the splendid multi-dimensional contributions of again the same remarkably dynamic 'Western-civilisation', to the entire humanity are indeed the greatest history-making 'factors' in this regard.

If that is so then, another 'weighty-question' crops-up, as to what all have really gone wrong for the entire-humanity; and then in turn, begins to pin-prick our conscience and also in addition at once, commences to challengingly encounter all the modern-era man's tall claims of having achieved all-round civilisational-progress as well as all inclusive multiply enriching improvements in his overall quality of life. Here it is nobody's 'brief' to deny all the civilisational advancements, mind-boggling scientific and technological progress as well as indeed, many multiple qualitative-improvements in the fast altering modern-era man's needs in his highly complex-life. Well in reality, we perforce find that the 'total-picture' is definitely not all that 'rosy' in the present 21st century. We further notice that in this century's fast-altering with enhancing complexities, both times and situations, where surprisingly almost everywhere perhaps now, the cultural-fixations for **"obsolescence"** has indeed, 'trippingly' become the central 'command-word' in the modern-era man's world and life-styles. **Consequently, man naively finds himself in a highly paradoxical 'catch-22' situation again due to man's own deeds/misdeeds, since many centuries apparently hidden but at the same time, indeed, historically visible with great 'embarrassment', for the present high-tech-cum-space-age era human-civilisations.**

So in the context of our presently on-going discussions, what all really has gone wrong and due to what dominating reasons? In the present fast-altering modern era, the newly developed or acquired, regressive socio-psychological-cum-cultural 'fixations' with respect to the above mentioned "obsolescence", as man's highly entrapping new 'command-word' in his demanding life have played 'havoc' personal as well as societal levels,. He is feeling at times even a 'lost-case' in his own environment, since, unable to break the **'shackles of that trap'**. Most ironically these in turn, have very damagingly been inducing the 21st century modern-man, to hopelessly continue throwing the **"babies along with the bath-water"**. What else perhaps, could be the greatest 'irony of ironies', isn't it so? Yes, irrefutably indeed, most probably it looks like so.

Logically now, first we come to what precisely are the 'shackles of that trap? Well, those 'shackles' in fact, paradoxically and so overwhelming for the man, are the ditches of blind **'materialism'**; which have entrapped and imprisoned the same thoughtlessly naïve man. This extremely irrational obsession with the blind-materialism, has then substantially liquidated modern-man's capacities for truly higher-thoughts, sound-reasoning and thereby, that of consistently adhering to the demands of visionary self-restraint. 'Babies' of that bath-water are in fact, man's richly cultivated and consistently having been imbibed for so many centuries, as since, belonging to the ancient-era; the enchanting wisdom of true **'spiritualism'**. In different contexts, we have noted earlier too that, the aspects as well as dedicated pursuits of spiritualism are themselves very vast and complex since they indeed lead to the spiritual vision of the Infinite with all the incomprehensible magnificence.

Having come to this now, we may precisely mention here that, with respect to the enchanting pursuits as well as captivating aspects of the same 'spiritualism'; the historically perhaps most colourful and, sometimes even 'mysterious', multi-faceted civilisation(s) of ancient India, appears to be much more focused and pronounced in this regard. Well, archeologically and also from the historic-evidence angles, the ancient Indian civilisation may not now have much left for the discerning scholars, as monumental ruins belonging to their extremely remote-past but the truly marvellous classical-heritage, in the form of so many unique and timeless 'Epics', belonging to that remote ancient-era, are indeed, priceless possessions of the whole humanity.

Then the **captivating ancient Vedas and other timeless epics, do contain multiply enlightening aspects of the 'concepts of spiritualism', as revealed to as well as visualised by the same ancient-era Indian peers, sages and great mystics, during many centuries, and they in turn have wonderfully crafted these unique Epics, during their respective times**. Naturally, these epics also contain wonderfully shown and explained 'paths' for the man to equip himself, for suitably reining-in that blind "materialism", and then to progress towards, fruitfully enhancing his 'value-systems' as well as enriching his overall 'quality of life' enduringly, in the world. Further therefore, it becomes precisely visible now, that the 'spiritualism', with its perhaps humanly possible comprehensiveness, got gradually conceptualised during the, earlier briefly touched-upon, indeed the marvellous Vedic and Epic ages of ancient India.

At this stage now, we may relevantly have few words perhaps, on the tenets of the same Spiritualism. First and foremost, we may here emphasise that spiritualism in its true sense and also indeed, having immeasurably infinite marvels; is definitely not particular religion-specific, since the same just cannot be so, in any manner. **Swami Vivekananda, the greatest as well as most learned and enlightened disciple of Ramakrishna Paramhansa**, has very enlighteningly explained to the man, by his voluminous writings and preaching, that "religion" is in fact man-made whereas, the "spiritualism", with all its captivating marvels and superbly divine-fragrance, are by itself 'infinite' and hence, just cannot be 'man-made'. **Therefore again in its real enlightening-sense it is a true Manifestation of God. Spiritual-masters the world over since ages, have enlightened us in their own distinctive ways that, Almighty God Is beyond any religion since,**

Being Universal and at the same time, also Being Infinite and then, having Infinite-Manifestations and Incarnations.

Further, God is also beyond any kind of complete or full comprehension by man or anything else since, Having Infinite Range, Dimensions and Divine-Light. In the same continuity of thoughts, we may also visualise that, as "spiritualism" being a true Manifestation of God, may also be conceptualised as a 'Manifesting Shadow', of again the Same Infinitely Divine God Himself. Now since, the Ultimate-Truth remains timeless, and so the highly accomplished sages and spiritual-mystics time and again, have emphasised that, 'Truth Is One', having infinite manifestations. No wonder that saints from all corners of the world since ages, have taught the mankind that the saintly 'pursuits of truth' is in fact also the enchanting 'path of spiritualism'.

Relevantly therefore, when we look back into very ancient times of the unique and fascinating Indian-civilisation when wonderful Vedas and various other timeless Epics were being gradually crafted, with great vision and poetic-skills, the overwhelming aspects of the same Ultimate-Truth, were also very enthrallingly, unfolded and immortally revealed before the man. Further, during the earliest of human-civilisational times, when initial advent of the truly multiple-unfolding of the aspects of spiritualism, gradually began before man; sages belonging to the ancient Vedic-age have very illuminatingly confirmed that, the Ultimate Truth is God and hence One, but saints have called Him giving different Names.

At this stage now, with an aim to further elaborate and suitably substantiate these thoughts, it seems appropriate to quote from some scholarly writings of highly contemplative thinkers. Referring to a very reflective and spiritually enriching 'article', "**Romain Rolland** wrote in his *Life of Sri Ramakrishna*: 'And it is Ramakrishna more fully than any other man not only conceived but also realised in himself the total Unity of this river of God, open to all rivers and all streams, but I have given him my love; and I have drawn a little of his sacred water to slake the great thirst of the world.'....*then further* '**Ramakrishna** is not a quaint person from an ancient culture, representing a particular religious background,' wrote the American writer **Lex Hixon** in the introduction of his book *Great Swan*—Meeting with Ramakrishna 'but an Einstein of the planetary civilisation of the near future, a greenhouse for the future evolution of humanity, the master key that opens to all the cultures and all hearts.' Following **Sri Ramakrishna's footsteps, Swami Vivekananda** gave the historic call for the harmony of religions in the Chicago Parliament of 1893, which became the zeitgeist for today's world, torn by terrible religious conflict and bloodshed: 'Upon the banner of every religion will soon be written in spite of resistance: 'Help and not fight'. Assimilation and not Destruction', Harmony and Peace and not Dissension', He declared; ... **The Vedic vision of India was '*Yatra Viswa Bhavati Eka Vidam* (the land where the whole world unites like birds in a nest)'.** It is **Swami Vivekananda,** again, who in modern times for the first time the need of a complementarity of Hindu spirituality and Western dynamism, in order to make the foundation of a global harmony and ideal world.... **Romain Rolland wrote that Ramakrishna and Vivekananda won his regard because they realised 'this splendid symphony of the Universal Soul. They are its Mozart and its Beethoven-Peter Seraphicus and Jove the Thunderer—Ramakrishna and Vivekananda.**' The highest mysticism and frequent flights to the super-conscious, of the angelic father **Ramakrishna** was combined with the most intense Universal love in dynamic action in his thundering son-apostle Vivekananda.... **Dana Zohar and Ian Marshall** wrote on servant-leadership in their well-known book on SQ: 'The most spiritually intelligent form of walking the path, we have seen, is servant leadership, which creates new visions and brings new possibilities into being.... A great leader serves something from beyond himself; a truly great one serves nothing less than

God. Each of us is a servant of God or the quantum vacuum, a servant of the manifold potentiality at the heart of existence.... They (the servant-leaders) make things happen that others have felt impossible, they create new ways for human beings to relate to one another, new ways for companies to serve society, new ways for society to be. **The Buddha, Moses and Jesus were such leaders.**" (Jitatmananda, 2007, pp. 42-44 & 57)

Well, we may even emphatically assert here in this context that, by no stretch of imagination or even through any kind of argumentative-perception, we would be able to 'conclude' that the above quoted thoughts; wonderfully enriching and, with universally deep spiritual-meanings, are any particular religion-specific. The fundamental reason in this regard being that, different 'religions' *(if we may perhaps 'insist' on considering various religions, somewhat conventionally different from each other.)*are just superficially different paths, taking different 'routes', always aimed at reaching the same Supremely blissful 'Ultimate-Destination'. In other words, paths may be only, worldly different, but the final-aim or the destination here always, remains constantly the same. Precisely that's what Swami Jitatmananda has tried to convey to the entire humanity, by citing certain highly spiritual perceptions of internationally well-known personalities with enchanting spiritual-vision. Interestingly in the ongoing discussions we have talked about the modern fixating perhaps, the 'regressive-culture of obsolescence', in a context, to stress-upon certain important points. Well, deep and spiritually captivating 'Divine-messages' in the just above quoted paragraphs, are timeless and universal too, and hence, beyond any kind of the limitation of 'obsolescence'.

Now retracing our thoughts again back to the weighty issues of monarchies and building of indigenous ancient empires within the Indian subcontinent, the history shows many civilisational upheavals indeed. What is of a particular interest here is that, with the 'turmoil-ridden' gradual beginning of the otherwise very colourful age of multiple monarchies the history has been witness to the repeated foreign invasions into the Indian sub-continental territories. Ironically further, as if these were not enough that, many a times even avoidable by means of obviously expected monarchial-pragmatism, mindlessly damaging and repetitive, both inter-clans as well as intra-clans violent clashes kept-on happening since the same ancient times, all over the Indian landmass.

And thereby, many a times even this irrational phenomenon of the recurring 'chain-reactions' of such conflicts/wars had been paradoxically transforming the entire Indian subcontinent into, both socio-culturally as well as socio-politically, an 'embarrassing-victim' of mostly those monarchies' own deeds/misdeeds. The 'aging-pages' of the history of ancient India, are full of the records of many such 'depressing' episodes and betrayals and they in turn, are 'pin-pricking' the conscience of many a conscious citizens of the modern-era India even now. Well in any case, the same history is questioningly staring into the 'wide-open' eyes of the modern Indian populace, as to what all could be the causative 'reasons' in this regard? We have earlier very emphatically even asserted that, history just by its own 'living-nature' as well as again by its own totally independent 'process of materialisation'; never spares or forgives any person, community/society or nation-state, in any manner and anywhere, at any given point of the same continuing living-history, always in the making.

As already talked about earlier, the 'chequered' history of modern India shows us that even before the beginning of the historically unique and glorious "Maurya Empire", foreign-invasions from the north-western side and through "Khyber-pass', into the Indian subcontinent had already started, having far-reaching consequences as well as bringing-in dramatic changes in the 'historic-courses' of the whole of India. Nevertheless, ancient India did witness advent of many monarchies and dynasties throughout the vast landmass of India from west to east as well as from north to south, even much before the beginning of Christ-era and then, all through to the modern times, right up to the end of British-rule in the year 1947 and subsequently, the historic commencement of liberal Parliamentary-democracy, as independent India's federation-based, genuinely democratic-system of governance.

Though according to the many thinking citizens in India, it remains a highly debatable-point that from very ancient-era to the present 'I.T.-cum-Space to the Nano-technology-Age', and in our discussions here as already emphasised that ironically rather India has been very 'embarrassingly' lacking the **"essential strategic-culture and thoughts"**. And then at the same time, during any given period of history again, the same India has been socio-politically, not even 'caring' to genuinely and gradually evolve, cultivate and imbibe the same, in her multi-dimensional socio-cultural, socio-psychological and geostrategic ethos. Well, whatever the 'debating-points' aspects concerning the experts, but for the laymen like us, certain hard historic 'facts' do keep staring into our eyes, without even a 'blinker'. While critically reviewing these aspects, we do notice that though the unique colourfulness of the ancient India's history apart, the hard historic-facts cannot be wished away that, the 'courses' of the same History, since its beginning and then down more than the already bygone three millenniums; would have been entirely different if the peoples, of the Indian subcontinent, historically would have had the elements of the same essential 'strategic-culture and thoughts'; as a number of times contextually talked-about earlier, during our these discussions.

Then we notice that while discussing ancient world-history, in his famous lucid style, late **Jawaharlal Nehru** has interestingly identified few 'reasons' for India's still lingering 'socio-cultural' historic 'handicap'. He writes that, ***"We have also said something of a number of great thinkers who appeared in different countries about 500 or 600 years before Christ. Buddha and Mahavira were in India, Confucius and Lao-tse in China, Zoroaster in Persia, and Pythagoras in Greece.*** We noticed that Buddha attacked priest craft and the existing forms of old Vedic religion in India; for he found that the masses were being imposed upon deluded by a number of superstitions and *pujas*. He attacked the caste-system and preached equality." (Nehru, 1934)

A very fascinating point emerges here that, regressively cultivating the 'sickening-culture' of nursing many kind of superstitions, blindly sticking to mindless rituals and in addition, callously remaining imprisoned by the demonic 'caste-system'; plainly under no circumstances, can ever co-exist with the many invaluable aspects of the repeatedly emphasised, essential 'strategic-culture and thoughts. Interestingly, the irrefutable historic fact remains that, the above mentioned aspects in reality, are 'uncompromising antidotes' to one another. Now it appears appropriate to briefly identify and narrate the many otherwise preventable but then, tragically disastrous multiple 'historic-consequences', that India had to perforce undergo all through a continuous period of last around two and half thousand years. Principally due to the almost total lack of the many invaluable attributes of the same, *"essential strategic-culture and thoughts"*.

The chequered Indian-history of the last two and half millenniums, has recorded details regarding many magnificent and though short periods of glory as well as human achievements and enrichment and then at the same time the highly embarrassing and painful saga of many mindless blunders and grave judgmental-errors of Indian monarchs/rulers, defeats, plunders of India's heritage and riches, cultural defilements and also then, the many consequential socio-political humiliations.

Journalist, strategic-analyst and author, **Mr. Abhijit Bhattacharyya** (an aluminous of the

National Defence College of India and a member of the International Institute of Strategic Studies, London) has written that, "An honest historical analysis will inevitably raise unpleasant questions pertaining to the State and the war machine. This will not be easy as the reality is bitter. For all the wonder that was India, its military history recounts a rather insipid track record of the garrison, stories of sporadic victory notwithstanding. The reality is that India has never been able to extricate herself from the clutches of the foreigners—from Alexander's invasion in 327–326 B.C. to the Kargil incursion in 1999. It is a country that has almost always been aggrieved as victim of aggression.... To cut a long story short, one would attribute the defeats to the failure of command, not necessarily of the military. It was a failure of the statecraft, lack of vision and unity to anticipate and act. To this, can be added the frequent acts of treachery; again not necessarily from amongst the military, but of the divided establishment of the day. Indeed, the recurring invasion of India stopped for the first time in 1000 years as the plethora of 'Armies in India' transformed into the 'Army of British India'. Since independence however, there has been a spurt of foreign invasions.... **One is inclined to quote HG Wells' assessment of the 3rd century B.C. king Ashoka's place in Indian history—'Almost the tens of thousands of names of monarchs that crowd the columns of history. The name of Ashoka shines and shines almost like a star.' In the context of Sam Manekshaw *(Field Marshal and India's Army Chief, during 1969–1972)*, Wells' assessment of Ashoka could be modified thus: Almost tens of thousands of names of defeated marshals and treacherous monarchs that crowd the columns of history, the name of Manekshaw shines and shines indisputably as the brightest star**." (Bhattacharyya, 10 July 2008)

It is perhaps one of the bitterest facts of long and still remarkably continuing saga of the colourful Indian civilisation since ancient-era, as reflected in the above quoted lines. Further, this also proves the fact regarding, periodic historical exceptions apart, almost the total absence of the same essential "strategic-culture and thoughts". But historically these facts don't in any case, constitute the whole 'history' of ancient India and her magnificently glorious civilisation.

And hence, any comprehensively objective-review of the captivating ancient Indian-civilisation and its historical records, precisely in the context of our present discussions; may not be ever able to ignore their glorious chapters as well. Relevantly, here we may also highlight another paradoxical fact of the ancient-era Indian-history that, India was never by herself militarily weak or so and thereby, incapable of successfully facing any foreign invader's menacing threats or battlefield challenges. But the irrefutable fact remains that in spite of always possessing sufficient military wherewithal, whenever India didn't remain or couldn't be kept united into a single, vibrant and powerful geo-political sovereign entity she perforce became weak and multiply handicapped.

Then being like a 'house-divided' consequently, in spite of she possessing enough military and allied valuable assets nothing could systematically be maintained and kept in place with requisite professionalism as well as abiding by the **essential principles of the "unity of command"**. And consequently therefore, mostly in such highly vulnerable situations paradoxically; India by her own Indians callously sitting in her lap, so mindlessly made themselves to look like a pathetic military 'push-over'. In any case ironically what could be any greater historic tragedy for her? Moreover as if all these were not sufficient, since down the ages, Indians from the top monarchs/rulers to the toiling-masses across the entire subcontinent, never perhaps even remotely paid any attention to the fundamental essentiality of prudently cultivating the same 'strategic-culture and thoughts', for fruitfully triggering the multiple remedial-measures. What an irreconcilable depressing-scenario concerning India have these historically been. But again historically in any case, facts remain facts and thereby, strategic prudence demands that these may instructively be taken with desired objectivity as well as period-based flexibility.

Now retracing our earlier steps and coming to the glorious Mauryan-era, humane imperialism of the **Mauryan-emperors**. The first noted emperor of this dynasty was **Chandragupta Maurya (321–297 B.C.)**, who ruled over his empire with his capital at the grand ancient city of Pataliputra on the southern bank of the magnificent Ganga, i.e. now the present City of Patna in the province of Bihar. He was not only a magnificently visionary emperor but a great military commander as well.

By the time he came to the Mauryan throne, **Alexander** had already returned from north-western India, very tired and exhausted and thereby, he just could not advance towards the then magnificent east of the vast Indian subcontinent, after his almost militarily 'crippling' victory over **gallant king Porus** of his kingdom, consisting of the modern day Punjab and Pakistani territories. Then later on, the vast empire of **Chandragupta Maurya** extended from almost whole of east to the west, some significant eastern part of the present Afghanistan.

Here historically there is a very interesting story after the demise of Alexander and his extremely tiring and 'disillusioned' return to Macedonia, his vast but short-lived Greek empire got divided into relatively smaller kingdoms and thereby, his one general named, Seleucus, became king of western Asian region. He perhaps unimaginatively following Alexander's footsteps, tried to invade the powerful Mauryan Empire from the west and, here he met much more than his match in the persona of great military commander and emperor, **Chandragupta Maurya**. In the process, **Seleucus** got a 'smashed bleeding-nose' in his hands. He made a hasty retreat to save himself also parted with a significant part of the ancient 'Afghanistan-region' in the hands of the Indian emperor. Later, during his unmatched and immortally famous grandson emperor Ashoka some of the unique 'edicts', so marvellously engraved in the age-old mountain rocks near the Khyber-Pass are still available. More about the historic military-exploits and unmatched generalship of the famous pair of grandfather and grandson, in our perspectives, would be dealt with in the next chapter.

Well in this context, another historically interesting point, particularly with respect to certain important aspects of the age-old socio-cultural ethos of the very ancient Indian societies also then clearly emerged **India's "caste-system", one of the most harmful socio-cultural as well as fallaciously-fixated social ailments.** The far-reaching praiseworthy historic fact is that after the decisive military-victory of Chandragupta Maurya over the Greek general Seleucus he as the victorious emperor, very magnanimously as well as amicably married the defeated Greek commander-cum-ruler's daughter. Further historically logical point appears to be that the emperor himself set an encouraging example, down the hierarchical line, a number of such marriages between the Greeks and Indians subsequently would have probably followed. Another interesting point is that historically, this happened even before the advent of Buddhism. Hence the fact emerges that did the **"caste-system"**, take root much later.

In this perspective we may emphasise that, when **Lord Buddha** came on the scene and after achieving his enlightenment, he took valuably educative initiatives to reform the then Indian societies; **the Brahmin-priesthood with selfishly vested-interests** got threatened and felt hysterically panicky. These so-called Brahmin-priests came out up in-arms against the Buddhist monks and monasteries to subjugate Buddhism's sway over the masses. They instigated Hindu nobles and warrior-class for the purpose, by very **sinisterly and craftily distorting as well as misinterpreting the ancient Vedic-Epics and other timeless treatises** for their most unscrupulously narrow-minded selfish-motives; with total disregard to the harmful consequences.

This was, perhaps the most damaging blow to the otherwise, very colourful Indian 'social-fabric'; indeed with a very long-term and still lingering crippling socio-cultural and socio-political fall-out. Another related historic fact is that all these socially most harmful processes took four to five centuries, and produced Brahmin-priesthood's sinisterly desired results decisively during later-age Gupta-dynasty period. Here we may add that, no doubt the '**Gupta-period**' otherwise, is very befittingly called a golden-age in the colourful ancient Indian-history; but this historic fact too, cannot be wished away.

Further these sinister machinations on the part of the entrenched **'Brahmin priest-class' for their vested-interests coupled with, naively aided and abated by many relatively 'foolish' and unthinking monarchs/rulers then** was a **'lethal-combination'** for the ancient-era Indian-societies, with many extremely damaging socio-political as well as other 'fallouts'. Needless to say that one obviously apparent and almost permanent socio-political 'crippling damage', has been inflicted to India's 'body-politic' in this regard; in the form of 'nipping-in-the-bud', the then magnificently evolving **"essential strategic-culture and thoughts"**. And then more so, whose 'tremors' are sadly being felt in the entire gamut of 'policy-formulation' as well as multiple-processes of governance regularly, by almost all the consciously thinking patriotic citizens even now; in the modern-era's socio-economically fast advancing and politically-independent democratic India.

We have earlier also noted that during the prolonged periods in the history of ancient India a number of royal-dynasties kept-on sprouting from time, in the entire length and breadth of the Indian subcontinent and then, independently ruled over their relatively big and small kingdoms. And among them repeatedly mindless both inter-cum-intra-clans based clashes too kept-on happening. Moreover, at the same time, periodical foreign-invasions kept-on taking-place into the territories of same Indian-subcontinent as well, by repeatedly following largely the same identical route through the north-western corridor of the well-known 'Khyber-pass'.

Well, what all could be the reasons to these humiliating as well as recurring and audacious foreign-invasions into the Indian-subcontinent, since thousands of years and at the same time, what all factors could have possibly motivated them? We have befittingly noted one of the principal complex reasons above, in the form of **selfishly irresponsible-cum-sinisterly scheming and machinations on the part of ancient-era Brahmin priest-class**, with their consequential damaging connotations.

John Keay (*India: A History)* says that, "But in the case of 'India' this demeaning connotations had lasted until modern times. 'Hindustan', 'India' or 'the Indies' (its more generalised derivative) had come, as if by definition, to denote an acquisition rather than a territory. Geographically imprecise, indeed moveable if one took account of all the 'Indian' in Americas, *(well, bracketing all the Indian-citizens of modern democratic and politically-independent India, with the Americas' 'red-Indians'; is neither historically correct, scholarly desirable, nor in 'good-taste' and hence, grossly misleading.)***'India' was yet conceptually concrete: it was somewhere to be coveted—as an intellectual curiosity, a military push-over and an economic bonanza. To Alexander the Great, as to Mahmud of Ghazni, to Timur the Lame as to his Mughal descendants, and to Nadir Shah of Persia as to Robert Clive of Plassey, 'India' was a place worth the taking.**"

Another rather 'amusing' point is that many Indian as well as foreign historians and scholars, since centuries, have with some strange logic been giving perhaps undue importance to Alexander's invasion of the then ancient India during earlier than 300 B.C. History has been able to somewhat-glorify **Alexander's name and personality**. He was no doubt very ambitious, with some outstanding attributes of a great military-strategist and also a highly inspiring military-commander in the field. He was only in his twenties, when he succeeded to the Macedonian throne, with all his irrepressible ambitions and multiple leadership attributes.

Plutarch says that, "...considering him to be of a temper easy to be led to his duty by reason, by no means to be compelled, he always endeavoured to persuade rather than command or force him to anything; and now looking upon the instruction and tuition of his youth to be of greater difficulty and importance than to be wholly trusted to the ordinary masters in music and poetry, and the common school subjects, and to require, as **Sophocles** says—'The bridle and the rudder too,' he sent for **Aristotle**, the most learned and most celebrated philosopher of his time.... **It would appear that Alexander received from him not only his doctrines of Morals and of Politics, but something of those more abstruse and profound theories, which these philosophers, by the very names they gave them, professed to reserve for oral communication to the initiated, and did not allow many to become acquainted with**." (Plutarch, 1930s) (*Well indeed very interesting and historic too, isn't it?*)

When Alexander marched on his empire-building campaign with full military-vigour, vision and essential preparedness across the Mediterranean, he did achieve a series of spectacular victories in the battles after battles. And then, after wandering across Persia, he entered into the Indian subcontinent through the Khyber-pass and via the ancient and highly famous city of internationally great learning and scholars, **Taxila**, near and north of the modern Pakistani city of Rawalpindi. Naturally he probably must have considered the then ancient India to be another 'cake-walk' and subsequently, he met the Indian king of the north-western region Porus in battle in the plains of Indus-river.

With extraordinary military difficulties and great loss of both lives and war-assets, finally he could achieve victory; but in the process he, his generals and soldiers were thoroughly tired and exhausted. He was given a costly taste of the 'military-metals' of ancient Indian armies. Further when he learnt more factual military details of the emperor **Chandragupta Maurya** in the vast Gangetic-plains in the East, he and his soldiers called it a day. His disillusioned retreat back to his homeland Macedonia was indeed very costly and full of hardships. The sad historic fact is that he finally could not even reach his home and died in Babylon itself in 323 B.C. at the age nearly 33 years only. Later, then what fate Greek **Seleucus** met in the hands of Chandragupta Maurya, we have already talked about little earlier, in this chapter itself.

So India was certainly not any kind of a 'military-pushover' from the beginning itself. However Indians have no grounds for being 'apologetic' in any manner, and also no justification for 'blaming' others for their own still lingering and highly embarrassing as well as glaring geostrategic and politico-militarily lack of imperative socio-political 'culture of visions', along with other allied and regressively recurring 'lapses'.

The pivotal-point in this regard is that since ages; different tribes, kingdoms, empires, Nation-states and so on, have been surviving/thriving and interacting among themselves in all sorts of scenarios, in a "state of nature"; with their own vital-interests always reigning-supreme regardless of any kinds of consequential fallouts. These phenomena, whether one likes/agrees or not, are even now perhaps unalterably perpetuating with time and situation, in different forms-cum-impacts in the international arena; and the same are likely to be so in future as well. What depresses anyone concerned is that even now in modern times, the peoples' mandate-based Democratic India's so-called highly 'shabby' "Power-Centres", still perhaps wilfully remain prisoners of the old and irresponsibly callous as well as highly damaging 'socio-psychological fixations' and thereby, regressively refusing to change.

If Indians, with all their civilisational magnificence, occasional soldierly-chivalry as well as cultural heritage and so on since ancient times' callously refuse to see reasons and prudently not to evolve, cultivate and imbibe the multiple aspects of **"essential strategic-culture and thoughts"** then, they are themselves to blame. Hence, it is all their self-created comical-cum-tragic 'funeral'. In fact, crucially important historic-truths should always be instructively accepted as such and also plainly be told with desired directness to fellow-citizens, without any malice-cum-confusion and at the same time, be similarly recorded for the future generations and posterity. More relevant details in these perspectives would be suitably dealt-with in subsequent chapters.

Now coming back again to the aspects of glorious Mauryan-empire, the unique and famously learned and visionary **Chanakya** was a minister in the magnificent court of **Chandragupta Maurya** and also his 'mentor', principal guide and adviser. **Chanakya's world-famous and timeless Treatise, named 'Arthashastra' remains unforgettable and of great historic value, for everyone concerned.** This 'epic' in its own right, contains many facets with valuably unique lessons, on the complex subjects of **'Statecraft', 'Diplomacy'** and also **'Matters-military'**. When studied with the modern-era contextually desired flexibility, the lessons are even now universally relevant and highly enriching in stimulating strategic thought-processes as well as multiple areas of human-knowledge; for the rulers, military commanders, decision-makers and scholars as well. Paradoxically it is a matter of great misfortune for the modern democratic India that most of the callous bureaucrats and politicians don't even care to imaginatively study and professionally learn from this timeless treatise of a truly ancient-era Indian genius.

Interestingly in fact, the **historically famous and learned Greek ambassador Megasthenes, who had come to India during the glorious rule of Emperor Chandragupta Maurya, then reigning over his vast and prosperous-empire,** also visited **Chandragupta's remarkable court in his capital city Patliputra.** He has given a vivid account of this 'dazzling' city, the over-all high quality of life of its residents, their high-taste, beauty, learning and glamour. He was especially impressed by their 'high-heeled' shoes and finery in clothes, food and houses. Further, **Megasthenes** has specially recorded his memorable meeting and learned discussions with Chanakya, since he was overwhelmed by his saintly brilliance, impeccable integrity, simplicity and multiple-range of scholarship.

Indeed most befittingly, world history has ranked **Chanakya**, a magnificent guide and mentor of his emperor; parallel to **Sun Tzu of ancient China. No wonder then, that both learned Indian as well as foreign historians have called Chandragupta Maurya, the Indian Julius Caesar.** He was succeeded by his son, **Bindusara**, not so brilliant and little is known of him and his rule. But then subsequently the wonderful golden-era of the Maurya-dynasty had again dawned, **with the crowning of Ashoka as emperor**, who was the son and grandson of **Bindusara** and the great founder-emperor Chandragupta Maurya, respectively.

At this juncture, history shows us that though before **Chandragupta Maurya**, Magadha-kingdom was not a vast and famous empire but, he too, had his predecessors in the names of king **Bimbisara** and his son **Ajatashatru**. In fact, **Bimbisara** had since predeceased **Lord Buddha** but, his son **Ajatashatru** was imperial-cum-militarily more adventurous and therefore, his repeated clashes with the unique and famously perhaps world's earliest democratic congregations of numerous **Licchivis**' 'small-kings', with their capital at **Vaishali** in the northern part of Bihar in modern India along with his romantic involvements with **Vaishali's superbly beautiful court-dancer Amrapali**; are some interesting historical episodes. **Ajatashatru** died around 330 B.C. or so and then, **Chandragupta Maurya** occupied the Magadha-throne in 321 B.C.

Ashoka succeeded **Bimbisara** and came to the Mauryan-throne in 268 B.C. and most famously ruled over his empire for glorious 35 years, i.e. up to 233 B.C. Volumes have been written about his talent, personality, military-genius as well as imperial-firmness and yet benevolent rule. It is not surprising that after **'Alexander the Great'** of ancient Macedonia it was he, whom history recognised as **'Ashoka the Great'**. Interestingly much later in a similar manner, **'Napoleon the Great'** and then truly greatest of the Mughal emperors, **'Akbar the Great'** were recognised. **Well then, the recorded ancient-history shows that, 'Ashoka the Great' is historically unsurpassed.**

No wonder that, **H.G. Wells** in his well-known book, *Outline of History*, while paying befittingly glowing-tributes to **Ashoka,** has also written that more living-men most respectfully remember him, cherishing his unsurpassed memories much more than even the great names of Constantine or Charlemagne and so on, in the pages of human-history. He has further emphatically recorded in his book that, he remains the only unique and perhaps the greatest monarch in the pages of man's living history, who after a decisive military-victory in the **extremely violent battle of Kalinga** wilfully with his supreme-desire, renounced warfare and violence against fellow human-beings forever. He became supremely humane, magnanimity-personified as well as a superbly saint-emperor.

No wonder that his truly sacred-dictates timelessly engraved in the huge mountain-rocks, since located in so many places, bear historic-testimony to his high-minded greatness. **It may also be affirmed that in history perhaps the great military-monarch, Ashoka of India's gloriously ancient Mauryan-dynasty was the first ruling emperor to have given the remarkably precise definition as well as meaningful elaboration of the many facets of 'Dharma' so sincerely practiced by him. He had since already become a most dedicated monarch-disciple of Lord Buddha.**

Jawaharlal Nehru has reproduced the few lines of the English translations of **Ashoka's** unique and historically unsurpassed dictates as, "Kalinga was conquered by His Sacred and Gracious Majesty, when he had been consecrated for eight years. One hundred and fifty thousand persons were thence carried away captive, one hundred thousand were there slain, and many times that number died. Directly after the annexation of Kalinga began His Sacred Majesty's zealous protection of the Law of Piety, his love of that law, and his inculcation of that law (*Dharma*). Thus arose, His Sacred Majesty's remorse for having conquered the Kalingas, because the conquest of a country previously un-conquered involves the slaughter, death and carrying away captive of the people. That is matter of profound sorrow and regret to His Sacred Majesty.... Moreover should anyone do him wrong, that too must be borne by His Sacred Majesty so far as it can possibly be borne with. Even upon the forest folk in his dominions His Sacred Majesty looks kindly and he seeks to make them think a right, for if he did not, repentance would come upon His Sacred Majesty. For His Sacred Majesty desires that all animate beings should have security, self-control, peace of mind, and joyousness." (Nehru, 1934, pp. 73 & 74)

The above reproduced few lines from English translations of the **immortal edicts of the Emperor Ashoka** do convey the magnificence of the divinity of his thoughts, their multiple range and vast depth. His thoughts were not merely philosophical but pro-active in perceptions, which he diligently practiced too; in the all aspects of his own personal life as well as, while carrying out his royal duties. He further in fact, superbly enunciated that the true conquest of a man is the conquest through his heart and thus remarkably practicing the same, he wrote unique history of peacefully expanding his empire throughout almost whole of South-East Asia, with magnificent weapons of 'love and spiritualism'. He was a strategic-genius, for he never took any initiative from any position of moral and any other kind of weakness. No wonder that during his magnificent reign, Buddhism spread very fast thorough out the South-East and Far-East Asia and in China too. **The most remarkable point is that he was the first person to show the world that a Religious-thought can be extensively as well as more enduringly spread, with honest love and then; without ever resorting to sword and scheming craftiness anywhere. These and much more have been Ashoka's unique greatness.**

Paradoxically some historians have been of the views that in the ancient Indian-subcontinent, a period of more than six centuries from the **Great Ashoka's demise in 231 B.C. to 320 A.D.,** when another **golden-era of mighty Gupta dynasty** began has been relatively a 'dark-period' in the long ancient-Indian history and thereby, the Indian-civilisation with all its multiple manifestations, went into a period of 'obscurity'. Well, certain scholars' individual perceptions apart, when this historically identified period of more than six centuries, is critically examined in a somewhat comprehensively strategic perspectives such opinion is not completely corroborated by multiple-advancements of the age-old Indian-civilisation thus taking place, triggering into multi-dimensional socio-cultural and economic developments and enrichment. **It is true that Kushans had not originated in India and had come to India from the foreign territories. But they adopted India as their home and became socio-culturally as well as religiously integrated with the Indo-Aryans.**

Well, **'every coin has two sides'**, and the same holds good in our discussions in specifically the present-context as well. Interestingly scholars and concerned experts too, seldom agree on various aspects of the multiple complexities of man's continuing history. And this is understandable but as far as humanly feasible, desired objectivity may always be adhered to. History informs us that Kushans were Mongoloid and gradually got racially mixed with other tribes in China and West-Asians, as they marched westward and on to Persia and then, to east towards the present Afghanistan and so on. In the process, they did build a big empire, established their capital in ancient Kabul. Interestingly, for around 300 years, almost whole of Central Asia remained under Kushan-rule. Historically, they did display a focused politico-military strategic-vision and appropriately in their own imperial-sense of priorities, they turned to the further east towards the huge Indian-subcontinent. By that time, Sun over the Mighty and huge Mauryan-empire had already set-in, and specially the northern India had already in a way, got 'Balkanized' into numerous smaller kingdoms.

After politico-militarily establishing their footholds in north-west Indian region, with their befitting strategic-vision; they shifted their capital to **'modern' Peshawar, which was then known as Purushapura.** In fact, with considerable historic justification, it can also be maintained that again Indians were found wanting, both geostrategically as well as militarily, in invading the Indian-subcontinent from the Hindu Kush sides and through the Khyber-Pass onward. Well, the over-riding socio-political and socio-religious reasons in this regard, have since been spelled out earlier in this chapter itself. However, that is no 'apology' on any pretext and historic facts do remain unalterable. Further, there is no record or any anthropological evidence or so for massacre of the then Indians in that region, in the hands of Kushans. They were certainly no ruthless invading herds, like later-day Mongols and Islamic Afghans. They came to India and got assimilated completely into the Indian societies. They wonderfully adopted Buddhism got integrated into ancient India. **In fact, the extensive Kushan-empire magnificently became a huge Buddhist-**

empire, in all its manifestations. Historically, emperor Kanishka was the most glorious among the Kushan-dynasty.

I notice that the phenomenon of 'aliens' coming to a particular ancient Nation-state, with a clear aim of politico-military expansion of their empires; did happen in feudalistic-cum-monarchist China as well. But any the then Chinese 'similarity' with India, historically ends at this point itself. In fact whatever the historic reasons, China never displayed any kind of paradoxical 'bankruptcy' regarding the various aspects of the earlier repeatedly mentioned the same **"multiple-aspects of essential strategic-culture and thoughts"**. As an example here we may appropriately mention that by the dawn of the seventeenth century, the clout of ruling Chinese Ming-dynasty had already declined as the regime became callously corrupt and weak and there by, China's southern-neighbour, the Manchus invaded with a similar aim of expanding their empire. Consequently by 1650s or so, the Manchus, interestingly as 'semi-aliens', got permanently settled-down integrating themselves almost fully with the indigenous Chinese-societies. Their regime had already taken firm roots and in addition, making China fairly strong.

Another strategically overriding historic difference between India and China has been that since very ancient-ages, China has had unique and unbroken political-legacy for having a strong and sovereign imperial authority for governing the entire country. Further, China was relatively fortunate not to get mindlessly broken into a number of geographically little bigger and smaller kingdoms and Chieftains. In fact, it was only during 19th century that the central dynastic authority in China got weakened; consequently resulting in the entrance of foreigners with their own selfish purposes, in the guise of traders and so on. Whereas in case of India in fact, history unfortunately has been entirely different. This pivotal difference had serious fallouts, causative to a number of geostrategic as well as politico-militarily historically humiliating injuries to the socio-political psyche of the Indian societies.

Now coming back to the ancient Kushan dynasty, historians say that not much is available as any kind of historical records giving details a of historical-value and also, about the kind of dynastic-successions regarding this the then huge empire. But some details about their geostrategic and military-prowess, have been historically worked-out and informatively written by the concerned experts for the interested peoples and hence; contextually now these indeed make great assets. In addition we also interestingly notice that, with respect to the extensive European Mediterranean coast-line and central Asia including the Indian-subcontinent, the vast Kushan-empire occupied geostrategically a sort of 'dominant' central-position. It was historically logical therefore, that the Kushan-emperors did maintain politico-diplomatic fruitful relationships with Persia as well as Romans.

Interestingly in fact before reaching the ancient India, when they began their sweeping march towards east; they overran the pastoral territories of Shakas. They were nomadic tribe of Turkish origin and then since pushed further east, they entered into the Indian-subcontinent. It is pertinent to note here that all these Turkish-cum-Mongoloid origin tribes such as Baktrians, Shakas, Scythians Turkish and Kushans were principally nomadic. Whatever they might have been, when the Kushans came to the ancient Indian-subcontinent their gradual inter-mixing with the Aryans, they became more civilised, settled-down here and, ceased to be nomadic. Since they adopted and became dedicated followers of Buddhism. Though they got totally 'Indianised' but at the same time, they maintained contacts with the areas of their origin and China and also kept visiting those places. In the process interestingly, they were historically instrumental, in spreading Buddhist thoughts in those distant areas and thereby, facilitating Buddhism to enduringly take firm roots in those regions in the north of Indian-subcontinent.

John Keay writes that, "Coins, plus inscription found in Taxila, bear early testimony to the pretension of the Kushana. 'Maharajah', 'King of Kings', 'Son of God', 'Saviour', 'Great One', 'Lord of all Lands', 'Caesar' and other such titles are reeled off as if the incumbent wished to lay claim to every source of sovereignty going. (*Well in those now bygone ancient ages and, later on during the medieval-periods too, these were part of many pretensions/practices the world over and hence, it was nothing peculiar to the Kushans.*). 'Son of God' is thought to be a legacy of the Yueh-chi' familiarity in China and its celestial rulers; 'Kings of Kings' was borrowed from the Shakas, who had imitated the Achaermenids of Iran; 'Saviour' came from the Greeks; 'Caesar' from the Romans. The coins are of the highest quality and show a switch to Roman weight standards; possibly they were actually recast Roman *aurei*."

Especially during emperor **Kanishka**, the historic Kushan-empire remarkably expanded covering vast areas of northern India as well as adjacent north-western territories, bordering Persia. Further during this period in India, art, architecture and literature, more so the Buddhist-literature, historically flourished. More and more interactions with western and southern regions took place, bringing in new ideas related to thoughts, literature and clothes as well. **Another important event took place, in the shape of Buddhism splitting into two sects of Mahayana and Hinayana.**

In our perspectives at every emerging stage now, we do notice ironically that in the long history of ancient India in each of the prominent ruling-dynasties coming on the scene one after another, historically barring one or two monarchs, others were largely of 'routine-cum-inconsequential' types. This whole highly complex as well as important socio-psychological and socio-political 'phenomenon' since, always crucially required to be live and continuing, remained largely personality-oriented and hence never as a 'system'. Therefore historically, this 'phenomenon' could never get transformed into any kind of "Institution-oriented Indian-system" in practical reality.

Further in fact, if at all any such process did ever 'materialise' in reality in the Indian-societies it remained mainly 'disjointedly-superficial', and hence largely ineffective. Then as a rather strange 'comical-cum-tragic paradox'; even now in our present modern-era in the 21st century precisely in this context, nothing much has changed and consequently therefore, independent India's leadership, mainly consisting of again a strange "politico-bureaucratic combine", amazingly remains a 'prisoner' of the same highly damaging regressive-past. Logically therefore, in this specific Indian-context paradoxically, what could be any bigger 'tragedy' for India and the Indians. Well, no wonder that the modern and fast developing so-called 'soft' Indian Nation-State, is also addressed as a country of many kinds of 'paradoxes' as well. Most ironically, India's embarrassingly lingering and highly regressive 'state of affairs' has many facets and fallouts as well as many 'causative-blunders'; so callously and also at times rather even willfully committed at many levels and in many segments in the Indian-societies/Nation-State, since ancient times.

Astonishingly again specifically in the Indian sub-continental context, all these in any case, do not constitute the complete story. In fact in addition to the above irrefutably lingering-realities, so disturbingly staring at the Indians, as part of India's continuing living-history in the making ironically, we also notice that there have been at times, incidents of 'betrayals' by some shamelessly selfish and unscrupulous Indians, in different situations/calls-to-duty, causing serious damages/ losses to the State as well as irremovable 'historic-blots'. Though even if highly embarrassing and unpalatable, these historic truths would perforce emerge in any maximum-possible comprehensively objective strategic-analyses, precisely in the perspectives of India's past, present and possibly the future.

Well understandably in this perspective, one may naturally raise perhaps a 'question'; that this author is displaying a strangely 'fixated-prejudice' against many important and also praiseworthy achievements of India's multi-faceted civilisational-aspects in their many past as well as present connotations. Well concerning this, the author with due respect and honest humbleness, may emphatically submit that nothing could be farther from the truth. **In fact the honest intention here with desired strategic-objectivity; has been to precisely emphasise that, most of the historic as well as contemporary India's geostrategic and politico-militarily embarrassing failings were, and at the same time presently have been, paradoxically preventable.**

Any critically worthwhile strategic-review of India's continuing living-history with all its multiple-aspects, would clearly establish this historic-truth, in no uncertain terms. In other words in fact, they pertain to the aspects of various kinds of challenges emerging before any nation from time to time and then making consistent efforts towards 'situation-specific' critical-pattern of successfully facing them, with expected strategic-prudence. Hence, multiple-aspects of any nation's civilisational 'glory' and/or the 'blots' are perforce linked and simultaneously also, hidden exclusively again into the historic-saga of those very consistent efforts with expected strategic-prudence.

Now again retracing our discussions back into the ancient Indian-civilisational aspects, interestingly with all their peculiar historic-paradoxes; we further learn from history that successful survival of Kushan-empire lasted for nearly 300 years. And then afterwards, of a totally different kind but independently glorious in its own rights, a new Gupta-dynasty began building a historic-empire in the vast territories of northern as well as central areas of the Indian subcontinent. This process in historic reality, was never a peaceful one but somewhat violent and gradual as well. After a passage of so many centuries of time, in objective retrospect, ancient India's living-history shows that the violent campaign of Guptas against the sinking Kushan-empire was perhaps misguided and also based on some 'short-sighted' notions and at the same time probably, it could have been preventable as well, up to a large extent. In fact the initial Gupta-king of relatively small kingdom in and around ancient city of Patliputra and his clan were sinisterly and consistently 'poisoned against the Kushans, by the entrenched Brahmin-priests for furtherance of their own selfish vested-interests, since Kushans were devoted Buddhists. Great ancient Chinese scholar and strategic-thinker, **Sun Tzu** has said that the best victory in a war is achieved without fighting it. This magnificent saying indeed, remains timeless and universally true as well.

Though to begin with, they did come into the Indian-subcontinent from foreign lands but, as they started settling-down, they got totally Indianised and also Aryanised in the process by adopting the Indian customs, dresses, food-habits and in addition, became dedicated Buddhists; interestingly a religion which to begin with took birth too, in India. In fact they never attacked the Hindus, since they did look-after them and also, their places of worship. But the 'rub' was lying somewhere else in the sense that, the Buddhism, based on Lord Buddha's preaching, sincerely tried to gradually bring-in many reforms in Hinduism and also, tried to persuade the Hindus to get rid of many superstitions, regressive 'caste-system' being a highly damaging socio-cultural evil as well as discard many mindless customs. Here a singularly interesting point to bear in mind is that Lord Buddha was himself born in a Hindu royal-family. **But these highly desirable historic truths were of no concern to the powerfully entrenched Brahmin Priest-class, since they naturally felt seriously**

threatened, as the then historically dynamic reforms were going to 'smash' their vice-like grip over the Hindu-populace and their ways of life. And as the Kushans started getting weakened and their empire began gradually to shrink as well as disintegrate, the same Brahmin-priesthood craftily saw the opportunity and started 'smelling the blood'.

Now at this stage while trying to critically look into the otherwise fascinating aspects of the chequered-history of rise and fall of various indigenous-empires we are likely to come across certain 'interesting-dimensions' in this regard, in the specific context of **Professor Samuel Huntington's** somewhat 'controversial' thesis on "the Clash of Civilisations and Remaking of World Order". It is also understandable that in such very complex aspects of specially the 'fluid' history of one of the oldest ancient-civilisations particularly India; multiple-shades of opinions are naturally bound to be engagingly different and captivating as well. But the imperial conflicts between the Buddhist-Kushans and the Hindu-Guptas; most probably do not appear to be any types of such 'Clash of civilisations', as discussed by Professor Huntington in his book.

The historic facts are that both the Kushans and the Guptas were largely belonging to the same age-old 'Indo-Aryan' civilisation, as peoples of the same Indian-subcontinent. In addition interestingly, both did belong to the two different but very closely allied ancient Indo-Aryan religions. Well at the same time, it remains also true that the Kushans were initially aliens and did come to India, with their own socio-cultural multiple 'baggage', to somewhat 'dominatingly' settle-down in the vast plains of northern India. But it is also a fact that in the course of time as centuries passed, the Kushans got gradually much more civilised from their own relatively primitive nomadic-past and simultaneously; got overwhelmingly submerged and assimilated into the then vastly superior and multiply colourful ancient Indo-Aryan civilisation. Hence consequently, their combined-life became quite integrative and also substantially one homogenous societal-whole. Moreover historically while critically looking into the otherwise, understandable imperial-ambitions of the Gupta-monarchs with more objectivity; the conflict-like situations in the then developing specific politico-strategic scenario; were deliberately instigated and sinisterly propagated by the regressively entrenched Brahmin priest-class, and certainly not for their any so-called religion-sake but for selfishly 'feathering their own nests', as already elaborated.

In fact to begin with, the main founder of the glorious Gupta-empire was a dedicated Hindu, **Chandragupta-I**, who came to the throne in 320 A.C. As a king he was young, ambitious and a truly capable military-commander as well as having strategic-prudence too. He married a licchvavi princess, named **Kumaradevi**. In the process he could secure friendship and support in realising his big ambitions. In the relentless conflict for more than a decade against his adversaries, he established his unchallenged sway over large part including heartland of northern India. His reign lasted for nearly 15 years, i.e. up to 335 A.C.

At this interesting historic-point, it appears informatively very appropriate to again quote few lines by **John Keay**, "The Licchavis had a distinguished pedigree which was doubtless highly desirable to the unknowns like Guptas. But the importance the Guptas attached to this union was of an altogether higher order. **Chandragupta I's** successor would style himself not 'son of a Gupta father' but 'son of a Licchavi daughter'. There are even coins showing king and queen together, an unprecedented development; they bear, as well as the king's name, that of '**Kumaradevi Licchavayah**'. It is known that the Licchavis had acquired territory in Nepal and it may be that 'they had taken possession of Pataliputra, the city which had been built and fortified many centuries earlier for express purpose of curbing their restless spirit.' **Certainly it is probable that the Guptas**

and the Licchavis ruled adjacent territories 'and that of the two kingdoms were united under Chandra-Gupta I by his marriage of Kumaradevi."

Though opinions can vary but, it may also be said that if, **Chandragupta-I** would have been able to think even more with strategic-vision and 'out of the box' as well as being 'befittingly' free from the 'clutches' of Brahmin-priest then he should have made suitable efforts to win the trust and support of the Kushans too, especially from a politico-militarily unchallenged position of strength; consequently then he and his empire would have reaped many enduringly far-reaching imperial-dividends. In addition, the many positive long-term impacts would have come and also, the subsequent history of India might have taken a different course. Further most likely their son Samudragupta subsequently then, would have been able to cultivate more far-sighted strategic-prudence and display more 'cool-headed' politico-military vision; while magnificently expanding his empire, on any specifically developing situation-based imperial calculations then; on the lines of above mentioned **Sun-Tzu's thoughts**. Well, in a larger historic perspective, all these may also be termed as 'matters of conjectures' only. In any case, ancient India's history has shown us that Gupta-dynasty too, after the magnificent Mauryas, was also a great one in many respects and also, Gupta-empire shiningly lasted for little more that the two centuries.

In fact later, it was **Samudragupta, son of Emperor Chandragupta-I and Queen Kumaradevi,** who came to the Gupta-throne in 335 A.C. who as emperor, vastly expanded his empire and ruled for nearly forty-years up to 375 A.C. In history for all this and much more, **he is also known as the Indian Napoleon**. Subsequently his son, named **Chandragupta-II**, was also a great military commander and emperor; who expanded the Gupta-empire even further, including the then Gujarat-region and also the adjacent Kathiawad. During his period; these regions of ancient India's upper south-western coast, were under the Shaka rulers, a 'clan' of the Turkish-origin. **It appears that motivated by his magnificent imperial achievements Chandragupta-II, assumed still a higher 'royal-title' of Vikramaditya; and so he is generally known in history by this name.**

Historically the golden Gupta-era of the 4th and 5th centuries and also onward for some time during the beginning of 6th century A.C. ancient India, is generally known as remarkable period of Hindu-imperialism. During this period, Indians were noted for multiply magnificent achievements. Besides art, linguistics, literature science and technology, geostrategic and multiple matters-military and so on; there were great advances in the field of metallurgy as well. Universally known the unique and historically dominating **'iron-pillar' situated near the Qutub-Minar in modern Delhi was built by the Emperor Vikramaditya**. This is timelessly a unique example of wonderful ferrous metal-casting, since even after nearly fifteen centuries or so, it has not lost its smooth shining, remains rust-free in the open and the beautiful engravings on its 'polished' surface remain clear. In fact even now in the presently fast-advancing modern high-tech era; the metallurgists are not able to scientifically evaluate completely, the truly wonderful foundry-technology and skills of that long bygone Indian era.

It is also important to note here that during the golden Gupta-era, a great Chinese Buddhist and scholar, named Fa Hien, came to India during the beginning of 5th century A.C. He stayed in India for nearly six years up to 411 A.C., and then returned to China, via Ceylon by sea; with his historically rich and huge 'intellectual-baggage', containing his memoirs and the voluminous Chinese-translations of many timeless Sanskrit epics and writings of so many Indian scholars of the ancient-era. During his long stay in India, he visited many important places of Buddhist-pilgrimage and remarkable seats of higher learning and multiply-enchanting Indian scholarship

of those ancient times and in the process, he also got intellectually mesmerised. He then very reassuringly found that during that ancient-era, Buddhism was flourishing in place of origin. His scholarly written historic-memoirs do give vivid accounts of the ancient India's civilisational, cultural and multi-disciplinary scholarly greatness as well as many magnificent achievements, for the universal human well-being. He found them covering many areas of relentless human-endeavour such as; art and crafts, literature, science and technology, architecture, medicine and matters-military as well.

No wonder that volumes have been written about the historically magnificent multiple achievements of ancient India, by many noted Indian and foreign experts and historians. Though **one of the world's most ancient and great universities, the Nalanda** was established much earlier; other great universities and excellent seats of learning were set-up and flourished during the magnificent Gupta-era, in ancient India. History notably informs us that, during this historic golden-era in ancient India the **Vikramshila University, near the modern Bhagalpur town in Bihar**, was established and further, we find that **another historic Vallabhi University, near Kathiawar in Gujarat**, came-up and flourished; and then there was also **another Amravati University down south**, as a noted seat of learning. Ujjain in central India, was the flourishing capital of the mighty Gupta-empire and hence befittingly therefore in this city itself; **another highly important ancient Ujjayini University** was established by the dynamic Gupta-dynasty, as a brilliant seat of learning as well as intellectually explorative debates and discourses.

Historically in fact, ancient-India's high cultural-tradition of establishing a chain of formal and efficiently organised seats of higher-learning as brilliant Universities, covering almost the entire vast landmass of the Indian subcontinent most probably appears to be unique and oldest in the global ancient-history of mankind'. In this specific context, if we care to conceptualise an enchantingly panoramic-glimpse of this unique 'chain' of the ancient "seats of higher-learning" as flourishing Universities of the distant past in our world; from Taxila in north-west to Nalanda in the east and in addition to the others since elaborated above, as equally brilliant centres of learning, exploratory-discourse as well as research and then, the same suggested 'glimpse' becomes attractively clear. Interestingly further history shows that the same phenomenon of brilliant intellectualism in ancient India did continue for many centuries in the post-Gupta era as well. These aspects would further be dealt-with little later, in this chapter.

Jawaharlal Nehru has written that, "The Gupta period of Hindu imperialism in India was a great revival of old Aryan culture and Sanskrit learning. The Hellenistic, or Greek, and Mongolian elements in Indian life and culture, which were brought by the Greeks, Kushans and others, were not encouraged....by laying stress on the Indo-Aryan traditions. Sanskrit was the official Court language. But even in those days Sanskrit was not the common language of the people. **The spoken language was a form of Prakrit, which was allied to Sanskrit**.... There was a great flowering of Sanskrit poetry and drama and of Indo-Aryan art. In the history of Sanskrit language this period is perhaps the richest after the great days which gave the Vedas and the Epics. **Kalidasa**, that wonderful writer, belonged to this period. *(Well, it may also be mentioned here that much earlier, even nearly a century before* ***Chanakya also known as Kautilya****; a great Sanskrit scholar called* ***Panini*** *in Taxila formulated and wrote a comprehensive* ***'Sanskrit-Grammar'****, in nearly 400 B.C., which even today remains the most authoritative treatise on the subject.).* **Vikramaditya is said to have had a brilliant Court, where he assembled the greatest writers and artists of the day.... Samudragupta changed the capital of his empire from Pataliputra to Ayodhya. Perhaps he felt that Ayodhya offered a more suitable background for his aggressive Indo-Aryan outlook—with its story of Lord Ramachandra**

immortalized in Valmiki's epic. The Gupta revival of Aryanism and Hinduism was naturally not very favourably inclined towards Buddhism.... But there seems to have been no persecution of Buddhism. Buddhist monasteries continued and were still great educational institutions. The Guptas had friendly relations with the rulers of Ceylon, where Buddhism flourished. **Meghavarna, the King of Ceylon**, sent costly gifts to Samudragupta and founded a monastery at Gaya for Sinhalese students.... What was happening in other parts of the world when the Guptas held sway in India? **Chandragupta** the first was contemporary of **Constantine the great, the Roman Emperor who founded Constantinople**.... Thus, just about the time when the Roman Empire was weakening, India had a very powerful State with great generals and mighty armies. **Samudragupta is sometime spoken of as the 'Indian Napoleon', but, ambitious as he was, he did not look beyond the frontiers of India for his conquests**. The Gupta period was one of aggressive imperialism and conquest and victory. But there are many such imperialistic periods in the history of every country and they have little importance in the long run. What makes the Gupta times stand out, however, and worthy of being remembered with some pride in India, is the wonderful renaissance of art and literature which they witnessed." (Nehru, 1934, pp. 121-123)

It may again be recalled that ironically concerning the ancient Indian-subcontinent, there have seldom been any ongoing reasonably detailed 'historic-records' with important-datelines, available in any manner for posterity; for the scholars and experts for their valuable research and writings. So perforce, they had to 'reconstruct' this history many a times; based on explorative as well as inter-connective 'historic-inferences' with their expertise and also if necessary then; duly guided by their scholarly 'logical-conjectures' in their efforts. Needless to say that in the process since last few centuries, the scholarly books so lucidly authored by them from time to time both in India and abroad; have certainly been invaluable for even any layman like this author, for their preliminary study and knowledge.

Here historically we also learn even before the final demise of the glorious Gupta-era, the masquerading as well as barbarian **'White-Huns'** of mixed Mongoloid-cum-European stock, had begun entering into the Indian-subcontinent, as waves of relatively 'uncivilised' human-hordes, through the north-western passages of the mighty Himalayas. In fact, Huns' barbarian intrusions into the ancient Indian-territories did create very unsettling times for the Indian-rulers. **But fifth in the ruling-dynasty of Guptas, Emperor Skandagupta militarily challenged the Huns, defeated them in battles and victoriously chased them away**. But they were also ruthlessly tenacious-lot and, they had no option but to try again which in fact they did, since they had nowhere else to go. Accordingly around nearly 15 years later then, they again violently intruded into India and gradually spread over now eastern-part of Afghan region as well as extensively over whole of north-western territories of the ancient India, including Jammu and Kashmir region.

In the process of gradually establishing their barbarian supremacy over the almost whole of the north-western ancient Indian-territories, the Huns gradually got intermingled with the local peoples there, who were already settled since long. This unfortunate phenomenon inflicted many serious damages to the Indo-Aryan socio-cultural ethos and magnificent polity, with tragic fallouts. These are not the matters of conjectures only; but in fact borne-out by the old historic-accounts of Huns' highly uncivilised responses and barbarous behaviour, as anywhere else and so in ancient India as well'. Their frightful impacts on the ancient India's high Indo-Aryan multi-faceted culture and more humane-systems of governance, so alien and harmful to them, had many long-term consequences; and the same unfortunately remain still lingering, even in today's modern-era India. **It is a different matter that subsequently, King Yasodharman of the Malwa-region of central**

India, inflicted military-defeat on Huns and hence, stopped their advances. In addition to this later, King Baladitya belonging to the same Gupta-dynasty, militarily collaborated with requisite strategic-vision; with the Mukharis' as well as the Vardhana clans for decisively taking-on the aggressor-Huns. Fortunately for India, they did engage the Huns in battles with their sizeable allied forces and finally chased them away. But Huns had already got enough ironically extended time in the Indian-territories for their cruel stay, and for inflicting unspeakable vandalism and harms in many ways to the ancient Indian societies and thereby damaging socio-political unique culture and high value-systems. Consequently in addition even a much bigger long-term damage came, that the repeated clashes with barbarian Huns had perforce weakened India in many ways; and that has been another extremely sad part of the prolonged and at the same time, otherwise still remarkably continuing, very ancient-Indian history.

In the Indian historical contexts, whenever efforts are made to critically analyse them as much as possible, in comprehensive geostrategic and politico-military perspectives as earlier too touched upon in some details; we do perforce notice a highly complex and mixed-scenario on India's long and continuing historic-canvas. In this respect as part of our presently on-going discussions we have tried to identify certain principal reasons responsible for 'blocking' India's paths towards gradually evolving and imbibing the same **"essential strategic-culture and thoughts"**, since already talked-about, even much before the barbarian Huns came on the scene. **We have also clearly discussed in brief the many damages caused by the violently vandal-Huns intruding into India and thereby perhaps many long-term fallouts, since visible in many ways in the present-era Indian societies as well. And we may also say that they as their permanently negative-consequences, have in effect sown the seeds of callousness, social-violence, corruption and much more. All these in totality, have so far probably 'crippled' India; making her incapable of progressing towards fruitfully acquiring the same "essential strategic-culture and thoughts". What a great historic misfortune, all these have been for India and the Indians.**

Now coming back to our specific-context at this point of discussion, we learn from those interesting pages of the history of ancient Indian-subcontinent; that as further centuries passed, the highly acclaimed golden-era of Gupta-empire gradually declined and faded-away. Consequently, during the process of the then emerging socio-political vacuum many dynasties cropped-up for asserting their temporary sway over much smaller territories in this vast-landmass, since typical to the irrationally feudalistic socio-psychological fixations of the Indians. And they naturally became prisoners of their own regressive imperial-illusions with hardly any requisite sense of history and geostrategic prudence. In addition, they also displayed mindlessly divisive-tendencies for destructively pursuing various kinds of inter-clan and intra-clan rivalries and violent-clashes; consequently then, they themselves being driven into oblivion. In this process of highly complex historic-churning as further centuries passed; gradual dawn of another historic dynasty, took place over the ancient Indian horizon. This dynasty was established by the Vardhana-family from Thanesar near Delhi and their extensive **"Vardhana-Empire"** was gradually established sometimes during the second half of 5th century A.C.

Here too as talked about earlier, systematic historic records have not been traced by the learned experts and historians and even, valuable archeological monuments and few heritage-related coins, royal-seals and other similar objects have so far been located. But History does show that **Emperor Harshavardhana** was the most powerful and glorious ruler in the **Vardhana-dynasty**. He was fourth ruler in the Vardhana-lineage. His elder brother was perhaps first to announce himself as the disciple of **Lord Buddha**, Then history further informs that emperor Harshavardhana was dedicated to both paths of Divine-worship, as his inner-faith and spiritual pursuits. He was born in 590 A.C. and, came to the Throne in 606 A.C. Though he being the second son of his father, he was declared heir to the throne by his dying father, since his elder brother was far away from the kingdom's capital. In addition, he also came to the limelight even as a teenager prince for having proven his brilliant military-leadership in his victories in battles with aggressor Huns as well as for having driven them away from his father's empire. But at the same time, succession to the throne was not violent in any way.

His empire's capital was later shifted to **Kanauj**, then a thriving capital-city in many ways. Kanauj is located near modern city of Kanpur in central India. How **Harshavardhana** shifted his capital to Kanauj is historically an interesting story, which further highlights the magnificent military-prowess of Emperor Harshavardhana. In fact, the Barbarian Huns had earlier attacked King of Kanauj, killed him and made his **queen Rajashri** a prisoner. **Rajashri was a princess from the royal Vardhana-family and sister of Harshavardhana**. On hearing this, his elder brother **Rajavardhana** proceeded to fight the Huns, drive them away and rescue their sister. But meanwhile, Rajashri escaped from the clutches of Huns to forests, undergoing many hardships and sufferings. Rajavardhana was treacherously killed by the Huns. On hearing this Emperor Harshavardhana himself proceeded on a military-mission to save his sister and decisively punish the Huns. Timely he was able to save Rajashri from committing *Sati* as grief-stricken widow and then, began to attack the Huns as per his well-formulated military-plans. **As expected from his brilliant generalship, Emperor Harshavardhana defeated Huns in battles and chased them away; and shifted his capital to the ancient city of Kanauj**. In the process with his remarkable strategic-vision, he was able to integrate whole of north and central India up to the Vindhya mountain-ranges, into his vast empire. Then he could not proceed further south, across the Vindhyachal mountains, as he was stopped by **another mighty empire of the Chalukyas over there**.

Besides his strategic-vision and capable military-leadership Harshavardhana was also a great scholar, artist and poet with high sense of poetic literature. Historically in his royal-court too, there were a number of notable scholars, artists and poets. He was a dedicated Buddhist and as a result Kanauj also became a great centre for Buddhist learning and monasteries, besides already being a thriving capital-city of the empire. In the ancient Indian-history, Emperor Harshavardhana was perhaps the last Buddhist monarch, in addition to having excelled in many areas of human-endeavour besides the matters-military. He magnificently ruled over his mighty empire for nearly forty-one long years, from 606 to 647 A.C. Emperor Harshavardhana led a pious life and he was also sincerely devoted to philanthropic activities. Though a dedicated Buddhist, he also participated in Hindu festivals and also regularly visited Hindu places of worship. His rule was superbly humane; social peace and tranquility, communal harmony and good education were guaranteed. And at the same time, owing to his farsighted strategic-vision and deep knowledge of the matters-military, security aspects pertaining to his empire too commensurate to the level of the then civilisational advancements remained in place with strength and vigour.

It was during Emperor Harshavardhana's time that the history informs us that another great ancient Chinese scholar, author and remarkable traveller, named Hiuen Tsang visited India. His lucidly authored voluminous memoirs are great source of many fascinating information about that era's Indian-subcontinent, and also an invaluable asset for posterity. In fact, after commencing his long-journey and visiting many important places in central-Asia and Persia as well as enduring many hardships he finally entered through north-west and broadly following the Alexander's route of almost a thousand years earlier, finally entered into the Indian-subcontinent in the year 629 A.C. During his highly exploratory-stay in India for many years as a learned Buddhist-monk, besides many important places, he visited Nalanda and Kanauj as well. He has left his wonderfully detailed accounts of his observations and experiences of his travels within the vast empire of Harshavardhana. In fact during his time, it were the early years of Tang-dynasty in China and also during that era, China was already the world's highly civilised country with great culture and scholarship. Logically therefore, having himself come from a great ancient country his praise of the multiple advancements in India then, does carry a special scholarly significance.

Jawaharlal Nehru quoting Hiuen Tsang from his historic memoirs mentions that while magnificently admiring India, Indian peoples and their highly advanced civilisation, the noted Chinese scholar and traveler records that, *'With respect to the ordinary people', he says,* "Although they are naturally light-minded, yet they are upright and honourable. In money matters they are without craft, and in administering justice they are considerate. They are not deceitful or treacherous in their conduct, and they are faithful in their oaths and promises. In their rules of government there is remarkable rectitude, whilst in their behaviour there is much gentleness and sweetness. With respect to criminals or rebels, these are few in number, and only occasionally troublesome. As the administration of the government is founded on benign principles, the executive is simple.... People are not subject to forced labour. In this way taxes on people are light and the personal service required of them is moderate. Each one keeps his own worldly goods in peace, and till the ground for their subsistence. Those who cultivate the royal estates pay a sixth part of the produce as tribute. The merchants who engage in commerce come and go in carrying out their transactions, and so on. (Nehru, 1934, pp. 143 & 144) *(These sentences are those of Hiuen Tsang's own from his memoirs, from their English translations; and precisely quoted by Nehru, in his book.).*

Further, **Jawaharlal Nehru** writes that, "He *(i.e. Hiuen Tsang)* tells us that when a person fell ill in India he immediately fasted for seven days. Most people recovered during this fast. But if the illness continued, then they took medicine. Illness could not have been popular those days, nor would doctor be much in demand. **A striking feature of India in those days was the great deference and respect shown by and military men to learned and cultured people. In India and in China a deliberate attempt was made, and with great success, to give the place of honour to learning and culture, and not to brute force and riches**."

Interestingly, well it would definitely be hard for anyone including Indians, to really believe that any such historically factual accounts about India (or for that matter for any other Nation.) would ever have been written by any 'on the spot' learned observer and traveler. In today's world, one cannot even dream of that sort of popular-character, quality of life and humane-governance in any Nation, in any context what so ever.

No wonder that sometime even learned-authors and historians specially some of those from the Occident, find it particularly difficult to believe them and in this particular case have even commented that Hiuen Tsang perhaps got carried away by 'Buddhism-based' sentiments and uncontrolled 'poetic-illusions'. Well, such critical observations appear to be misplaced, since Hiuen Tsang was no novice, and also he came from no less civilised, learning-oriented and advanced country of his times. Moreover in any case, historic-facts do remain irrefutable and timeless as well. After his long stay in ancient India for many years and then subsequently, on his return to his homeland, he was warmly received by the then Chinese Emperor of Tang-dynasty; and then the emperor himself encouraged him to write his detailed memoirs based on his personal diaries and notes, for posterity.

Emperor Harshavardhana, with great governing-skill and remarkable geostrategic as well as politico-military vision ruled over his huge empire covering almost whole of ancient Indian-subcontinent for nearly 42 years and died in 648 A.C. In fact, during his time itself, the march of living global-history in distant Arabia, had begun preparing the 'platforms' for far-reaching multiple revolutionary developments as birth of a new socio-religious prophet, later on to be named as **'Hazrat Mohammad'.** These historic aspects have many dimensions, which triggered the course of entire world-history in many ways globally; such as in socio-cultural, socio-religious, geostrategic and politico-military aspects and so on.

But even earlier to them, again as centuries passed, many other civilisational developments had already taken place in huge geographical expanses of both Asian and European continents. These historic aspects too, would be taken-up even earlier to the 'Advent of Islam'; in the immediately forthcoming chapters. These exercises would further bring-in the required 'expansion', in continuity, of the 'perspectives', in their overall contexts; precisely with respect to our on-going this "Strategically Critical Journey". The naturally interlinked 'points' specifically in this regard, are the historic-complexities of these civilisational-developments covering many areas such as, multiple social-aspects concerning historic 'power-game' and so on. At the same time, this multi-dimensional 'power-game' among numerous historic international and intra-national politico-military 'centers of power'; since we do notice as always perforce getting 'cropped-up' by themselves with their own complexities.

Now coming back to the last-phase of historic Vardhana-empire again, we do also notice that with the passing-away of Emperor Harshavardhana the sun had started setting on his magnificent empire. History also records this as the beginning of the end of extensive "Hindu-imperialism" of the northern India in this bygone era's Indian-subcontinent. The genius of Emperor Harshavardhana also historically lies, at least temporarily, in liquidating many damaging-fallouts of the sinister machinations of entrenched 'Brahmin priest-class', for promoting their own selfish vested-interests. These have since been beautifully recorded in the detailed historic memoirs of the Chinese scholar and traveler **Hiuen Tsang**, who visited India during his rule.

Ironically as historic imperial-glory of northern-region of the ancient Indian-subcontinent had begun fading; its southern-region, encompassing the vast peninsular territories, south of the mighty Vindhyachal and Deccan mountain-ranges was already on rising-trajectories, towards their unaffected imperial civilisational-glories. These complex historic processes, fascinatingly did throw-up many **southern-empires with their own magnificent imperial-dynasties**. These empires had many unique historic features and in addition, their exclusive imperial-histories have been no less magnificent. They had their own glorious saga, covering many centuries, of remarkable multi-faceted achievements as well as civilisational-advancements; which were definitely not only military campaign-based but, rich socio-cultural-based as well. One distinctly visible feature in their case has been then, that their far-off location in relation to the north-western Himalayan-passes, facilitating passages for entering into the ancient Indian-subcontinent and their geography too, geostrategically remained in their favour. Further the above mentioned formidable mountain-ranges gave them many politico-military as well as a number of other long-term advantages.

Historically, who can forget the magnificent **Chalukyan**, **Pallavas**, **Pandyan** and **Chola** and also the **Kalinga-empire?** These historic southern-empires, enduringly turned-out to be the great protective-shields against the newly converted and ferociously violent as well as ruthlessly destructive Muslim Turkish-hordes whose invasion, through the same age-old north-western passes into the northern-India, had already started.

Even going back to the much older-times when the Aryans had started coming to the Indian-subcontinent and thereby, begun the gradual process of 'Aryanising' the ancient indigenous population and started settling-down in fact; much later only, they were able to gradually come in contact with peoples already living in the southern Indian-peninsula. These peoples already had their own well-rooted and distinct Dravidian-culture. Here then started a very fascinating historic-process of mutually enriching cultural interactions. This was a sort of historically unique gradual process of socio-cultural as well as civilisational-synthesis. In this historic context, we also notice a remarkable socio-cultural fusion and at the same time interestingly, both retained their certain important distinctions as well.

In the middle of South-India, there was an independently thriving Chalukyan Empire. Chalukyan's imperial territories largely consisted of the modern Maharashtra and other adjacent areas. As already mentioned earlier in fact, they were the same Chalukyas; who were successfully able to militarily prevent the imperially ambitious Emperor Harshavardhana's march from north to south for further territorial-expansion of his mighty-empire. In addition we have also noted earlier that unlike ancient-China, India never enduringly had one single central imperial-dynasty, with any such vast empire covering whole of the then Indian-subcontinent.

The Chalukyas also had a dynamic 'Naval-culture' and consequently therefore, they did develop commendably powerful Naval-fleets of those times and in addition with their mighty military-prowess; they were able to expand their empire from eastern to the western sea-coasts, in the northern part of India's southern-peninsula. Further south, another historic 'Pandyan-empire' flourished. Later on then another famous and vast 'Chola-dynasty' flourished and they too, were able to build-up significantly powerful Naval-forces, and commensurate to their historic era; they were able to effectively develop blue-water Naval-fleets.

No wonder that with their mighty 'blue-water' imperial Navy and allied with their noted military-prowess, they were able to magnificently colonise almost whole of far-eastern Asian-islands in the Pacific and, across the Bay of Bengal. This process had far-reaching socio-cultural, political and religious fallouts. In addition, these developments also triggered significant civilising-process of the then indigenous societies in these islands. Interestingly even earlier to this, historically non-violent 'Buddhist-expeditions too, were sent to these areas, including Burma, and they were instrumental in extensively spreading Buddhism in these far-eastern Asian territories. But later-on the mighty Hindu Chola-Kings sent militarily conquering imperial-military expeditions with well-conceived plans to these areas. And consequently therefore, those southern Hindu-empires too; socio-religiously as well as geo-politically began to colonise them with their overwhelming military-might and well-developed civilisational assets. More related historic details in this regard, would be appropriately dealt with, in the next chapter.

In spite of the larger socio-cultural and religious similarity, even from very ancient to epic-ages onward the Indian sub-continent never remained geo-politically a single-entity and politico-militarily united. Throughout the otherwise multiply magnificent process of civilisational-advancements of the ancient-Indians, but at the same time, territorially remaining irreversibly divisive; have most paradoxically never allowed them to achieve meaningfully gradual-evolution of the earlier already talked about the same, **"essential strategic-culture and thoughts"**. Consequently, numerous mighty Indian-dynasties naively remained short-sighted and with hardly and long-term larger and highly imperative 'geostrategic as well as related politico-military' vision. Paradoxically therefore, history also clearly shows that these perforce crippling-shortcomings triggered many tragically historic-fallouts. The central reasons have been the same age-old lingering regressive culture of inter-clan as well as intra-clan conflicts and short-sighted frictional competitiveness as well. It may appear relevant to quote few lines by Jawaharlal Nehru, "The Chalukyans had to hold Harsha in the north, the Pallavas in the south, and Kalinga (Orissa) in the east. They grew in power and spread from sea to sea, and then they were pushed away by the Rashtrakutas *(of the Deccan's of south-western India.)*. And so the big empires and kingdoms, flourished in the south—sometimes balancing each other, sometimes one of them growing big and overwhelming the others. **Under the Pandyan kings Madura was a great centre of culture.... Most of the classics of Tamil date from the beginning of the Christian era**. The Pallavas, whose capital was Kanchipura—the modern Conjeevaram—also had their day of glory. They were largely responsible for the colonization of Malaysia. Later, the Chola Empire grew in power, and about the middle of the ninth century it dominated the south.... Vijayalaya was their first great ruler....the Rashtrakutas suddenly defeated them...soon under Rajaraja, who restored the Chola fortunes.... He conquered Ceylon, and the Cholas ruled there for seventy years. His son Rajendra was equally aggressive and warlike. He conquered southern Burma, taking his war-elephants with him in his ships. He came to northern India also and defeated the king of Bengal. The Chola Empire thus became very extensive, biggest since the days of the Guptas.... But more important than the kings and their conquests is the cultural and artistic record of those times. Artistically, there are far more remains in the south than the north has to offer. Most of the northern monuments and buildings and sculptures were destroyed during the wars and Muslim invasions.... But a more remarkable arose in the south, destined to play a more vital part in India's life than all the kings and emperors. This young man is known as **Shankaracharya**. Probably he was born about the end of the eighth century. He seems to have been a person of amazing genius.... **It is difficult for an appeal to the mind and to the intellect to go far. Most people unfortunately do not think: they feel and act according to their feelings**. Yet Shankara's appeal was to the mind and intellect and to reason.... Shankara's choice of the four corners of India for his *maths*, or the headquarters of his order of *sanyasins*, shows how he regarded **India as a cultural unit**. And the great success which met his campaign all over the country in a very short time also shows how intellectual and cultural currents travelled rapidly from one end of the country to another....the colonies and settlements of peoples from south India in Malaysia and Indo-China...these settlements were deliberately organised and arranged.... There must have been frequent journeys across the seas and a sufficient mastery over the seas, to permit of this deliberate colonization simultaneously at several places....these colonies began in the first and second centuries of the Christian era. They were the Hindu colonies bearing South Indian names. After some centuries Buddhism gradually spread, till nearly the whole of Hindu Malaysia had become Buddhist.... For 300 years or so, up to the eighth century, there were three different Hindu States in Indo-China. In the ninth century a great ruler arose—**Jaya-varman**, who united all these and built up a great empire. He was probably a Buddhist.... As empires go, it was supposed to be splendid and powerful. The royal city of Angkor Thom was known all over the east as 'Angkor the Magnificent'. It was a city of over a million people, larger than the Rome of the Caesars had been. Near it was wonderful temple of Angkor Vat.... Across the sea, not very far from Indo-China, lay the island of Sumatra. Here also the Pallavas from South India had established their earliest colonies in the first or second century after Christ.... This Sumatran Empire is therefore known as the Buddhist Empire of **Sri Vijaya**. Sri Vijaya went on growing bigger and bigger till it included not only Sumatra and Malay, but Borneo, Philippines, Celebes, half of Java, half of the island of Formosa (which belongs to Japan now: *interesting—that was to the year 1945 only and at present, Formosa is an independent State with her capital at Taipei.*), Ceylon and even a port in the south of China near Canton.... The Chinese and Arabian writers of the time give us long lists of ports and new colonies subject to the Sumatran State. The lists go on increasing.... This Singapore, which is a great city now, was originally a settlement of the Sumatran colonists. The name....is a typical Indian name: Singhpur.... So the empire of Sri Vijaya was not unlike the British Empire, though of course it was much smaller. But it lasted longer than the British Empire.... Sri Vijaya Empire spread from Ceylon to Canton in China. It included most of the islands in between.... Thus while western Java was under Sri Vijaya, eastern part was independent.... Hindu State of East Java.... Thus there was rivalry between Sri Vijaya and East Java, and this developed into bitter enmity.... It was a cruel war and there was great destruction. Both the cities of Sri Vijaya and Singapore were destroyed. Thus ended the second of the great empires of Malaysia—the Empire of Sri Vijaya...in Java that the Hindu State had attained a high degree of civilisation....what it excelled in was building and specially and specially...the temples...and among these are said to be some of the world's finest and most artistic specimens of stone architecture...that both Borneo and the Philippines learnt the art of writing from India, through these Pallavan colonies. Unfortunately many of the old manuscripts in the Philippines were destroyed by the Spaniards...that the Arabs had their colonies all over these islands from the early days, long before Islam. They were great traders, and wherever trade was to be found, the Arabs went." (Nehru, 1934, pp. 147-149, 150, 154-158)

The above quoted lines from Nehru's book, do highlight many commendable facets of ancient India's remarkable history. But at the same time, almost total lack of few of the otherwise crucially important 'strategic-attributes', on the part of Indians and their very ancient civilisation so prominently 'staining' the pages of well-accepted glorious ancient Indian-history too, cannot be wished away. Contextually here Nehru's well-reasoned contention regarding **Adi-Shankaracharya**, in his own 'Divinity-based' enchanting spirituality considering "India as a Cultural-unit" nearly 1200 years back, indeed perceptively remains well-accepted as well as a historic-fact too. But this historic socio-cultural-cum-spiritual unity on the part of Indians alone, does not even remotely compensate for their paradoxically lingering and highly damaging lack of the "**essential sense of history as well as territorial-geography**" and thereby; triggering persisting absence of the same **geostrategic and politico-militarily vital, "socio-political unity"**.

And hence, history shows that since the very ancient times these lacuna in turn, have perforce been causing many inglorious-fallouts, tainting the historic images of the earlier-era Indian monarchs/rulers; and then later-on, including post-1947 democratic-India's 'politico-bureaucratic class' so naively 'arrogant'. Then separately on their own, modern India's military-commanders mostly found lacking in the higher-level 'professionally perceptive-vision'. In this specific context, ironically an obvious question in the larger-perspectives, arises that even now in the beginning of 21st century as to have these embarrassing situations been substantially altered for India's larger long-term multiple-betterment and if so then, how much in the areas of top-level 'decision-making process' as well as strategically in what precise-manner concerning the methodologies?

Now again coming-back to above topic of our discussion here we may also say that, whatsoever happened has become part of history and naturally therefore, they are needed to be taken as such since, historic facts in any case, will always remain unchanged. This aspect of fruitfully accepting the concerned historic-facts would always perforce require the same "essential sense of history as well as territorial-geography" in this regard as emphasised earlier. In continuation therefore, if these imperative aspects are adhered to in the spirit of due pragmatism and with open-minds then gradual multi-dimensional and far-reaching positive-processes of simultaneously evolving, cultivating as well as imbibing the same "essential strategic-culture and thoughts" would get effectively triggered, since vital for the Indians and their now fast developing Nation-State. The logical points in this perspective, appear to be that their long and historic absence from the otherwise remarkably unique manifestations of the ancient Indian-civilisation as well as historic-processes of governance in so many Indian-empires/kingdoms, down the centuries; have caused both geostrategically and politico-militarily, multiple harms/disasters time and again.

As emphasised earlier, the moot-point remains that history never spares any human-societies or Nation-State and hence it would naturally be naïve to think that the Indians would be any exceptions anywhere. In this context paradoxically, the more embarrassing-reality so clearly visible, remains that even after the year 1947 in modern-India, and now as a liberal-democracy nothing much substantially appears to have changed. Here it appears relevant to refer to the irrefutably 'factual-lines', from a scholarly and thought-provoking article of Shri Abhijit Bhattacharya, on "The Victor of 1971: Glorious Triumph of a historically Inglorious Military"; since already quoted in the pages 157 and 158 of this book, naturally bringing-in these otherwise factual-perceptions in a more 'focused-perspective'.

It may appear to quote few relevant by a well-known former Indian-Army officer, political leader, scholar and diplomat **Shri Jaswant Singh** wherein he so informatively writes that, "Two axioms, seemingly contradictory, open our enquiry on India's strategic culture. The first, that warfare is not the concern of the military alone. This is a self-evident and oft-repeated truism. But not so the second: the proposition that strategic decision-making is the function and responsibility of a small political-military class alone. Thus for this class of leaders a culture of strategic thinking emerges as an irreplaceable attribute, no matter what the form of governance. But was of them alone? Principally, yes; for though the civilisational, cultural and societal influences shape the thought processes of the leaders, entire societies or a people can hardly be expected to be endowed with a culture of strategic thinking.... There is another factor: of geography, of a sense of territory. Indian nationhood, being largely cultural and civilisational, and Indians being supremely contented with what was theirs feared no loss of it, for it—the civilisational—was as unconquerable as is the spirit. **Thus both were absent: a territorial consciousness, and a strategic sense about the protection of the territory of residence.... This absence of a sense of geographical territory persisted from ancient India down the ages, to the medieval period and even later.** Of none of these periods does there exist even a suggestion of a reliable map, even though there are descriptions, but with only the unmistakable landmarks being fixed—say, of rivers, the mountains, or the ocean. It is only during the British reign that territory, its survey, settlement and mapping came recognizably in India." (Singh, 1999, pp. 1, 16 & 17)

Well, what could be paradoxically and also singularly more 'historic-irony' of India and the Indians, as so fascinating and colourfully ancient-civilisation; since the dawn of various human-civilisations in this world? Further, democratic India's "politico-bureaucratic-class", always arrogantly intoxicated by power do not hesitate to gloss-over them for 'comically facilitating' perhaps their own fixations for wishful-thinking, and also coupled-with fictitious 'make-believe world'. Bertrand Russell is understood to have also said that, power corrupts and absolute-power corrupts absolutely.

Well, contextually can't we again assert that historic-truths, whether sweet or bitter, remain nothing but undeniable-truth. But modern India's political-class, whether ruling or in opposition, have also amazingly cultivated a typical habit of mostly remaining in an unchanging 'denial-mode'; since as dictated from time to time, by their respective situational 'vested-interests'. For them what all matter, are their own personal-interests, Party's interests as dictated by their own high-commands, and in addition also, interests of their lobbyists as well as hefty financers. With such leaderships always in place, it remains an uphill task to initiate a multi-dimensional process of gradually cultivating and imbibing the same all encompassing "essential strategic-culture and thoughts".

Interestingly in this specific context, it appears befitting to also add that, these aspects of multiply lingering extreme-negativities are not only 'India-centric'; but in addition, very much present in different countries too, especially in certain African and other Asian countries as well. It is altogether another matter that, they are there again in different situation-specific degrees, compositions, and complex-manifestations, having different socio-psychological and cultural aspects, as well as processes of governance. These aspects are indeed, highly complex in many ways. Their detailed analytical-studies would naturally be much beyond the scope of the sub-matters of this book.

(f) Semitic and Aryan: 2800 to 1925 B.C. onward

During the very distant-past, migrating-Aryans from Central-Asia and also now from the modern southern-Russian territories, moved in different columns and in different directions; for colonising and settling-down in different ancient geographical-territories, since already discussed in some detail earlier in the beginning of the sub-chapter No. 5 above, on the ancient Indian-civilisation. It is also well-known that in these far-reaching historic-processes one column of the migrating-Aryans moved towards the ancient middle-east Asian territories. And here they came in contact with Semitic-races, since perhaps 'original' inhabitants there. Their mutually intense interactions, historically resulted in gradually transforming the whole region in such far-reaching manner that, many noted scholars call it the 'cradle of early human-civilisations'. No wonder that many of those scholars and experts call them as a distinct 'Semitic and Aryan' civilisation.

They were also gifted with certain specific natural-barriers, readily available formidable defenses to act as effective deterrent; especially against any possible invasion by the moving Mongolian-tribes and storming into their geographical territories. We notice that this particular natural-deterrent was in fact, the mountain-ranges extending from Kashmir to the Caspian-sea; since already located so strategically on their eastern-side. **Their geographical territories also valuably included the fertile-valley of the well-known Mesopotamian twin-rivers of Tigris and Euphrates. In addition, these valleys are appropriately called as fertile-crescent as well.** Further their extended-territories, so strategically located between Asia and Europe, have also been found to be of great politico-military importance later by the competing European imperial-powers during the modern-era period of their far-reaching colonial-ventures for their respective empires-building. These by themselves are a combination of historic-developments with multiple-fallouts, which ultimately changed the entire course of continuously advancing man's living-history. Interestingly, the historic Industrial-Revolution also, commenced during this period itself. In addition, two highly violent and destructive World-Wars, indeed unprecedented too, took place during this historic-period as well.

Interestingly, historians have mentioned in their scholarly work that, if an 'imaginary-line' is suitably drawn from the Caspian-sea in their north to the Persian-gulf in the south then the eastern-areas of this 'line' may be identified as the Indo-Caspian belt, historically acting as a civilisational nursery then, breeding the ancient Indo-European race and also consequently, inhabited by the Medes and Persians historic off-shoots of the Aryans. And then the western-zone of this substantial territory, indeed commanding great strategic-values; consisted mainly of Asia-Minor and Mesopotamia. At this juncture of our deliberations, another fascinating historical aspect is noticed, that is, the western-zone has been territorially shared by a number of 'sub-branches' of the same Semitic-race. And at the same time, each of them having distinctly different culture, religion and ways of living with separate socio-political groupings and methods of their governance. Historically, one may also say that they had then, constituted a highly complex as well as fascinating civilisational mosaic in that strategic-region so much so that far-reaching multiple epoch-making fallouts were gradually triggered, in course of many centuries. These aspects are again by themselves vast subjects of historical studies and research, meant for experts with their own specialised-programme in this regard.

Now coming back to the various Semitic-tribes inhabiting their own controlled areas as part of the same earlier identified 'western-zone'; according to Professor S.R. Sharma, we find them to be **Semitic-Hebrews, Armeans, Phoenicians and, then the Aryan: Hittites, Medes and Persians** and so on. These historic-phenomena had been taking place much before the advent of Islam and interestingly further, they also had their own distinctly different developmental-contributions towards the remarkably multiple advancement of a number of ancient human-civilisations; the then taking birth in their age-old strategically situated geographical-region, since already explained earlier. Then we also notice that the historic-birth of Islam having taken place in the same Middle-eastern region of Saudi-Arabia during little more than one and half centuries before and this in turn had many complex socio-cultural, socio-religious, geostrategic and politico-military fallouts, with long-term and far-reaching historic consequences. **These aspects would be covered in the later sub-chapter on 'The Advent of Islam'.**

We then learn from the history that as Islam spread in this strategic region as well, most of the Semitic-population in this strategic-region got converted to Islam and were also made to convert, except one distinct die-hard, tenacious and proud minority known as the same Semitic-Hebrews. This is another distinctly remarkable feature of the ancient civilisational-history of this strategic-region. This particular historically ancient-tribe is now also known as the **Jews or Israelites**. With all their modern-age progressive dynamism they too, are deeply religious and highly proud of both their ancient language and religion. Their distinctly ancient and separate language with different alphabets is known as **Hebrew** and then, their religion is called **Judaism,** which is the world's second oldest religion after Hinduism.

In the present modern-era, historically one highly complex and disturbing fact for the entire humanity has emerged more pronounced that, after the creation of the independent sovereign State of Israel in this extremely volatile strategic-region in the year of 1949 A.C. under a specific and internationally 'binding' U.N. Security-Council Resolution, unfortunately there have been a number of wars between Israel and the middle-east Muslim-States and in addition, a perpetual state of animosities and conflicts as well. Moreover the problems are becoming as complex as days keep passing by, that they are defying any internationally worked-out solution. Interestingly both sides historically belong to the same ethnic-stock, but there ends the similarity; and then a highly complex and seemingly unending tragic-scenario of all sorts of again appearing to be irreconcilable-conflicts, begin appearing on the historic horizon of this strategic-region; since already having a chequered-history with a saga of great human-sufferings and inhuman-tragedies. Long before the advent of Islam, in fact around 2000 years back, the Jews' territories in and around what happens to be now the modern sovereign-State of Israel, they were overpowered and driven-away from their settled homes and dispersed.

During that ancient-period, when the Jews were violently chased-away from their homes; they dispersed in almost all corners of Europe and Asia, a long period of their great sufferings, persecutions, violent killings by others in those regions (of mainly Europe), and extreme humiliations, perforce began. Their forced dispersal from their then ancient homes, resulted in great migrations into the new locations with indescribable-hardships for all those Jews including their women-folk and children as well. As would have been naturally inhabitable under such compelling circumstances for anyone so in case of those Jews; many perished on the way, many were killed by the rival and/ or hostile ancient tribes, during their journey wherever they became vulnerable as well as on arrival in their various new locations. In any case, whatever the cost thrust-upon them from time to time, both in terms of lives and their meagre-belongings they never gave-up; in fact, always preserved and carried with them with their unshakable faith and respect, wherever they went—their sacred books and symbols of **Talmud, Torah (a set of five books)** and **Cabala (the book of mystics)**. Therefore at this historic-point, it appears to be contextually important to bear in mind an exclusively unique age-old aspect regarding the Jews' mental, religious-cum-spiritual as well as their socio-cultural 'make-up', both individual and collective unwavering tenacity and determination, irrespective of the any kind of forced-circumstances thrust-upon them. These qualities have certainly have hardened them in many ways and at the same time substantially contributed in their gradual multi-dimensional evolution socio-politically as well as imbibing the earlier mentioned "essential strategic-culture and thoughts". As if such adverse situations and unprecedentedly prolonged and unimaginable sufferings were not enough that during the present modern-era their hardships and persecutions, especially before and during the second World-War, further increased manifold unfortunately for them in Europe, more so in hands of Nazis in Germany.

During that period, the war history chapters on **German politico-militarily fanatic Dictator, Adolf Hitler** and his terrifying concentration-camps and 'Gas-Chambers'; meant for the demoniacal mass-killings of the Jews, including their women-folk as well as innocent children are well-known and authentically available. These are indeed, unsurpassed 'black-spots' on the long and continuing saga of the human-history. The logical-point is that factually they were cold-blooded and mass-scale murders and hence obviously, cannot be bracketed with the war-related military and civilian casualties. In this respect, volumes of history on geostrategic, political, and also military are available for the interested readers but in addition, for maintaining the comprehensive-objectivity in such studies; contextually here it is no less important to also critically look into the desirable Jewish-versions as well. **In this historically context, one may like to read a very 'spine-chilling' and engaging book giving a detailed account, interestingly titled "EXODUS", written by a well-known American author, Leon Uris.**

No wonder that by the time the World War-II was ending in Europe, with the forthcoming final defeat of Nazi Germany, **Hitler's suicide** and unconditional surrender of the top-level Nazi military-commanders, Jews had already got organised from all over the world for sustained and relentless struggle for getting back their holy lands on the eastern-coast of the Mediterranean-Sea and north of the adjacent African coast. Jews were since disciplined with an unshakable-faith in their religion and hence they had already started moving towards their above identified ancient holy-lands, facing all sorts of obstacles, both natural as well as man-made, and again deaths, destructions and untold sufferings. Their organisers had since already evolved a daring strategic-plans and put them in place, with prudent-vision and tenacious will-power; as they had begun sensing the emerging opportunities. These processes ultimately resulted into the second modern-era colossal 'exodus' in the reverse direction, after more than 2000 years, since known as another unique historic-episode; and finally resulting in the formation of a new independent sovereign State of Israel there, under a U.N. Security-Council resolution with a revolutionary success.

In *EXODUS* the author regarding the very **ancient city of Jerusalem**, when Israel was just born in the year 1949 and then only may be a couple of years after, writes that, "How strange the place is.... Here so far away in those barren, the merging point of a hundred civilisations in its thousands of years. Of all the earth, why this place, this street, this wall, this church? Roman and crusaders and Greeks and Turks and Arabs and Assyrians and Babylonians and British in the city of maligned Hebrews. It is holy, it is sacred it is damned. Everything strong and everything weak, all that is good in man and all that is evil in him are personified. Calvary and Gethsemane, and the room of the 'last Supper' happen to be there as well. The last supper of Jesus....to the holy Sepulcher, the site of the crucifixion and the tiny chapel lit with ornate hanging lamps and perpetually burning candles over the marble tomb of Jesus Christ....northward into the Galilee....through the timeless Arab villages into the fertile carpet of the Jezreel Valley, which the Jews had turned from swamp into the finest farmland in the Middle East. As the road wound out of the Jezreel towards Nazareth again....moved backward in time. On one side of the hill the lush lands of the Jezreel.... Nazareth was much as Jesus must have found it in His youth.... The task of regenerating this miserable land seemed nearly impossible, but it could be done if there were only more Jews with a fighting spirit...that more Jews had to Palestine come sooner.... Everyone recognised that something was missing that was not in the Talmud or the Torah or the Midrash or the Mishna." (Uris, 1959, pp. 333, 334 & 218)

Amazing, isn't it? **That is the unique and ancient city of Jerusalem and in addition the surrounding territories, now constituting the modern sovereign State of Israel; and also decidedly one of the most dynamic Nations in the present 21st century world**. It is indeed an unparalleled saga of a relatively small but, fast advancing Nation; now so modern, so assertive for their self-defence and geostrategic aspects of their multiple-security, so westernised and yet, so devoted to their own ancient-religion as well as age-old values. In addition, more focused as per compulsions of the fast-altering their region-specific perpetual game of international power-politics, again so high in their scientific-temper coupled with wonderful spirit for technological innovations.

Multi-dimensional highly praiseworthy results achieved by them in a very short period of the birth of Israel have indeed, now placed them in a relatively high-pedestal in the modern-era comity of Nations. These are no mean achievements. No wonder that Israel always maintains herself in a multiply commanding strong position, with a high degree of strategic-confidence. For India and the Indians in the present-era, Israel has become very important, since is unreservedly providing certain high-tech military hardware and software as well as similarly valuable support in dry-farming and drip-irrigation and so on, at mutually very advantageous costs. Hence, internationally it appears imperative and also in the larger interests of the humanity, since prudence demands that Modern Israel may be multiply-assessed and accepted with modern-era strategic vision and objectivity; rather than living rigidly in the past with perhaps any "time-warp fixations". So in any case, that remains that.

Globally well-accepted norms and necessary politico-diplomatic checks and balances do remain equally applicable to the all-big and leading as well as small nations of the modern world and at the same time; honestly justice-oriented fair yard-stick may be necessary to apply to all nations with equal norms. But at the same time, international compulsive situations always do remain perforce delicately-complex and never simplistic and also, each Nation's own politico-economic as well as geostrategic compulsive situations too, are different and therefore, they always need to be handled with care and impeccable objectivity. Administration of justice not only must be fair and impartial, but seen to be so and, same goes to the international-arena as well.

Interestingly, history of the Jews goes back to even the much earlier times in the distant-past, pre-dating the Bible's Old-Testament. If the Jews' ancient history as well as certain important and equally ancient civilisational aspects are even briefly reviewed from the historic days of Moses to Solomon and onwards dating back to nearly 1480 to 1020 B.C. and further as many centuries passed we would notice that according to **Professor S.R. Sharma**, the Jews started facing many types of hostilities and violence. Historic date-lines may not be always possible to get authenticated with precision but historians do anticipate that Judaism, as a distinct religion, might have made an initial beginning nearly 3000 B.C. or so. Similarly Hittite as a tribe too might have come on the scene from nearly 2800 B.C. or so.

In any case, besides such ancient-datelines being largely matters of conjecture; one aspect remains clear and as already emphasised earlier, that all these historic-tribes belong to the same stock, except the Persians, and, i.e. the Semitic-race. Persians were from the Aryan-race. In addition then even in those ancient times, their trades followed with distant lands, including the ancient Indian-subcontinent and so on. In this regard, even during Jewish monarch Solomon's reign, the Jews were permitted to do mining of precious metals, and also they were required to pay taxes to **King Solomon's royal-treasury**. Regarding their trade activities, and 'so-called' ancient Biblical *(Old Testament)* warnings, **Prof. S.R. Sharma,** in *A Brief Survey of Human History*, mentions that, "....besides creating state monopolies of trade in yarn, horses and chariots. He also (i.e. King Solomon) imported articles of luxury—ivory, apes and peacocks—from distant countries like India. But it was all a nine days wonder: for on the tenth day came death and devastation: 'vanity of vanities and all is vanity'."

Interesting, isn't it? Now coming to the present modern-age, while contextually scanning through some Israeli publications, with some strategic-criticality; one may come across a very informative official-publication, titled, '**NAHAL**' containing lot many interestingly revealing photographs, along with text prepared by **Mordechai Noar** and published by Israel's Ministry of Defence, Tel Aviv in the year 1970: rebuilding of modern and fast developing their sovereign Nation in mid-twentieth century onward. With precise narrative and revealing photographs, this booklet informs about the successful execution of their projects through unique on the job experimentation of putting young soldier-farmer as well as simultaneous nation-builders of youths consisting both highly motivated and patriotic young boys and girls. Israel's these all-inclusive disciplined youths set a unique example by simultaneously undergoing military training, devoted farming activities; they have been pursuing academic-education as well.

Then another booklet giving short and crisp, bird's 'eye-views' on Jews' long history from 10th century B.C.E. onward, is interesting and informative too. The title of this booklet is 'Secure and Recognized Boundaries: Israel's Right to Live in Peace within Defensible Frontiers'; A Carta Publication, 1971, Jerusalem, Israel. Though this too is again an Israeli-publication, but this also does contains a number of geographical-maps along with some factually relevant data as well. Hence Israel's neighbouring Arab-Nations may even 'counter' these perceptions, but the counter-arguments are expected to be historically authentic duly backed again, by relevantly factual-data as applicable. It is also true that such publications anywhere, may also contain both some propaganda or counter-propaganda matters and therefore, they have to be discretely set aside with necessary strategic-criticality and prudence here for our own contextually desired study and analysis. Further, it may also be considered imperative to emphasise here again that, in our ongoing 'Strategically Critical Journey'; it remains always essential to take the utmost care to proceed in our studies, avoiding any even remotely possible 'controversy' but at the same time; while consistently adhering to ensure objectivity, one may never feel shy of 'calling a spade a spade' and remain unwavering in explicitly expressing the facts, with no subjectivity at all.

In its 'Introduction' (above referred booklet shows that, "Israel's history in the Middle East goes back to the dawn of recorded history. It is the only country in the region whose people live in the same area, speak the same language and maintains tradition and memories of three thousand years ago. It is only under Jewish rule and Jewish sovereignty that the area in question maintained its uniqueness or, indeed its own geographic entity. It was conquered and reconquered no less than fourteen times in thirty centuries, yet no conquest—with the brief and solitary exception of the Crusades—resulted in more than administration from without. Throughout this period and even in spite of enforced dispersions the Jewish community maintained its presence in the Land while Jews dispersed abroad kept an unbroken link with their Homeland throughout the generations, praying for the redemption and celebrating its seasons and remembering its map and geographic features. The Land as an entity thus goes back to biblical times." (***Magnificent historic facts indeed—Jews are great!***)

In this context now, we may also need to bear in our thoughts that Jews were not the only earliest people of that extended-region, known as the 'cradle of earliest human-civilisations'. Further, now these geographical-areas are since commanding great importance; as per both the geostrategic and politico-military considerations, for all the leading-Powers and global-Economies in the present modern-era. In this respect besides its unique location the Suez-Canal, the near-by Mediterranean-Sea, and consequently trans-Euro-Asian very busy sea-lane-passages and then, also the substantial **'oil-wells'** located there, happen to be the overriding geostrategic as well as global-economic reasons for these phenomena of multiple-importance. In addition though, there may not be unanimously identical perception now, particularly on one specific 'line-of-thought' and interestingly, that happens to be the oil-wells' related dimensions of importance. And that would remain so, till that region's oil-stocks are not exhausted and at the same time till the multi-disciplinary sciences and technologies are not developed and perfected–leading to the revolutionarily viable and globally eco-friendly alternative as well as relatively perennial sources of energy for the multiple uses of the entire humanity at large.

Now again coming to the aspects of earliest peoples belonging to the both Semitic and Aryan-stock, inhabiting the above-mentioned ancient and geographically extended region; we have since noted that besides the **Jews, Armeans, Phoenicians, the Aryan Hittites and Persians** too, had settled there. As earlier noted, like the ancient Hebrews or Jews in fact belonged to the Semitic race, so were the **Armeans** and **Phoenicians**. Since as mentioned earlier, any ancient particular civilisational advancement, naturally have had very complex-roots as well as causative-reasons in this remarkably slow and gradual-process and at the same time fascinatingly, they have certainly not been identical and common, for all different civilisations in this world. Hence, if one may care and try to even somewhat observe and study them with desired strategic-criticality then the astonishing and engaging complexities of that unique, slow and gradual multiple-process, with respect to any specific ancient-civilisation would fascinatingly begin emerging. Contextually again, we may also say that any such ancient human-civilisation in the process, does simultaneously evolve certain special and exclusive attributes as well. This too, is another remarkably interesting-facet, so special as well as different; again with respect to any particular civilisation or even a part there of, as visible in any geographical region of our world.

Professor S.R. Sharma mentions that, "**They were noted as carriers of trade—the former *(i.e. Armeans)* by land and the latter *(the Phoenicians)* by sea**—rather than for anything creative of their own. Damascus was the most important centre of Aramaeic activities: They carried on trade between Mesopotamia, Phoenicia and Egypt. They are credited with having supplanted the cuneiform and hieroglyphic writing by the alphabetic. The Aramaic dialect developed into the international language of the Levant: and that too was the language which Jesus spoke. The Phoenicians were better known to the ancient world as their activities were spread over a wider area. Their land was enclosed between the Lebanon Mountains and the Mediterranean Sea, but they created ports wherever they traded.... The antiquity of such activities may be determined from the fact, that the Tell-el-Amarna tablets mention Tyre as already of great importance in the fifteenth century B.C. and it was also flourishing under **Hiram I**, a great friend of Solomon *(as already mentioned earlier, a famous Jewish king of pre-Christ era)*, in the tenth century B.C. The Phoenicians are mentioned in Bible, and the ancient Greeks believed they owed their alphabet and weights and measures to them. Perhaps the earliest specimen of Phoenician writing is that found on a bronze cup of **King Hiram I (c 960 B.C.)**.... Indian scripts, both Sanskrit and vernacular, are equally supposed by some to have originally derived from their writing. These strange symbols are the most precious portion of *India's* cultural heritage. Indians gave them ivory, cotton fabrics and spices in exchange. The introduction of these and pen, ink and paper into Europe are attributed to the Phoenicians." (Sharma, 1948, pp. 111, 112 and 113)

Interestingly further in this respect, within specific-framework of the principal subject-matter of this book we also learn from the same living-history that, the larger dimensions of **'matters-military'** have been continuously developing in tandem with gradual advancements of various ancient human-civilisations in this world. Then in addition, history further shows that these multi-dimensional continuous developments of the matters-military, especially in precise contexts of the larger-humanity, have positive as well as negative aspects, or perforce both simultaneously being visible in these on-going complex processes. We have also seen that their allied and complex geostrategic as well as politico-military aspects too, have been multiplying manifold with expanding consequences. Again at the same time, they were present during ancient-era as well, but in relatively much more simplistic as well as limited pattern, and also with different manifestations, as would now be seen. Of course, these multiple-issues would naturally be dealt-with in much more detail, in the next chapter.

Then very valuably we learn from the same earlier referred book of Professor S.R. Sharma that, another learned-expert, **Winwood Read**, most probably a British national, gave a highly informative 'graphic-account' of the ancient-era multiple-activities of Phoenicians. Since probably from the earliest of times of their civilisational existence these ancient Phoenicians were fond of carrying them through their suitable sea-routes and hence at the same time, for significantly facilitating their trading-activities they strategically built many posts along their sea-routes, wherever they traded. From all possible critically evaluative concerns, commensurate to those phenomenal historic-times; these were definitely no mean achievements. In addition, for them to suitably equip themselves in this regard they had to be highly skilled sailors as well navigators and also, necessarily having developed the desired-expertise in building the then, their sea-worthy boats.

In this respect then, we also notice another negative attribute in those ancient Phoenicians that from being remarkable sailors, they gradually got transformed into sea-pirates. This likely happened somewhere around 1500 to 1000 B.C. and could even be the earliest historic-record of the culture of sea-piracy. But at the same time the ancient Phoenicians then, couldn't have indulged in sea-piracy, if they didn't possess the requisite naval military-fighting expertise, commensurate to their mission. Then we also notice another interesting dimension that from simple sea-based traders, they became competitive innovators-cum-manufacturers as well. We are also informed that, naturally in spite of certain obvious constraints since always imposed on man, the ancient Phoenicians were remarkably able to develop some captivating dyes and colours, from natural sources as befitting raw-materials: bringing-in handsome returns and wide publicity resulting in multiple-expansion of their trade.

They were perhaps the first in historically evolving and practicing the novel concepts of innovating 'sales-bazaar' for their new and attractive goods; wherever they sailed and set-up camps for their substantially-profitable trade. Further, on their arrival to a new location, they used to light-fire for attracting potential customers, who mostly then lived naked. For their multiple-goods, they even set-up their primitive 'factories'. All these they did with remarkable vision and in addition, they always sailed for well-planned trading with their 'need-based' military-might.

But then the 'silver-lining' regarding Phoenicians' ancient commercial activities ended there. As per above cited-source, their accompanying so-called 'need-based: military-might' was not always so: as this also had their certain evil-intentioned sinister dimensions as well. Paradoxically sometimes while returning they did indulge kidnapping and carrying selected women and children with them.

Now coming to the Hittites and Persians, history shows us that they as tribes belonged to the Aryan-race. Though, there happens to be an opinion that, the Hittites were of mixed-blood; between Semitic and the Aryans. Well in this too, there could be some truth; since as many centuries passed and migrating tribes, belonging to different tribes, began moving into various Geographical regions of the world naturally there had to be some gradual intermixing of races, and consequently distinctly different tribes got evolved in the concerned locations. Interestingly, based on certain fairly 'conclusive' evidences learned historians and archeologists have opined that the ancient-Hittites had already achieved a remarkable feat of firmly establishing a distinctly their own 'great-civilisation', in their region of decisive geostrategic as well as politico-military dominance since a very distant-past; and that could clearly be around 2700 to 3000 B.C. These aspects have since been firmly established by the leading archeologists by well-planned programme of unearthing their magnificent monuments, massive-constructions as well as other singularly great valuable relics. These in turn admirably speak of their special engineering-design acumen and great construction skills by expertly using very large stone-blocks, without applying any kind of mortars and they have successfully withstood the ravages of nature over thousands of years.

Further, whatever historically may be said and also have expertly been opined, one aspect admirably stands out that, commensurate to their very ancient civilisational-era they had marvelously evolved and imbibed their own outstanding 'essential strategic-culture and thoughts'. And these in turn naturally made them politico-militarily very ambitious for territorial expansion for imperialistic-dominance, even during that very remote-past. Naturally, as a consequence, these uncontrolled ambitions made them militarily-aggressive as well. They came in violent clashes with equally ancient Egypt as well as Assyria. Some informative-maps displaying the geo-political state of the ancient-world, indicates the region located somewhere around south-west of the Caspian-Sea and northern side of the well-known Tigris river. And obviously the then location too, gave them strategic-importance.

In their relentless military-mission aimed at maximum possible territorial-expansion; the Hittites even Babylon quite early in 1925 B.C. itself. Facing repeated Hittite military-offensives in that distant ancient-era; had perforce transformed the Assyrians into a militarist-State for its effective defence, as feasible. The Hittites continued with their insatiable ambitions and invaded the ancient Egyptian-empire and, temporarily did make quite aggressive in-roads, around 1370 B.C. During that fateful period for the Egyptians, since militarily-weak **Emperor Akhnaton** was ruling and therefore, strategically situation turned quite favourable for the aggressively-militarist Hittites, as the emperor happened to be soft and 'naively' idealistic Pharaoh. But that is only one-side of the story in a context, since Akhnaton a great king and reformer as well. Whenever, we review the historic monarchs and rulers since ancient times, with requisite strategic-criticality; we notice their personalities having many facets and also, this interesting aspect is universally true.

At that point of time, the Hittites then did achieve military-victories and, they captured fairly substantial territories and brought them under their control. But in any case, all this didn't continue for long and geostrategically tide turned against the Hittites. They met more than their match in **Rameses-II**, who forced the Hittites to enter into a Treaty on his terms, after their clash in 1295 B.C., since historically known as the battle of Kadesh. Further, one singularly important development were already taking place even much earlier to these historic happenings, since Indo-Aryan columns had moved deep into these regions and settled there and consequently Hittites too, are principally known as of Aryan-race. No wonder that around 1400 B.C., the very ancient Hittite-inscriptions now unearthed revealingly mentions names of some Gods since originally belonging to the Indo-Aryans, such as Indra, Varuna, Mitra and so on and in addition, many Hittites spoke a dialect, again belonging to the already historically quite developed and fascinating Indo-Aryan languages.

Now coming to the matters military of ancient Hittites, we learn that even during those days they had tactically fine-tuned many facets of their 'profession of arms'. These imaginatively included the aspects of command and control, administration, training, discipline and areas of logistics as well. They may even sound 'odd' today, but they may befittingly need to be viewed in the context of man's long-past ancient-history. Then ancient Hittites' even more remarkably striking-features in this respect are noticeable in their astonishing military acumen in building tactically well-conceived network of situation-specific and well laid out field-defenses. They imaginatively included military-engineering dimensions as laying-out camps, multiple-fortifications, tactical design of ramparts, ditches and trenches and catering for crucial compulsions of guard-duties, cataloging and indexing of books with details of the concerned authors, laying of obstacles and their multiple strengthening, for effectively facing enemy-assaults. Here we may wish to note that **both the Persians and Medes were Aryans**. Historically it remains a different matter that, the ancient Greeks didn't differentiate between them and so for their own civilisational as well as geostrategic purposes, they indeed kept them 'bracketed' together. The Medes had established their capital in the ancient Ecbatana; and the Persians made the capital-city of their Empire, at ancient Susa and then subsequently, they shifted their capital to now more historically famous more famous Persepolis. At this juncture, we may like to mention here that interestingly, modern Iran as a sovereign-State, now carefully preserves and maintains the same ancient Persepolis and its relics, as a valuable heritage-site and understandably now, it has become a place of great international tourist-attraction.

The ancient Persians also had a couple of great emperors such as the Cyrus, David Cambyses and then, Solomon Darius; and they all ruled over the ancient Persian-empire much before the historic beginning of the Christian-era. In fact, they all were also noted warrior Kings and believed in the territorial expansion of their empires and in addition, also they were well-versed in the multiple-aspects of geostrategically crucial '**matters-military**'. Even at the risk of repetition, we may wish to emphasise that, these crucial aspects of the same 'matters-military, are indeed always timeless and therefore; even in the present modern-era, they remain the same.

The **Persian-king Cyrus** during his regime at the opportune time integrated the Medes, both socio-culturally and geo-politically, into his dominating empire. He attacked Babylon and also in course of his campaign, as per his own strategic-vision; helped the ancient Jews to reclaim their socio-psychologically as well as religiously the most dear **region of Zion**. This is located on the east-coast of Mediterranean and hence, now also known as the Palestine and where the modern sovereign-State of Israel has been founded; under an historic UN-charter, as per the referendum held in the then UN General Assembly in the year 1949. Well, during the ancient-era, Cyrus achieved this feat by then militarily defeating the proud ruler of the region in the year 546 B.C. and consequently, the ancient Jews too got magnificently benefitted in their own way. At the same time, this successful military adventure of the Persians, got them in direct contacts with the ancient Greeks and their own magnificently powerful Greek-empire. In fact, that's the way, the living-history keeps marching ahead over the all-encompassing and unending 'canvas of time' and then simultaneously; many far-reaching events keep taking place, all over the world.

In fact, as a fairly common factor then, during the ancient Persian imperialistic-era too, it was all an age of imperial ruthlessness and human-cruelty of man against the fellow human-beings. No wonder, that when the mentally unstable **Persian-Emperor Cambyses** committed suicide, a sort of revolt took place in the various provinces of the empire. The successor Emperor Darius then, ruthlessly suppressed the revolts, with no mercy shown to any one at all. Even earlier, during the sacking of Babylon, more that 3000 of important Babylonian citizens were ruthlessly done to death by cruelly putting them to the crucifixion: those were the days and such were the mindlessly inhuman imperialistic practices. **But the point is that why alone blame the ancient world? Can the man, even in the present high-tech modern-era, ever claim to be faring any better? What about various deeds/misdeeds happening today? Therefore, the same weighty question keeps coming back that can the man even remotely and objectively claim to be truly 'civilised'?**

Since the Persians were racially the same as Indo-Aryans; the ancient Persian religion too, was largely akin to the ancient Hinduism. All these were long before the advent of Islam. Their ancient most sacred religious treatise is Avesta written in their own ancient language Zend, and the same is similar to the ancient-India's Rig-Veda. Their great and highly revered Prophet happens to be Zarathushtra. Present-era's Parsis and Iranis are the only survivors of this ancient religious-faith. Now, again coming to the historic contacts between the ancient-Persians and the Hellenic-world, i.e. the Greeks, as spelled-out earlier; these contacts were never for any peaceful purposes and hence there were many wars too, between them. Many noted and learned historians have given vivid accounts of these military-campaigns on the part of both the Persians and the Greeks. Most famous one is the valiant defence of the ancient Greek-empire by their most brave and fearless warriors known as **gallant 300 Spartans**; who magnificently fought the Persian invaders, in the year 480 B.C. This is a unique episode in the long military-history since ages indeed, with few parallels anywhere in the world.

Jawaharlal Nehru has narrated this, as a vivid 'picture', "The Hellenes or Greeks were frightened at this great host, and forgetting all their quarrels, they united against the invader. They retreated before the Persians tried to stop them at a place named Thermopylae. This was a very narrow path, with the mountain on one side and the sea on the other, so that even a few persons could defend it against a host. **Here was placed Leonidas with 300 Spartans to defend the pass to death**. Right well did these gallant men serve their country on that fateful day, just ten years after Marathon. They held the host of the Persians while the Greek army retreated. Man after man fell in that narrow pass, and man after man replaced them, and the Persian army could not advance. Leonidas and his 300 comrades lay dead at Thermopylae before the Persians could go ahead....and even today one's heart thrills to think of this unconquerable courage; even today the traveler to Thermopylae may see with tear-dimmed eyes the message, engraved in stone, of Leonidas and his colleagues:

Go tell to Sparta, thou that passesth by, that here obedient to her words we lie."

So that is that, and the magnificent words coming out of the 'mighty' pen of Nehru. But what happened to him in the post-1947 times of independent India, during his period as the Indian Prime Minister when he, as the chief-executive of his sovereign-State, was historically required to face the challenging security-threats? Modern Indian-history is still asking this question.

(g) Ancient Greece: 3600 to 1200 B.C. onward

We may even emphatically say that the ancient Greek civilisation is in no way less magnificent and also multiply advanced than any other equally famous and historically noted ancient human civilisation in the world. So far as the ancient Geeks' 'grip' over even the multi-dimensional aspects of 'matters-military' are concerned, historically they certainly are in no way less magnificent than any other similarly great ancient human civilisation. A very-brief glimpse of the historically factual narration of so famous and unsurpassed bravery of 300 ancient Greek-Spartans in the face of invading Persians in very large numbers have already been given above. Historians have also admiringly called ancient Greece as the great Hellenic-World. Indeed, ancient Greece was unique in many ways and, so were the multiple-facets of their remarkable civilisational-advancements. Their world-famous also beautiful heritage-sites, art-effects, magnificent statues carved in dazzling white-marbles and, beautiful monuments, buildings and palaces, so ornamentally carved and built again in selected white-marbles are indeed timeless and naturally enchanting objects even today in the modern-era.

No wonder and as reproduced by the learned **Professor S.R. Sharma** in his book **G.A. Dorsey has so befittingly said that, "For weal or woe, the Greek hand touched every phase of this civilisation of ours that was born in the Nearest East—and, touching left a mark." Dorsey's this observation has understandably been regarding the far-reaching historic impacts of Greeks upon various European-civilisations**. In fact, Greece too is a part of European continent and, located on the northern coast of the strategically well-known Mediterranean Sea. It may even be interesting to recall here that some of the noted scholars and historians have mentioned that, areas of river-basins and valleys Nile, Tigris and Euphrates have been called the 'fertile-crescent' And then the vast geographical-regions, all around this 'fertile-crescent', extending to the ancient Persia on one side and then the modern Israel, West-Bank and so on the other side; have interestingly be noted as the 'cradle of civilisations'.

There is an interesting and very ancient historical-story on the 'vast-canvas' of human-history, about the earlier name of the 'Hellenic-world' given to the geographical areas of modern Greece. But in fact, the ancient Hellenic-world was much larger than the present-era Greek Nation-State. In fact, the legendary Hellas are supposed to be the ancestors of even the ancient Greeks and at the same time, names of Hellas are to be found even in the Biblical stories.

From the noted books of History; we also learn that these Hellenes belonging to the very remote past were indeed from the same Aryan-stock. They were known to have migrated from the ancient Caspian-Sea region, during even earlier to 2100 B.C. or so, towards their later areas of settlement. Logically therefore in the same historical-perspectives, many learned historians have referred to them as, ancient European-kin or cousins of the same Indo-Aryans, Persians or even the ancient 'war-like' Hittites of the ancient Mesopotamian and adjacent regions. In our specific context, we have already briefly deliberated upon all of them earlier, in a sequence.

Professor S.R. Sharma mentions that, "**The missing links between the Asiatic (including the African) and European civilisations were discovered in the buried cities of Tory, Mycenae, Tiryns and Cnossus, by Heinrich, Schliemann and Sir Arthur Evans, in 1870 and 1900 respectively**. The former was an American born German grocer who shipwrecked on the coast of Holland, but later amassed a fortune all of which he invested in the hobby of seeking for **Homer's Troy**. The later was an English archeologist. **Their excavations have done for European history what Mohenjo-Daro and Harappa have done for India**. *(Fascinating, isn't it? Yes, it is indeed so.)* Prior to their wonderful discoveries Greek history was begun with Homer: now it goes back to about 3600 B.C.... Cnossus is on the island of Crete which is centrally situated in the Eastern Mediterranean, thus forming a link between Asia, Africa and Europe. The excavations carried out on this island, by Sir Arthur Evans and others after him, have revealed a number of sites the chief of which are Cnossus and Phaestus. **They represent a civilisation which are contemporaneous with Egypt, Babylon and Mohenjo-daro, and had spread as well over the mainland of Greece as across the Aegean to Troy in the north-east and Sicily in the west. For this reason it is called the Mediterranean or Aegean civilisation. It is also called Cretan on account of its island centre and Minoan because of its legendary connection with King Minos who had his home in Crete**." (Sharma, 1948, pp. 139 & 140)

And for the first time, the light of civilisation arrived on the southern-coast of Europe when the ancient southern European-State of the same Hellenic-world or the Greeks, came in contact with Asia-minor region and the migrating-Aryans. But it is equally true as also already discussed above; the ancient Greeks were already fairly civilised, when they came in contact with the glorious old-era Asian-civilisations. But the same cannot be asserted at all with respect to the other European-areas and the peoples then residing over there. Interestingly in fact, historians have even called them as barbarians belonging to the Europe's ancient 'dark-age' era.

At this juncture, interestingly we notice the ancient Greek-civilisation had a number of socio-cultural commonalities with the very ancient Indian sub-continental civilisation. On the one hand, when the migrating-Aryans arrived in the ancient Indian sub-continent, they found themselves contacting the glorious Harappan-civilisation of the north-western India and subsequently then slowly coming in touch with the Dravidians of Central and Southern India. Therefore, interestingly as mentioned earlier **if as a consequence, Aryanisation of India began then simultaneously, fascinating historical parallel-process, the Indianisation of Aryans too, slowly but consistently commenced**. We may even say that these, in fact, are unique illustrative examples of the phenomenal interaction between the two different and also both being highly advanced as well as enriched civilisations, coming in contact on a common-platform; and triggering great civilisational activities. Identical historic phenomena took place with far-reaching consequences in the ancient Hellenic-world too, when they came in contact with the ancient and powerful civilisations of north-African as well as middle-eastern Asian regions; specially the war-like Persian-Monarchs, resulting in violent-clashes.

In any case, this weighty aspect perforce assumes a much greater significance and perhaps an exclusive importance, whenever we are discussing the historic Greeks along with their multi-faceted enchanting civilisation; being so multiply enriching for the entire humanity, since ancient times. The ancient Greeks have not only displayed great valour in the battlefields, while dedicatedly defending their motherland with unhesitatingly supreme-sacrifices, but much more with numerous such remarkable achievements in many other fields of human endeavour. Display of magnificent courage and fearlessly facing powerful enemies in battlefields with desired military acumen at both tactical as well as strategic levels since ancient times, is one aspect of their so many civilisational achievements.

In any case, whenever we do begin a 'critical-review' of the Greeks' long history; we do also notice that on certain occasions, they had also been politico-militarily found 'wanting' sadly in fulfilling the 'call of destiny', historically at certain critical moments and thereby; inevitably coming to grief as a Nation-State. These being so at times, whenever they had turned somewhat prosperity-oriented complacent and hence consequently, perhaps going callous in the essential aspects of the matters-military. But while reviewing these historic issues, specifically concerning the ancient and unique Greek Nation-State as well as their magnificent civilisation; we encouragingly do also notice that unlike perhaps 'unfortunate' Indians, these Greeks did display the essential strategic 'sense of history' and seldom found wanting in valuably learning from the same living-history, as demanded by the dictates of times and thereby; bringing-in the requisite 'course-correction' for their own good.

As we proceed further in our ongoing discussions, contextually always keeping in mind the main subject-matter, as our essential 'compass' in the overall path of our same 'strategically critical journey'; we do prominently notice the many other magnificent civilisational achievements on the part of the unique Greeks since ages. Again while looking into man's same exceptionally long and living-history we also interestingly learn that the Greeks earlier seldom believed in launching well-planned military-campaigns for acquiring extended geographical territories and building their own powerful empires, like the ancient Persians, Hittites and the Indo-Aryans and so on. But in a sort of haste, we may not think that the Greeks' truly magnificent-saga ends here, as we would be able to notice later, in our continuing discussions.

An exception being the Macedonian warrior and outstanding military genius named **Alexander the Great**, who came on the horizon of human-history with a bang, sometime during 300 to 200 B.C., who shook the world of his ancient times, with his ruthlessly relentless march of military-conquest with a sort of lightning speed. Well, while viewing in larger historic perspectives; we do also notice that this was indeed a sort of 'cyclonic' but a short-lived phenomenon and at the same time on man's much larger multiple civilisational canvas, this hardly left any enduring footprint except in the domain of the matters-military.

In the volumes of man's multiple military-history since very ancient times, Alexander's life-sketch can be defined as: outstanding military-leadership qualities, unique attributes of a young warrior's unprecedented art of conducting his military-campaigns, professionalism and magnificent genius of his historic generalship, are indeed even now the important subject-matters of studies and professional-research in almost all the leading military institutions around the world.

Now again coming back to the other equally magnificent civilisational aspects of ancient Greeks, we find that during that bygone era, Greeks believed in small and extended, but beautifully organised, self-sufficient and well-governed independent **'City-States'**. Historically this idea might have been pretty attractive as well as more efficient in many ways in the beginning of its applications and actual practice on the ground; but as time naturally moved on, its many dark-sides began emerging. Unfortunately, these independent 'City-States' were just not mutually co-operative and supporting but instead they became each other's bitter-rivals and there-by, getting locked at times in violent clashes too. In any case, what could paradoxically be more ironical when they all have had same culture, religion, ways of life and socio-psychologically as well as socio-politically identical value-systems and in addition, and belong to the same ethnic-stock as well?

Now looking into these aspects in overall geostrategic and politico-military perspectives, with specific reference to that very ancient-era Greek socio-cultural as well as socio-political institutional arrangements for affecting the governance of their various City-States, we obviously notice many chinks in their armour. In fact, in a very simplistic-term, we may even perhaps with some accuracy say that these, otherwise multiply-magnificent Greeks; strangely ignored the age-old saying that, 'divided we fall and united we stand'. At this point specifically in their ancient historic-perspectives, we may even emphasise that their at times 'regressive' obsessions with their rather fixated theory and practice of the concepts of 'City-States' perforce made them militarily highly vulnerable, and consequently, humiliating military-defeats as well as great misfortunes visited them. No doubt that in many other civilisational aspects, their great creative achievements and magnificent advancements, are indeed well-established and do remain highly praiseworthy even in today's modern high-tech era but their earlier mentioned strategic 'negligence', did cost them very dearly in those ancient times.

In the above discussed specific context, one glaring example of the Persian King **Darius's** tornado-like military 'sweep' over the ancient Greek territories during nearly 499 to 490 B.C comes in our minds for some review in this regard. We very informatively learn that even the military-adventures of the royal predecessors of Persian-King Darius, then ruling over the ancient Persian-Empire had resulted in substantial territorial-expansion of their empire.

And in the process, their military-campaigns swallowed a number of Greek-colonies spread over Asia-Minor and then touching the southern coastal-belt of the Mediterranean. These geostrategically and politico-militarily highly threatening developments then, even touching the marine-borders of the main-land Greek territories should have naturally sounded serious alarm-bells in the Greek-ears. But historically as a great and otherwise avoidable tragedy for them, nothing of this sort happened and thereby as an unfortunate irony the ruthless military-devastations of their magnificently multiple civilisational achievements along with tragic sacking of the historic marvels of unique Greek City-States; became pre-destined.

These extremely tragic developments perforce resulted in so many adverse fallouts for the great ancient Greek-civilisation. In fact, when viewed with requisite strategic-criticality; we notice that in relatively high prosperity of their otherwise well-organised and shining City-States, Greeks got used to easy-going comfortable life-styles and in the process; they perhaps developed a sort of foggy-vision resulting in highly dangerous negligence of their crucially important strategic-compulsions, to their great peril.

Now returning to our contextually specific-point of discussions at this stage, we the royal-predecessors of the **Persian-King Darius**, the big Persian-Empire, under the ruling Monarchs Cyrus, Cambyses, and then Darius himself, had considerably expanded. All these resulted in a sort of dangerously bleak-scenario for the Greeks and at this historic point it looked as if almost all their small Hellenic City-States would be swallowed by the advancing Persian armies, and brought under the Persian rule. The relentlessly marching military-forces of King Darius overwhelmed those small City-States and decisively crushed their relatively weaker military. In fact, Persian King Darius was just not satisfied with his ruthless military exploits and subsequently then, he re-organised his forces for two years for his renewed military-campaigns and started having an eye on the **magnificent ancient Greek City-States of Athens and Eretria**. As per his operational-plans, he came knocking on the doors of these two City-States sometime during 490 B.C.; naturally creating grave military threats to the Greeks.

But as the ancient Greek-history shows us, this time **King Darius** was not going to have just a 'walk-over and interestingly; too much of 'surprises' awaited him. As a great development of militarily-historic astonishments, this time far-reaching good-sense prevailed among the Greek ruling-classes as well as dominating influential-aristocracies, belonging to their numerous City-States and befittingly took imaginative-steps to forget their petty-differences and quarrels and, got united; creating magnificent geostrategic and politico-military synergy, for facing their common and powerful Persian-invaders. **Consequently, when the historically decisive military-engagements did take place between the united Greek warriors and the armies of Persian-King Darius; on 21st September 490 B.C. at Marathon, 'Lady-fortune' had already changed sides.** The unique saga of this historic battle between the Greek and Persian armies on that fateful day of the distant-past at Marathon indeed had far-reaching influences, in further shaping man's living-history for many centuries to come.

Otherwise probably the entire shape of man's continuous history and in addition, the multiple courses of their socio-cultural developments, during many centuries yet to come would have been quite different. If we may care to look into these aspects and begin critically examining them in a comprehensive historic-perspectives; **first and foremost we notice that this decisive military-defeat of the Persians at the hands of the ancient-Greeks then at Marathon in fact, totally stopped any further hostile Persian military-misadventure in Europe**. Further, this historically remains no small matter, since historically this perforce remains a great turning point. We may like to bear in mind that with time the way European-societies gradually developed along the highly complex paths of socio-psychological, socio-cultural and socio-political spheres, could never have been comparable, if instead of the Greeks, the Persians would have achieved victory in that battle at Marathon, during more than two thousand years ago.

Interestingly, in anespecially informative manner in these specific perspectives, learned Indian Professor S.R. Sharma writes that, "The history of Hellas from the defeat of the Persians to rise of Macedonia 9470-338 B.C.), constitutes one of the brightest epochs in World History: it was certainly the glorious period of in the history of the Greeks.

Richard Overy, who was Professor of History at the University of Exeter, writes that, "The 8th century B.C. was a period of great transformation in Greece. It saw the appearance of the first monumental buildings and with them other indications of the emergence of new communities, including changes in burial practices and artistic styles.... From the middle of the century, following in the wake of the Phoenicians, groups of Greeks began to create settlements around the

Mediterranean. The earliest were in Italy and Sicily, but by the middle of the 6th Century there were numerous Greek communities in North Africa and, to the east, along the Black Sea coast. These colonies were set up for a variety of reasons. Some of the earliest were trading posts, which over time developed into permanent settlements. Others were formally dispatched as a response to land shortage in the mother city.... From its earliest existence, decision-making in the Greek *polis* lay with an assembly of adult male citizens. Leadership however would have been in the hands of the wealthy elite. Increasing wealth and overseas contact in the 7th and 6th centuries led to the emergence in many city-states of powerful individuals, known as tyrants, who were able to impose their will on the community, usually with popular support." (Overy, 2008, pp. 58 & 59)

The above quoted paragraph from a book of a learned British professor and author of History, does in fact, very revealingly show many important aspects of the socio-political culture as well as geostrategic inclinations around the then Mediterranean regions, of ancient Greeks and that too again; during the pre-Alexander era. In addition, such far-reaching multiple-historic initiatives on their part did bring-in substantial riches into the mainland Greece, triggering multiple long-term fallouts. These were in fact, caused by the expanded trade opportunities and avenues for trade and commerce for the Greeks, with other peoples and also, empires.

Such considerable swelling of Greek wealth and resources in turn greatly facilitated the Greeks during that bygone era in crafting their superbly captivating statues, many kinds of art-effects, magnificent monuments, ornamental-buildings and at the same time, creating timeless master-pieces of the work of Greek literature. This was truly a great-era for the magnificently captivating multi-dimensional socio-cultural, art and architecture, literature as well as civilisational advancements and enrichment as well. No wonder, that even during the presently on-going highly high-tech modern-era; peoples from all over the world keep flocking into the Greek Nation-State for having enchanting glimpses of them and enjoying their unique magnificence. These global activities also have another highly beneficial outcome for modern Greece, in terms of substantial revenue-generation from their booming tourism-industry and so on.

Jawaharlal Nehru very befittingly writes that, "Yet these few *(i.e. books)* are such as to fill us with admiration and to make us wonder as the **many-sided greatness of the men of Hellas. How rich their minds must have been and how deft their hands, to produce their beautiful statuary and their buildings. Phidias was a famous sculptor of those days, but there were many others of renown also. Their plays—tragedies and comedies—are still among the greatest of their kind. Sophocles and Aeschylus and Euripides and Aristophanes and Pindar and Menander and Sappho and others can only be names... now...but... their works...*are*...something of the glory that was Greece**.... Athens especially became famous during this period. It had a great statesman for its leader. Pericles was his name, and for thirty years he held power in Athens. During this period Athens became a noble city, full of beautiful buildings and great artists and great thinkers. Even now it is spoken of as the Athens of Pericles and we talk of the age of Pericles.... **In life Socrates served the cause of truth and knowledge well, better still he served it in his death.... When Plato was getting old, another Greek who has become famous, was coming to the front. His name was Aristotle. He had been the private tutor of Alexander the Great, and Alexander** helped him greatly with his work. Aristotle did not trouble himself with problems of philosophy, like Socrates and Plato. He was more interested in observing things in Nature and in understanding the ways of Nature. This is called Natural Philosophy or, more often now, Science. **Thus Aristotle was one of the early scientists.**" (Nehru, 1934, pp. 51-53)

We do also notice certain, may be, some 'weak-spots' in the truly magnificent Greek civilisational-canvas. In our specific context and also as far as possible in fact, meticulously adhering to the duly mandatory objectivity in this regard we are able to notice certain 'gaps' in the otherwise crucial awareness-levels of the then certain aspects of the geostrategic and politico-military compulsions, staring in the eyes of the ancient Greeks.

So far as the aspects of military-acumen of both the Greeks and Romans are concerned, nearly four to five centuries before the advent of Christ-era, their praiseworthy military tactical-formations for their battle-plans oriented deployments in the fields for engaging their enemy's forces had become fairly famous. Every student if military-history is well-aware of the ancient-era skillful art of forming-up of their troops as **'Greek Phalanx'** and **'Roman legion'** for fighting with the enemy in battles. These along with many other allied matters regarding the above mentioned multiple compulsions, indeed always of paramount importance would be more befittingly discussed in some detail, in the next chapter.

(h) Ancient Rome: 1000 to 753 B.C. onward

While perusing the pages of very long living-history of man, we find that the noted and glorious Roman-civilisation in fact, was preceded by the unique Greek one. If we may care to review the otherwise glorious history of fascinating ancient Roman-civilisation, then we find that though in many ways both civilisational as well as geostrategically they were bitter rivals; the advent of Romans succeeding the Greeks since already commanding great and multiple civilisational-advancements; fascinatingly brought in numerous kinds of advantages to the Romans. One point appears to be obvious that historically there was already an interesting kind of multiply civilisational 'road-map' available to the Romans for some kind of need-based strategic reference. In addition, both geostrategically as well as politico-militarily the Romans had an access available to fruitfully draw from the Greek experiences as well.

Many well-known scholars and historians have in their own enriching ways written about the magnificence as well as interactive saga of strategic-encounters between both the Greeks and Romans, over the great history-making periods of a number of eventful centuries. In this context, we learn that Earnest Barker, with great historic intellectual-precision, has said that, "While it was the Greek genius which, in its later days, rose to conceptions of the unity of humanity, it was the Roman genius which translated these conceptions....into an organised system of life." Indeed these have been so and hence no wonder that they timelessly do form shiningly glorious-chapters of the vast recorded living civilisational-history of the mankind, since a remotely distant ancient times over the same unending 'canvas of time'.

Then very relevantly in our specific context, we notice that Jawaharlal Nehru mentions that, "....perhaps remember the City-States of Greece. Wherever the Greeks went, they carried this idea of City-States with them, and dotted all over the Mediterranean coast were Greek colonies and City-States. But now in Rome we are dealing with something very different. To begin with, perhaps Rome was not unlike the Greek City-States, but soon it spread by defeating the neighbouring tribes. The territory of the Roman State thus grew and comprised the great part of Italy. Such a big area could not be a City-State. It was governed from Rome, and Rome itself had a very peculiar type of government. There was no big emperor or king; nor was the modern type of republic. Still the government was a kind of republic, dominated over by the rich families owning land. The Senate was supposed to govern, and this Senate was nominated by two elected persons called

'Consuls'. For a long time only the aristocrats could become senators. The Roman people were divided into two classes; the patricians or the rich aristocrats, usually landowners; and the plebeians, who were the common citizens. The history of State or Republic for several hundred years is one of conflict between these two classes.... The plebeians go on struggling and fighting to gain power and slowly some crumbs fall to their lot. It is interesting to note that in this long struggle the plebs successfully tried non-co-operation of a kind. They marched out of Rome and settled down in a new city. This frightened the patricians, as they could not get on without the plebs, and so they compromised with them and gave them slight privileges. Gradually it became possible for a plebeian to attain high office and even to become a member of the Senate." (Nehru, 1934, pp. 82-83)

In this chapter on 'Advent of Human Civilisations and Ancient Era'; we have also very interestingly and historically seen, fairly understandably, that in case of certain highly acclaimed ancient human-civilisations, with respect to their initial beginnings; two streams of perceptions do prevail rather somewhat amusingly and hence, one being legendary and naturally the other one being acceptably historic. And interestingly here we do find the same in case of the glorious ancient Roman-civilisation as well. Hence man's history from its inception, shows us that the Romans' traditional legendary beginning is understood to be largely perceived by the modern sovereign Italian-State as around 750 B.C. from the highly acclaimed Roman emperor Romulus or Remus coming to the throne. And interestingly, noted historians have broadly recorded its beginning from the year 500 B.C. when the ancient Roman-empire was administratively as well as both geostrategically and politico-militarily organised into a dynamic Republic.

Understandably to great amazement of the historic-coincidence as per one specific perspective, we do notice that as in case of both the ancient Indian and Greek civilisations, interestingly the ancient Romans too, coming on to the same 'platform' of the then praiseworthy pursuit of civilised human-endeavour. And that specific perspective in this regard, has historically undoubtedly been making of the socio-cultural as well as socio-economic pursuit of 'agriculture' as their national-occupation. If viewed rather narrowly, we may find that as in case of still more ancient and also pre-civilisational advent of hunting as possibly a feasible means of food-gathering, relatively the later era advent of agriculture too, has been then another source of growing the food-grains for the man's consumption. Not the least, this would in any case be a great civilisational misnomer of man's same living-history. As already elaborated earlier in this chapter itself, history very vividly shows that certainly the magnificently gradual advent of agriculture in man's life during the bygone very ancient-era, had multiply enriching fall-outs.

Historically this phenomenal gradual development is considered to be a great initial landmark in the otherwise cumbersome path of man's passage into the domain of phenomenal dawn of these very ancient and marvellous different human-civilisations in their respective far-off geographical regions in this world. In fact, they were all off-shoots of the same Aryan-race. As the Romans and their fellow Latin the Italians were primarily of the same stock of Indo-Aryans as the Iranians; they all did display certain identical civilisational similarity such as, dedicatedly maintaining agriculture as their principal pursuit of life, along with fruitfully rearing the cattle, always having great attachments to their land and remaining a good landlords with high degree of perseverance. Then interestingly the above mentioned now commonly well-known human attributes have been historically identical to all the three great ancient Indian, Greek and then subsequently the Roman civilisations.

Interestingly, as per Indian learned Professor S.R. Sharma, both the Greeks and Romans attached a great dedicated value to their live-stock, as the Aryan Indians did the same to their cattle in the ancient India. Their Divine deities were identical too, to that of the ancient Indians. Probably the greatest of ancient Roman Deity was Jupiter; as in the case of the Greeks was Zeus and so in case of the ancient Indians was the Vedic Dyauspitar. Well, as the saying goes that all that glitter is not gold; and so there were pretty darker-sides of the otherwise fabulous ancient Roman-civilisation too. In fact to begin with, ancient Rome was socio-politically a City-State and then, as the few centuries passed, Romans geostrategically started expanding broadly in all the four directions, with their consistent politico-military initiatives and simultaneously, along with the gradual unfolding of opportunities unfolding in front of their rulers and the military commanders. In their geostrategic context and as per the strategic planning of the Roman commanders; they first secured their supremacy over the adjacent ancient Italian territories and then, proceeded to conquer the Sicily.

Further the Romans embarked on their subsequent conquest of ancient Sardinia, Corsica, North-African region, and then proceeded towards East to politico-militarily subjugate Macedonia, Greece, Asia-Minor and subsequently Syria. Moving towards their West, the Romans established their supremacy Spain and ancient Gaul. When we begin drawing the outline of geographical map encompassing the above mentioned countries and territories, we do find the picture of a vast ancient Roman-empire. **Interestingly in the precise historic perspective, whenever building of any empire has materialised anywhere in the world at any given period of man's continuously ongoing living-history since a very distant-past, with all its magnificence and splendor, simultaneous sprouting as well as growth of the imperialism cannot be far behind**.

In fact, even during the fundamental inception of any kind of thought-processes with respect to the initiation of any kind of multiple-process of empire-building anywhere with a rather well-planned geostrategic as well as politico-military oriented relentless endeavour on the part of any conquering power, perforce have seeds of inseparable and simultaneous germination and multi-dimensional growth of imperialism with all its numerous shades and connotations. These in reality are part of the basic make-up of inherent societal and nation-based human-nature, under the supreme command of mother-Nature as the truly as well as unassailably Sovereign-Creator. In fact, we have already delved-upon these aspects in some detail in earlier chapters. As already mentioned above, imperialism has many shades and varieties and thereby, it gets manifested accordingly; as the empire as part of the history-making process, gradually proceeds with again multi-dimensional activities of its own consolidation in the territories and societies of the conquered areas, conclusively establishing the expansion of the same empire.

In other words, as time keeps moving relentlessly, old era and ages keep becoming part of the man's same living-history of the past and thereby; as civilisations advance, man's progress in the multiple areas of continuously developing science and technology, keeps enhancing and consequently triggering many kinds of mind-boggling as well as enchanting fallouts. They in turn multiply affect various advancing human-civilisations, human-societies, life-styles, multi-dimensional capabilities and man's value-systems. But, as mentioned earlier, the fundamental causative-ingredients and natural tenets shaping the attributes of any historic period-based empire and its accompanying imperialism have perforce remained the same.

Now again coming back to specific aspects of the magnificently glorious ancient Roman-civilisation; we do notice that as it dramatically did expand in different territorial-directions, naturally a stage came when one man assumed the charge as Head of State, exercising governing sway and control over the whole of the vast Roman-empire. As per the imperial governing policies, under which all residing citizens were to avail equal rights and facilities and were also, charged with equal obligations towards the Empire. But in practice were they really so? Well, ironically the answer cannot be emphatically in the affirmative. History shows us that the countries other than Italy and also territories other than Italian, though a part of the same Roman-empire were being considered and also treated as the conquered subjects of the Roman-empire and its ruling elites in administering the matters of State.

Since historically in this specific context, that being the reality on the ground, any manifestation of any kind of 'sense of equality' naturally got washed-away and in the process, paving the way for a kind of hegemony-based imperialism to set-in within the empire. With different kinds of socio-cultural and socio-political thinking and practices over the years, they became a basis for formulating relationships between the ruling-elites and the ruled in any particular empire, in different civilisational-territories. These are historic facts pertaining to the ancient human-history.

We then come to notice and as recorded in the pages of history that such norms and practices have not been true only during long periods of ancient-era in our world but even during the present modern-era, larger socio-political, as well as geostrategic picture broadly remains the same and only external appearances and methodologies of maintaining the empires and practicing the tenets of imperialism have changed. But while reviewing these aspects with some criticality in our specific context; we do find that many changes too have taken place. For example, earlier during the long bygone ancient-era, taking men as slaves and women too from the militarily conquered territories by the victors were the prevalent norms, which were both largely as well as ruthlessly followed.

And as man's same living-history moved-on over many centuries, historic-era too changed, bringing-in many alterations in many such prevalent norms and practices as well. Interestingly during the present modern-era, if we may care to look, geostrategically as well as politico-militarily, a critical-look over the historic-saga of Japanese, Chinese, French, Spanish, Portuguese and even once the massive British Empires and also, enforcement of their own different kinds of harsh Imperialism then we may not come across multiple narrations of their colonial-era histories in anyway softer. We may further befittingly add that in this context, historically for the colonised peoples anywhere, the socio-cultural, socio-economic and even socio-political **crippling-lethality** has largely been the same. These again remain moving-stories of great historic-tragedies. Well, historically with respect to different empires belonging to different periods of history and also in addition, regarding different European and Asian colonial-powers; degrees of the above mentioned lethality as applicable to their respective colonies, may vary and even at times just marginally; but the larger aspects of tragedies then being forced to visit upon their colonised subjects, have definitely been crippling in various ways

Now again coming back to our present specific context pertaining to the ancient Roman-civilisation, interestingly, we find that, **Richard Overy** writes that, "The emperor's figure was central to the empire: everywhere statues and coins were constant reminders of his presence. In the former Hellenistic kingdoms the kings had been objects of religious worship, a practice which continued with the cult of the emperors. In the western provinces, temples and altars dedicated to the emperor became focuses of Romanization. The early 2nd century saw important cultural developments: Greek and Latin literature flourished; and the distinction between Italy and the provinces dissolved as rich men from all over the empire were admitted to the Senate, with some, such as **Trajan** and **Hadrian** (117–38), even becoming emperor. For the poor there were fewer benefits, and differences in rights and privileges of rich and poor grew. By the time **Caracalla** extended Roman citizenship throughout the empire in AD 212, it gave little advantage to the newly enfranchised citizens. Later in the 2nd century, pressures grew on the frontiers. **Marcus Aurelius** (161–80) spent much of his reigning days at war with barbarian invaders, and his successors faced threats both from the north, and after 224, from rejuvenated Persian empire under the Sasanids." (Overy, 2008)

Before quoting the above paragraph from Richard Overy's book; we have been discussing in the overall historic-perspective, certain aspects of man's inherent civilisational-legacies with respect to the actual imposition and practice of **hegemony-based imperialism** along with its many consequential connotations over his fellow-beings; whenever they have faced each other as the victor and the vanquished. Further in our present context, historically wherever any conquering power has been able to establish its politico-military sway over the subjugated peoples and their areas, this has naturally resulted in territorial-expansion of the victor's empire. In fact, it won't be erroneous to emphasise that the principal geostrategic aim behind the launch of any military-campaigns by any country or its empire against its rivals or neighbours has largely been towards its territorial-expansion as well as gains of substantial riches and at times, for capturing women too.

Now in the context of our ongoing discussions on the ancient Roman-civilisation and its vast historic-empire; interestingly if we may take the trouble of reading in between the lines and also try to dig their certain 'hidden' meanings in the above quoted paragraph from Richard Overy's book then, we may even somewhat fascinatingly find rather somewhat ruthless history of imposing 'practices' of the victor's *(in this particular case, the Romans')* imperialism; over their conquered countries or territories and their inhabitants. More interestingly in this specific context, we may even see that all these had been phenomenally turning a 'full-circle' and thereby; especially for the scholars and historians, engagingly triggering a fascinating process of further writing in continuity; a kind of new living-history in this regard, and they all in turn naturally forming parts of the overall legendary history of ancient-Romans and their multiple achievements.

We also quite engagingly notice that since the historical times any fast-expanding empire of a militarily strong Imperial Power naturally have had perforce to face both geostrategic and politico-military opponents or any other competing imperial-power. And at certain point of time, powerful Romans had also to face the might of **'Carthage'** who had already established their supremacy over the areas on Western-Mediterranean. Strategically, the Roman ruling-elites then had finally taken a decision that in the larger imperial-interests of their mighty empire the Carthage had to be militarily engaged with decisiveness with the singular aim to destroy them once and for all. As history shows us that both from geostrategic and politico-militarily perceptive-analyses, these were long-drawn affairs of violent military clashes for more than a century.

In fact, this was no 'cake-walk' for the Romans and their clashes resulted in **three Punic-warslasting over a period from 264 to 146 B.C., with a complete destruction of the Phoenician *(i.e. Carthage)* military-sway and rule over their areas in the Western-Mediterranean region**. At the end of **first Punic war during 264 to 146 B.C.**, the major military engagement took place in shape of Naval-clashes resulting in their decisive defeat of the Carthage at the sea with destruction of their naval-ships along with the crippling loss of the islands of Sicily, Corsica and Sardinia in the hands of the victorious Romans. But this was principally a naval-battle between the Roman fighting-ships and those of the Phoenicians and naturally this didn't result in their final military-annihilation, which the Romans then were aiming at.

The historic reality then was that the Carthage was not at all weak in their land-territories across the Western-Mediterranean coast-line and hence, militarily they were in any case no push-over. Further, after having been dislodged from the Mediterranean-sea they did display their sound expertise over the matters-military, both in tactical and strategic domain. After their major reversal at sea, they began consolidating their strong land-border defence under their military-plans as per their own threat-perceptions and hence; militarily occupied the Iberian Peninsula for establishing their firm-base over there to meet the threatening Roman challenges. In fact, their military build up against the invading Romans subsequently resulted in the **second Punic-War (219 to 202 B.C.)**. From the ancient military history points of views, this war was indeed historic in many ways and also, history saw a great military-leader and brilliant commander coming on the centre-stage to face the Roman military-might: in the persona of **Hannibal**. Hannibal was son of **Hamilcar Barca**, the ruler of the Phoenicians. In addition, **Hamilcar Barca's son-in-law, by the name Hasdrubal**, also came-up on their side, to face the Romans.

So the battle-lines were drawn-up for the above mentioned second Punic-War, resulting in extreme bloody clashes during those times. History of this war then shows us that during the first phase of the war itself, Carthage suffered greatly; as except Hannibal, the other two above mentioned Phoenician top-commanders fell and consequently it was left to the destiny of Hannibal to further lead the campaigns against the menacing Romans. It is a different matter that ultimately the mighty Romans finally at the end of this second Punic-war, did prevail with military-decisiveness; but it is no small matter of historic importance that the this entire war established Hannibal as a and great military-commander and magnificent warrior in the pages of long global military-histories, for all times to come. But we may note at this point, that this second Punic-War did not end so simply and so soon and it would be appropriate to give some more space here for narrating the magnificent saga of Hannibal's valour and brilliant generalship in the campaigns against the attacking Romans.

In fact after the first battle, Hannibal got some time to reorganise his forces for further taking-on the Romans. History teaches us that he organised his infantry consisting of more than one hundred thousand soldiers and significant number of elephants; and trained them precisely for the forthcoming battle. Hannibal then marched along with his forces; via Spain and then through the southern coast of ancient France. He then crossed the mighty Alps mountain ranges for entering into Italy. It is interesting to note that at this stage, as per the emerging situations there; Hannibal did also display his brilliance in conducting the crucial strategic-diplomacy in winning over the **war-like Gauls**, who were found to be settled in the Po valley of that region. Consequently, with his precise survey of multiple situations facing him as military-challenges and assisted by his visionary generalship; Hannibal's military-strategic 'recipe' was complete and, he was ready to take on his adversaries with a series of lightning strikes. Hence, total surprise was achieved by him since the city and the military-camps in front of him in the peninsula, were taken totally off-guard.

Hannibal made a series of swift moves with brilliant military maneuvers as if, he was waiting for this opportunity and also; he had organised, trained, equipped his fighting-forces and then, brilliantly moved them over long distances through tough terrain and situations with stealth and speed; never once losing sight of his aim and also direction of his march. This was his finest-hour as a warrior and as a historically-noted commander in the battlefield. 'Die was cast' and Hannibal was on the offensive. History has already recorded that with a series of quick attacks on his totally surprised enemy, he inflicted a number of crushing defeats on his enemy-forces. Otherwise valiant Romans and Italians were militarily out-maneuvered by Hannibal and cornered, to be ruthlessly killed by his forces. **In this military-campaign of Hannibal, the most decisive battle took place**

in the year 216 B.C., since known as the battle of Cannae in the annals of ancient military-history; resulting in a disaster for the Romans and they were in fact, routed and massacred. But this was not the end of Hannibal's story and to his great misfortune soon the tide begun turning against him.

In the above mentioned battles, Romans were no doubt defeated but not finally finished. And more so, they certainly didn't lose their heart and military-grits, both geostrategically as well as politico-militarily so vital specially, in such extremely adverse situations, in spite of a severe blow to them, with a tragic but valiant death of one of their top commanders and consul-generals, named **Claudius Nero**. Strategically so far as the aspects of military command and control in the battlefields are concerned, Romans had a ready and great line-up of competent generals, always capable of transforming any adverse-looking military-situation into that of even decisive advantage, before it was too late. **No wonder, as to why great strategic thinkers, military leaders with proven generalship in war and learned authors since ages, belonging to this specialised professional-domain have always laid great stress on the decisive importance of strategic-vision and also unwavering professional-competence coupled with leadership-potential of the military-commanders, leading his forces in the battlefields.**

History so knowledgeably informs us that this was what had happened at the point of history favouring the Romans, which as a consequence, had decisively sealed the fate of Hannibal once for all and to his great misfortune; he was finally pushed into politico-military oblivion and thereby, his extremely cruel destiny perforce ensured him to meet his lonely end. In fact, as already talked about earlier, after the death of Roman general **Claudius Nero** another great Roman military commander named, **Scipio Africanus** immediately emerged on the scene at a most crucial moment and took-over the reins of military-command, even before morale of the Roman soldiers began to sink low and, saved the situation for Romans in the battlefield. Scipio innovatively regrouped his forces with sound tactics in the field and launched repeated attacks on Hannibal's forces, forcing defeat after defeat on him. Hannibal stood isolated and exhausted as a disillusioned commander; since Phoenician had no choice but to abandon Italy to its fate and rush to organise defences of his own principal-city as it was already getting threatened; but ironically there he met with his decisive defeat in the **historic battle of Zama in the year 202 B.C.** in the hands of attacking Romans.

As the saying goes that any calamity brings-in many damaging 'fallouts' and, this stands equally true universally in the battlefields as well. Hence finally Carthage too, was forced to leave control over ancient Spain to Victor Romans and then they in turn forced down the throat of Carthage, an extremely humiliating and also, politico-militarily a totally crippling treaty. This naturally brought the Romans in the region then, in an unassailably commanding as well as controlling authority. Now viewed in our perspective, we notice the Romans then geostrategically making a brilliant move to their great advantage. In fact, here history shows that the financial amount as fines, levied by the Romans upon their adversaries was next to impossible to fulfill, with an intention to ensure their total financial collapse. On top of it, Romans further laid down another one-sided condition on Carthage that under no circumstances they could ever go to war against anyone, whatever might be any pressing and even otherwise legitimate circumstances; without their advance permission. What else could have been any more ruthlessly crushing-trap to be laid-down by the Romans, as conqueror?

But, there was an intentional 'design' to all these, on the part of Romans. No doubt, tenacious Carthage was down but certainly not yet finally finished. His slow and steady process of inescapable economic recovery to start with, was a very long-drawn process; but they did recover after more than half of a century. But the scheming Romans were not going to leave them alone, to substantially re-build in their own way. In fact, Romans had not even once allowed the Carthage to move away from their sharply monitoring a kind of 'Radar-screen'; since it was all a patiently calculated move, with a focused aim just to bid for time and appropriate opportunity. Meanwhile, Romans relentlessly kept on prodding and coaxing to attack Carthage. When the attack came, obviously Carthage perforce had to fight a self-defence war. This gave the Romans much sought after chance they were craftily waiting-for and hence, Carthage was charged by the Romans, for willfully breaking the treaty they had earlier entered into.

The Romans then ruthlessly launched attacks and totally destroyed their city. History then, has recorded this as the **Third Punic-war, having duration of nearly three years from 149 to 146 B.C. and, historically this resulted in the then final eclipse of Carthage**. Powerful Roman-empire was too obsessed with their intoxicating imperialism for achieving universal-domination, regardless of their own certain inherent both geostrategic as well as politico-military limitations. Well at this juncture, it may appear to be a matter of conjecture but interestingly; one may even say that perhaps Romans wished to follow the footprints of the earlier Macedonian conqueror, Alexander the Great But with all their mastery of generalship as well as their other multiple civilisational achievements, Romans were in any case no Alexander.

Then in addition, it may also appear appropriate to contextually add that man's living-history too, makes it clear, that it doesn't ever seem to be repeating itself; i.e. the way Romans then, perhaps so naively had wished for. In this regard in the broader context, Prof. S.R. Sharma writes that, **"This vast Empire (i.e. Romans) endured in prosperity until Rome herself was overwhelmed by the barbarians after the death of Marcus Aurelius, the last of the great emperors. Between 200 and 250 A.D., that empire was already tottering. Constantinople, which was to be the seat of the Eastern Empire, was founded in 330 A.D. By 500 A.D., Gothic kingdoms were established in Spain and Italy; a Vandal kingdom arose in Africa, a Frankish kingdom in Gaul, and a Saxon kingdom in Britain."** *Hence, we notice that the history in the above mentioned regions during that era was indeed moving in a much faster pace. And naturally as a consequence, many far-reaching changes as well as major upheavals too, were taking place. By having a closer look with some strategic as well as politico-military criticality; we also notice that ironically, the Romans and their subsequent Emperors had already lost touch with their ground realities and in addition, they had lost the essential 'sense of history' as well, paving the tragic path for their downfall and ultimate withering-away of their once glorious empire.*

The above mentioned fait-accompli for the Romans, had since already germinated in the process of reckless expansion of the Roman-empire without ever giving any thoughts to the crucially important strategic-considerations in this regard. As the empire had already expanded to far-flung areas it was obviously not possible to ensure a direct control by the central authorities with the much needed close-watch, for effective governance and essential imperial revenue-collection. So mediocre military-governors were appointed in those far-off provinces to do the job and more so, it also needs to kept in mind that since bygone era communication as well as multiple connectivity, were perforce quite primitive giving enough scope to all sorts of corruption to fast take deep-roots everywhere and at all levels in that historic empire. The once famous shine and glory of the powerful Roman-empire began fast withering-away making it a stinking breeding ground for slaves, to be forced into the service of since already turned corrupt Roman officials and arrogant ruling-class. As the saying goes that one malaise since creeping-up, if not checked in the beginning itself, create causative-effects for other malaises to thrive regardless.

No wonder that as the once mighty and glorious Roman-empire began decaying and weakened; Romans and their societies too, started losing their values and character. In fact, they became more and more maliciously corrupt and cruel as well. They gradually stopped treating their slaves as 'lowly' humans and hence; simmering frustrations, acute sufferings as well as contempt against their masters among the slaves also, began getting stronger. Roman elites never cared since for them, what mattered was to make money, lead criminally luxurious-life and corner more and more wealth. They even indulged in forcing the slaves and criminals to fight in pairs, as **gladiators**, for public-entertainment and called this as a type of martial-sport. Historically it was no surprise that all these led to violent-revolts against the Roman-rulers; resulting in a kind of **'Social-war'** spreading in the already tattering empire, shaking the grip of the authorities and this turmoil lasted for nearly two years, during 91–89 B.C. Results were that slaves too were granted Roman-citizenship, with rights and privileges for keeping them somewhat pacified within the empire.

But all these came too late and were unable to bring-in the desired results, which the already decayed Roman authorities had wished for. Roman imperialism over the empire was sinking fast and, it became apparent that a strong and competent 'dictatorship' had to be brought on the scene, to somehow save whatever was left of the empire. All this were part of the Roman-elites' extremely desperate thoughts, and in any case the march of man's continuing living-history was moving rather too fast then, for the Romans to be able to maintain even workable 'grip' over the process of governance. Hence as per the strategic-initiatives then taken by the Roman-Senate a kind of dictatorship to be exercised by a team of triple-personalities consisting of **Julius Caesar**, **Crassus and Pompey** with their own distinct attributes, were brought-in for the purpose around 60 B.C. But the team hardly remained cohesive, as the inevitable mutual rivalries and also inevitable clash of personalities among them became stronger; especially between Caesar and Pompey and, in spite of all such initiatives, nothing was working for the empire since it appeared that it was already too late.

Among the above mentioned trio; history rates Caesar since gifted with some qualities of charismatic leadership and attributes of a great warrior and outstanding generalship, as a historic genius. He did achieve outstanding military-victory over the Gauls, i.e. modern France and then, the **great scholar and orator Cicero did bestow glorious praise on Caesar**; since the facts remain that he was victorious in every battle. But his soldierly qualities didn't work against then the reigning-Queen **Cleopatra** of Egypt, who was able to otherwise also overwhelm him with **her captivating beauty, charm and rare intellect**. Well, we may also say that all these attributes along with the politico-military sway over her domain, being unquestionably supreme; together do manifest as another kind of "**lethal-combination**". No wonder that Julius Caesar along with his trusted comrade **Marc Antony** later too, were victims of this lethality in a woman and moreover, she being an outstanding reigning-Queen.

History does inform us that long historic saga of the Roman-empire and its multiple-dimensions with many ups and downs, have a highly chequered and absorbing narration. Romans during their magnificent civilisational past had displayed a number of brilliant examples of both geostrategic as well as politico-militarily innovative initiatives and indeed with ruthless and at times even cruel single-mindedness. But both geostrategic and politico-military critical-review of the many marvels of multi-faceted Romans and their prominently ancient empire is certainly not the whole story. They have left many magnificent marks on the pages of their captivating-history, in their socio-cultural, art and literature, religion, architecture and philosophical spheres; and their enduring marvels cannot be ignored. As befittingly quoted by Prof. S.R. Sharma, **H.H. Asquith** says, "Alike in literature, in art, in philosophy, and in religion, Rome built the bridges over which many of the best thoughts and finest models of antiquity found their way into the medieval and hence into the modern world." (Sharma, 1948, p. 203) The above sentence by itself, coveys, volumes regarding Romans and their ancient civilisational achievements.

Subsequently, further in the same book, the learned Professor himself writes that, "The Romans were unlike the Greeks in most respects. But though they lacked the versatility of the **Hellenes**, they had a genius for construction and organisation. Hence they were content to develop such virtues as were useful in building up a great society, and for the rest borrowed from their more gifted neighbours. In art, science, philosophy, the Romans made few contributions of any distinctive merit; but in law, military organisation, government of different peoples, and other practical matters there have been few to equal them. Consequently, we shall find that the Romans, though lacking in intellectual and imaginative faculties, particularly in that grace which lends charm to life, made their own contributions to civilisation as men of action. **H.G. Wells** is a too uncharitable when he roundly declares, 'The reputation of Rome has flourished through the prosperity of her heirs. The tradition of Rome looms far greater than the reality' *(It is indeed a great irony that such comments have been recorded by one of the extra-ordinarily learned historian of international fame, i.e. H.G. Wells.).* **The defects of any people should not blind us to their virtues. The evil that men do lives long after them; and the good is often interred with their bones. But the duty of the historian is to present both aspects in their true proportions.**" (Sharma, 1948, p. 203)

But historically perhaps the darkest spot in the history of Romans; has been their one singular and most uncivilised barbarian act and heinous crime and i.e. the act of the *crucifixion of JESUS CHRIST.* And history can never forget this, nor forgive the Romans and their civilisation for this sinful crime. In fact, Romans and their otherwise magnificent civilisation too, got regressively fixated ironically with rather one of the most uncivilised idea that, the 'worship of Emperor was supreme'. Well, this too, became one of their biggest undoing. It would naturally be anybody's guess that, in such a 'suffocating' environment, no tolerance remains for a saint or more so for a Divine-incarnate.Later, the **same ancient-city of Rome became the most sacred place, after Bethlehem, for the Christians worldwide; in the Divine-name of Jesus Christ**. (***Indeed, so captivating—an unique historic saga!***)

(i) Europe: 476 A.D. onward

History further informs us that in the ancient-era Europe, the ultimate fall-out of the once mighty and so glorious Roman-Empire had in fact multiple and far-reaching fallouts for the whole of Europe. Then in our context, whenever we begin to look into the historic fall of the Roman-empire with certain degree of geostrategic criticality, we also notice that the principal reasons for the same have substantially been both internal as well as external as its causative-factors. No wonder that, as the historic Roman-empire began decaying gradually, from its one imperial corner to the other perhaps simultaneously and consequently; another parallel phenomenon too got triggered, as a distinct and historically unstoppable process. And this process was commencement of the **Dark-Age in Europe**. This had been indeed a major historic development and then; this dark-age in Europe did last for a few centuries.

In any case, history of the dark-age in Europe by itself is somewhat long-drawn and also socio-politically, a complex-development. Roman-empire always had to put-up with the various kinds of military-threats, mainly from barbarians from across their borders in different areas. Over centuries as the empire began expanding and flourishing in many ways the threats from the barbarians too, got more intensified.

Interestingly, history also shows that no mighty empire down the ages; has ever been free from many kinds of threats, both internal and external. And as might of the empire begin growing with its expansion; threats also start cropping-up and growing with many dimensions. Distinctly termed as **'threat-perceptions',** many modern nations today very befittingly, adopt inter-disciplinary approach while discerningly studying and conceptualising the aspects of threat-perceptions as valuable inputs into their processes of policy-formulations.

Now, again coming back to the ancient-era Europe, interestingly we do also notice that both in civilisational as well as in the imperial perspectives the principally vast-regions for the multiple activities and adventure of the Romans, have been the continental Europe, and also some areas on the eastern Mediterranean coast especially such as Jewish homeland particularly Jerusalem and Palestine-regions. Naturally, therefore, the multiply enchanting history of the ancient Romans; do also remain inseparably overlapping, from history of the ancient Europe. For the historically noted Roman-empire, first threats did crop-up from the barbarian Gauls and therefore, in order to protect themselves and their mighty empire they adopted some effective geostrategic and military measures. In this context, history also informs us that even during the days of the Romans in spite of being geostrategically as well as militarily so powerful; peoples of the ancient German-territories since they themselves being barbarians, were quite war-like and also militarily very strong. Moreover, these ancient-era Germans happened to be well versed in conducting raids and gradually penetrating in others' territories where ever opportune-openings they found; with a military-strategic aim to occupy for aggressively settling down.

In these struggles in the ancient-Europe, besides other factors, the clashes then taking place among various intra-cum-inter-tribes, the always highly influential racial-considerations also triggering the conflicts happened to be not uncommon. In addition, though the Romans had taken innovative steps to militarily strengthen their defences against such raids and border-violations; defensive fortifications were sometimes being breached too, especially by the ancient Hadrian and Trajan and then; both tribes in fact being of non-Roman origins. Contextually here, one more important notable point emerges that the Roman-emperor Augustus then, while militarily engaging the Germans for overpowering them and driving them away from borders of his empire lost around ten Roman-legions in the bloody clashes.

Further as the process of decline of once mighty Roman-empire gradually started; the Germans seized the opening opportunities for more and more aggressive raids to capture additional areas for their fresh settlements. In course of few centuries, such opportunities too began expanding and naturally then, Attila the Hun and even the Vandals were not far behind. On the ruins of Roman-empire, a number of ancient barbarian-kingdoms began sprouting. History shows that Nordic kingdoms began coming-up such as Visigoths in the Spanish areas, Vandals in the north-African territories since earlier being part of the empire, Franks in the ancient France and so on. Interestingly, during those turbulent times one prominent name emerges on the pages of history and that is of **Theodoric the Great**. His grip over the art of generalship in battles was definitely noteworthy, and he was also a brilliant military-tactician. He decisively achieved victories over Odoacer, in three historic battles between 489 and 490 A.D and; he did so since taking initiates in the name of Eastern Roman Emperor. Since he himself got converted into Christianity but as a strategic visionary, he tolerated peoples of other religious-faiths too, within his imperial domain. In those ancient times, this was indeed remarkable.

But historically, the die was perhaps unalterably cast, for final eclipse of the empire and in addition tasks facing him were too difficult for the final success. Ironically then glorious city of ancient Rome too, was not spared with any mercy and consequently Italy got overwhelmed and ruthlessly converted into come-up, leave their inedible-marks on the pages of history and, then gradually fade-away. But these historic processes don't take place suddenly since, centuries roll-by and they gradually begin to happen in a highly complex manner. And also, by the time all these do come to a somewhat 'visible-conclusion' as multi-dimensional phenomena with a prolonged time-frame; the concerned period becomes a history of highly turbulent and at times, even violent historic-saga. All these indeed were much more in case of the prolonged and highly violent eclipse of the Roman-empire, indeed having multiple ancient era civilisational achievements. No wonder that many noted scholars of history, have called this turbulent period, covering a number of centuries, the **'Dark-Age'** in the history of Europe.

In this context we may also bear in mind that during the golden age of Roman-empire spanning over a number of centuries, in fact, geostrategically had expanded in almost all directions; not only in the ancient continental Europe but also in north-Africa along the vast Mediterranean-coast and then touching even the Asian-border. Politico-militarily, having flexibility with scope for need-based changes in the systems of governance in those far-off territories but yet; always ensuring maximum possible imperial firm-grip over the vast-empire and hence obviously, these visionary concepts were the fundamental strategic guidelines for their own built-in system of policy-formulations. Based on these historic-points, as we begin having some 'review' of the empire we notice that, when due to the prolonged barbarian-invasion as the European part was lost, the Romans still maintained their better fortunes in their African-territories.

Then, **H.A.L. Fisher** has written that, "Over against the Greek world ruled from Byzantium and the Saracen world governed from Bagdad and Cordova, the vast territory of Latin Christianity stretching from the Ebro to the Carpathians acknowledging rule of the Frankish empire and the Pope of Rome." (Sharma, 1948, p. 237)

In fact interestingly, during 4th and 5th centuries A.D, in this eastern part, a new kind of civilisation had flourished: historically known as **Byzantine-civilisation**. This in fact, was a harmonious-blend of both the ancient Greek as well as Roman magnificent thoughts and culture; and hence contextually giving some noted strategic-importance to **Byzantium**, which is not so commonly well-known. Whatever, this was no small achievement on the part of Romans; when viewed with the important compass of the same **"essential strategic-culture and thoughts"** *(where multiply even the civilisational aspects are integral-part.)*, especially during the times when the Sun was slowly and finally setting-in over once the mighty Roman-empire.

As then further times past, another historic-era of the gradual as well as far-reaching rise of **Christianity** began, across the same vast regions of the earlier-era magnificence of the Roman-empire. In this context, **Richard Overy** writes that, **"Christianity began as a small sect within Judaism, but gradually established itself as significant religious and intellectual force throughout the Roman-empire. It offered both a promise of eternal salvation to individuals and, from the 4th century onwards, a powerful new vision of an empire united under a Christian ruler *(i.e. Theodoric the Great, as already mentioned earlier)* which was to be of enormous significance for the future of Europe. The earliest Christians did not see themselves as founders of a new religion but as witnesses to the fulfillment of God's promise to provide his people, the Jews with a Messiah or redeemer. By raising him from the dead, they believed that God had shown the**

Jesus of Nazareth was the Messiah and that the risen Jesus had commissioned his disciples to preach the good news of God's kingdom. Though the earliest Christians were Jews, they did not interpret the message in political or national terms. Christianity was attractive to people besides Jews for its promise of eternal life and for the spiritual benefits of membership of a close-knit, supportive community.... By this time the Church had grown from a network of small communities, meeting in the homes of richer members, into a well-organised body owning buildings and burial grounds, and led by a ministry of bishops, presbyters and deacons. It was well placed to benefits from the changes that began in 312... From the roots in Jewish monotheism; early Christianity underwent a deal of doctrinal development. Many theologians from the 2nd century onwards attempted to combine the teachings of Christianity with the intellectual assumptions of Greek philosophy." (Overy, 2008, pp. 79-81)

In fact during the momentous advent and gradual spread of the Christianity, Europe was certainly not what it happens to be today; even after giving the universally common factorial-considerations pertaining to the ancient times of the distant past. But as further centuries rolled-by; what all were the central reasons, for making then the Europeans since already having landed in the so-called 'Dark-age'; into civilised-communities as well as marvelously dynamic, ensuring their gradual and steady multiple-advancements and in addition, also along with multiply nationalistic-cultures in different regions of the continent? This appears to be an engaging question. But when we critically begin reading in-between the lines in the above quoted paragraph with the same earlier mentioned valuable 'compass' of the 'essential strategic-culture and thoughts'; the fascinating answers to this weighty question also, begin surfacing. In the process then, we further significantly notice that it was the **Christianity with its many influential attributes**, so capable of triggering the far-reaching dynamism for the later-era progress of Europe and the multi-cultural Europeans. Well, if at all it was so then, it perhaps happened that way in the beginning only and that too we may say so at the most in retrospect.

As we move on, we notice that in those ancient times, overall socio-political as well as socio-cultural conditions in the continental Europe happened to be far from being peaceful for ensuring stable tranquility. Civilisational value-systems since considered to be essential pre-requisites in this regard, were still in the process of gradual evolution and then in any case all these perforce had to be very prolonged affairs, covering a long span of many centuries. Historically, all these definitely have not been particularly Europe-specific, since those turbulent phenomena were no less unsettling and turmoil-ridden in Asia and North-Africa along the south and eastern Mediterranean coast-line and deeper; specifically in our context, as we begin critically to have a glimpse of their ancient histories. We may add further that in any case in larger perspectives, all these have been very much part of the overall mighty Nature's process of human-evolution with its multi-dimensional attributes; covering man's long-history over very many centuries. Naturally, their vast and deeper studies are domain of specialised subject-matters of different disciplines and hence outside the scope of our subject in hand.

Now as Christianity began gradually spreading and taking deeper-roots all over the Europe, with its many socio-cultural as well as socio-political connotations, we also notice that with the eclipse of the classical Roman-civilisation, barbarians found more and more territorial space for spreading their own kinds of imperial-tentacles. Among them the most noted and with significant strategic-vision, was the **shining Charlemagne; since more commonly famous as Charles the Great**, who ruled over his kingdom from 768 to 814 A.D. He and some other barbarian rulers in Europe had already got converted to Christianity. In the West, they were also historically known as the Franks. Significantly even earlier, another great ruler happened to be **Charles Martel** and he was military-visionary and a great warrior, with great tactical-skills in battles. His particularly brilliant feat was defeating the invading Arabs in a keenly contested battle of Tours in the year 732 A.D. and thus saving Europe from the Arab-domination and their intended rule over the West. Geostrategically then, this was no routine politico-military achievement since otherwise, the entire subsequent course of the European history would have entirely different, with many far-reaching fall-outs and also not the least; even affecting possibly, the further survival as well as expansion of Christianity.

With respect to the above mentioned aspects, there may possibly be some opinions that historically, in our geostrategic and politico-military perspectives are largely matters of conjecture. Interestingly, we notice that Christianity initially originated in Jerusalem, much after the **birth of Jesus Christ in Bethlehem** but subsequently to start with, it gradually spread all over the European continent. Subsequently then in this context, it may appear to be another point of special significance that though the birth-place of Jesus Christ happens to be ancient Bethlehem now in Israel, the **'Mecca' of Christians from all over the world, is Rome in south-western Italy near its international border**. This is not only fascinating but, does have some far-reaching historic significance and also has been playing role since inception; in decisively shaping the globally influential European-history in the making. Therefore contextually, in a larger political and geostrategic terms we may also notice though may be inadvertently and as already mentioned earlier; the ancient Roman-civilisation multiply contributed historically, in dynamically shaping-up the European-nations.

The important specific points are that Europe was not only civilised but in a way; Rome subsequently gave the religiously most significant Christian "Mecca" to the world. But these developments were not fast and hence, took long period of time. In the history as such, nothing remains isolated either as a 'causative' or 'effect' factor and so inter-relations or interconnectivity is always a 'binding-law' in this regard. Hence, though once glorious and greatly powerful the Roman-empire was sinking then; various European-nations had gradually begun getting formed and advancing in their own distinctly different ways. But again, these gradual and historic phenomena were never smooth and peaceful but always full of turmoil and ruthless violence. Interestingly therefore, in our specific global geostrategic and politico-military perspectives; we may also view them again as historic legacies of both the ancient Greeks and Romans.

And then subsequently they in civilisational-terms, formed substantial segments, both socio-politically as well as imperially; of various European-societies and their nations. Then as a number of centuries passed these were causative in also triggering the gradual birth of various European Monarchs, Royal-dynasties and their own kingdoms/empires in the European-continent. At the same time again, these historic-phenomena were anything but smooth and peaceful. Contextually again in macro-perspective this has been one of the most important nature of the man's living-history since its inception. Well, further in our context, if examined with some strategic criticality we may interestingly notice the 'seeds' of the much later-era turbulent birth of their different Empires and then, some of these empires did acquire powerful global-reach and also multiple-influence on their foreign subjects.

Appropriately therefore, more so in our specific context, learned historian and author **Paul K. Davis** has quoted **Robert Heinlein**, that, **"Anyone who clings to the historically untrue,—and thoroughly immoral—doctrine that violence never settles anything I would advise to conjure up the ghosts of Napoleon Bonaparte and Duke of Wellington and let them debate it. The ghost**

of Hitler would be referee. Violence, naked force, has settled more issues in history then has any other factor, and the contrary opinion is wishful thinking at its worst. Breeds that forgot the basic truth have always paid for it with their lives and their freedoms." (Davis, 2001) (*Well, the above quoted perceptions, especially during the present 21st century and perhaps even beyond may not be agreeable to all and which is also quite understandable. But historically, the point remains valid and well-taken since, the irrefutable facts of man's vast living-history of the thousands of years; conclusively and precisely establishes the above mentioned perceptions regarding the man's world. And at the same time even today, they are certainly not totally obsolete. Hence the point precisely remains that as per the global geostrategic and politico-military considerations with respect to the past, present and possibly even future; the above quoted perhaps timeless perceptions may perforce to be contextually considered.)*

Now again referring to the 'controversial' book of **Professor Samuel P. Huntington** writes that, "Religion is a central defining characteristic of civilisations, and, as **Christopher Dawsonsaid, 'the great religions are the foundations on which the great civilisations rest.' Of Weber's five 'world religions', four—Christianity, Islam, Hinduism, and Confucianism—are associated with major civilisations. The fifth, Buddhism is not. Why is this, the case? Like Islam and Christianity, Buddhism early separated into two main subdivisions, and like Christianity, it did not survive in the land of its birth**." (Huntington, 1977) (*Perhaps, everybody would not fully agree to his perceptions as since earlier, we already noted. But these happen to be his perceptions. Now, one may like to know as to what precisely persuaded him to write and, how his book in question, developed in to a shape and form; the way it has been brought before the world?*).

In this specific context, we may like to refer to the highly analytical USI National Security Paper 1997 by Shri J.N. Dixit, IFS (Retd) where he has mentioned that, "First Huntington's thesis on the Clash of Civilisations was not a result of any objective or detached interest in the contradictions and conflicts afflicting the world in the post-cold war era. It was a specially commissioned thesis in which Professor Huntington was asked to analyse the changing security environment and to assess how US national interests would be affected by the changes which he could discern. It was more a prognostical position paper on the strategic challenges that the US would face and the options it would have to exercise to meet these challenges, than a scholarly attempt at understanding the social, cultural, political and economic processes which have emerged after the end of the Cold War and which may affect inter-state relations and human conditions in general, well into the 21st century."

From the above mentioned facts, it becomes clear as to how Professor Huntington's well-known and also somewhat controversial book came into shape and, for what specific purpose. Moreover, he has been part of US Strategic and Security Establishment and then; after the end of second World War and subsequent disintegration of the USSR triggering the demise of Cold-War era; this American Establishment had also developed a perception that US had to be responsible for and also take initiatives for maintaining world order by restructuring it for international peace and discipline. Well, these would always understandably be debatable in certain quarters. Therefore, it may be prudent for taking this book in desired perspectives and thus, enabling us to perhaps avoid viewing it in a rigid 'black and white spectrum'. Further, so far the credentials of Professor Huntington are concerned there can be no doubt about them and in addition, his academic accomplishments and scholarship have definitely been extraordinary and also widely well-known.

Now coming back to the ancient-cum-medieval era in Europe, it is widely known that though initially Christianity had begun as a middle-eastern Semitic religion, subsequently it had become predominantly European and thereby, almost whole of Europe had adopted it in some form or other; with many far-reaching fall-outs. But during the European medieval-era, one aspect stands-out and that, Christianity as a religion was principally responsible for comprehensively triggering the socio-religious as well as socio-cultural vital unity among the then multiple and highly feudalistic European-societies. In addition, historically viewing the same contextually in our specifically critical geostrategic and politico-military perspectives; we may notice this aspect to be no small matter. And in order to substantiate the same, we may find that when prolonged **War between the Cross and the Crescent** had begun and then subsequently **known as a historic 'Crusade'** this particular sense of valuable identity and unity among the Christians came in handy and in great favour of the Cross for meeting the threatening challenges of the Crescent.

Interestingly, on the historic-legacies of once mighty Roman-empire; various Monarchies and feudal War-Lords gradually commenced cropping-up in different European-societies, carving-out their own highly ruthless and feudal empires/principalities as different political-entities; a great turbulent-age of wars and violence then, had dawned on medieval-Europe lasting for many centuries. In basic human-terms, history of this extremely violent European-era was ruthlessly dictated by the swords, spears, arrows and soldierly-horsemanship in military-maneuvers and in battles and hence written with human-blood on its pages. All these got shaped into a prolonged saga of chivalries, unhesitating sacrifices and fighting and also at times, brilliant or otherwise records in the history-making military command, leadership at different levels in battles.

Thus as centuries further passed, and Europeans gradually got socio-politically organised into different nations; they then inherited the same history-making soldierly-attributes, military-leadership qualities as well as empire-building potential from their mighty Roman ancestors. Whenever proceeding with a review of the history of medieval-Europe, with some geostrategic and politico-military criticality; we do notice the above mentioned multiple military-attributes, imperialistic-traits coupled with their determined perseverance towards the multi-dimensional efforts for empire-building, on the Europeans' past, both within the European-continent and subsequently during much later-period in Asia and Africa. The concerned segments of the history, specifically dealing with them, are again medieval Europe's chequered-saga of wars and all sorts of highly intriguing-conflicts. We would be suitably dealing with these aspects, along with their multiple causative-factors in later concerned chapters.

In our specific context at this stage of our discussions; one historic-fact we need to always bear in mind is that the modern concept of democracy as the most preferred system of socio-cultural and political governance with all its multiple-attributes as well as heat and dust; were yet to get born in any of the ancient human-civilisations and also even during the medieval-age anywhere. The point is that in our specific geostrategic exploratory studies this important aspect does bear a special significance, as would be seen subsequently. Now the historic-saga of the multiple human-civilisational progress as well as those of relentless wars and feudalistic-conflicts in medieval-Europe won't have taken place at all but, for the gradual dawn of the age of far-reaching feudal chieftains, war-lords, emperors and their empires in all noted countries in the medieval-era Europe. Logically therefore, even advent of the European Monarchs and the various monarchies all-over in medieval-Europe too, have been certainly no less violent, conflicts-ridden and ruthlessly turbulent. These were nothing peculiar to Europe but in fact, historically common to almost whole of Asia, Africa, the Middle-East and in the similar other places, in distinctly different ways and with different culture during those times.

Since the historic-spread of the Christianity along with its few different Sects in the then Europe did give the Europeans a re-assuring sense of identity and valuable socio-politically conditional unity, for facing any threats from other religious-regions but; they also kept fighting among themselves, as their different competing interests kept-on relentlessly clashing regardless. Added to this, the mutual power and supremacy-oriented rivalries between the Monarchs and the Pope, further increased all sorts of complications and many kinds of other problems, in the medieval-Europe. Precisely in this context, **Prof. S.R. Sharma** has written that, "The cause of the struggle, though it might appear trivial now, was considered most vital by both parties in those times. In effect it was the question whether the Pope or the Emperor should be regarded supreme in Christendom. Both were trying to arrogate to themselves the myth of a bygone age, viz. the ideal of Theocracy. There may be little doubt about the secular sovereignty of the Emperor and the spiritual sovereignty of the Pope. But unfortunately society, especially in medieval Europe, was incapable of such dichotomy. Each party was desirous of exclusive authority and none prepared to accept a compromise. Indeed, conflict was inevitable owing to overlapping jurisdictions, and an impartial tribunal was lacking." *(Interesting indeed and more so, historic-facts are nothing but facts and unfortunately; even these clashes too, caused highly avoidable blood-sheds during those now bygone medieval-age in Europe.)*

To substantiate the above; we may wish to recollect the pre-planned and cold-blooded murder with extreme cruelty of Thomas Becket sometime in the 12th century, who was then learned, courageous, highly-principled and visionary Archbishop of Canterbury in London. And most ironically, this was arranged and ordered by none other than the British King Henry II. The King earlier had spotted him as a highly capable, possessing great administrative-skills and visionary person and accordingly, since recommended by the earlier **Archbishop Theobald of Canterbury**; whose secretary happened to be Thomas Becket himself. King then appointed Becket as his Chancellor in 1154 AD, who during those times was like the British Prime Minister. In this important position too, he carried out his responsibilities with rare distinction; making him the King's great and trusted friend as well. King Henry II then wished him to be the Archbishop of Canterbury. Visionary Thomas Becket had warned that this may lead to irreconcilable clash of interests; and the rest is a piece of great historic tragedy. In addition, this also happens to be a non-washable 'black-spot' on the pages of medieval-era British-history; since those times the British too, were no less ruthlessly obsessed with socio-cultural feudalism in many ways.

Interestingly, **Rodney Castleden** writes that, "In 1162 Henry decided.... Given the charge of looking after the interests of the Church in England, **Thomas** did just did that, and steadfastly resisted Henry's attempts to reduce the Church's great power. Within a short time, the two men were enemies, and Henry wanted to remove Thomas, who was now an obstacle to his reforms. **The men who were to remove this obstacle were four barons: Reginald Fitzurse, Richard le Breton, William le Tracey and Hugh de Morville. These were the men who were demonised as the four 'knights' when Thomas was canonised. Perhaps the most peculiar about these men is that in spite of committing the most notorious crime of the middle ages, very little is known about them. They appeared out of nowhere, rode to Canterbury to insult and murder the Archbishop, and vanished again**. Rather more is known about Reginald Fitzurse than the others.... Obscure though these men are—and were, even in their own day—they are remembered for one encounter: their single act of phenomenal brutality and sacrilege, the murder in his own cathedral of Archbishop Thomas Becket, the greatest saint of the middle ages." (Castleden, 2009, pp. 331-332)

Again the history of the birth and subsequent development of European Monarchs and their kingdoms/empires too, are no less fascinating. **Whether it is China, India, the Middle-East and then in Europe; history quite emphatically is able to display that, first there had been all sorts of clashes and wars between chieftains of different tribes/clans and then gradually leading to the historic formations of small, medium-sized or relatively big kingdoms**. Interestingly, as a parallel and in certain cases perhaps, even much earlier historic-phenomena of the birth and gradual advancement of various ancient as well as medieval human-civilisations too, began taking place. In addition long with them, man was also gradually able to evolve, develop and imbibe the already discussed **"the multiple-aspects of essential strategic-culture and thoughts"**. We are largely aware that since, all these historic-process have been overlapping, and hence, certainly not compartmentalized, and then triggering them to be highly complex and also human blood-ridden.

Subsequently then, some of the comparatively more powerful kingdoms were able to expand, as opportunities came their way leading again to the gradual evolution of empires and also, simultaneous formation of various Nation-States in different continents.

Nehru writes that, "The Holy Roman Empire was not a continuation of the old Western Roman Empire. It was something different. It considered itself the Empire, the Emperor being boss over everybody else in the world—except perhaps the Pope. Between the Emperor and Pope there was for many centuries a contest as to who was the greater. But this also was to come later. What is interesting to note is that this new empire was supposed to be a revival of the old Roman Empire, when this was supreme, and Rome was said to be the mistress of the world. But to this was added a new idea—that of Christianity and the Christendom. Hence the Empire was 'holy'. The Emperor was supposed to a kind of Viceroy of God on earth, and so was the Pope. One dealt with political matters, the other with spiritual. This was the idea, at any rate, and it was from this, I suppose, that the idea of the Divine right of kings arose in Europe. **The Emperor was the Defender of the Faith...that the English King is still styled the Defender of the Faith**." *(This is indeed a historic magnificence of an idea; still in continuation in Europe. Whatever may be, as we further critically look into it in our specific context; we may very fascinatingly notice the seeds of the causative-factors for the rivalries between the Emperor and the Pope in medieval-Europe, and they did tensely survive for a number of centuries—since how could smoothly 'two swords in any case be accommodated in a single scabbard'?)*

Historically, we are informed from the well-received books of noted Indian as well as European scholars and writers that around 476 A.D., the demise of Western Roman Empire took place during the times of **Romulus Augustulus**; providing the space for coming of the New Empire. In fact, this process was in a sort of continuation, socio-politically, in many ways. Further by this time, socio-culturally, one new and the other in continuity; the two factors stand out rather prominently. One was the phenomenal spread of Christianity gradually on firm foundations, all over the medieval-Europe, and other was perhaps the continuation of the same **age-old and all-pervasive culture of feudalism.**

We have already noted earlier that earlier, that the Arabs had begun menacingly threatening some of the medieval European Empire, during the reign of **Charles Martel**. This historically resulted in two wars, during 717 and 732 A.D. respectively. Militarily both the battles were clearly decisive and hence, in both Arabs were defeated. And in any case, the outcome of these two military engagements between the Europeans and Arabs later on were also instrumental in maintaining a particular course of the history of medieval Europe, where the times then, were moving at quite high pace. History then further leads us to another noted development in West and Central Europe, between 768 and 814 A.D., and that happened to be the gradual creation of the Frankish Empire.

Historically this was during the reign of Charles the Great (i.e. Charlemagne. Living-History of the medieval-Europe was indeed moving fast during that age, since historic division of the Frankish Empire took place between 843 and 870 A.D. and during those times throne was occupied by Lothaire, Charles, and then Louis. Further when Otto the Saxon was ruling, sometime in 962 A.D. Holy Roman Empire got consolidated and firmed-up.

As mentioned earlier during our discussions; the above development naturally resulted in relentless struggle for decisive supremacy, between the Emperor and Pope; and no side was even remotely willing to accommodate the other with any reasonable adjustments. During the period between 962 and 1056 A.D., as there were long such struggles and this then resulting in the Emperor prevailing over the authority of Pope, on the powerful and rich Christian religious Institutions, with vast estates then in medieval Europe. Then again during 1066 and 1254 A.D., a fresh struggle began between **Henry VI and George VII**; with another outcome of **Pope's supremacy**.

In our specific context at the present juncture of our continuing discussions, we are further informed by the European-history **that intermittently from 1122 to 1226 A.D there have again been rivalries for supremacy at Worms between Henry V and Pope Pascal and then with Fredrick Barbarossa and, Fredrick II with Pope Innocent III respectively; encompassing over a time period of 104 years**. History has also recorded the Pope Innocent III as the greatest of Popes. By this time, Christendom had already expanded extensively almost whole of medieval-Europe with firm roots. And simultaneously then we notice that Islam had already got born earlier in the Mid-East, and this too had expanded there with firm-roots. Then the Turks under its fold and naturally, had the irresistible urge to expand by raising their swords. Consequently then, not much the Arabs but then even more so, the Turks had begun threatening and harassing the European Christian-pilgrims, while visiting Jerusalem and the Bethlehem. Even further then, the Turks had also started invading the southern European territories, across the Mediterranean; resulting in the Islamic subjugation of Spain.

European Christian-world was in no more in any mood to tolerate such relentless Islamic incursions and therefore a long-drawn historic process of extremely violent clashes began between the Christians and the Muslims first in southern Europe and then in the Middle-East at various locations, particularly in religiously highly important Jerusalem, and then in Constantinople as well. We have earlier already noted that both socio-religiously as well as from war considerations, especially geostrategically and politico-militarily, the superbly motivating Christendom had already emotionally united various distinct European Nations. These aspects had since usefully triggered in evolving and imbibing the crucially important **'essential strategic-culture and thoughts'** too and thereby, further ensuring invaluable synergy in making the war-preparations and necessary both strategic and tactical operational-plans for launching the military-campaign in this regard. History has recorded them as ruthlessly violent and prolonged **'Crusades between the Christians and the Muslims'; intermittently lasting for nearly 400 years**. Then, history further informs us that, in these momentous crusades the certain noted personalities actively involved and leading them, were **Peter the Hermit, Pope Urban II, Conard III (Germany), Louis VII (France), Saladin and Richard Coeur de Lion** respectively.

In this respect, **Richard Overy** writes that, "**A Crusade was a Christian war concerned with recovery and defense of lost lands. For those taking part it was a means of salvation. Starting with the First Crusade in 1096-99, the crusaders altered the political balance in the eastern Mediterranean for many centuries. Christian holy war also remained a recurrent feature of northern European politics. Christians had made pilgrimages to the Holy Land since the 2nd**

century AD. The combination of religious tourism and spiritual inspiration offered by pilgrimage had greatly the number of visitors after the conversion of Constantine and the discovery by his mother Helena, of the supposed True Cross in Jerusalem. Many new churches and monasteries were built in the Holy Land, often on newly identified holy sites of both Old and New Testament events. Jerusalem in particular held pride of place in the imagination and hearts of Christian everywhere, which not even its conquest by Arab forces in the 7th Century, could overturn. Indeed, access thereafter far from denied and throughout the 8th, 9th and 10th centuries there are reports from travelers from Britain, France and Italy who prayed at the holy sites. A constant stream of souvenirs and holy relics, furthermore, reached with great reverence in churches across Europe. Allied to the importance of pilgrimage to the Holy Land were the missionary endeavours to convert pagans conducted in Western Europe. Between the 6th and 11th centuries this saw the conversion of the English, the Frisians, Saxons, Slavs, Scandinavians, Bulgars and Rus by Irish, Roman, English, Frankish, Saxon and Byzantine missionaries. Within the Christian Church, moreover, there were intense bouts of eradication of heresy. Pilgrimage, the striving to achieve orthodoxy and conversions of pagans, coupled with the religious piety and devotion of the laity, came together in the crusading movement of the central and later Middle Ages."

We learn from the history that in fact, the two principal causative-factors since irreconcilably and gradually emerging in the medieval-Europe as well as in the ancient historic-territories in and around Jerusalem were, indeed, highly volatile and powerful enough for triggering this historic-crusade. The first one was highly persuasive but mainly peaceful Christian-missionary efforts to convert the peoples to Christianity; and the second one being the emergence of highly intolerant militant-Islam, especially Muslim-Turks; who fanatically believed then, in **ruthlessly violent Jihadi-Islam.**

In fact as quoted above, conversion of Russia and other Europeans was complete; much before the beginning of crusade. But we learn that at first the Bulgarians were quite hesitant to convert to Christianity, and also half inclined towards Islam; but the overwhelming attraction of Constantinople became finally decisive in favour of Christianity. Here, it would also be relevantly interesting to note that, **Byzantium happened to be much earlier the older name of the Constantinople.** Then as further time passed, many European Christians along with their Kings/ Emperors, while visiting Jerusalem on pilgrimage; were harassed and even some were killed and looted by those mindless Islamic Jihadi Turkish-gangs.

Then so far as the some historically influential saints and apostles are concerned, they were indeed many; but among those, a number of them prominently increased the magnificence of the Christianity as well. Among them **St. Paul** and **St. Peter** are, indeed, the extraordinarily shining apostles. There are weighty historic reasons, as recorded by momentous Christendom-history of the medieval-Europe; pertaining to raising these two Saints to such exceptionally high religious pedestal, and that is understandable.

In this historic context, **Professor S.R. Sharma** writes that, "**St. Paul and St. Peter, 'the two glorious apostles ', may be considered the founders of the Catholic Church even as Jesus Christ was the founder of the faith. In the estimation of Lord Birkenhead 'Of all men who may claim to have changed the course of the world's history, St. Paul must surely take the first place. He altered the basic ideas of Western civilisation: the whole of our history bears the marks of that busy career of impassioned teaching which the Jewish tent-maker undertook after his conversion to faith in Jesus Christ.' About the importance of St. Peter we have the testimony of Christ Himself:**

'and I say also unto thee, that thou art Peter, and upon this rock I will build my Church: and gates of hell shall not prevail against it. And I will give unto thee keys of the Kingdom of Heaven: and whatsoever thou shalt bind on earth shall be bound in Heaven: and whatsoever thou shalt loose on earth shall be loosed in Heaven. St. Peter was the First Pope (Latin, *papa = father*) or Bishop of Rome. Both on this account and because of the prestige that Rome enjoyed throughout Europe as the imperial capital, the See of St. Peter; became naturally the Head the Roman Catholic (Universal) Church.... Next to the Popes, who were the head of the official hierarchy of the Catholic Church, there the unofficial army of monks who greatly influenced the shaping of Christian life in the middle ages? On account of their lives being very strictly regulated, they were called the 'regulars', and the official clergy were distinguished from them as the 'seculars' or persons still connected with the world (*speculum*). **Monasticism was a philosophy which considered the normal life in the world miserable and sinful, and therefore to be redeemed through severe discipline. It was however, not peculiar to medieval Europe. It corresponds to the Hindu idea of *sannyasa* and the Buddhist ideal of asceticism which was carried to excess by the Jains in India**. It is better, some thought, to undergo voluntarily the maximum suffering in this world and earns merit in Heaven, rather than sinfully enjoy here and earn the torments of hell later as the wages of sin." (Sharma, 1948, pp. 246-248) *(These are indeed very thoughtful analyses in a context and at the same time, they do convey some important facts regarding certain socio-cultural as well as religious aspects of the then societies in medieval-Europe; which in turn, were principally instrumental in bringing-in many far-reaching changes in the later-period course of man's living-history. Historically, these are no small matters, though they may appear so, when one may at the most care to have just a fleeting-glance only.)*

Hence, when we may try to again look little deeper into the above quoted paragraph from a scholarly book, and many momentously causative-factors may come, otherwise since hidden, to one's notice. **Significantly, St. Paul is considered to be a pioneer who innovatively gave the basic ideas of the Western civilisation which in turn subsequently, were decisive in multiple-ways.** We are aware that Judaism, as a religion, has been much older than the Christianity and at the same time, among the historic Semitic human-races; the Jews have also been singularly more tenacious, scientific in multiple-approaches in life and also extra-ordinarily industrious, more focused, absolutely non-feudal, dynamically progressive, and then innovative as well. Specifically here, we also learn that **earlier in his life St Paul himself was a Jew and a hard-worker**. In this context, the precise-point as in earlier chapters too since brought out, remains that the multiple aspects of human-civilisations have been singularly instrumental in triggering the magnificently multi-dimensional continuing human-progress and simultaneously ensuring the many parallel advancements of those civilisations as well.

In totality as well as in our context, when viewed again in critical geostrategic and politico-military perspectives St. Paul's pioneering contributions have been marvelously enduring and far-reaching, indeed. Further in the larger perspectives, as additional centuries gradually passed we may also add that these aspects of the dynamic European societies, with all their positivity; again were decisive more so, in evolving and enriching their same **"essential strategic-culture and thoughts"**. In other words, with the Europeans; the religious aspects acts as, at time hidden, 'catalytic-agents' for multiple-growth and meaningful progress—and not hindrances on their way. But ironically perhaps even remotely not so in the case of some others; particularly with respect to the entire Middle-East and the Arabian-region (of course, excluding Jews and then Israel.), the Indian sub-continent since distant ancient-times and certain other areas as well.

Further viewing the above as well as earlier mentioned aspects regarding Christianity, again in totality, especially in the context on our continuing deliberations with the requisite strategic-criticality; it is all the more understandable as to how and why, to begin with, nearly the entire European-continent enduringly and with vision-oriented flexible-dynamism; became historically invaluable part of **gradually expanding and enriching Christendom**. These aspects facilitate our efforts to develop some understanding as to again—why and how some civilisations went on flourishing and some gradually 'perished', faltered and continued perhaps 'limping for survival'. (***How precise and true!***).

(j) Rise and also 'negation' of Islam: 570 A.D. onward and beginning of the medieval-age, as well as its modern-history in the making

It appears to be appropriately more relevant in our context that, before taking on the various aspects of Islam; we may briefly touch upon the interesting history of the pre-Islamic Arabia. **In fact, Hazrat Mohammed was born in Mecca, in Arabia, in and round 570 A.D.** History interestingly informs us that during ancient times, i.e. a number of centuries before the advent of Islam; the vast and largely desert Arabian peninsula was mainly populated with highly non-developed nomadic tribal-groups, and also population was thin and deprived of even relatively primitive 'human-amenities'. This was naturally understandable; since this vast Arabian desert-oriented geographical-region, did not possess any kind of green worthwhile forest-cover, rivers and other valuable natural resources. Logically therefore, there was hardly any rainfall and hence, water in fact was the most precious commodity. Life was hard, leading to perpetual struggle for survival; from birth to the sandy-grave. The tribes were organised in groups and, these numerous groups shared many common customs and some systems of kinship. But again such kinship was unable to create any socio-economic bond. Moreover, in the absence of even any primitive law-enforcement system these tribes remained hostile to each other and indulged in frequent violent and mindless clashes and then, all sorts of scarcities added 'fuel to the fire'.

Their dialect was mainly primitive Arabic and these nomadic tribes are principally known as Bedouins. Nehru in his well-known book, covering world-history says that since, deserts and mountains breed tough, brave and freedom-loving peoples and hence these Bedouins too have been so, which historically hardly gets substantiated. But there are some interesting social-facts about them, for example, a special and highly valued friend was compared to their desert-donkey, without any kind of even remotely-intended insult. Their donkey was always a very faithful friend and a precious beast of burden and therefore, their friends to be compared with a donkey was especially taken as a high tribute and salutation. In addition, their swift camels and famous horses were their almost constant companions. (Interesting: isn't it?))

Besides these Arabian Bedouins; the ancient Jews too, were residing as their neighbours. These Jews were staunch followers of their ancient and well-organised religion of Judaism. Then in addition, the Jews were already relatively much more sophisticated and developed; living in their well-organised and nicely managed villages and ancient towns. Therefore, in socio-cultural and politico-economic as well as even in the matters-military aspects; the Bedouins were even distantly no match to their co-habitant Jews. We may recall since earlier too noted that, both the Jews and Arabs; interestingly to begin with, belonged to the same Semitic racial-stock but subsequently differences historically developed between these two tribes, resulting in a drift.

In this respect, **Richard Overy** writes that, "While most of the peninsula was arid desert, there were towns, villages and quite large settled areas. The most important of these was Yemen, in the southwest corner of the peninsula, where the monsoon winds bring water to a mountainous landscape heavily populated with towns and villages. In antiquity, Yemen had supported a thriving urban and mercantile community that prospered from the trade in incense between the area of Suhar, where the frankincense trees grew, and the Mediterranean. Cities such as Marib and Shabwa boasted fine stone-built temples and palaces and a developed language and culture.... There were other cities and settlements in Arabia. There were Jewish communities, notably at Yathrib (later called Medina) and Khaybar, and Christian ones at Najran and in the areas bordering the Gulf. **There were others which were pagan shrines. Arabia had a number of ancient holy places where tribesmen who would normally be at war could meet together in the sacred enclave (*haram*) to arrange treaties and trading arrangements. By the late 6th century, the most important of these was the Ka'aba *(Interestingly, it is believed that the Kaaba, in fact a cubical black stone, has fallen from Heaven at that spot as a Gift of God to the Earth.)*, the shrine at Mecca, whose guardians were the Quraysh tribes. The Prophet Muhammad was born into this tribe in around 570, and the reputation and skills of the Quraysh were an important ingredient in his success**..... Arabia lay on the borders of the two great empires of late antiquity, the Byzantine and the Sasanian Persian. Until the late 7th century, both had supported client Arab rulers, known in Greek as phylarchs: the Ghassanids in Syria and the Lakhmids at Hira in Iraq. The collapse of these kingdoms left the way open to nomad advance." (Overy, 2008, pp. 87-88)

In our specific context regarding the prelude to the advent of Islam, the above quoted paragraph does bring to us some fundamentally important historic facts with precision and clear objectivity and without even remotely falling for any 'mythical-entrapment'. This is indeed highly valuable for our on-going discussions; and also contextually in this regard, with the desired strategic criticality. If, we try to little look into some basic grass-root realities even with a preliminary glimpse in Indian societies today; one perforce can't miss typical regressively fixated and self-negating religious-fanaticism among our Muslim fellow-citizens. Modern India is a secular liberal-democracy with a well groomed and truly free society-based composite-culture. In this healthy scenario; if a sizeable portion of ill-informed and half-educated Muslim citizens still in 21st century India; perforce living in a kind of mental-level 'cocoons', like powerful dark-holes then, it does become a matter of some national concern.

This author has come across many elder Muslims who have since already returned from Hajj, and when asked about the conditions in Mecca and Medina, during pre-Islamic era; they emphatically claiming to have deep knowledge and saying that it was all like an extremely 'dark-age' inhabited by primitive sort of 'sub-humans' and they were worshiping all sorts of senseless and ugly stone and mud images. These had to be broken and cleared by the dedicated followers of the Prophet, for establishing the magnificently enlightening and most revered holy relics of Islam. **And again referring back to the above quoted paragraph from Richard Overy's book; we find that the well-established historic facts happen to be entirely different. And he is not the only historian but many others too, have written on similar lines; for example, even Nehru, in his book, have substantiated the same facts.**

But, this sort of stagnation and regressive naivety is not typical to the Muslims alone and even now, many ill-informed Hindus too in India; are mentally 'stagnating' in similar kinds of mental-darkness. In this respect, this author again may please be permitted to cite a similarly amusing example—some years back he came across quite a literate Hindu middle-aged man, who was not a Brahmin but was proudly masquerading as a 'Hindu-priest' and in any case now, in no way there was anything wrong in his doing so and also he was not a professional and even he couldn't be objected to at all since, he was capable of conducting 'Puja-rituals' as well. So this author happened to ask him that was anything perhaps 'improper' in offering Arabian or Iraqi dates as offering, since this a very delicious and nutritional fruit—and to my utter shock, he typically posing as 'knowledgeable' as feasible and objected to doing so; as this nice fruit was grown in a 'Muslim-land'. Naturally it was not at all agreeable to any right-thinking person and hence later-on; this author advised his wife not to pay attention to any such 'foolish-advice' and she still prefers such dates too along with other fruits/dry-fruits for any such religious-ritual. Moreover, it would be quite amusingly interesting to say that the same dates, when totally dry and hardened, are quite preferred along with other dry-fruits by any Hindu-priests. And at the same time, such naïve Hindu-priests do know that their highly preferred raisins, largely come from Afghanistan and Iran and both happen to be 'Muslim' nations as well. So, there is no dearth of such irrationalities even now; and in any case, the world is like that.

So as since noted earlier, **Prophet Hazrat Mohammed was born around 570 A.D. in Mecca** (now in Saudi Arabia). **And the Prophet in the beginning was in no hurry to give 'birth' to a New-Religion and, the commencement of Islam, since as initiated and guided by him, did gradually begin later. In fact, the history of the birth and subsequent growth as well as somewhat rapid expansion of Islam is no less violent and turbulent too.** The highly revered tenets and teachings of Islam are based on selfless love, peace and universal brotherhood; but the history of the many fanatic Muslims and their various kinds of activities, as the so-called self-styled dedicated followers of Islam; regarding many a times forcible/coercive conversion of some members of the other communities and also relatively the fast expansion of Islam in different geographical directions of Arabia and Turkey; have been anything but peaceful. It was largely based on ruthless sword-wielding. Many have also rather called it historically, as the "Swords of Islam". Well, with prudence and also in any substantially important and relevant context; these rather not so proud facts of Islamic-history just cannot be accepted as matters of any pride in any manner, whatsoever. **More so it is a matter of great irony of ironies that, when the highly enlightening tenets and teachings of Islam since already said earlier; are irrefutably based on totally selfless love, universal-peace and universal-brotherhood.**

Like most Prophets, Mohammed too, in the beginning was shockingly persecuted by his own people. In fact, he was even threatened and hence, had to perforce take shelter in another place. That city, where he went to escape such persecutions; is now named as **Medina, i.e. Medinat-un-Nabi or the Prophet's city**. Then in order to regain his entry into Mecca and take control; he had to fight a violently contested battle at Badr and achieve a decisive victory over his adversaries. So the important point to note is that it was no 'cake-walk' for him, to become an unchallenged master of the entire Arabian Peninsula. In the process, his unparalleled historic achievement remains as being singularly instrumental in giving the Divine-Concept and birth of a new religion, known as Islam.

Now, for getting an initial glimpse of the Spirits of Islam Professor S.R. Sharma's mentions that, "*Islam* meant 'submission to God, and those who accepted this creed had above all to submit to the five disciplines of the faith: (1) Belief in one God and Mohammed as his prophet; (2) the duty of praying five times daily; (3) giving alms to the poor; (4) fasting in the whole month of *Ramzan*; and (5) making a pilgrimage to Mecca. The teachings of Mohammed were in a volume called *Koran*. This and the Hadis or 'traditions' constitute the scripture of the Muhammadans. The prophet during his lifetime was both then spiritual leader and temporal ruler. Hence Islamic society to begin with was a *Theocracy*. Soon after Mohammed's death a dispute arose as to the succession. Some were for the Prophet's son-in-law Ali; others were for electing the

Caliph. The former called the Shiites, were defeated; and the later known as the *Sunni*, triumphed. This rent Muslim society for all time to come into two hostile camps, though there are no fundamental differences of creed between them. Arabia represents the latter, and Persia the former. In India there are representatives of both the sects." (Sharma, 1948)

Now, these basic aspects of the above mentioned code and spirits of Islam are indeed, magnificent, universal and timeless and also when truly taken with most sincerely heart-felt sense of universal peace and love for the entire humanity with not even any remote feeling of hostility towards any other religion or community. **After the death of Prophet Mohammed, the first successor was Caliph Abu Bekr**, who himself was a thoughtful man with high learning and simple living. His sermons have substantial point regarding the limitlessly magnificent Mercy and Divinity of Allah; and his preaching are, indeed, great.

Now, again coming to the preaching of **first Caliph Abu Bekr** and **according to Professor S.R. Sharma's book**, the Caliph further says that, "In the name of most merciful God, to the rest of the true believers. Health and happiness, and the mercy and blessing of God are upon you. This is to acquaint you that I intend to send the true believers into Syria to take it out of the hands of infidels. And I would have you know that the fighting for religion is an act of obedience to God. *(God Knows, how these perceptions can be truly rationalised and ever justified as humane in totally unprejudiced manner.)* Remember that you are always in the presence of God, on the verge of death, in the assurance of judgment, and the hope of paradise. Avoid injustice and oppression: consult with your brethren, and study to preserve the love and confidence of your troop.... As you go on you will find some religious persons who live retired in monasteries, and propose to themselves to serve God that way: let them alone, and neither kill them nor destroy their monasteries: and you will find another sort of people, that belong to the synagogue of Satan who have shaven crowns: be sure you cleave their skulls, and give them no quarter till they either turn Mohammedans or pay tribute." (Sharma, 1948) *(How these too can ever be rationalised and justified in the best interests of universal-peace and tranquility for the entire humanity?)*

While proceeding further in our analytical study and review with requisite criticality in geostrategic and politico-military perspectives, historically in this specific context; learned **Professor S.R. Sharma** writes that, **"With the conversion of the Turks and other savage peoples, Islam tended to follow more and more the closing part of the message rather than its nobler portions.** *(That is concerning the above quoted preaching of the first Caliph Abu Bekr)*. The tribute extracted from the infidels was called the Jiziya. The Jews and the Christians were treated with consideration as 'the people of the book' as Mohammed drew much of his own theology from their traditions. (But, a great irony is that the Muslims everywhere neither say this historically momentous truth openly; nor even silently or internally acknowledge it.) 'We believe in God', declares the Koran, 'and in what hath been sent down to us and what hath been sent down to Moses *(Prophet of the Jews)* and to Jesus *(Prophet of the Christians)* and to the prophets from their Lord. We make no difference between them; and to Him we are resigned; and who so desireth any other religion than Islam, it shall no means be accepted from him, and in the next world he will be among the lost.' (iii, 78-79)." (Sharma, 1948, pp. 261 & 262)

Well, with respect to the above 'underlined-part of the sentence'; the obvious question comes in one's mind as to why and how: and any way, that may be perhaps Prophet Mohammed's perception—but, the most noted and magnificent point remains irrefutable for the posterity that, Prophet himself never, even indirectly, never preached any kind of violence on religious-grounds;

and also mindless torture or killings of fellow human-beings; under any kind of blind-obsessions of religious fanaticism. And these remain the foundational tenets of the Islamic religious-concepts of being a Religion of the inviolable 'Universal-peace and the universal-brotherhood'; and these were strictly adhered to; both in letters and spirit, by Prophet Mohammed, till he breathed last. But alas, the conduct and the recorded history of his followers, i.e. the Muslims; never largely and sincerely cared for this most magnificent message of Islam—which can very justifiably be taken as a Divine-Piece of Timeless Message indeed, for the entire humanity; without any distinction/ differentiation on the grounds of religion, caste, creed or class of any kind whatsoever.

No wonder, that the Muslims, in spite of indeed their praiseworthy magnificence of the multiple-civilisational or other achievements in different contexts; they sadly got ashtray rather very early, from their directed-path from the Prophet and consequently; their history remains perhaps 'unacceptably', soaked in human-blood of innocent human-lives the world-over. And rather very painfully, one perforce observes that in this specific context, even mow in the present modern-times of the 21st century; same kind of medieval-period Islamic history is being written and rewritten, regardless of most tragic fallouts for the Muslims themselves and also the larger humanity around the world. Naturally in the same context, when these aspects are viewed with the desired criticality, we may also notice that tragically in the whole process of this ruthlessly demonic series of rather 'unending' violence; more and more Muslims are being killed by the so-called Muslims and their fanatically-blind leaders themselves.

Then as these extremely tragic-aspects are further contextually examined as per the 'guiding-compass' of the geostrategic and politico-militarily critical-perspectives; we just cannot fail to notice that such madness of unending violence being triggered by Islamic-terrorism and their fanatic-masters and at the same time, being thrust-upon the larger innocent humanities, including the women and children, without any constraint of various national-boundaries; are indeed, posting grave dangers to the whole world. These most dangerous and disturbing aspects of the dehumanized and so-called Islamic Jihadi-violence then; have perforce acquired another kind of now unprecedented and mind-boggling new dimensions of destructive lethality. Now, since some irrational Islamic nations so uncontrollably obsessed with a rather mad kind of Jihadi-fanaticism and in addition, are possessing nuclear weapons with different types of delivery-systems; since got hold of, by all sorts of dubious international methods.

Then extending these arguments further with requisite strategic criticality; we may clearly notice that the real dangers of these weapons and their delivery systems already being 'unsafe' and 'unsecured' and thereby, somehow falling into the hands of various Islamic terrorist-organisations, with their own kinds of dubious international-reach, just cannot be wished-away. And more so, in today's overall on-going highly complex-scenario; any such possibility cannot be ruled-out, in any manner. In addition God forbid; if any such real possibility ever arises and, this may in reality, arise anytime; then its multiply destructive consequences would be unimaginatively long-term and catastrophic, for the entire humanity. In addition then, we also find that so far, the mankind with all its command over the ever advancing high-tech capabilities as well as internationally organised and efficiently managed multifarious systems and Institutions; have yet to find any assured and effective preventive-systems to these unprecedented and highly lethal dangers to the same mankind. And then added to them, the many kinds of chemical and biological weapon-systems of warfare; we may also emphasise that, can there any more grave and worrisome situation for the humanity today? More about these aspects, we would be dealing with, in later chapters in greater details.

Now, coming to the indeed, enchantingly larger aspects of Islam; **French scholar Maurice Bucaille** has mentioned that, "In his objective study of the texts, Maurice Bucaille clears away many preconceived ideas about the Old Testament, the Gospels and the Qur'an. He tries in his collection of writings to separate what belongs to Revelation from what is the product of error of human interpretation.... **At the end of a gripping account, he places the Believer before a point of cardinal importance: the continuity of a revelation emanating from the same God, with modes of expression that differ in the course of time. It leads us to meditate upon those factors which, in our day, should spiritually unite—rather than divide—Jews, Christians Muslims.**" (Bucaille, 1982)

Maurice Bucaille (pages 242 and 243) further writes that, "Some of the collections were made public in the decades following Muhammad's death. Just over two hundred years to pass before some of the most important collections appeared.... As regards their origins, some of the Hadiths and Gospels have one point in common which is that neither of them was compiled by an author who was an eyewitness of the events he describes. Nor were they compiled until sometime after the events recorded. The Hadiths, like the Gospels, have not all been accepted as authentic.... Very early in the history of Islam, masters in Islamic thought exercised a thorough criticism of the hadiths, although the basic book (The Qur'an) remained the book of reference and was not to be questioned."

In the similar larger context, the second President of the Sovereign Indian Republic and a noted scholar and philosopher of great repute, Dr. S. Radhakrishnan writes that, "The Hindu *Gayatri* prayer is a perpetual quest for illumination. The scientist knows that truth is always provisional and hypothetical and what matters is the search. The followers of different religions should regard themselves as fellow-seekers of truth. Gandhiji wrote to an American missionary who claimed that the Christian way was the best for all: 'You assume knowledge of all people which you can do only if you were God. I want you to understand that you are labouring under a double fallacy: that what you think is best for you is really so: and that what you regard as best for you is the best for the whole world. It is an assumption of omniscience and infallibility. I plead for a little humility.' (*Well, without any even unintentionally remote offence to anyone or any religion in any manner whatsoever: since all religions the world over, are seekers of the same Supreme Truth. May be that, they perhaps appear to be trying to lead the believers to advance apparently on 'different-paths' and hence; any truly enlightened and knowledgeable preacher belonging to any religion, fully realises that God Is One and the Only One; since the Same Almighty is always Indivisible and again Supremely Commanding Absolute Non-Duality.)*.... The attitude of absolute possession of truth is not consistent with the catholicity of the great religions at their highest. Besides, every great religion has learnt from others. If religion is to gain the dynamic power which it once possessed of shaping society, rivalries of religions should give place to co-operation among them. The world can no longer afford to be divided on this vital matter. **God is active everywhere in Christianity or in other religions. If we believe that God has revealed truth not to all mankind but to a chosen few, Hindus, Buddhists, Jews, Christians or Muslims, strife and hostility start."** (Radhakrishnan, 1970, pp. 72-73) *(Therefore, precisely the broader point remains that, whoever in a truly rational and right frame of mind anywhere, would not wish any such irrationally mad hostility to begin at all.)*

Now, coming to indeed, the magnificently multiple contributions of Muslims in various countries around the world, since last one thousand years or so; books on various spheres of the world-histories, as authored by many learned historians and scholars do give us very vivid and enriching accounts of them. In this specific context historically, India does command a highly distinguished place; in captivating horizon of Islam, with its multi-dimensional manifestations. The pivotal reasons for these have been that, historically since the ancient times of distant past and hence, very many centuries before the advent of Islam India as one of the oldest and uniquely advanced, dynamic and multiply grand civilisations of the world was superbly placed to receive many contributions from the Islamic-civilisation and in these prolonged interactive historic phenomena, it was also capable of enriching the Muslims and their societies in many ways and in so many spheres of human-endeavour. Hence, whenever and by whosoever, these aspects are critically studied and researched with honest intellectual-objectivity; these magnificent aspects do stand out as irrefutable facts of history.

In this specifically on-going context, **Abdul Qadir** has written that, "That modern Indian Civilisation has developed from the action and reaction of so many different races and creeds upon each other that it is extremely difficult to say which of its features is due to a particular influence. Hardest of all to assess is the influence of Islam, for the various Muslim incursions into India brought comparatively few people of an alien race into India. Even the great Babur, when he 'put his foot in the stirrup of resolution' and set out to invade India, in November 1525, only took with him some 12,000 soldiers and merchants. Of the eighty odd million Muslims, who today form a quarter of the population, the great majority are descended from Hindu stock, and retain certain characteristics common to Indians as a whole.... Muslim culture in India, being a blending of two civilisations, is something *sui generis*, and as such has its special contribution to make to the Western world, as well as to the rest of Islam. The process by which the blending took place is of special interest. A passage in the *Cambridge History of India*, by **Sir John Marshall** (vol. iii, p. 568), well describes the influence of Hindu and Muslim culture on one another. He observes:

> 'Seldom in the history of mankind has the spectacle been witnessed of two civilisations, so vast and so strongly developed, yet so radically dissimilar, as the Muhammadan and the Hindu, meeting and mingling together. The very contrasts which existed between them, the wide divergence of their culture and their religions, make the history of their impact peculiarly instructive.'

The historic 'rainbow' of the multi-dimensional marvels and the enchantingly resultant creativities of such prolonged interaction and blending of the civilisational-aspects of these two unique civilisations have definitely not been confined to the multiple cultural-aspects alone. In fact, when reviewed with the requisite strategic-criticality in our precise perspectives in this regard; we do notice that the magnificence of such historic aftereffect have been much more vast, enduring and far-reaching.

In fact, the Muslims coming to India during the medieval-period for hundreds of years brought with them certain new military-technologies and, tactical-concepts as well as fresh and more flexibility-oriented and specific situation-based the then innovative techniques of forces' deployments and also, more effective positioning of the heavier weapon-platforms. In turn then, these historic developments did work as decisive causative-factors in further advancing composite India's matters-military. Hence, all these need to be perceived and prudently reflected upon with desired sense of historicity and scientific intellectual objectivity.

Now, we further interestingly find that **Abdul Qadir** in his earlier referred chapter itself, in continuity has mentioned that, "In the time of the Mughals the study of the Persian language was eagerly taken up by Muslims as well as non-Muslims. The Hindus, who possess a great capacity for adaptation in matters intellectual, took kindly of Persian literature, just as they are now eagerly studying the English language and literature (*Indeed, historically how precise and true.*).... Things have so changed now that, in spite of the fact that orthodox Hindus have still got idols in their temples, their attitude towards the worship of idols is very different from what it used to be. The intelligent and the educated among them declare that idols are only meant to serve as aids to concentration of thought, and those who appear to worship them are, in reality, offering worship to Him to whom alone it is due (*Misguided and regressively fanatic Muslims suffering from hopeless darkness of minds, should care to read with coolness and openness of attitudes; the above and the following scholarly statements; to clear their own entrenched foggy thought-processes in this regard, since it is always best in India's national-interests. More such quotations would follow subsequently, to substantiate these irrefutable truths even further; specially regarding the gradual multiple 'negation' of Muslim-societies the world over.*).... **The mingling of Persian, Arabic and Turkish words and ideas with languages and concepts of Sanskritic origin is extremely interesting from the philosophical point of view, and this co-ordination of unknown resulted in the origin of the beautiful Urdu language. That language in itself symbolized the reconciliation of the hitherto irreconcilable and mutually hostile types of civilisation represented by Hinduism and Islam. The language thus developed by the combined efforts of Hindus and Muslims now boasts of fairly varied and wide literature, which may be claimed as a common heritage by both, and is gaining every day in importance and strength.** Urdu literature is rich in poetry.... The most popular form of versification in Urdu was the *ghazal*, consisting of stray thoughts on such subjects as love, beauty, and morality. Each line was in the same metre, and endings of each line rhymed with one another. This style of writing has found numerous votaries among Muslims as well as Hindus. In the collection of the ghazals of many of our eminent writers you can find literary gems bearing comparison with some of the best pieces of literature in other languages (in fact, the world over.), though for the bulk of this kind of verse no merit can be claimed. Hence it was that some of the second half of the nineteenth century who realised the limitations of the *ghazal* and its shortcomings felt the need of literary reform. In Delhi, **Ghalib** was the first to realise this, but it fell to the share of **his distinguished pupil, Hali**, to inaugurate the reform. He started a new school of Urdu poetry, which has had many adherents among the contemporaries and successors." (Qadir, 1937, pp. 287-290 & 293)

Well, in the larger interests of our on-going strategically critical journey, we learn that Islam's contributions have definitely not been exclusively in the fields of literature only. We have earlier noted that as Muslim-invaders and their armies crossed India's western-borders and entered, from foreign lands their military conquests had been perforce very violent and ruthless but, vast majority of them settled-down here and became Indians. Some invaders were just demonic with no dedication to the true teachings of Islam and were criminally betraying the Prophet himself.

Further in fact, as a medieval-era process of the Muslim-invasions began, India was not only materially quite rich but, she was also commanding the very rich traditions of multi-dimensional socio-cultural, socio-political as well as many spiritual kinds of civilisational-marvels and multiple-enrichment. And the same were enchantingly noted by their many accompanying Islamic scholars and thinkers from time to time, for their own scholarly understanding and fruitful assimilation. Subsequently, these in turn, gradually triggered another magnificently historic-marvel of the unique blending and also the mutual enrichment of otherwise; two substantially different civilisations, on the territories of medieval Indian-subsequent. **And hence, about these very historically unique civilisational-wonders; Abdul Qadir has talked about in his above referred essay and also contextually, a significant part of the same has since been already quoted above.**

The language, literature and the poetry are definitely important segments of the rich culture of any ancient society; commanding very high multiple developments historically in civilisational perspectives; since already earlier noted and substantiated in case of the ancient India's civilisation.

For more precise elaboration of these aspects, it perhaps appears more appropriate to again quote from the highly educative writings of some of the well-known experts-cum-authors in these highly scholarly areas of man's relentless pursuits of learning.

Hence again referring to **Professor S.R. Sharma** as he writes that, "In the field of philosophy alone two names are famous throughout Europe, viz. Avicenna and Averroes. The former whose real name was **Abu 'Ali-al-Husayn ibn Sina** (980-1037 A.D.) was one of the greatest scholars of the Islamic world. Though primarily a philosopher he made valuable contributions to medicine and science as well. His *Canon of Medicine* was an encyclopaedia dealing with general medicine, simple drugs, and diseases affecting all parts of the body, special pathology and pharmacopoeia. It was greatly in demand in its Latin translation down to the seventeenth century.... The name of Averroes was '**Abu'l-Walid ibn Rushd** (1126-98 A.D.). Averroism continued to be a living factor in European thought until the birth of modern experimental science. The golden age of Islamic science and medicine was from about 900 A.D. to about 1100 A.D. The *al-Hawi* or 'Comprehensive Book' by **Rhazes** (c. 865-925 A.D.) may be cited for illustration. It is considered as perhaps the most extensive work ever written by a medical man. **For each disease Rhazes first cites all the Greek, Syrian, Arabic, Persian, and Indian authors, and at the end gives his own opinion and experiences, and he preserves many striking examples of his clinical insight**.... The name of **al-Biruni** (972-1048) is familiar to readers of early Muslim history in India. He came to India with **Mahmud of Ghazni**. But few, perhaps realise the nature of his contributions to various branches of knowledge. Familiarly known as 'the master' (al-ustadh) he was a physician, astronomer, mathematician, physicist, geographer and historian).... But, by far the most important of Muslim scientists of this age was **Abu 'Ali al-Hasan ibn al-Haytham (Alhazen) of Basra** (965 A.D.). Though his original work in Arabic is lost, it has survived in Latin translation. In it he opposes the theory of Euclid and Ptolemy that the eye sends out visual rays to object of vision. He discusses the propagation light and colours, optic illusions and reflection, with experiments for testing the angles of incidence and reflection. In architecture they produced a wonder of the world like the Alhambra in Granada. In the minor arts too they made valuable contributions too numerous.... 'In manufacture', one writer has pointed out, 'they surpassed the world in variety and beauty of design and perfection of workmanship. They worked in all the metals—gold, silver, copper, bronze, iron and steel. They made glass and pottery of the finest quality. They knew the secrets of dyeing. Their many processes of dressing the leather and their work were famous throughout Europe. They made tincture, essences and syrups. They made sugar from the cane and grew many fine kinds of wine. They practiced farming in a scientific way. They had good systems of irrigation. They knew the value of fertilizers. They fitted their crops to the quality of the ground. They knew how to graft and were able to produce some new varieties of fruits and flowers. They introduced to the West many trees and plants from the East... They produced a number of versatile and profound scholars. They made permanent contributions to European science and vocabularies (e.g. words like *sofa, tariff, algebra,* etc.). **Omar Khayyam,** who is celebrated as a Persian, wrote in Arabic a book of the first rank on *Algebra*." (Sharma, 1948, pp. 265-267)

Peoples everywhere may like to be objectively aware that those medieval-era superb geniuses (since just by the way belonging to the Islamic-religion), were so creative and brilliant for the whole humanity and also, they were indeed far above any kind of extremely dehumanizing religious-fanaticism. Moreover so far in any case, these are just one-side of the story.

At the same time, since the above aspects in this extra-ordinarily important perspective, now under discussion in its requisite historicity as well as strategic-criticality, are equally relevant and true for so many different peoples inhabiting this world, since ancient times. And with respect to the similarly brilliant and superbly creative geniuses belonging to the other ancient religions as well as civilisations; preceding Islam and during the different historical-ages and also, in different geographical-regions in the world too; identically belong to the entire humanity. Important point in our same on-going perspective and also since historically emphasised earlier and hence, may always be borne in our mind is that no civilisation howsoever great, has ever been able to advance in isolation; or even grown on its own and without any valuable and external inputs of any kind. We all are also aware that the civilisation is a very composite word, meaning that it is truly inclusive of man's so many exclusive and captivating creative-excellence; such as multiple-cultural-aspects, matters-military and then, man's enchanting religious and spiritual pursuits apparently showing different paths but also, leading to the same destination towards One and the only One Divinity.

One apparently 'hidden' but, crucial difference comes in our minds and i.e. between the Muslim-empire and other so many historically prominent Empires, preceding it and then since following it as well. **And interestingly, this crucial-difference in this regard has been that the historic Muslim-empire was exclusively Islamic religion-centric and based on at times even extreme religious-fanaticism and intolerance, contextually in almost all-inclusive manner. But, other historic Empires have not been so since, they were principally and only socio-politically imperialistic, and even at times, harmonistic or politically, following again the historic age-based culture of hegemony and at the same time; as these being suited to their own geostrategic and politico-militarily larger imperialistic-interests of their Empires as well as their own projected greater strategic-needs for the future.**

But again coming to the above mentioned points; we are unable to ignore that the important aspects of Islamic-societies and in their different kinds of kingdoms since the medieval-age itself, have been rather radically different. Even in today's modern super-scientific era, things have largely remained the same and stagnant. Moreover ironically, the Muslim-societies even now have largely been 'forced' to continue living in a rather pathetic-kind of 'time-warp' of their own making. **Precisely in this perspective, the central overwhelmingly important and unassailable reason has been their now, totally obsolete and regressive religious-fixations, always being based on their one-track mindedness and in addition, even now; powerfully being controlled by their dictatorial ruling-class, extremely fanatic Mullahs and then, aided and abetted by their corrupt and also perhaps incurable Islamist-military**. These three segments of Islamic-societies are indeed, unchallengeable and thereby, have always been able to form mutually supportive and also their vested-interests' based 'Nexus' indeed, as their own society/State specific **'lethal-combination'**; definitely though, in different 'tailor-made' manners in various Muslim-States; but their contents everywhere, have principally and perforce remained identical. In this regard the earliest seeds; as since already having been discussed and quoted in earlier paragraphs above, had been sown by the first Caliph of the Muslims then; **Abu Bekr** himself in his preaching as cited by **Gibbon**, perhaps immediately after the passing-away of Prophet Mohammed, and is all well-known history itself.

No wonder, that many independent-minded, progressive, without any kind of regressive prejudices and also, highly learned Muslim scholars the world over; have been forcefully writing brilliant and thought-provoking articles as well as well-received books, fearlessly criticizing these unbecoming realities in their own societies and, still largely living in the dark and also, trying to inspire the Muslim-masses to show courage for coming out of their suffocating cocoons and into the bright lights of new scientific-ideas and, relentlessly strive for achieving the modern-era multiple progress; like the other societies and communities. **In this respect, the first and foremost need is the scientific modernisation of the education-systems of both the Muslim boys and girls; with special emphasis on the girls' education, in maximum possible Muslim States, and with true-consistency and perseverance.** In any case, the crucial facts here remain that there are no substitutes to honest intellectual-objectivity, analytical comprehensiveness with the desired depths and dimensions and hence thereby; thoughts/ideas consistently being progressive as well as scientifically dynamic as well. In these specific contexts, we may wish to refer to some relevant writings of the few noted Muslim-scholars and thinkers; to suitably substantiate the above mentioned contention.

A well-known Indian Journalist and writer **M.J. Akbar** has mentioned that, "In the beginning was a miracle and the miracle was a word: *iqra*. Read. Read, said the angel of God, Gabriel, and the illiterate Mohammed began to read the word of God out to the world in an enchanting prose that was later compiled and called the Holy Quran.... The sheer excitement of the achievement *(This is in the exclusive context of medieval-age Muslim forces' speedy conquest of various other regions/kingdoms; for bringing them forcibly under their rule and then conversion of subjugated peoples into accepting Islam, by any means possible)*, has tended to eclipse a basic impulse of the Muslim mindset. Jihad is not a sanction to empire-building, though empires did emerge in its wake. **The most powerful manifestation of jihad is not when all is won, but when all seems lost: that is the spirit of Badr.... Sufi derives from *saf*, an Arabic word meaning wool; for the only possession of these inspired men and women was a piece of coarse woolen cloth. Their lives were dedicated to a mystical union with Allah, inspired by such verses of the Quran as the one in which Allah says that he is closer to man than his jugular vein and knows more about his innermost thoughts than even the ego. The wisdom of these masters is divine, but they neither offer any explanation for nor glorify their powers. They treat life with the disdain of one who can see nothing but the Beloved.**" (Akbar, 2002, pp. 23, 25, 26 and 155)

If we ever read the above quoted lines and also in between them, with true objectivity and without any kind of bias then; a fundamental point with respect to the building and expansion of any particular totally religion-based empire; so exclusively obsessed with mindless religious-fanaticism and extremely irrational intolerance of any other much more ancient religion and since also, they definitely being equally magnificent and gloriously enriching too—need we say or accept anything more; specifically with respect to the birth, growth and expansion of the Islamic-empires? Well, that is a different story and, not very soothing as well as humane part of the man's on-going and vast living-history, always in the making. Unfortunately, it does remain such a sad-chapter of the history; written ruthlessly with human-blood on the unending canvas of time and more so then, attacking, killing as well as committing all sorts of crime against the innocent non-Muslim civilians in the conquered territories and, exclusively driven by the mindless religious-fanaticism and inhuman intolerance and hence; the same cannot be erased and wished-away.

Now, coming to the 'subject' of **Jihad**, M.J. Akbar, in the above quoted paragraph has indeed, magnificently as well as in a very enchanting manner, has brought out the true spiritual-meaning of and great substance behind the word: Jihad. In the identical spirit of true Islam and as also mandated by the Holy Quran; noted Indian scholar and with pious living based on Sufi-thoughts, **Maulana Wahiduddin Khan**, in his book 'Jihad, Peace and Inter-Community Relations in Islam' and as translated from Urdu and edited by **Yoginder Sikand**; the true-concept behind Jihad, has been enlighteningly brought out, not only for the Muslims but for the entire humanity. The English version of the book has been published in 2010, by Rupa & Company, New Delhi.

And accordingly, **Maulana Sahib** has written that, "**The Arabic word Jihad is derived from the root Juhd, which means 'to strive' or 'to struggle'**. It denotes the extension of oneself to the utmost, to the limits of one's capacity, in some activity or for some purpose. This is how the word is understood in Arabic grammar. **As fighting against one's enemies is also an act of striving, it is also sometimes referred to as Jihad. However the actual Arabic word for this, is *qital* (i.e. killing) not jihad.** Fighting with one's enemies is something that might happen occasionally or exceptionally. **However, jihad, properly understood, is a continuous action or process that animates everyday and night of the life of the true believer. Such a person does not let any hurdle his life, including the desire for gain, the pressure of customs, the demands of pragmatism, and the lust for wealth. All these things serve as obstacles in the path of doing good deeds. Overcoming these hurdles and yet abiding by the commandments of Gods the true jihad, and this is the essential meaning of its concept**.... The true Islamic way is through addressing others while being inspired by the spirit of love, compassion and concern for their welfare in spite of their opposition. In contrast, polemical debates aim at defeating and demeaning others. Instead of love and understanding they produce only more hate and conflict, thereby creating even more problems.... **The terms *dar ul-kufr* ('the abode of infidelity') and *bilad al-kuffar* ('the land of the infidels') are not found in the Quran. They are a later invention, which emerged after the demise of the Prophet and date to the Abbasid period. They were not in use among Muslims before this. In my opinion, these terms are not proper**." (Wahiduddin, 2010, pp. 1, 2, 51, 52)

Hence, when we make objective efforts as per our parameters of a sort of comprehensive strategic-criticality; with an aim to visualise the above highlighted 'gems of wisdom indeed, in those enchanting words' in a truly elevated and humane perspectives for our own enlightenment, in the larger as well as multi-religious contexts of our on-going discussions; we are perforce unable to find any sense in the mind-set, regressive thought-processes and the extreme acts of criminality; on part of the present modern-age Muslim terrorists, Islamic-fanatics and extremists, the world over. In addition the moot-point emerges here that, on what basis and wisdom (*well, are they capable of having any wisdom?*), these perhaps incurable fanatics and the many kinds of criminalized and sadistic Islamic terrorist-organisations; now violently active as armed-gangsters in various parts of the world, can ever justify their such extremely demonic mind-sets and misdeeds? The many known and hidden dimensions of perhaps, this singularly the most destructive as well as threatening international phenomenon of 20th and 21st century; would be discussed in greater detail as we proceed further, in the subsequent chapters.

In these contexts, in the ending pages of this sub-chapter, we may just point-out that now; it is widely believed that **Pakistan** is generally considered to be the pivotal-centre of the mindless international Islamic-terrorism, with piles of irrefutable evidences in this regard; continuously being gathered in India and also other nations, including the sole super-power the USA. Further, there happens to be a widely held view that, Pakistan is fast turning into a failed State. This remains highly disturbing not only in Pakistan's neighbourhood but, in other distant countries as well. Contextually, we would take-up Pakistan, her birth as a Nation-State, 'condemned' to perhaps in a state of perpetual turmoil and many other aspects of her later, in the eighth chapter; as multiple points of our precise discussions. At present in Pakistan's precise context at this stage, it may appear to again come to Maulana Wahiduddin Khan's already referred book and interestingly, we notice that respect to the menace of Islamic-terrorism; he himself has sharply talked about Pakistan's multiple-negation and her irrational obsession with it and specially so, against India.

Maulana Sahib says that, "**Pakistan-based radical Islamist groups such Lashkar-e Tayyeba and Jaishe-e Muhammad often refer to a saying attributed to the Prophet Muhammad in which, so they claim, the Prophet is to have prophesied a battle against India, the *ghazwat ul-hind*, that would be fought by a group of Muslims who would be saved from the fires of hell. This alleged Hadith is routinely used by these groups as a major tool for recruitment and to justify war against India, which they describe as a Jihad, promising their fighters that this would guarantee them a place in the heaven....groups like Lashkar deliberately twist the term *ghazwat ul-Hind* to give it a wrong meaning, claiming that it refers to a violent war against India by some Muslims. They translate the world ghazwa to mean a physical war. Actually the original and literal meaning of word is simply 'to move or to shift from one place to another'. In the more contemporary period of Muslim history, however, the term came to be used synonymously with violent war, or what is also known as *qital* in Arabic. This is how the term *ghazwat ul-Hind* is misinterpreted by terrorist groups such as Lashkar**." (Wahiduddin, 2010, pp. 67 & 68)

During the presently on-going modern-era and especially in the 21st century; having read the above quoted highly scholarly as well as informative words of wisdom sincerely and also contextually with requisite strategic objectivity, since indeed totally irrefutable; does anybody could ever feel any need to say anything more in this regard? Now, to bring-in certain important aspects with respect to the truly enchanting meaning and magnificent dimensions of Jihad, as already elaborated little earlier here, and by referring to the writings of a couple of noted modern Islamic scholars; especially keeping in mind the contemporary man's level and range of understanding and awareness, are indeed, something like an eye-opener and also, highly enriching as well. In addition, one may also appropriately maintain that **Jihad by itself is not Islam-centric at all; but factually secular-concept as well, soaked in the Divinity of spiritually highest-level captivating-fragrance, in the true sense**.

With respect to the above idea, there is no need to get somewhat perhaps 'alarmed' and, try to make efforts towards creating some kind of an unnecessary 'controversy' too; more so probably by the relatively fanatic members from among the Muslim-societies, anywhere in our world. The most sincere intention of this author is to go-in for geostrategically all inclusive objective-thinking with vision and pragmatism; in the highest interests of global peace, since already relentlessly being demanded by the modern-era humanity from all over the world; irrespective of the class, caste, creed, culture and religion. In these precise perspectives as well as if we may care to view them in the requisite geostrategic and politico-military contexts in this regard then; we may globally be in a dispassionate position to prudently visualise the urgency for meaningfully fulfilling the demands of the entire humanity, from the rulers and highest-level power-centers from all over the world.

Now, coming back to the subject-matter of Jihad again, we do learn from the highly educative narrations by certain noted Indian Islamic-scholars in this regard that, jihad in fact, means dispassionately doing one's duties without any kind of greed-based worldly attachments, in the service of the mankind from all over the world, and obviously where there is no place of any kind of violence at all. In addition, simultaneously and perpetually remaining attached to the one's Beloved, i.e. The Almighty, Who always timelessly belongs to everyone without any distinction whatsoever and further in this captivating spiritual path, one remains fearlessly ever prepared to lose all without any grief. In that sense all the great Saints from all over the world since the very distant ancient times; were indeed great and highly respected Jihadists. Further, All the Divine Incarnations coming to this world for good of the entire humanity and perhaps belonging to different religions since as recorded by man; were also the Great Jihadists. That precisely is the true spirits of jihad and hence; Jihad may also indeed, be appropriately called an enchantingly secular and totally non-violent concept with all inclusive universal comprehensiveness as well.

*

CHAPTER 5

Military Aspects and their Parallel Progress

In the otherwise necessarily long previous fourth chapter, we have had fairly engaging and detailed discussions on multiple aspects of the advent of human-civilisations and the ancient-era; with special emphasis on the specific demands since as dictated, in the geostrategic and politico-military perspectives. In the process, we have referred to many relevant as well as scholarly writings of various prominent and learned academicians and writers; we have taken liberty to quote certain paragraphs from them for essentially substantiating our analytical deliberations and possibly enriching them too, as well. **We are also aware that man's history in a way, is substantially the history of war itself**. Here we may recollect that in our first chapter on 'Introduction' itself, we have noted that with beginning of the evolutionary process of man on this Earth and; as man initially got in some shape, the simultaneous and gradual birth of some sort of very primitive 'arms' since obviously needed by him for his own protection as well as hunting in the wild too commenced in the right earnest and, their slow but continuous process of improvement also began.

This entire phenomenon of man's evolution under the command of mother-Nature perforce became highly complex, but captivatingly fascinating as well. We may again perhaps befittingly say that at the root of this fascinating phenomenon of the very slow but consistent evolution of man's **'arms, aims and aspects'** and though quite primitive to begin with; was again his same "**analytically thinking-cum-creative mind**".

It has been quite correctly said by the learned thinkers and historians the world over since ancient times that, **war is as old as the history of man himself**. Contextually in the specific perspectives of our now continuing strategically critical-journey too; we do notice that war by itself, is as old as man but, as times kept multiply changing with all their complex manifestations the nitty-gritty of war too, kept on changing alongwith gradual enhancements in its complicities as well. And at the same time, with changing situations as well as unending march of different human-civilisations the world over, man has also realised that war cannot be sole concern of the soldiers alone. History, since the dawn of humanity in this world, has shown us that the civil life and societies (*whatever might have been their form and contents of existence during any particular age or* *period*) have always been perforce extensively affected by the relentless clashes, conflicts and warfare. It may also be observed with the requisite objectivity and, it appears to be a great irony of history that since the beginning man's multiple-progress and multi-dimensional techno-scientific discoveries as well as many revolutionary innovations have largely been caused by the primitive armed-struggles and subsequently by the organised military-conflicts. And further very interestingly, we do notice that the same phenomenon is rather predominant in today's highly dynamic modern high-tech age, with all their very fluid as well as fast changing geostrategic and politico-military situations and substantially expanding complexities too.

Late Field Marshal Viscount Montgomery mentioned that, "**Why do wars happen? Some will say that war is the child of civilisation, others that war stems from raw human nature. But war has always been the arbiter when other methods of reaching agreement have failed.**" Whatever may be one's perceptions or line of thinking in this regard but, Montgomery's above quoted precise observation has always timelessly been substantially valid in the man's continuing living-history. Though it is also equally true that as centuries after centuries have continued rolling-by over the unending canvas of time, the warfare's larger nature and also its multiple manifestations have certainly been changing along with the changing times as well as situations. In this specific context, we may very appropriately recall **Paul K. Davis**'s scholarly considered opinion quoted earlier substantially coinciding with the British Field Marshal's above perceptive reflections.

Clausewitz has defined war as continuation of political transactions intermingled with other means, i.e. furtherance of the state-policy with multi-faceted military means and resources in action. Warfare has inherently been the most powerful guiding-factor of the evolutionary process of man as well as the various human-civilisations and therefore, the pivotal-theme of man's continuing living-history and in addition simultaneously, this living-history has again substantially been also the living-history of warfare as well. This is in a way interestingly, an unique historic fact in the sense that within the man's larger 'macro-level' living-history encompassing almost all aspects with their maximum possible comprehensiveness, in fact, also principally contains man's living military-history as its substantial segment. No wonder, that **Liddell Hart** has opined that if you want peace then, understand war. Similarly, the unique words of wisdom of **General George Washington**, the illustrious and visionary founder of the modern United States of America, have also been that if you want peace, always be prepared for war. It is another way sad and embarrassing fact, that in post-1947 India, its first P.M., Mr. Nehru, never cared to understand this hard reality in its desired perspectives along with its essentially requisite proactiveness.

In the beginning of the third chapter of this book itself, we have noted from Professor S.R. Sharma's scholarly book that **man's life is a moment between two eternities and, it is in fact a hyphen that establishes the essential connectivity between those two eternities; i.e. the past and the future. Simultaneously then, he also brilliantly mentions that this so important a hyphen is derived from the present itself, i.e. in fact, that very moment which has been talked about just above**. Contextually then, viewing this brilliantly thoughtful perception appropriately in a larger perspective; we also do notice that this highly captivating phenomenon by itself just does not get triggered in any kind of isolation.

In addition then and as noted in third chapter itself, Prof. Sharma also says that most people throughout life remain confined to a sort of animal satisfaction by living in the present. But getting shackled by this stagnant as well as regressive fixation; no civilisational or other kind of multiple progress for the man is ever possible, in this world. Well at this point, man is also pushed up to a much higher level of dynamic human-existence by means of his pro-active as well as analytically thinking and creative mind. And at the same time interestingly, we may also notice in our perspectives that, during 'strategically critical journey' **the multiple aspects of the matters-military and their parallel progress too, owe their birth and so on; to the same comprehensive 'causative-factor'**, as mentioned in the above paragraph.

At this stage in order to befittingly further enlarge our discussions, we do find essentially enriching to quote a very thought-provoking paragraph from learned British author **Arthur Birnie**'s historic book where the author says that, "The popular notion that man is by nature a fighting animal receives little support from researches into the earliest history. Most archeologists are agreed that for hundreds of centuries warfare played no serious part in the life of mankind. Among the remains neither of Paleolithic nor of Neolithic man have weapons of war been found. This covers a period of 28,000 years (30,000-2000 B.C.). If war had been a regular or organised activity during these times, it would inevitably have led to the invention of tools or instruments designed for fighting and for nothing else. The absence of warlike weapons entitles us to believe that though occasional skirmishes or scrimmages may have occurred amongst and between primitive communities, these were no more than occasional interruptions of a normally peaceful existence, like occasional drunken brawls that disturb the quiet of a modern village. **War did not, indeed could not, become an organised institution until man had acquired offensive and defensive weapons, and for these he had to wait till the discovery of the metals (about 2000 B.C.).**" (Birnie, 1942)

In the context of the above referred book, it may appear rather important to bear in mind that this book remains very scholarly and highly informative in many ways on the subject, and at the same time **Arthur Birnie** too is fully entitled to express his perceived opinions on certain highly relevant pre-historic aspects belonging to a very distant and ancient times in man's long and living-history. But one may perhaps be somewhat 'compelled' to contradict his thoughts in the above quoted paragraph. In this context here, it appears to be rather contradictory to certain well-accepted pre-historic facts, since brought on the pages of history, after extensively logical research and inferences in this regard. On that basis, it may be maintained that fighting rather in different forms has always been an inseparable part with the nature of all the animals and mammals including that of the pre-historic humans, since the very beginning of their creation by the continuously creative, altering as well as destructive evolutionary-process of the mighty Nature itself.

It may also appear important to note that so far as the arms were concerned to begin with, they were somewhat crudely shaped out of the suitable pieces of the branches of trees and strong bones of very large animals for their use. Moreover, these pieces were not only used for themselves but also for hunting of animals for food and their skins as well. It appears very relevant to quote from **Major General Gurcharan Singh Sandhu's book**.

On **Indus-Valley civilisation dating back to around 3500 B.C or so** General Sandhu writes that, "At higher excavated levels specimens of better built and stronger spearheads were found. These blades have a midrib as well as a tang for riveting to the shaft. The later versions were a great improvement on earlier flimsy heads and weapons fitted with them withstand the wear and tear of the battle. Two other blades excavated at the same level are $1^1/_2$ feet (0.45 m) in length --much too long to be spearheads. It is more likely that these blades for short swords or dirks. Some authorities opine that these blades are imports from Mesopotamia while others regard them technically very Indian.... Daggers have also been excavated from Harappan sites. The most common specimen has a long tang for insertion into a handle.... One specimen of a double-edged dagger with a midrib was also found at Harappa. Similar daggers made of bronze and with a midrib were also found at Mohenjodaro but from later levels only.... Slings were an important Indus weapon. It fitted well with the concept of battle of the Indus people. It had the maximum range of all available weapons and was particularly effective when shot from a height with a clear field of fire, as would be the case when fighting a defensive battle from the ramparts of a fortified Indus town.... These are balls of clay compressed by hand and then well-baked. A heavier specimen weighs about 12 ounces (34 grams) and lighter version is about half that weight. 98 sling shots of the lighter variety were found heaped together on the parapet walk between two southeastern towers at Mohenjo-Daro.... It seems unlikely that the Indus people invented the catapult.... The Indus people do not seem to have made use either shields or body armour...far from clear evidence there is no reason to believe that Indus warriors made use of any defensive equipment.... There is no indication that they expected to take offensive action even in their own defence...body armour which had by this time, come into use in West Asia. There, copper breast plates and helmets and helmets were in general use along shields of raw hide, wickerwork or metal.... The economic prosperity of the Indus people makes it probable that a military class of some kind could have come into being. The main accomplishment of such a class, if it existed, was the vast extent and uniformity of the Indus culture; it could not have spread over such a large area without the use of some form of coercion.... On the other hand, it is evident that the armed force of the Indus people did not enjoy a high social status nor were adequate resources were allocated to make the army into an efficient organisation.... This would be an indication that whereas the trading community enjoyed social pre-eminence, it ensured that the minimum resources were diverted to the upkeep of the army.... Spearheads were so feeble that a good strong thrust would either break or bend them. This weakness was not due to the lack of metallurgical skill or poor workmanship.... The simple explanation is that a merchant-dominated state saw no reason to use scarce tin or use enough copper or bronze for a spearhead. The discovery of stronger spearheads at upper levels may denote a change in the threat perception and priorities after the arrivals of the Vedic Aryans, if so, it came much too late." (Sandhu, 2000, pp. 63-64)

Though it has specially focused on the otherwise captivating aspects of matters-military, historically pertaining exclusively to Ancient India, but the we do also notice that it does throw substantial light on much larger gamut of similarly certain historic and even pre-historic global-issues with respect to the man's long living-history as well. In this context, first we may find that Arthur Birnie's particular perception that before the acquisition of more effective and lethal weapons both offensive as well as defensive in nature and that too, before the man's discovery of metal were largely peace-loving, appears to be not in coincidence with historic facts in this regard.

In fact, even before the advent of metals and the initial art and craft of metallurgy, inherently the pre-historic man did perforce have had to engage in fights, as per the specific demands of different situations or certain kind of alluring opportunities confronting him. In any such situational-setting, these might have been in the form of an alluring animal for hunting, a young female or any kind of perhaps unavoidable threats emanating from any quarter; probably including anybody of his own kind. Interestingly, the weapons used by him in such clashes were definitely crude and principally made of timber, animal bones and stone-pieces. Moreover, the inherent instincts to engage in fights and some kind of clashes/conflicts both individually or in groups have always been present in almost all species, including man. This is as per the nature of evolution itself, since already discussed in the beginning by us in some detail, in the first three chapters. Contextually here in addition, in our critical-review, we do learn from the man's long living-history that his inherent culture to fight or respond to any kinds of war-like situations and, certainly not necessarily with peaceful intentions, have seldom been triggered by some sort of path-breaking scientific or technological inventions, but it has been mostly the other way round.

Now again coming back to certain very important historically factual details in this regard, since as contained in the above quoted 'paragraph' from the Indian General Sandhu's book at this point; otherwise do also clearly hints towards a very ancient glaring failures regarding even primitive imperatives to pragmatically take necessary initiatives and also at community-level move for evolving and cultivate the essential strategic-culture and thoughts. The important aspects to bear in mind are not about trying to go tangential and then, indulge in some kinds of 'tall-orders'; without even giving some thoughts to the historic limitations concerning a very ancient-era in this regard. In this specific context, we may perhaps justifiably emphasise otherwise very laudable historic-reality that the Indus-Valley civilisation belonging to the remotely past-era of around 3500 B.C. and that the peoples belonging to it were quite developed with well laid-out semi-urban settlements having high sense of hygiene and sanitation, nice system of agriculture, metallurgical-skills and craftsmanship and also a well-regulated community-life.

In continuity, we may further add that the multiple-degrees of the enchanting advancements of so ancient as the Indus-Valley civilisation were truly a marvel of man's long and checkered living-history on this Earth, since its beginning. Again in this specific context as we move on; we may also not be able to wish-away this particular case of perhaps failing to bring-in the effective awareness on the part of the then their community-level leadership, for taking the requisite pro-active measures for evolving and imbibing the **'essential strategic-culture and thoughts'**; since these aspects always being the keys for comprehensively ensuring their security and well-being. It is needless to repeat in detail here since earlier, we have perhaps fairly discussed the aspects of man's exclusively inherent-instincts for his **survival** coupled with the compelling urge for relentlessly making **multiple-progress** in this world; since his beginning itself. These interesting aspects again have triggered the gradual advent of various human-civilisations, since the remotely ancient times.

Hence, as civilisations were born and as centuries have kept-on rolling-by; these civilisations naturally moved on their paths of multiple advancements. And they again further intensified the man's urge for progress and also his struggle for survival as well. Hence, any particular human-civilisation, as it firmed-up in its territorial-region of birth and its subsequent advancements since the beginning in turn; also brought-in the imperative necessity for the cultivation, refinement and fruitful adoption of the 'essential strategic-culture and thoughts'. Interestingly further, the overwhelmingly multi-dimensional vibrant-powers of the same strategic-culture and thoughts in turn; gave birth to the strategic-concepts for any situation-specific process of the **threat-perceptions**, and this again in turn; gave birth to the multiple aspects of the **concepts of the organised-defence also offensive-action** against any threatening adversary, as deemed necessary by the man in his any given situational framework. Further in this context we may also notice that all these above mentioned crucial as well as interestingly living-segments collectively in-turn; give birth not only to the **matters-military**, but in addition also, they are the principal causative-factors for growth of the meaningfully effective and pro-active **conceptual-framework for the all-round developments of the larger civil-societies and their Nation-State** concerned.

Before we further move-on the engaging 'road-map' for this strategically critical-journey and at the same time, since in this precise context; we may always remain aware that the 'compass' for this journey have already been defined in the beginning itself, as the 'geostrategic and politico-military perspectives'. Having said that, it may perhaps also be needed to bear in mind that the highlighted 'segments' in the above paragraph precisely with respect to the above mentioned comprehensive 'conceptual framework'; since they both individually as well as collectively, are always phenomenally complex in their own manifestations and substantial fall-outs with far-reaching consequences; for any given human-civilisation in this world. These universally important aspects have always been true during any given historic-age and hence the same do remain equally true during our presently on-going modern-era, and further in the similar perspectives; they would perhaps definitely be so, during any foreseeable futuristic-times as well. Therefore, in the total context of these on-going discussions and their purposefulness; we may again always remain critically aware of the above narrated important aspects.

Now come backing to our earlier point regarding the inherent attributes of the pre-historic man regarding their clashes and conflicts/war; we may again note that **Jericho, located very near the ancient-city of Jerusalem towards its north-east; was then strongly fortified with praiseworthy tactical-acumen, to effectively defend against any possible threats of enemy-attack; even during around 7000 B.C.** We are also aware that birth of the highly checkered ancient-history of Israel, goes back to the antiquity of the dawn of man's history itself. The great historic exodus of Jews from land of their origin and which being the land of the Holy Bible as well; was militarily enforced with extreme ruthlessness, since they were chased-away into almost all-corners of the world, by the threats of lethal-arms and man's ancient military wherewithal more than a thousand years before the birth of Jesus Christ. And we do know that this episode is mentioned in detail in the Holy Bible.

Man's very ancient history shows that the **discovery of wheel during 4000 B.C. and probably in Sumer** was indeed a revolutionary breakthrough in the path of historic-progress of man. And at the same time, in the marvelously multiple-advancements of many ancient human-civilisations in almost all regions of the world and since thousands of years, wheel has played astonishing-roles in many ways. In addition therefore, wheel was also instrumental in another revolutionary discovery of the magnificently looking horse-drawn chariots not only then the most efficient means of an ancient collective-transport but, also for the carriage of various kinds of goods. **No wonder, that with respect to the otherwise very enchanting multi-dimensional civilisational-aspects of man since the dawn of his history itself; wheel perhaps has been most influential as well as a path-breaking tool of his progress.**

Further, the chariots then became the most effective means of military-mobility and in addition; a wonderful weapon-platform as well. Chariots in the ancient-era then, proved to be very handy for achieving highly beneficial elements of surprise over the enemy and also greatly helpful in launching quick and decisive attack for subjugating enemy, especially in the plains. As additional centuries rolled-on over the continuing living-history of man; further many new innovative-ideas were triggered by the wheel in multiple-areas of the matters-military, much before the another historic dawn of the Industrial-revolution. As an example in this regard, we may cite the suitably designed wooden wheels of efficient haulage of again the wooden chassis-mounted heavier weapons such as cannons and their shells' carriage etc., by suitable beasts of burden like horse, ox and trained elephants to the vantage points in the battlefields. But, these were indeed, much later developments.

Well in this context, **Field Marshal Montgomery** has said that, "What traces we can discover of the primitive peoples of ancient times show that, from the first, men have had to organise themselves for defence. It was because their dwelling places had to be refuses as well as shelters that they so frequently lived in cliff-side caves, on hilltops or in houses built on piles in lakes and marshes so as to be moated. Among the Paleolithic instruments which could be used as weapons of war were stone cleavers and knives, and blades and lance-heads of bone, flint and horn. Since it is virtually impossible to discover the causes and modes of warfare among these earliest peoples, the most instructive approach for us is to look at the primitive peoples of other continents, some of whom in our twentieth century are passing through their stone ages. (Montgomery, 1983) (*Interesting indeed, isn't it? In any case, what he most probably means are the peoples in remote corners in Africa and also perhaps; tribal-belts situated in the not so easily accessible hilly and jungle areas in certain countries in Asia, as well as those of the South-American continent as well.*)

The much larger facts in this regard, happen to be that the birth and also the fascinatingly multiple-advancements of the various ancient human-civilisations as well as later-on those of the medieval-age civilisations and in addition subsequent historic-decay of some of them too **won't have materialised the way they have, on the pages of the same living human-history, without compulsive resorting to the use of arms and also at times, just display of the military-might coercively in the process**. Hence, this fundamental fact since the pre-historic age in this regard, has perforce remained unchanged and interestingly therefore as old era gave way to the new ones in a historic sequence here, and only the technologies and the modalities concerning these matters-military; along with the lethality of the accompanying arms as well as weapon-platforms kept-on changing.

In this context, it is perhaps also relevant to bear in mind that interestingly, the multiple-concepts with respect to various ideas of democracy as complex-systems for the process of governance have been relatively much later developments. In fact, the advent of these ideas concerning democracy, as we perceive and understand today have been a very remarkable development and, only during few centuries back in our presently ongoing modern-era itself. Hence, the fascinating and more humane ideas concerning democracy were totally unknown to the very ancient as well as medieval civilisation man. Further therefore, as man's continuing living-history informs us that the various initial concepts of tribal-cum-feudal lordships and subsequently then, those of kings, various forms of monarchies and dynasties; were largely the socio-political systems for governance, available to man during those bygone historic ages of the past. Looking them contextually concerning man here and, in the larger evolutionary-perspectives; we may also say that in fact, these complex-processes form themselves relatively small historic-segments of the overall and much more complex evolutionary-phenomena of the mother-Nature on this planet-Earth.

Further in this specific context and already mentioned in the earlier chapters as well that, as various ancient civilisations began sprouting as well as developing, the gradual birth and growth of a number of monarchies and also royal-dynasties, in the different locations of the world; naturally led to the founding of kingdoms and empires. In addition then **the establishment of any kingdom or empire, would have remained historically inconceivable without a king having any credible and effective military-force for the purpose** and also, either the king himself directly assuming command of his military in action or even may be, he appointing a suitable and professionally competent commander for his campaign; towards the intended founding of his kingdom in his area of imperial domination. Though to begin with, man had since already mastered gradually even his primitive fighting-skills during the primitive-era itself; for his security and also meeting his other needs.

But as many centuries kept passing-on; and with the advent and growth of various ancient human-civilisations, along with a prolonged passage of time the man's overpowering obsession with royal-dynasties based imperialism, began expanding and taking-over command of the era-based history-making process itself. Hence in continuity then, **in the process of founding of many kingdoms; simultaneously triggered the gradual birth, growth and consolidation of relatively more organised and disciplined military-force; for the assured security of the concerned kingdom and also for its opportunity-based any possible expansion. At this fascinatingly historic-point, we may also notice sowing of seeds and the sprouting of the civilisational-era based multiply more engaging and gradually enlarging matter-military**. In addition then, we may also notice that along with these military-aspects; simultaneously the various ancient human-civilisations too, kept on multiply advancing with the continuing passage of man's living-history in the making, and at the same time; all these complex-processes also do fascinatingly remain mutually complementary as well as pro-actively supportive and also, as part of the overall highly complex phenomena of man's gradual-evolution itself, in this world. Naturally therefore, all these evolutionary-processes perforce do remain inseparable from each other and hence, that is one of unaccountably so many marvels of the continuously evolving supreme mother-Nature herself.

At this specific point of our continuing discussions, another remarkably fascinating point may not be ignored since that pertains to the man's parallel and fast developing multi-dimensional engineering-design as well as outstanding construction-skill, with path-breaking multiple-applicability indeed, both in vast civil and military sectors. The countless monuments, invaluable historic-sites and construction-works, even today enduringly improve his quality of life, since their completion during ancient as well as medieval ages, all over the world. Hence as the civilisations took birth and commenced advancing, with respect to the matters-military during those historic-ages of the past; many developing processes of the construction of more and more strong forts, tactical-fortifications and strong-points, many kinds of logistical infrastructure, advancements of armaments and weapon-platforms with gradually increasing lethality and so on too got more intensified. And obviously at the root of all these aspects were the man's above mentioned same continuously developing scientific and technological knowledge as well as practically applicable multiple skills, indeed making path-breaking differences. Again these scientific and technological aspects in their own simultaneous advancements have also triggered many kinds of decisively new discoveries, with dual applicability in both civil as well as military sectors. Interestingly since ancient times as centuries have been rolling-bye multiple urgencies of the man's military needs have perforce ensured compulsively accelerated process of these new discoveries and innovations, indeed making remarkable differences along the way. And that's precisely the way that man's fascinating as well as highly checkered and also multi-dimensional living-history since his advent on this planet-Earth is so-far unendingly being formed.

We also recall the specific case of preparing then a remarkable defence for the ancient Jericho during nearly 7000 B.C. or so. It was indeed then a brilliantly memorable example in this regard, of man's creative-genius and engineering innovative-skills in the fields of military-construction works for countering enemy-threats there. For effectively defending Jericho then, and as per its authorities' though limited but perceived geostrategic and politico-military perceptions; a strongly fortified mud-cum-stone wall having a height of nearly 21 feet was prepared with innovative tactical-alignments. Another important aspect in this context is that even during ancient times, locally available suitable materials were imaginatively identified and efficiently made use of. Well interestingly during the present high-tech modern-era particularly in the military construction works as well as allied fields, the same practices are always been examined.

During the bygone ancient-era itself, two important discoveries with far-reaching and multiple consequences for man indeed decisively changed the course of continuing as well as living human-history. **First, the discovery of wheel during 4000 B.C. or so, and then subsequently that of iron during nearly 1500–1600 B.C**. Even before the advent of iron, man did have good knowledge of the art, science and practices of metallurgy and of course; commensurate to those ancient-times and, with their multiply fruitful applications. In this specific context, the moot point historically remains that even before the discovery of iron; man was well-versed in the multiple use and including in the fields of the many kinds of military-applications, of copper and bronze as well. That's precisely the reason, why archeological excavations have discovered well-established remains of the war-like implements, made of copper and bronze, in the different ancient human-civilisational locations at many places in the world.

Major General G.S. Sandhu of India mentions that, "The earlier view that iron came to India from the West around the beginning of the first millennium B.C., however, can no longer be sustained. Iron was apparently discovered independently in Central India. Iron weapons which can be dated to the 14th century B.C. have been excavated in Eran in the Nagda area. These are double-edged daggers, socketed axe-heads, and a flat axe with broad cutting edge, spear heads and arrow-heads (*all made of iron.*). In Malwa iron replaced copper tools and weapons around 1300 B.C. and about 1000 B.C. was in widespread use in Karnataka.... In Northern India use of iron commenced around 1200 B.C. after the arrival of the Rig Vedic Aryans there. Excavations in Eastern Haryana and Western Uttar Pradesh have revealed iron implements in association with PWG (painted grey ware) pottery which dates the use of iron to around 1200 B.C. At Hastinapur and other sites in the area iron arrow-heads, spear-heads, daggers, knife blades and other implements have been found. The use of iron in the extreme northwest of India, at Swat, can be placed around 1000 B.C. thus making it probable that use of iron spread in India from south to north. From Central India, it went to the Ganga valley and there it spread to the northwest; it did not come from West Asia." (Sandhu, 2000) (*Interesting, isn't it? Well in this respect, some substantial clarifications have been given in the following paragraph*).

All these 'controversies' or rather certain differences in the opinions regarding some issue-based perceptive-differences among the archeologists the world over, are understandable. In fact, it is all the more so if the specific issue/subject concerned, is carrying some weighty importance in the context of the any kind of path-breaking breakthrough/discovery along the advancements of various human-civilisations. In any case we may need to always bear in mind that, it remains an important segment of the inherent-nature as well as temperamental-constitution of the man that, he is basically rather fixatedly resistant to the any kind of change. As an appropriate example, we interestingly recall the important issue concerning the revolutionary discovery in recent-past during the British-rule in India; of the undeniable evidences as well as antiquity of the remarkable existence of Indus-Valley civilisation, in the North-Western region of the very ancient Indian subcontinent.

Now, coming back to the multiple fall-outs in the specific context of matters-military, we may notice that the first and foremost outcome was in the fields of man's achievements regarding improvements in design, strengthening and also assured increase in the durability as well as lethality of various arms, weapon-systems and weapon-platforms. In fact, historically to begin within our precise context here, we may also befittingly say that these aspects have been singularly the most important contribution of the advent of iron along the paths of the multi-dimensional advancements of various human-civilisations. No wonder that beginning with revolutionary-discovery of iron during nearly 1500 B.C. and then, for many centuries and up to perhaps end of the medieval-era; this long historic period is also called the **'age of the iron'**.

Then as the medieval-age began approaching its end and simultaneously there was the gradual advent of the wonderful scientific-age and the 'industrial-revolution', again multiple contributions of iron towards the continuing advancements of human-civilisations, in the form of various kinds of steel and its alloys with many new kinds of applicability; cannot even be precisely measured. These and many allied aspects in this regard, would be dealt with in much more detail, in the later chapters.

Then, besides multiple-contributions of iron in the areas of arms and so on; iron had played a key role in enhancing the decisive military-aspects of the ancient 'Cavalry-arm' as well. Field Marshal Montgomery writes that, "During this period arms began to be diversified and more value became attached to cavalry and lightly armed troops. In 415 (B.C.) Nikias took only thirty horses to Sicily; but he soon found that he had to protect his foragers, and so he raised more cavalry. But the old limitations on the use of cavalry still existed. Fodders and water were scarce during the Summer campaigning months, and the Greeks had no horseshoes or stirrups—**horseshoes were invented by the Celts in the fourth century B.C., and the stirrup appeared in India in the first century B.C**. The Greek horses were not strong to support armour; hence they were vulnerable to spears, particularly at the moment when cavalry wheeled away after the charge. But the obvious value of cavalry in the broken countryside of Greece was increasingly recognized." (Montgomery, 1983)

Now, in the precise context of the matters-military here the gradual evolution of cavalry and its far-reaching applications in both defensive as well as offensive operations of war since the ancient-era itself, and especially for achieving surprises by ensuring decisive mobility of own troops and then in addition, thrusting upon the enemy at times a crushing shock-action we may term them perhaps collectively, as the '**dawn of cavalry-age**'. Then over the ages, innovative military-commanders then, were quite imaginatively able to gradually organise their valuable cavalry-troops into highly advantageous, swiftly mobile and flexible **cavalry-formations** as well.

In fact, in the above mentioned historic-process, resulting in the cavalry gradually achieving a kind of commanding prominence in the battlefield; iron too, played no less a decisive role. First, instead of copper and bronze, various arms and weapon-systems commenced being made of iron; since naturally with so many decisive advantages. Then the design as well as making of especially the sword and spears witnessed a sea-change with improved effectiveness and lethality in such a manner that, a horse-mounted soldier was able to carry both the arms together. He was able to keep his sword in a kind of cover and tie it up in his waist and thus making his both hands free. Swords made of iron and later-on of steel were much tougher in fact, facilitated soldiers to fight in duals; both on horseback as well as while on foot. Further then, all the essential accessories needed to suitably equip the horse for cavalry were made of iron. In addition, these were to the chariot much more effective and versatile in war and also achieving enhanced mobility.

So far as ancient India is concerned, cavalry has been an important arm for battle indeed, since a very distant past. In this precise context, the Indian Major General G.S. Sandhu, "The first probable reference to the cavalry as a distinct arm occurs in the *Yajur Veda*, its introduction thus predates the use of war elephants in battle within the country. The Veda refers to the 'dust raising horses', a possible reference to a formed body of mounted sowars (i.e. cavalry-men). The Chandogya Upanishad's reference to an *asvalaysana* (a leader of cavalry is unquestionable.... The efficacy of cavalry prior to the introduction of stirrups may be discussed here. It is true that as a weapon platform a horse's back without stirrups for the rider is somewhat precarious. Both the chariot and the elephant provided more stable weapon platforms for the efficient use of weapons in battle. This fact does not, however, distract too much from the efficacy of the cavalry on two grounds. Firstly, well trained cavalry did manage to the handicap. The Assyrian cavalry introduced in the middle of the second millennium B.C. was an effective force. Its riders are depicted hunting lions with bows and arrows at full gallop. Alexander of Macedonia's cavalry is another example of how effective such a force could be even when fighting from a rather unsteady platform. A part of his (Asian) cavalry charged shooting arrows at full gallop." (Sandhu, 2000)

Well, the above mentioned points in rather a kind of critical-praise of the cavalry especially in the context of those bygone ancient times, is indeed well taken. But at the same time, we may also bear in mind that since the beginning of the history of man at no given point, cavalry has ever commanded a totally unassailable superiority in battles. In fact at times, **various other relevantly important factors** in a combination have been able to achieve crippling surprises over an adversary's cavalry-formations in action. That's indeed a different interesting point that the today, the horse-based cavalry of the earlier-era, has been totally replaced by the **armoured-formation consisting of the armoured fighting vehicles, i.e. AFVs or Tanks as well as the infantry combat vehicles; ICVs, carrying the always unquestionably essential infantry-soldiers, and in addition also, the modern self-propelled artillery-guns** and hence in a way forming a composite and quite mobile fighting-formation, with substantial military-lethality in an offensive operation.

At this specific-point of our interesting discussions, it is relevant to bring-in another highly important point regarding the multiple military-aspects of the unique **'Infantry-arm',** being the oldest, most versatile as well as capable of moving anywhere and into any kind of terrain-cum-weather conditions for any kind of both defensive and offensive military-operations, as per the concerned commanders' plans. Last but not the least, in any size of contingent-cum-composition. No wonder, that infantry is also befittingly known as **'Queen of the Battle'**. Certain important aspects in this regard, have already been discussed in the initial larger civilisational contexts in the earlier chapters. And hence otherwise too, it remains well-known that some kind of different manifestations of the primitive 'infantry-arm' have always got evolved in some perhaps parallel-manner simultaneously; along with gradual-evolution of the different ancient human civilisations, in different regions of this world. In fact, so far as other military-arms of the ancient armies are concerned, they got gradually evolved relatively much later and at the same time more so when the corresponding ancient human civilisations had got more firmed-up and also historically, in addition, was able to claim somewhat more multiply-advanced.

At this juncture of our on-going discussions on the vast and multiple matters-military of topical-importance regarding 'Military-aspects and their parallel progress' in our specific perspectives of the ancient-era, it is crucial that we may have a look into the pattern of the constitution of certain historic fighting-formations. Indeed, a few number of these distinct fighting-formations; did in fact belong to the different and mostly competing socio-political entities as part of their own those very different ancient human-civilisations. In this respect, both the **ancient 'Greek-Phalanx' and then the 'Roman-Legion'**, do stand-out clearly as two important historic examples. In fact, when reviewed with the requisite strategic-criticality; the relevant pages of military-history have revealed certain distinct strong-points as well as weaknesses too, concerning both these two ancient fighting-formations.

On the **'Greek-phalanx'**, old British scholar Arthur Birnie mentioned that, "The typical Greek soldier was the heavily armed pikeman or hoplite. **He carried a six-foot spear and a short sword.** (*Interestingly, military history for the first time perhaps distinctly mentions the advent of a specially designed and crafted 'short-sword', being advantageously adopted by the ancient Greeks in their one exclusive pattern of their military-formations; while going into battle. This is definitely a historic advent of the distinctly designed new type of sword, and that along with a long-spear and also a long-cum-sleek shield; gave the ancient Greek soldier in his phalanx-formation, a kind of simultaneously both protection as well as offensive capabilities*). For defensive armour he had a helmet, a shield, a cuirass, and greaves. Sometimes they were supported by forces of cavalry and archers, but till the time of Alexander the Great these auxiliaries did not count for much. The hoplites fought in the famous phalanx formation—a dense hedge of spears, normally eight files deep. It was a stiffened and reinforced line which crashed easily into thinner formation. In a charge the phalanx was almost irresistible. But it had its weak points. It lacked mobility. On rough ground it was easily thrown into disorder. And it could not quickly change fronts, so that its flanks were particularly vulnerable. In earlier wars against the Persians, in which Greeks had fought as auxiliaries or mercenaries, it was mostly the weak points of the phalanx that were exposed. Its real value as a tactical unit was first revealed on the historic day of Marathon." (Birnie, 1942, pp. 13 & 14)

We may recall that the ancient historic **'battle of Marathon'**, having taken place even few centuries before the beginning of Christ-era has been touched upon in a brief-detail in the last chapter itself. In fact, **King Darius of Persia** during around 490 B.C. had launched an invasion of the ancient flourishing city-States of Eretria in Greece with comparatively large force of nearly 26,000 infantry and 4000 to 5500 cavalry. During those times, Darius was definitely in command of much more forces but sensibly he had no plans to conquer the whole of Greece and he wanted to punish those two city-States and teach them a lesson, not to mess-up with the large Persian-empire; since according to him, they had audaciously dispatched military-assistance to the Ionian-Greeks, then being part of the Persian-rule but having rebelled against the Persian King. Darius was also politico-militarily perhaps 'over-confident', as per his own military appreciation of the overall geostrategic situation since; his Persian army consisted of much larger infantry, archer-units as well as relatively well-equipped cavalry and at the same time, he almost had and an unchallenged naval-supremacy in the sea.

But ironically, in his calculations and operational-planning he had perhaps injudiciously ignored and not cared for certain unpredictable as well as in advance unknowably and also otherwise, very decisively 'crucial-factors' in any war anywhere since the beginnings of human civilisations themselves. In fact, any of those unknown factors of war, either in single or in combination of more than one, if allowed to dominate the battle-situation in any given scenario in war then they can considerably upset any commander's plan. This is a universal truth since ancient times and also, the same holds good during even the present high-tech modern age. Interestingly, these factors may be any kind of surprise suddenly achieved by the enemy, at times not fully in advance known his military wherewithal, suddenly changing weather-conditions, sudden thunder-storm and/or even powerful storms in the sea as applicable and at times, perhaps some hidden fighting-qualities and military-tenacity of the enemy and so on. In the beginning of the Marathon-war, all the military-strategic parameters were totally in favour of the Persians but informed military-history shows us that though not crushing but the Greeks had finally won the day.

In the recently published highly informative book titled as *Great Battles*, with General Editor being **Christer Jorgenson** the concerned contributor, writes that, "A typical hoplite of the Persian war era: His primary weapon was long iron-headed spear, which could be between 2-3m (6-10ft) in length. It was usually held over arm in combat, and under arm when maneuvering. He also carried a short sword about 60cm (2ft) in length, made of iron with bronze fittings. The sword was used in both a cutting and thrusting motion. For protection he carried a hoplon (shield) made of wood with a bronze face and leather inner lining. The hoplon was secured to the hoplite's forearm by a band, and he held a grip in his left fist. His Corinthian helmet is topped with a plume of horsehair, which could be dyed for effect. His torso is protected by stiffened linen with metal scales for greater protection. On his shins are moulded bronze greaves, while simple leather sandals are worn on his feet. *(Naturally in those pre-Christ ancient times this way the hoplites were fully equipped, suitably armed and then rigorously well-trained in advance and also, whenever required; they were immediately able to form-up as historic phalanx-units, under their commanders for the battle.)....* **Popular myth has it that following the Greek victory, the Athenian runner, Philippedes, delivered the news to the worried Athenians 42km (26 miles) away. The modern marathon race celebrates his feat ..."**(Jorgenson, 2011) ***Interesting, isn't it.***

Now coming to the important aspects of historic **'Roman-Legion'** in fact we interestingly learn that in the beginning, the legion was simply a kind of ancient Roman-militia, which consisted of relatively higher-class Roman citizens and, available for service to the empire only during summer months and again that too as an auxiliary or perhaps mainly as an amateur fighting Unit. But the Roman-rulers were in fact not persistently so naive and hence, did not continue for long with such a 'showpiece' militarily half-baked army. Naturally, if that would have been their imperial-history then certainly, they won't have succeeded at all in so successfully building such a historically magnificent Roman-Empire; with indeed a powerfully and dominating and conquering military-might.

Arthur Birnie writes that, "By far reaching reforms, adopted probably in the fourth century B.C., the structure of the legion was entirely recast. It was divided into three sections, the *hastati*, the *principles*, and the *triarii*. The hastati were the young soldiers; the principles the men of mature age; and the triarii the veterans whose liability to serve had almost expired.... The normal strength of the legion was 4,500 men, but on occasions it was enlarged to 6,000. Usually four legions were raised for a campaign, two commanded by each consul, but in times of emergency the number and size of the legions were increased. The consul was assisted by a staff of military tribunes, six in each legion. Each maniple was officered by two centurions. The legionaries were equipped with helmet, shield, cuirass and short sword. Formerly all ranks carried the *hasta* or pike, but this was now retained only by the triarii. The hastati and principles carried each two javelins or pila.... The chessboard formation of the maniples made it easy for the rear divisions to pass through the frontlines if necessary. Moreover, each maniple was an independent unit which could be manoeuvred separately. If a gap appeared in the front line, a maniple could be pushed forward to fill it. If part of the legion was overwhelmed, the remaining maniples closed up and carried on the fight. Unlike the phalanx which manoeuvred as a single mass, the legion did not break up when penetrated by the enemy. It was indeed a flexible phalanx, combining solidity and divisibility, pliancy and strength, weight and manoeuvring power." (Birnie, 1942, pp. 37-39)

With some contextual strategic-criticality, we have briefly reviewed certain particularly important military-aspects of both the **'Greek-Phalanx'** and the **'Roman-Legion'** concerning their different systems for troops-formations; before they being launched into the battle by their commanders. In the process, we do also notice that in fact both the systems have had their distinct advantages as well as disadvantages in any kind of given situation with respect to a particularly recorded historic battlefield scenario. Moreover during those ancient times interestingly, in these two historic-systems for their own kinds of armaments as well as body-armour and helmets, and so on; for equipping and dressing-up the their troops for battle, did contain lot many similarities and at the same time, some differences as well. We have since seen that in certain challenging situations, the Greek-phalanx did suffer limitations in mobility and also, in their much needed flexibility and speed in wheeling-around and changing their directions for timely responding to their enemy's surprise initiative in a rather fast-moving battle-scenario.

But in the above sequence at the same time, if the ground is favourably flat and also if the war is comparatively not very mobile, as was generally the case in the very distant past then troops in the phalanx-formation was able to march forward quickly and hence at the scene of actual engagement with the enemy-troops, their sudden attacking thrust en-masse was always mostly irresistible and crippling for the enemy. In addition, at this point, we may also say that this indeed was perhaps the first historic conceptualization as well as any situation-specific military-application of the **'concentration of force'**; as clearly pronounced one of the now well-known **'Principles of War'**. Military-history also informs us that in fact, the formal a kind of military 'treatise-based' any conceptualization and listing of a number of now so familiar "Principles of War" to the military-officers as well as military-historians the world over have indeed been done much later and perhaps, during the beginning of the present modern-era or may be little later. Interestingly further, during our discussions at this specific point itself, we may have also noticed that another familiar principle of war, i.e. the **'Surprise'** has also been understood and variedly practiced, since indeed very ancient times.

At this specific point of our on-going discussions, we may further notice that to begin with perhaps **Alexander** was the first ancient-era monarch and an outstanding military commander; who was able to brilliantly synthesise main positive military-attributes of both the Greek-phalanx as well as the Roman-legion with the then maximum possible practical applicability in his historic-battles to achieve decisive victories, during his long and relentless campaign across a number of continents. The saga of his long military-campaign is indeed a unique chapter in man's vast living military-history of the whole world since the beginning, which is fascinatingly as well as very knowledgably still engaging the attentions everywhere, of the military historians, scholars and military personnel. In this perspective we may also say that another four 'principles of war' namely; **'maintenance of aim', 'unity of command', 'initiative' and then 'logistics'** too; perhaps further got sharpened and conceptualised during a series of military-victories in war for phenomenally expanding his empire. In fact, according to noted British historian Arthur Birnie: Alexander was among the first to recognise that a commander's proper function is not to fight himself but to direct the fighting of others (*last sentence of p. 36 of his earlier referred book on the "The Art of War"*).

Field Marshal Montgomery writes that, "After Gaugamela, Alexander marched on to take the great Persian centres of Babylon, Susa and Persepolis, all of which surrendered easily. In 330 (*B.C.*), he then set out to conquer the East. In the early summer of 326 the Indus (*one of the well-known rivers in the ancient Indian-subcontinent*), after four years of fighting, he and his army after performing a number of remarkable military feats, in the pursuit of **Darius** (*King of ancient Persia*) towards the Caspian he covered 400 miles in eleven days, and then he mounted 500 phalangsits on horseback to ride another 50 miles at night. He defeated the nomads at the Oxus in 329 by bombarding them with catapults from boats (*this appears to be perhaps the first brilliant tactical-use in attack, of the catapults en-masse in the ancient-era military-history with such decisive and remarkable success.*), and then attacking with a force of heavy cavalry interspersed with javelin-men and archers.

In the winter of 328 **Alexander**'s men captured the stronghold of the Sogdian Rock, climbing 300 feet in the snow with the aid of ropes. In 327 he crossed the Hindu Kush with 27000 men, and after very hard fighting reached the Hydaspes-river in the following year. He had the first to cross the river while the enemy (*i.e. King Porus of the then ancient Punjab and adjacent territories in the North-Western India.*) were massed on the other bank, and then fight against the elephants which put out his cavalry out of action (*For the first time in his ruthlessly-cruel but, brilliant-saga of military-victories in such vast territories; Alexander was given a taste of fighting with another brave and great fighter and also, was forced to realise that the supremacy of his magnificent cavalry could not be guaranteed in a highly tough battle with a remarkable King in the then ancient Indian sub-continent: kind courtesy—the age-old tradition of the Indian Kings' having large contingents of the highly trained war-elephants. In fact, much later in the medieval-period; the historic and far-reaching advent of the cannons was able to trigger gradual-eclipse of the unique and highly dependable Indian military-tradition of maintaining sizable segments of war-elephants, along with both infantry as well as cavalry; for deployment in the battlefields.*) But shortly after, at the river Bias, Alexander's army mutinied and would go no further. It was eight years since they had left Greece, and they had marched 17000 miles. He led them back to Susa, which they reached in spring 324. That last phase of the war was the toughest of all. Alexander was wounded, and he died in June 323 at an early age of thirty-three." (Montgomery, 1983)

In this specific context of the 'study of Alexander', with our somewhat focused domain of the critical geostrategic and politico-military perspectives; we do further notice that Alexander also proved himself to be his age-based, a master of the theory and practice of the **'manoeuvre warfare'**. In fact, very judiciously as well as with situation-specific based military-innovativeness, while adhering to the above mentioned important principles of war; Alexander always moved his troops brilliantly in manoeuvring his forces with the required as per his any particular strategic-plan of campaign. And then in addition, also manoeuvred his troops with speed in the battlefield itself while maintaining his **initiative** and then simultaneously achieving decisive **surprise** over his enemy. Hence no wonder that, Alexander was able to achieve such military-feats; as recorded in the annals of the man's living military-history in those ancient times.

William S. Lind writes that, "**Maneuver warfare is not new**. It probably dates from the first time caveman surprised an enemy from behind instead of meeting him club to club. The first clear case in recorded history was the battle of Leuctra in 371 B.C. The Theban won that battle, thanks to a surprise strike against the right flank of the rigid Spartan phalanx. Hannibal's defeat of the Romans at Cannae in 216 B.C., one of the most decisive tactical victories of all time, was an example of maneuver warfare. Why are all these cases of 'maneuver warfare'? What is 'maneuver'? A traditional definition is offered by **Soviet Colonel F.D. Sverdlov** in a recent study, Tactical Maneuver, during combat operations to a new axis (line) and region for the purpose of taking an advantageous position relative to the enemy in order to deliver a decisive strike." (Lind, 2005, pp. 6-7)

In this context, we do also notice very interesting aspects of the author of the just above referred book, William S. Lind, regarding his remarkable scholarship as well as his rare talents for writings on particularly certain highly complex subject-matters concerning the matters-military we like to refer to 'Foreword' of this book itself. In the beginning of it, retired U.S. **Colonel John C. Studt** of the United States Marine Corps very revealingly writes that, "The author of this book has never served a duty of active military duty, and he has never been shot at, although there are no doubt some senior officers who would like to remedy that deficiency. Yet he demonstrates **an amazing understanding of the art of war**, as have only a handful of military thinkers I have come across in my career. I served over 31 years active duty with the Marine Corps, saw combat in both

Korea and Vietnam, and attended service schools from The Basic School to the National War College. Yet only toward the end of military career did I realise how little I really understood the art of war.... **When I first heard Bill Lind speak, I must confess I resented a mere civilian expressing criticism of the way our beloved corps did things.... But what he said made sense.** For the first time I was personally hearing someone advocate an approach to war that was based on intellectual innovation rather than sheer material superiority." (Lind, 2005) *(Simply remarkably rare and indeed wonderful: isn't it?)*

But, as William S. Lind too confirms that, **'manoeuvre-warfare' does indeed mean much more than all these and hence, it's no wonder that the pivotal professional-concepts regarding the same do also appear to be quite confusing as well.** Historically, this has indeed been so; since beginning of the highly engaging living military-history of the mankind. In continuity, the author in his book further says that, "Tactics are defined as 'how to' of things like a penetration, an attack, coordination of supporting arms and so on.... Maneuver warfare is not very concerned with names of stoves and skiing techniques. It looks at tactics differently. A useful definition of tactics in Maneuver warfare is: Tactics is a process of combining two elements, techniques and education, through three mental 'filters' or reference points—mission type orders, the focus of effort or Schwerpunkt, and the search for enemy surfaces and gaps—with object of producing a unique approach for the specific enemy, time and place."

We also notice that **engineers**, along with the combat-service support, are indeed crucially important, for a commander's operational-plan in maneuver warfare. These have been so since the ancient-ages and of course forms of expression might have gradually got altered over centuries, but the principal concepts in this regard, have remained largely the same. In the same continuing context, man's living military history also informs that as sometimes in other more conventional or classic forms of warfare, in manoeuvre-warfare, no two different combat situations have ever remained identical. Therefore logically in this particular perspective, so far as the important aspects of manoeuvre-warfare are concerned the theoretical definitions of certain obvious military-terms like tactics, larger strategic-parameters, movement of troops towards making contacts with the enemy for intended combat where he least expects so, tactically moving the 'supply-train' to keep in constant touch with own troops and so on' do always look very fine in the military-manuals.

But in these continuing perspectives of manoeuvre-warfare, what remains of crucial importance is the basic intellectual-cum-professional ability of commander in the field to think innovatively; especially with respect to the challenges confronting him in that given situation and simultaneously, keeping in mind his intention/aim for success. And hence, here comes the personal ingenuity of the commander concerned. In fact, the main point is **that any situation-specific precisely correct thinking and then seizing the initiative in time for innovative manoeuvre would most likely always lead to success** and similarly any faulty steps, are doomed to failure. In addition interestingly, correct concepts and undertaking innovative steps for manoeuvring troops with confidence on the part of concerned commander; may also act as a very beneficial force-multiplier as well. In this respect, timely valuable field-intelligence inputs too are no less important. **Alexander the Great and then Chandragupta Maurya of the ancient India** and then, few other historic military-personalities too were indeed pioneers in manoeuvre-warfare. Further, the history of both world-wars during the 20th century have shown historic examples of the brilliant practices of manoeuvre-warfare, on the part of various commanders on both sides of the frontlines.

In terms of the overall historic-totality, all the more so in our specific geostrategic and politico-military perspectives, and our presently on-going strategically critical journey we need to always bear in mind that the history is never static, as an essential phenomenon. And at the same time, it becomes naturally obvious that since the beginning itself, man's living military-history too is perpetually on the multi-dimensional moves. This march of man's military-history in continuity is perforce triggered, guided and then controlled by the man's inherently constant urge to multiply-progress, including again in the multi-dimensional spheres of the matters-military as well. Here interestingly, we may also notice that the just above mentioned specific as well as continuous multiple-progress since the ancient-times itself; has been rather sometimes relatively slow and then faster, during different historic-ages; depending upon various other allied causative-factors.

But historically, this progress has historically been always there in the man's overall socio-cultural as well as politico-military perspectives, and simultaneously also then with its own multiple-fallouts. At the same time interestingly then, these 'fallouts' do act as multi-layered and powerful causative-factors in decisively altering the different paths of the continuous multiple advancements of various human-civilisations, historically in the different geographical-regions of the world since the very ancient past. **At this point of our ongoing discussions, we do also notice that this entire gamut of appearing very different but always being interrelated processes together; do dynamically constitute in totality, very complex-phenomena themselves again, in the vast canvas of the man's matters-military**. We may also call these different but always remaining interconnected processes; in fact, constituting a number of concentric-circles maintaining their own in-built synchronization and simultaneously then, continuously expanding as well.

Hence it becomes clear that since the beginning itself, man's living military-history is perpetually in the complex process of being evolved and also on the move and never being static in any manner. Then, how can any historic tactical as well as strategic level concepts can ever remain static; and the same time, these also since, emphasizing concerns for the strategic composition of forces, pattern of their tactical-deployment ensuring essential flexibility with respect to meeting the vital needs of an operational-plan for any specific battle/campaign in a given situation; and simultaneously also, all these in totality ensuring necessary synergy with singular purpose towards fulfillment of the larger highest-level 'strategic-aim'? In this specific context, the fundamental basis for **Chanakya's own multi-dimensional strategic-thoughts**, during the ancient-times even before the advent of Christ-era were indeed valuable during a relatively short duration in that historic-age specially when India was yet to get exposed to the basic initial concepts as well as applications in different kinds of live-situations, of the 'mobile-warfare'.

Hence in this continuing context, larger reasons for India remaining almost perpetually rather in a kind of strategic-logjam; were certainly not at all the purpose of the thoughts of Chanakya and his teachings. But ironically since the ancient times the Indian rulers and their top military-commanders, down the different historic-ages chose to remain prisoners of their own kinds of regressive mental as well as professional fixations. Naturally in these kinds of extremely damaging, lingering and also over-powering situations; the first casualty is the inescapably essential need to fruitfully maintain the mental dynamism as well as that of the thought-processes of those Indian authorities, during any given changing periods of time and situations. **Why blame Chanakya at all when faults glaringly lie in those elite Indians' own self-made as well as all round regressive-fixations; as mentioned above. And then at the same time, the elite-Indians in power naively chose or even perhaps 'foolishly' falling repeatedly again and again in the 'conspiracy-traps' of the multi-layered, selfish, powerfully entrenched and also, the colossus-nexus of the age-old 'Brahmin priest-class'**.

Sadly, these regressive aspects down the ages and then even in a way now, in the present modern-era have been singularly perhaps the greatest negation of the Indian societies as well their Nation-State. We have since discussed these aspects in a perspective, in the earlier chapter itself while dealing with the ancient Indian-civilisation. Hence so far as the principal aspects of the matters-military are concerned; the main question arises as to why and how these have always been a kind of very 'shamefully' unfortunate legacies of the age-old Indian-civilisation as well as her various political-incarnations, since the very distant-past itself? **Perhaps the main reasons for these age-old socio-cultural as well as geostrategic and politico-military 'ailments' of the Indians, have historically been the regressive-fixations of again the 'lingering closeness of their minds'.** Well at an initial glance, this perception may perhaps appear to be a kind of over-simplistic statement, but if we may care to pause and give some cool thoughts over it with desired criticality, some hidden but highly relevant meanings behind it; may appear to be interestingly unfolding.

Hence they in turn, do indicate to us the above mentioned 'lingering closeness of minds'; and this in turn may have perforce been triggering many unsettling negative fallouts. First and foremost, it ensures cropping-up of the irrational socio-psychological arrogance of the targeted peoples and in turn; it unfailingly brings-in **a kind of very unfortunate 'intellectual-blindness'**. Further in continuity, this 'blindness' sets-in another **very tragic and regressive self-righteous attitudes**; preventing meaningful considerations of any fresh-ideas with due open-mindedness, since these ideas in turn may be highly helpful with the **evolution of far-reaching and strategically advantageous socio-political culture**. Here at this specific point, we may also add that **it is immaterial whether those newly fresh-ideas are internally generated or even coming from foreign lands**. In the larger comprehensive strategic-perspectives, what perhaps timelessly important is in fact; the depth and range as well as the degree of permanence of those fresh-ideas.

The multiple-importance of any such worthwhile new-ideas cannot be over-estimated at all. In fact, at this specific point of our on-going discussions, we may recall that earlier we have mentioned a number of times in different perspectives that, the mother-Nature has exclusively gifted **'analytically thinking-cum-creative mind'** only to one mammal, and that happens to be the man. Hence **man's this unique mind; which relentlessly keeps thinking analytically, always remains as a highly creative 'causative-factor' at the roots of the fascinating germination as well as fruitful evolution of any such valuable new-ideas**. Therefore in other words, we may also say that the above mentioned 'intellectual-blindness' in addition very harmfully crushes the multiply vital-aspects of the invaluable strategic-prudence of any otherwise historically highly vibrant and dynamic civilisation. More so in our specifically relevant geostrategic and politico-military perspectives; these aspects which ensure the continuously fascinating and multiple-advancements of various human-civilisations since ancient-ages are again, all the more important.

In addition at this point, we may also recollect that in the overall perspectives of our on-going discussions here, concerning our principal subject-matters; the **multiple-importance of the 'essential strategic-culture and thoughts'**, have been repeatedly emphasised. Hence interestingly, while critically reviewing a number of very important points since raised and also illustrated in brief in the immediately above few paragraphs in continuity; we may also very revealingly notice that those 'important-points', both individually and collectively and also very encouragingly are indeed responsible in a decisively pro-active manner; for the initial germination, fruitful cultivation, evolution, multi-dimensional development as well as simultaneously fascinating imbibing of that 'essential strategic-culture and thoughts' by the peoples concerned; of any particular noted ancient human-civilisation in the world.

If we further may thoughtfully observe and study the above mentioned points, with their own inherent mutual-connectivity and comprehensiveness we may then notice that what an overall highly complex as well as wonderfully magnificent phenomena historically they indeed are, with such unique and again multi-dimensionally far-reaching consequences. Ironically, we also notice that India's ancient, medieval and also modern-era long histories are full of a series of 'shameful' saga of such consequences. In this context tragically, this irony of ironies persists even today; in spite of India now being geographically a vast independent democratic entity in the modern comity of nations; she as a sovereign democratic State is still refusing to learn these most valuable geostrategic and politico-military lessons from the pages of man's long living-history. In this specific perspective, we are all aware of independent India's post-1947 times' recent embarrassing history of her first Prime Minister, **Late Jawaharlal Nehru's uncalled-for arrogance and extremely damaging 'intellectual-blindness' with respects to the multiple aspects of the matters-military as well as along with the essentials of vital geostrategic and politico-military considerations** and thereby bringing-in great and yet avoidable harms to the Indian-Nation's international prestige and also, internal well-being. We would be coming to these and also other allied historic-episodes more appropriately in the later chapters.

Now, again coming back to the highly valuable aspects of Chanakya's unique ancient treatise on his multiple-teachings; it is not disputed that his principles and parameters of forces' composition and their deployments in the battlefields, were principally based on **concepts of the static-warfare**. But ironically, due to the some of the reasons having been discussed earlier, **vital concepts of the mobile-warfare** were yet to dawn on the minds and thought-processes of the Indian-rulers and their military-commanders. Every valuable treatise of great thinkers, is needed to be studied with openness of mind and due strategic-prudence and at the same time; valuable gifts of other thinkers and authors too, are required to be studied with desired strategic-criticality. Moreover, one thing is historically true that besides the multiple-aspects of matters-military, Chanakya's teachings valuably cover much wider aspects of intelligence-services as well as those of still much more vast **crucial subjects of the Statecrafts** as well. We would be discussing these aspects of **Chanakya's thoughts along with those of still more ancient Chinese thinker Sun Tzu**, little later, in more details.

In these ongoing important contexts of our discussions at this point, regarding India, it may appear to be very relevant to **refer to another valuable volume of Indian General G.S Sandhu,** where he writes that, "India, like every other part of the world, was subjected to several foreign invasions during the ancient period of its history.... The indigenous Indus people (*i.e. pre-Aryans*) who had evolved to a markedly higher scale of civilisation than the invaders (*i.e. the incoming Aryans then*) lost to them because they did not accord to the defence of their land the importance it deserved nor did they spare the resources needed for the purpose.... The Defensive has never proved effective by itself except as a part of the other measures, particularly a counter-offensive capability. The Persians who captured the Northwestern fringe of the country during the 6th century B.C. made no impression on the Aryans in Madhya Desha who had practically disowned their sister Aryan tribes which had remained behind the Sapta-Sindhu.... The next invasion was by Alexander of Macedonia during the 4th century B.C. It was blunted by the stiff resistance put up by a small state inhabited by the gallant Pauravas and ruled over by a brave king, mistakenly called Porus by the Greeks, but the real name of this Indian hero is lost to posterity.... The Paurava resistance was also instrumental in the refusal of Alexander's Greek troops to proceed beyond the Beas and his exit from the land. **The claim of western historians that their refusal was occasioned by the home-sickness of his troops is untenable because after Alexander's death in West Asia a**

majority of them settled down and never went back to Macedonia. This invasion and India's response to it as an event is much misreported and misrepresented by Western historians.... He (*i.e. Porus*) had Alexander so worried about a possible unfavourable outcome that he initiated negotiations for a truce.... The Shakas and the Kushans were hinduised and later in the middle of the 4th century A.D., deprived of their kingdoms by the Guptas. The Central Asian tribes gave the country's armies an invaluable lesson in mobile operations and convincingly proved the superiority of the new system over a set-piece form of static warfare with all arms groups.... The Guptas practiced the newly learnt art of mobile war so well that Skandagupta defeated a force of 300,000 Huns and chased it all the way back from Yamuna to the Sindhu. This valuable lesson, alas, was deliberately ignored a few centuries later by the country's armies which reverted to their fascination with elephants and ponderous set-piece battles. It was an unfortunate triumph of conservatism over progress." (Sandhu, 2000, 39-40) (***Mainly due to 'foolish' Indian-rulers and their military-commanders naively falling into the entrenched as well as extremely damaging 'trap' of the powerful and age-old 'Brahmin priest-class'***: *what a tragedy indeed, for the Indian peoples and their magnificently glorious and in many respects, unique ancient-civilisation.*).

Indian General and a scholar-cum-author Sandhu, in the above quoted paragraph, has indeed substantiated clearly that the **inhabitants of the Indian subcontinent since the very ancient-ages have been lacking proactively strong tradition of the "essential strategic-culture and thoughts" in all their traditions of socio-cultural and socio-political lives, in their societies as well as in their processes of governance of their numerous Nation-States**. These glaring negations with all their historicity on the part of Indian-States, their rulers and their military-forces have already been explained and highlighted earlier in this book, a number of times in different contexts during our on-going discussions; as we have been proceeding along on the 'road-map' of our this continuing 'Strategically Critical Journey'. To suitably enrich our deliberations further, in the subsequent pages and chapters here as we reach them, we would appropriately be emphasizing the same again essentially in different contexts. However at this point it may appear rather appropriate to again quote few lines by Field Marshal Montgomery, "It will be understood that the warfare of the ancient Hindus is not very fruitful area of study. The main weaknesses in the composition and organization of their armies were excessive reliance on elephants, poor cavalry and a feudal method of recruitment which militated against the unity of command and standardisation. Even the more successful commanders, such as Chandragupta Maurya who drove out the Greeks, and **Skandagupta Yasodharman** who repelled the Huns, seem to have lacked vision and had little strategical or tactical sense. The passes of the north-west frontier were not held, and armies moved sluggishly. Hindu wars were petty affairs, fought with reserves—affairs of politicians, which were ignored by the peasants cultivating the fields. Hindu civilisation by A.D. 1000 was complacent and conservative. These weaknesses were exposed by the Islamic invaders.... Turkish-Islamic conquest of India developed a pattern. It was a gradual process which began in the tenth century and was completed only in the seventeenth...and the Hindu territories were gobbled up successively as the forces of Islam progressed south and eastwards. Only in the seventeenth century did the tribesmen of the Assam jungles halt the then decadent Moghul forces. As the invaders of the earlier waves settled and became Indians, they themselves were swamped by succeeding waves from the north-west. Each separate advance was short, but occasionally an outstanding Moslem conqueror swept all before him in a devastating tide. Four name stand out: **Mahmud of Ghazni** (997-1030), who is said to have led seventeen campaigns in India (*and ironically as well as 'shamefully' for the then Hindu Kings in both Gujarat and Rajputana with all their military valour and armies, smashed* *famous temple of Somnath in Gujarat.)*; **Shihabuddin Ghori**, victor in Tarain in 1192; **Timur, the Mongol**, who swept through Hindustan in five months during 1398, sacking Delhi but returning to his capital at Samarkand; finally Timur's descendant, **Babur**, the Tiger, who invaded India in 1525, defeating his enemies at Panipat and Sikri." (Montgomery, 1983, pp. 128-130) (*We would be coming to these episodes of the medieval-era Indian military-history later on, suitably in more details. In any case, the above narrated, not at all palatable historic-facts, concerning the making of the Indian 'military-history are a sad commentary. The principal reasons for them have already been spelled-out, in different context earlier too, and the same would be taken-up contextually later-on as well. Hence that remains obvious here that since ages; India has been unpardonably callous in evolving, cultivating and imbibing the inescapable aspects of* ***'essential strategic-culture and thoughts'****. History has never pardoned anyone and, that would never do so in future as well and that's what today, even the modern-era sovereign Indian Democratic Republic perforce needs to always remember the same and make it prominently an integral segment of her socio-political culture as well as process of her national governance.*).

In the similar context in the last chapter itself, while discussing the aspects of the interesting advent of medieval human-civilisations in the European continent we have quoted timelessly relevant 'golden-words' of **Robert Heinlein**. He has said with his great strategic-prudence and vision that, **"Violence, and naked force, has settled more issues in history than has any other factor and the contrary opinion is wishful thinking at its worst. Breeds that forgot this basic truth have always paid for it with their lives and their freedoms."** In any case, whether one remains 'comfortable' with this irrefutable timeless universal fact or not but truth just cannot be wished away. Sadly for the Indians, even in today's modern high-tech era; and hence precisely, this remains perhaps all the more singularly applicable to them and their modern Indian democratic-polity as well as almost all the important segments of the otherwise, multiply dynamic Indian Nation-State.

Comparing two outstanding as well as truly immortal and great ancient-era historic personalities, namely **Sun Tzu of the ancient-China and then Chanakya of the ancient-India** contextually here in our own multiply-strategic and politico-military perspectives are never easy and also simplistically linear, especially when **they happened to be both geniuses in their own rights independently as great thinkers, philosophers, strategists, tutors par excellence and then; great writers of their own invaluable and timeless 'Treatises' as well**. In addition, this difficult assignment becomes all the more daunting, when they both belonged to more than 2200-2300 years in the past and at the same time, their firm historic dates of birth as well as other important personal-data, and also along with the **original-copies of their manuscripts too (except perhaps 'copies from the original' of the those of Chanakya.)** are not authentically surviving for posterity. Naturally therefore in the absence this crucial data-bank in this case; since many centuries and more intensely during the last couple of centuries, many thoughtful as well as intellectually argumentative perceptions and conjectures even leading to some controversies in this regard, had to gradually perforce crop-up internationally among scholars and historians; broadly in the process of the unending making of the man's living-history in this world.

It would certainly be worthwhile to visit the **Oriental Research Institute (ORI) located in Mysore**, one of the historically famous erstwhile princely cities in India in the south. This valuable ORI was set up in 1891 by the then Maharajah of Mysore, medieval India's since a very famous princely kingdom. ORI, Mysore, has been set up there in a very magnificent heritage building; blending architectural styles such as Gothic, Corinthian and Romanesque as well. This Institute houses nearly 60,000 palm-leaf manuscripts, indeed from antiquity. Interestingly then, while coming back to our again ongoing discussions with respect to the **ancient China's Sun Tzu and the ancient India's Chanakya we learn that, historically it appears to be scholarly inferred by some learned historians that Chanakya was around during 321 B.C. and Sun Tzu around nearly 473 B.C.** Hence, Sun Tzu historically pre-dated Chanakya by nearly 150 years or so.

In this precise context, we may refer to a highly informative article (famous Indian News Magazine, '*The Outlook*', 'The Years of the Guru', 27 July 2009) 'The Years of the Guru', since written by **Sugata Srinivasaraju** where the author writes in her article that, "What brought fame to the Institute, however, was the discovery among them of **Kautilya's (i.e. Chanakya)** Arthashastra some 100 years ago. A manuscript of the treatise on politics and governance, believed to have been written in circa 4th century B.C., was found and identified by **Rudrapatna Shamashastry**, a refined scholar of Sanskrit who was the librarian and later curator of the Institute. Shamashastry came across the work in a heap of manuscripts he was going through. This was in 1905. But it was in 1909 by the time he transcribed, edited and published the Sanskrit edition, making the current year (*i.e. 2009.*) the centenary of his landmark mark publication. He then painstakingly rendered the work into English, publishing it to astounding ovation in 1915, by which time excerpts had already made appearances in Journals like Indian Antiquity and Mysore Review, preparing Indologists across the world for the watershed appearance of the English edition.... **Prof Jagannath**, an expert in manuscriptology at the ORI, put it all in perspective. 'Don't expect an autograph of Chanakya on these palm leaves,' He says. 'This is perhaps only a recopy of a recopy made some 500-600 years ago. It was with a pandit in Tanjore, who handed it over to the Institute not knowing what was written on it. Other such copies were found elsewhere in India, but that was later, after our discovery.' He explains that the manuscript is in Sanskrit, but written out in the Grantha script, not Devanagiri. Since the Tamil script couldn't accommodate certain sounds from Sanskrit, Grantha was created to allow for the representation of those sounds in a script accessible to those who know Tamil."

In this specific context at this point; any praise bestowed on the ORI, Mysore (India), won't be excessive at all; since it may be perhaps rightly considered, to be a sort of an Indian 'incarnation' of the **"Department of Oriental Printed Books and Manuscripts in the British Museum, London"**. We may also rightfully say that one of the so many invaluably unrivaled contributions of the ORI, Mysore, can be gauged in this regard; from the contents of the above quoted paragraph, from the above referred article; published in India in 2009, by Sugata Srinivasaraju. Further with respect to our own project since undertaken, as our on-going 'Strategically Critical Journey', within its already precisely laid-out focused perspectives; since the above highlighted a jewel-like piece of invaluable information, along with again so many other similar jewel-like 'pieces of information' as well as vast store-house of the fascinating historical-facts; naturally constitute highly useful quantum of raw-materials for us in this regard.

Since as broadly accepted, Sun Tzu has preceded Chanakya by around 150 years, it would be desirable in our context, to begin a sort of critical studies of these two well-known historic personalities; first with Sun Tzu of ancient-era China. To begin with, we may recall that Sun Tzu is known to have remarkably said that, **'the art of war is of vital importance to the state. It is a matter of life and death, a road either to safety or to ruin. Hence, it is subject of inquiry, which can on no account, be neglected by any one of us today...and tomorrow'**. Well, it remains indeed amazing as to during such distant ancient times of historic-antiquity; this famous Chinese thinker and definitely the human-civilisations' first genius to have said so, with indeed such unique clarity of thoughts and then in addition with politico-militarily unsurpassed multiple strategic-vision.

Since historic-antiquity and then, in the vast arena of the man's matters-military; Sun Tzu's above reflected statement are indeed the first words of wisdom.

So far as traces of the original manuscripts of Sun Tzu are concerned, the persisting uncertainties are yet to be clearly solved. But in any case, man's living-history since its civilisational inceptions; perpetually keeps-on moving and being formed as well. In the process therefore since ancient-ages, brilliant scholars from China as well as around the world have succeeded in somewhat zeroing-on to, with some reasonable historicity; towards throwing appreciable lights on the engaging subject-matter regarding invaluable manuscripts of Sun Tzu's treatises. These are indeed matters of some reassurance. **In this respect, a British scholar, Lionel Giles's book may substantially be very usefully referred to, in this regard**. Interestingly, Lionel Giles happens to be an Assistant in the Department of Oriental Printed Books and Manuscripts in the British Museum.

The author writes that, "I have found it difficult to glean much about the history of **Sun Tzu's text**. The quotations that occur in early authors go to show that the '13 chapters' of which **Ssu-ma Ch'ien** speaks were essentially the same as those now extant.... During the **Ch'ien and Han dynasties** Sun Tzu's **ART OF WAR** was in general use amongst the military commanders, but they seem to have treated it as a work of mysterious import, and were unwilling to expound it for the benefit of posterity. Thus it came about that **Wei Wu** was the first write a commentary on it.... Thus until the end of 18th century, the text in sole possession of the field was derived from **Chi Tien-pao's edition**, although no actual of that important work was known to have survived.... Sun Tzu can boast an exceptionally long distinguished roll of commentators, which would do an honour to any classic. **Ouyang Hsiu** remarks on this fact, though he wrote before the tale was complete, and rather ingeniously explains it by saying that the artifices of war, being inexhaustible, must therefore be susceptible of treatment in a great variety of ways." (Giles, 2008, pp. 24, 25 & 27)

We have earlier too, emphasised that Sun Tzu's main thrust appears to have been to give to the humanity then, to multi-dimensionally think on various aspects of matters-military and formulate comprehensive perceptions regarding them; which subsequently got gelled into the first ever 'Art of War'. These Treatises on the art of war, for the first time, had so scientifically covered almost all the related aspects concerning the matters-military during those times of the distant past of man's history and thereby, shining as an original work of a true genius. Here another remarkable point getting noticed is that they are not at all obsolete in a context of the geostrategic and politico-military criticality even during the presently ongoing times of the super modern-age. Therefore no wonder even today, they do create great interests among the international community of strategic-thinkers, military-commanders as well as the military-historians and writers from allied areas of learning-cum-research.

As per Lionel Giles' book, Sun Tzu's treatises on the art of war have thoughtfully been worked-out broadly covering the following main war-related areas: 1. Crystallizing and formulating the 'war-plans'. 2. Actual deployment of troops in the field and waging of war. 3. Finding-out the tactical-dispositions of first the enemy-forces and then, imaginatively positioning own troops along with supporting arms; for the fulfillment of own politico-military aim. 4. Militarily estimating the weak as well as strong points, of the enemy and own forces. 5. Professionally maneuvering of own troops to gain situational advantages. 6. Launching of attack on the enemy forces as per well-conceived stratagems. 7. Initiating variation in Tactics by the own commanders as per developing situations; with clear perceptions to gain and maintain Initiative as well as achieving sudden-surprise over the enemy, i.e. now; may also be explained as per militarily known 'Principles of

War'. 8. Aspects of Terrain in question. 9. Strategic aspects concerning the Army on the march. 9. Sun Tzu's nine situations, conceived as the Art of War. 10. His thoughts on launching attack on the enemy by making use of 'Fire', i.e. fire which burns. 11. Deployment of well-trained and well-briefed Spies, with focused purposes. 12. Sun Tzu's thoughts regarding the matters-military guided strategic evaluation of 'Energy' related factors.

We then may notice that during those bygone ancient times; it is intellectually remarkable indeed, that Sun Tzu was so visionary with his unique brilliance of mind that he was able to elaborate in substantial as well as focused details, his main 'factors' listed above constituting his concepts of the 'Art of War'. With respect to the formulating and spelling-out the 'War-plans' in a given situation being faced by the concerned monarch and his military-leaders; Sun Tzu says that while doing the same, perhaps five conditions mostly present always, may have to be borne in mind. And they are: 1. the moral-law, i.e. coming to the complete understanding by the troops with his monarch and always remaining totally and with unquestioned loyalty to him. 2. Heaven and 3. Earth: by the 'Heaven', Sun Tzu means daily time-frame in terms of day and night and then the weather-conditions based on daily/monthly and also on the seasonal-basis; and further, by 'Earth' he means relative distances between two points/locations, details concerning the terrain as well as related security aspects and threat-perceptions. 4. The Commander: and according to Sun Tzu, that means the commander's personality-aspects, such as his levels of professional competence and wisdom; along with his nature; instrumental in shaping his courage, maintaining morale, as well as his grip over the command and control aspects, in all kinds of situations and challenges. 5. Discipline and systems of organising the troops in appropriately different groups/sub-groups for the training and deployment of troops in the battlefield, as dictated by any given situation-specific conditions. During Sun Tzu's era, these aspects were highly fluid and hence; they principally depended upon the concerned Monarch and his top commanders and on their military-ingenuity.

We have listed broadly twelve principal constituents previously, which in fact formulate and precisely shape Sun Tzu's 'The Art of War'. Of which the first one is crystallizing the 'War-plans', since elaborated in some details in the above paragraph; as per Sun-Tzu's own concepts, since also available in the books on these subject-matters. In this respect, the remarkable points are that Sun Tzu has indeed elaborated over them and also explained them with great professional precision and outstanding clarity of vision. It emphatically establishes Sun Tzu's genius and capacity of his unique intellectual prowess over the multiple-aspects of the matters-military (around more than 400 B.C.); which were indeed much ahead of his times. He has specially emphasised the needs for attacking the enemy, with clear strategic expertise and military-precision; which are so relevant universally and even today, with cool and unwavering confidence for achieving victory in war. For militarily to ensure these, he stresses the crucial importance of putting logistics-management in place, appropriate organisation and composition of troops for the war, their befitting training, morale, professional competence of the commanders/officers at different levels, and so on.

As dictated by any particular given situation; he further stresses that the commanders should be imaginative enough to ensure maximum feasible mobility of his troops and simultaneously also, they must tactically as well as promptly be able to manoeuvre their troops with the desired co-ordination and synergy and hence, always remaining confident of the final victory. These words may not be directly coming from the later-era formulation of 'Treatises' on Sun Tzu's concepts on his 'Art of War' by the subsequent scholars; but whenever we may study those treatises with the requisite strategic-criticality, we may find similar meaning always getting surfaced. He then in addition, goes on to caution regarding some relatively common mistakes and military-naivety as well. **Like, he says that avoidably prolonged war is always very costly, along with tragic negation of military as well as the Kingdom's valuable resources and thereby, perforce landing in a jam and playing into the enemy's hands**. In addition, if these crucial politico-military aspects are not effectively taken-of and ironically then the otherwise assuredly forthcoming militarily decisive-victory, may even tragically passed-on into the enemy's hands.

It is commonly well-known that, **Sun Tzu was the first military thinker as well as strategic-genius, who is reported to have said that the finest victory for any General is to decisively achieve the same without even fighting that war**. In fact, he emphatically and even rightly maintained that the successful fruition of this does indeed universally command the gloriously highest-point in the art and science of the 'generalship' in war. Sun Tzu further goes-on to say that the General is indeed a kind of bulwark of the State and if that bulwark is complete in all the crucial politico-strategic aspects, then the State is bound to be strong and thereby; always displaying great deterrent, for any potential enemy of the State (*Paradoxically and even ironically even today in the 21st century sovereign democratic Indian-Republic, the ruling political-elite as well as members of the political-opponents, objectively and also with essential strategic-prudence are refusing to accept their highly damaging regressive-fixations and ignorance, for their own reforms and simultaneously, for their politico-strategic enlightenment; in the highest national-interests.*).

Another outstanding feature of Sun Tzu's concepts regarding his 'Art of War' is that with his typical strategic-brilliance, he has also clearly spelled-out the five essentials of the victory in war. And as we learn from the noted books on Sun Tzu's 'The Art of War' these have been spelled out as: 1. He, with all broadly probabilities, should be assured of the final victory against his enemy; who confidently knows when to engage in fight and precisely when not to do so. 2. He will win the war, who strategically knows how to handle both superior and inferior forces facing him. 3. He will win whose military forces are animated with identical spirit and spirit de corps throughout the all ranks. 4. He commands a much assured chances of victory, who has strategically prepared himself well for that war and then, tactically waits for a chance to swiftly and decisively engage his enemy; giving him no scope to escape or to siege the initiative in any manner. 5. That military commander will win, who has been suitably facilitated to acquire the requisite military wherewithal and at the same time he is not unnecessarily interfered with, by the sovereign or his political bosses; since that is a great botheration for him and his profession of soldiering.

Interestingly in the present modern-era too, the above narrated timelessly remarkable and politico-militarily brilliant strategic perceptions of Sun Tzu, do remain all the more important and relevant in a context in the crucial domain of matter-military. The unique genius of Sun Tzu lies in the fact that he was the first visionary, great strategic-thinker and unprecedented teacher of the matters-military; as per the man's the then civilisational-awareness in the world and also, as relevant to profession of arms as well as the level of military infra-structure, prevailing during those very distant and ancient historic-ages. But as per the details available in the first ever historic treatises of Sun Tzu and also, as illustrated from time-to-time, by the various noted scholars and thinkers in this regard; we learn that Sun Tzu's intellectually brilliant thoughts, mostly remained confined to the aspects of soldiering and crucial segments of the military-command and then all important, both strategic and the tactical levels jobs of the commanders and their own aspirations, regarding any situation-based vision including their professional thought-processes.

But, the larger fact too, just cannot be ignored that **the entire gamut of the governance of any**

Kingdom or the Nation-State; perforce contains many other aspects than just the 'profession of arms'. Though at the same time, the aspects of matters-military with all their vastness and complexities; certainly do remain always crucially very important, since they are continuously involved with both the internal as well as external security-aspects of the concerned Kingdom/ Nation-State. And at the same time since very ancient-ages, again all those segments of the entire gamut of the process of governance, along with the crucial matters-military are all certainly always multiply inter-connected as well. Now at this specific point of our ongoing discussions, we may even very befittingly say that the above highlighted aspects are the central-fulcrum of another allied as well as crucially important aspects of the gradual evolution and cultivation the same **'essential strategic-culture and thoughts'**; whose far-reaching and multi-dimensional decisiveness-cum-advantages in this regard, have already been discussed and in addition, stressed-upon earlier. And the same would again be discussed and highlighted in different as well as interconnected contexts subsequently, as we further proceed along with our continuing deliberations.

Interestingly, now coming back to the **ancient-India's immortal legend: Chanakya, his history and his unique Treatises of such great antiquity**, according to the noted scholars of different ages Chanakya came in our world nearly 150 years later than Sun Tzu of the ancient-China. In this respect, besides other source-materials thereis ***Kautilya: The Arthashastra*, edited, rearranged, translated and introduced by Mr. L.N. Rangarajan** for our valuable information, guidance and reference. Since it is well-known that 'Kautilya' was also another name of the same legendary Chanakya of the ancient India. In the beginning of his book, Rangarajan has reproduced a very important verse in Sanskrit, reported to have been written by Chanakya and which conveys a highly learned as well as comprehensive message, concerning the paramount duties of a King. According to Rangarajan, its meaning in English is that, "In the happiness of his subjects lies the king's happiness; in their welfare his welfare. He shall not consider as good only that which pleases him but treat as beneficial to him whatever pleases his subjects" (Verse No. given as: 1. 19. 34). The multi-dimensional comprehensiveness so meaningful, of this above mentioned verse can only be visualised if, we may go deeper into it and give a little careful thought on it.

Similarly, we do learn that there are so many verses written by Chanakya with deep thoughts as well as his unique genius remarkably, during those very ancient times. One highly learned Indian professor in the University of Bombay, **R.P. Kangle** too, has done lot of deep studies and research on the vast treatises of legendary Chanakya; and has written a book consisting of three scholarly volumes on his 'Arthashastra', and understandably he also owes a deep gratitude to Dr. Shamashastry and his monumental as well as historically unforgettable work in this regard; during his long stay at the Oriental Research Institute, Mysore (India). The work of Dr. Shamashastry has indeed attracted a number of foreign scholars of oriental-studies, with great interest.

All those international scholars and authors do befittingly owe great respect and academic gratitude to the legendary **Dr. Shamashastry of the ORI**, in this regard. **L.N. Rangarajan** writes that, "Subsequently another 'original' manuscript and some fragments, in a variety of scripts, were discovered as well as old commentaries of the text. In addition to Dr. Shamashatry's translation, there is a complete Sanskrit commentary by **Dr. T. Ganapati Shastri**, a German translation with voluminous notes by **J.J. Meyer**, a Russian translation and translations in many Indian languages.... The actual number of verses, (sutras and shlokas) in Kangle's edition is 5348. The fact that the text does not contain 6000 as claimed in (1.1.18) has led to the speculation that a part, about one-ninth of the present length, may have been lost over the centuries.... The figure of 6000 is an approximation, not meant to be an exact count. Barring any new discoveries, the text, as we now know of it, can be taken to be complete and faithful a rendering of the original as is possible for a work of this antiquity." (Rangarajan, pp. 21-22)

Hence if we may delve deeper into Chanakya's above conveyed meaning behind those 'words of wisdom', we may notice their very enriching as well as much wider connotations, concerning the earlier mentioned entire gamut of the process of governance of any Kingdom/Nation-State. In the above mentioned meaning of Chanakya's one verse originally in Sanskrit what is apparent is that Chanakya conveys the principal and foremost duties of any King, with respect to his kingdom but that is certainly not his only message. In fact contextually with respect to a king's duties, the meaning becomes much larger as well as multi-dimensional in range. To anyone, these would be quite clear who happens to read translations of almost all the verses in a perspective, since all the messages contained in them would be found to be conceptually interconnected. Here the welfare and happiness of the people do not at all mean even remotely to adhere to their short-sighted and irresponsible appeasement; but these would always be pragmatically implemented in practice, keeping in view the essential and futuristic security-concerns regarding the kingdom as well as its people. Here in a context, these multiply larger connotations are universally relevant even today, and all the more so to the present-era's sovereign Indian Democratic-Republic. Since immediately after the year 1947, almost all the Indian political-leaders, certainly with few exceptions like India's first deputy prime minister Late **Sardar Ballavbhai Patel**, have been mostly irresponsibly callous and lacking in essential vision, regarding the paramount geostrategic and politico-militarily national-security concerns. We would be more appropriately discussing these aspects in some detail, in the subsequent chapters as we come to them.

Then with respect to the King's as well as his kingdom's multiple economic-activities in his verse: 1.19.35-36 historically indeed with great vision during those very distant ancient times, Chanakya has precisely said 'Hence the King shall be ever active in the management of the economy. The root of wealth is (economic) activity and lack of it (brings) material distress. In the absence of (fruitful economic) activity, both current prosperity and future growth will be destroyed. A king can achieve the desired objectives and abundance of riches by undertaking (productive) economic activity.' One may even very rightfully say that these historically timeless words, as available in the treatises on Chanakya's great visionary thinking and writings are indeed beyond any kind of time-bound obsolescence.

Similar to the above, Chanakya of the ancient-India has further written with great vision and detail, almost on all the principal segments constituting the entire processes of the gamut of the governance of a Nation-State. They include even his perceptions on the matters-military and war though his concepts on warfare were somewhat 'linear' in visualization and also, laden with rather excessive 'conservatism'. But at the same time when reviewed contextually, with the desired politico-strategic and allied military perspectives interestingly we may notice that the ancient-China's Sun Tzu's treatises too were mostly China-centric, as per those very ancient times. In this specific context, in addition we may also say that Sun Tzu's grand strategic-perceptions were no doubt visionary but they were principally confined to only some strategic aspects of the matter-military as well as brilliantly timeless tactical-concepts, concerning the different operations of war. Understandably these are such historically perceptive observations on which, opinions may differ and hence any such matters in addition, may remain debatable as well.

As per the ancient highly complex Indian-history of Chanakya's times of those by-gone era he too, displayed far-reaching vision and scholarly great analytical acumen while historically formulating his comprehensive treatises. In that he has brilliantly spelled out that there are seven principal constituents of a Kingdom/State and, according to him they are:

> 1. The Ruler or the King. 2. King's 'Council of advisers, ministers in-charge of different 'departments' or the areas of governance. 3. The armed-forces to defend the kingdom's borders, including fighting both defensive and offensive wars, always making best military as well as strategic efforts to ensure the king's decisive victory. 4. The 'geographical' territory of the Kingdom; including the entire population inhabiting it, as the king's subjects. 5. The populated and effectively fortified cities and towns, located within the kingdom's territory. 6. The treasury, along with the entire valuable assets/wealth of the kingdom. 7. Strategic and politico-military 'Allies' of the Kingdom.

As already discussed earlier, since Chanakya in his timeless treatises has perceptively written about the essentials of the whole gamut of the aspects of governance of any Kingdom/State, as expected from the King, it was in the fitness of his visionary thoughts that first he spelled-out as to what were the principal constituents, which gave cultural as well as socio-political 'shape and life' to any Kingdom/State. Therefore, he did the same as a brilliantly visionary teacher-cum-mentor, in the beginning itself and precisely those elements of the principal 'constituents' as seen in his *Arthashastra*, have also been listed here in the above paragraph. At this precise point of our on-going **comparative-study of both Sun Tzu and Chanakya** we may always bear in mind that with regard to Chanakya, he too as per his own bygone ancient-era, was India-centric and hence; he wrote his treatises as relevant to then prevailing societies and kingdoms specifically in India. And logically therefore, treatises of both these independently ancient Chinese as well as Indian genius-thinkers respectively, may be critically studied in the requisite geostrategic and politico-military perspectives; in their own distinct contexts.

Now coming back to **Chanakya's perceptions regarding the then entire processes of governance, and as mentioned in the earlier already referred book of L.N. Rangarajan (Kautilya: The Arthashastra) we learn that the governance consists of these main segments: 1. First, the king or ruler of any Kingdom or State, 2. A well-organised State, 3. Treasury: including State's sources of revenue and Accounts and Audit, 4. Matters-military: Defence and war, 5. Covert operations 6. Civil-Service's Regulations 6. Government's Various Departments of the Government 7. Law and Justice Departments, 8. Foreign-Policy Aspects—Subsequently in the Treatises, each one of them has been analysed.**

Well in addition, as learnt from the voluminous commentaries on Chanakya's treatises, we also say that chapters on each one of the above mentioned segments of the complex and vast processes of governance, has been analysed as well as thoughtfully dealt with in great depth and detail; consisting of a number of sub-segments or sub-chapters. No highly relevant point with respect to any of the above listed principal segments of the entire gamut of the governance possibly perceived based on ancient-India's overall conditions during Chanakya's times has perhaps been overlooked or ignored by him, in his legendary treatises. Logically therefore, it is indeed remarkable and a matter of historic wonder, that broadly similar to the ancient-China's legendary great strategic thinker Sun Tzu, but in different and more comprehensive visionary context, nearly 150 years later but totally independent in the ancient-India, another great thinker lived. Interestingly, he became a historic-legend in the name of Chanakya or Kautilya in the ancient Indian-history and preceding the beginning of Christian-era by nearly 150 years and, since discussed above.

Let us further discuss some historically irrefutable 'chinks in the armour' of otherwise brilliant and visionary-thoughts as contained in his legendary treatises on the entire processes of the gamut of governance of any Kingdom/State. The plain facts do also remain loud and clear, on the pages of very long and chequered living-history of the ancient Indian-State with far-reaching and indeed highly damaging as well as 'shamefully' embarrassing consequences for her. And **unlike Sun Tzu, those facts are that Chanakya's thoughts and writings on the complex subjects of exclusively important matters-military, in the same total gamut of the governance, have been found to be inadequate and wanting; in the perspectives of the geostrategic and politico-military criticality**.

At this specific point concerning the matters-military, in order to cut the long 'story' short, it may therefore appear to be highly relevant to quote few lines from **General Sandhu's** book wherein he writes that, "A feature of the Arthashastra which is particularly striking is the author's (*i.e. Chanakya's*) belief that war is an intellectual exercise. He distrusts initiative or innovation in battle he deliberately plays down the role of commanders in war and denigrates the art of generalship (***ironically and totally unlike his predecessor in ancient-China, Sun Tzu***). He goes to great lengths in laying down how troops should be deployed; he leaves no room for deviation or error. It is only at the lowest level of command (*patti*/platoon) where his instructions are left vague. His clear message is: 'Battle is a mathematical exercise. I as an intellectual Brahmin, have the duty to tell kshatriyas how to conduct a battle. All they need to do is to follow instructions and if other factors, that I have enumerated, are in their favour they will win'. He warns kings repeatedly to be on their guard against treachery by his commanders.... Such a suggestion can be made only by a person with no experience of war. No successful battle can be fought by a commander who is not even sure of the duration of his authority. **The distrusts of generals by the civil authority, apparently has a long history in the country (*i.e. India, here*)."** (Sandhu, 2000) *(In any case, without any intention to even remotely create any kind of debatable 'controversy' in this regard one point perhaps appears to be very apparent that notwithstanding, legendary Chanakya's truly remarkable multi-dimensional intellectual-acumen, his unsurpassed genius as well mental-reach; his path-breaking any visionary strategic-grip over highly complex 'matters-military', ironically happened to be rather 'naïve'. Hence in the multiply-futuristic and timeless, as well as very consequential geostrategic and politico-military perspectives; these aspects have indeed repeatedly resulted in greatly strategic-cum-militarily embarrassing reverses for the Indian-State, since ancient-ages onward. We would again be discussing these very important aspects suitably in more detail, in the subsequent chapters).*

Well, it is also historically very enriching and true that, indeed Chanakya was the mentor as well as multiply great guru of the brave and highly talented warrior-king Chandragupta Maurya, who marvelously succeeded in establishing the vast and famously glorious Magadha-empire in India, during those very ancient times. Some learned historians are of the considered opinion that Chandragupta Maurya, as a great General as well as a visionary Emperor, was instrumental in establishing his mighty and highly famous empire but, behind the whole process, it were indeed Chanakya's genius and thoughtful guidance in ensuring his glorious success. But in the larger perspectives of perpetually changing historic 'time-frame', we may also say that during those times of the enduringly mighty Mauryan-empire; there were indeed different kinds of situations prevailing in India.

Nanda King, though not very competent and a relation of Chandragupta Maurya, was on the throne of Magadha-kingdom. And as per their tradition, Nanda King too, has massive army with large number of war-elephants. Moreover, Alexander came to India much earlier than the coronation of Chandragupta Maurya, in the ancient-city of Patliputra, being the Capital since located on the southern-bank of the Ganges in the eastern India. In addition, it is also historically well-known that after Alexander's closely fought battle with King Porus, then ruling over a small kingdom in the far-off north-western India, Macedonian soldiers were extremely exhausted and militarily very tired and the back of their fighting-spirit stood completely smashed by the brave and highly resilient soldiers of Porus. Hence the Macedonians refused to commence their conquest-march towards further east, for facing one more fresh and very large military-formations of the ancient Indian soldiers, along with another vast number the war-elephants.

Moreover, though those ancient-era Indian-armies were not at all well versed with the highly beneficial military aspects of the mobile-warfare and, the Indians were experts in only the art of static-warfare and hence, they lacked flexibility as well as art of then the manoeuvre-warfare. Though Alexander was himself highly innovative and a pioneer in the battlefield arts of the both mobile as well as manoeuvre warfare he had to move back home, highly dejected and militarily frustrated, for the reasons cited above. Reasons for this in no way, historically goes as a 'feather in the cap' of Chanakya but to the unshakable military-valour and leadership of the King Porus. Well, one very predominantly important segment of the multiple-aspects of matters-military, i.e. areas concerning warfare apart; but otherwise, Chanakya's multi-dimensional vision and range of strategic-thoughts, regarding vast and highly complex areas of the then prevailing practices of the systems of governance of any Kingdom/State, as existing in the ancient-India along with its overall socio-cultural environment and traditions are indeed unique as well as in many ways his timeless treatises do contain highly valuable lessons for us, even during our present fast-moving super high-tech modern age.

In this specific context and in addition especially, in the presently ongoing modern-era; what is important as per the imperatives of strategic prudence is that today, we need to critically study Chanakya's treatises in such a way that we are judiciously able to filter-out those still relevant valuable lessons and at the same time; keeping in mind the aspects of our geostrategic and politico-military perspectives. **Who was historically Chanakya** of India, during those very ancient-times? There are many opinions in this regard. Some scholars say that he was originally a Keralite Brahmin and interestingly, some maintain that he was a learned Brahmin hailing from the Magadha-empire in the eastern India. Well, that is understandable concerning any very ancient-era famous thinker in any part of the world and also since many legends too, historically in course of time do tend to grow around him.

Interestingly, **Jawaharlal Nehru** has written that, "Chandragupta appears to have been a very clever, energetic and ambitious person, and the Nanda king, thinking him too clever, or not liking something he had done, exiled him from his country. Chandragupta went north to Taxila, attracted perhaps by stories of Alexander and the Greeks. With him was a very able Brahmin named Vishnugupta, also called Chanakya. The two of them, Chandragupta and Chanakya, were not meek and mild persons bowing down to fate or whatever might happened to them. They had great and ambitious schemes in their heads and they wanted to go ahead and succeed. Perhaps Chandragupta was dazzled and attracted by the glory of Alexander and wanted to follow his example. In Chanakya he had an ideal friend and counsellor for his purpose. Both kept their eyes open and watched carefully what was happening in Taxila. They bided their time.... Chanakya, the great minister of Chandragupta Maurya, tells us something about the navy in his Arthashastra.... Thus it appears that even at the beginning of the Mauryan period shipbuilding was a flourishing industry in India. And ships are obviously meant to be used. So quite a considerable number of people must have crossed the seas in them." (Nehru, 1934, pp. 58 & 117) (*Nehru's well-known books are not only scholarly in pleasant flow of good English but indeed historically highly informative as well.*)

Interestingly, well-known Indian scholar and the third President of the Sovereign Republic of India, Late **Dr. Zakir Hussein** is known to have asked that after all: **'what is specialization?'** And then he further enlightened us to say that **'specialisation means, knowing more and more of less and less.'** If these golden words of wisdom looked into deeper and with desired criticality, one may indeed find them quite thought-provoking and engaging as well. Now coming back to the various branches of man's living-history since its inception; we may notice that gradually as time rolled by over the unending 'canvas of time' and, as various human-civilisations in their respective geographical-areas; simultaneously and also multi-dimensionally their own exclusive as well as in addition, interactive among themselves, commenced achieving multiple-progress and the same history too, began advancing and then, gradually got 'segmented' into a number of branches and much more; towards continuously expanding 'store-house' of human-knowledge. Their list perhaps may not be that easily completed; but contextually for our specific needs, they may be listed globally per as below:

> And they are: (1) Man's civilisational history, (2) Political history, (3) Economic history, (4) Cultural history, (5) History of Religions, (6) History of Cuisines since the ancient-ages, (7) History of clothes, garments and dresses, (8) History of Fine-arts and their many fascinating sub-branches, (9) History of Music and musical-instruments, (10) Military-history and so on. And thereby; they may appear to be adequate to serve our need-based discussions.

In an allied context, **Professor E.H. Carr**'s compilation of his lectures delivered in the University of Cambridge in 1961, in a book-form and titled as to 'What Is History?', and then later on, also published in 2008 by Penguin Group, Australia, may appropriately be referred. Hence, here on page 7, as quoted from the 'Personal Anthology' of **J.L. Borges**; we find his interesting reflection that the "Ultimate history we cannot have in this generation; but we can dispose of conventional history, and show the point we have reached on the road from one to other, now that all information is within reach, and every problem has become capable of solution. And almost exactly sixty years later **Professor Sir George Clark**, in his general introduction to the second Cambridge Modern History, commented on his belief of Action and his collaborators that it would one day be possible to produce 'ultimate history'."—*G. Lucas is quoted as saying that,* "**What is historical fact? This is a crucial question into which we must look a little more closely**. According to the common-sense view, there are certain basic facts which are the same for all historians and which form, so to speak, the backbone of history." (The Historical Novel, p. 10)

Then on page 11 of the same book, **Professor Clark** himself, very knowledgeably comments that, "It is a necessary condition for his work, but not his essential function. It is precisely for matters of this kind that the historian is entitled to rely on what have been called the 'auxiliary sciences' of history—archeology, epigraphy, numismatics, chronology, and so forth. The historian is not required to have the special skills which enable the experts to determine the origin and period of a fragment of pottery or marble, to decipher an obscure inscription, or to make the elaborate astronomical calculations necessary to establish a precise date. These so-called basic facts, which are the same for all historians, commonly belong to the category of the raw materials of the historian rather than of history itself."

How precise, thought-provoking as well as scholarly beautiful indeed, the above quoted paragraphs are for we laymen in the contexts of the clarity of concepts of history and its multiple-aspects. We may notice an important point in this regard that, in general perhaps due to lack of any such crucial clarity naturally there remain many kinds of confusion regarding history as a whole. And thereby, concerning history's multi-dimensional aspects along with its many 'branches' and in addition then; many kinds of 'raw-materials' too, since explained above by a noted scholar of history are also available for further studies and research. Ironically, the same perhaps may be so in some cases; among the concerned people, both soldiers and civilians having interests in the areas of matters-military as well as military-history too. If that is so, then perhaps in a context, the same is understandable as well. In these perspectives, it may appear somewhat appropriate to again emphasise that the main subject-matters of this book; don't pertain to history at all, in its classical meaning and sense. **In fact, as already mentioned and elaborated in the beginning itself that it is precisely a 'Strategically Critical Journey: Through Humanity's Past, Present and Possibly Future: In Global Geostrategic and Politico-Military Perspectives'.**

In our context at this point of our continuing discussions, the valuable details given in the little earlier quoted paragraph, from General Sandhu's book on 'A Military History of Medieval India' since already referred, are indeed highly informative and well-taken; but while reflecting on them, we may notice two far-reaching points concerning India; and one is positive and another quite negative for the ancient and definitely glorious Indian-civilisation and its many age-old socio-political imperial entities. The positive one is that in one way, **India's ancient glorious legacy of the matters-military as well as traditions of brave-soldiering** have been continuing since very distant-past of the fascinating living-histories of the advent of various human-civilisations in the world and also, perpetually continuing multi-dimensional advancements of some of those ancient and medieval-age civilisations. As obvious examples, we may name ancient Chinese, Indian and Christendom-guided civilisations and then medieval-era Islamic-civilisations. In this specific context here, another negative-point since mentioned just above and hence, on our detailed and historic identifications we notice the extent of negativities of this particular point as extremely damaging and, very unsettling as well, which perforce did multiply change the course of India's long living-history, with many lingering adverse effects. And naturally thereby as their still felt after-effects, **even in the present 21st century of the presently continuing highly dynamic modern-era, and also as a vast democratic independent Nation-State, ironically India has not yet decisively succeeded in coming out of this dark and deep ditch**. Frankly to be specific now, that negation in this context since the ancient-age has been India's highly regressive fixations to live in a sort of pathetic 'dark-cocoon' of her own making.

And hence further as a consequence, India has not been able to see and read the invaluable and also continuing writings on the wall; highlighting historic-lessons more crucial in the paramount spheres of the matters-military, from time to time in different ages and, as they sprouted and gradually got shaped-up for the larger humanity and also, in their own historic-sequence. Paradoxically, India's age-old historic civilisational-failure, which has been her own negation in many ways and sadly therefore, as a most glaring consequence has been India's greatly embarrassing failure to fruitfully cultivate, evolve and imbibe the multiply crucial the same and earlier already emphasised 'essential strategic-culture and thoughts'. **No wonder that in the present 21st century, with all her noteworthy economic progress and also well-recognised democratic-polity, India is yet to overcome numerous hold-ups in the paths of her complex-governance**, which are indeed otherwise differently unique too, as per the different kinds of situations-specific compulsions of different Nation-States the world over, in the present modern-era. We would be more appropriately dealing with these aspects and their multiple connotations in more detail, in the subsequent chapters.

In course of time, medieval-age slowly dawned into the pages of the man's living-history naturally being a fascinating segment of the supreme Nature's own totally independent and incomprehensively complex and continuing phenomena of evolution. Logically then it is obvious that man's living military-history too, entered into the same medieval-era. Hence as a part of that, as man's own multiple 'store-house of knowledge' further slowly kept-on expanding and along with it, man's multiply experimenting as well as innovative-skills too continued to improve and sharpened. Naturally therefore, these interesting phenomena triggered further gradual advancements in the areas of arms-making and those of weapon-platforms as well as other allied military-technologies as well. In this specific perspective interestingly further, when we may study and review these processes being parts of the man's same living-history with our desired strategic criticality, we do also notice that as their consequential-results then prevailing entire gamut of the warfare too, progressed further and simultaneously got more complicated.

During the early years of the many centuries belonging to the highly chequered and blood-soaked **medieval-age** history shows us that as the western part of the once mighty ancient Roman-empire, began gradually disintegrating and thereby getting sandwiched between the expanding as well as consolidating Christendom in their north and west and then the advancing Arabs from the south and across the Mediterranean. Interestingly the eastern-part of that Roman-empire at Byzantium proved to be better survivor, displaying much greater fighting-qualities historically and simultaneously, more tenacious and enduring military-acumen. In fact, since the beginning of Islam in Arabia; during 622 AD, **Prophet Mohammed** undertook a mission along with his followers from Mecca to Medina and then relentless expansion of Islam and Arabianism, further into various other regions of the medieval world. Among Mohammed's followers, few more capable military-leaders got evolved; historically more prominent were Khalid ibn-al-Walid and Amr ibn-al-As and so on. Within only a one or two centuries these Islamic-warriors in turn, were instrumental in bringing in vast territories of the Middle-East, Africa and Asia in north-western fringes of India and some southern parts of medieval-China; into the fold of Islamic-empire.

In these contexts, **Field Marshal Montgomery** writes that, "The feature to observe in history of western European warfare between the seventh and eleventh centuries is the continued race of cavalry. The Arabs made their conquests by **intelligent use of camelry and cavalry in suitable terrain**, the open lands of North Africa and western Asia.... It was the number and appearance of Arabs which made them so formidable. As a Byzantine commander observed, 'they are very bold when they expect to win: they keep firm in their ranks, and stand up gallantly against the most impetuous attacks.... The infantry was mostly inefficient and ill-armed; their strength lay in cavalry. In the early part of the seventh century their cavalry was very lightly equipped and highly mobile; but in succeeding centuries the Arabs learnt most from their most stubborn enemies, the Byzantines, relied in increasingly on mounted archers and lancers, protected by chain-mail shirts, helmets, shields and greaves. The best qualities of the armies of Islam lay not in equipment and organization but in morale, created by religion, mobility due to camel transport, and bred of hard living in the desert. The crusading idea of the jihad, the holy war, was very real to the close followers of Mohammed.... The Arab explosion of the seventh century the forth, last, and the greatest of the Semitic migrations. In the traditional and natural pattern, the migrant went first for the Fertile Crescent, before over spilling towards and beyond the valleys of the Euphrates and the Nile.... The culmination and turn of fortune in the struggle between the Arabs and the Byzantines came with the siege of Constantinople in 717-18...**before the age of gunpowder**.... Since Constantinople was surrounded on three sides by water, all seemed to depend on the relative strength of the opposing fleets, and the Arabs had a vast superiority in numbers. Leo, however, led the twelve months defence of Constantinople with courage and brilliant resource, and when the siege was raised the Byzantine fleet pursued the enemy down the Hellespont; there into a storm and only a small portion of the Arab forces survived.... The maintained the classical Roman practice of regularly building fortified camps and a corps of engineers always marched with the vanguard of the army. **Every unit of 400 men had its medical officer and six or eight stretcher bearers**.... The keystone of the Byzantine military system was tactical organization: they fought cunningly and efficiently. The Byzantines rightly considered that the methods to be applied in battle must be varied according to the tactics of their enemy, and they studied carefully the methods of probable opponents. Their most important military writings are the *Strategic on* of Maurice, *c.* 580, the *Tactician* of Leo the Wise, *c.* 900 and a manual on frontier warfare by Nicephorus Phocas (conqueror of Crete and Cilicia from the Arabs, and emperor 963-9)." (Montgomery, 1983, pp. 79-82)

In varying contexts, we have since observed during our discussions earlier that, historically it remains a big and indeed very embarrassing irony for the man that since his inception as well as gradual development over many centuries, and also the continuing march of his otherwise magnificent civilisational-history too continuously in the making; have ruthlessly been human-blood stained and thereby, an extremely painful saga of the events of all sorts; over the last many thousands of years. And at the same time in reality, the same phenomena are still continuing unhindered. In other words, peace in real and fruitfully enduring terms; in the man's individual, socio-cultural as well as socio-political lives, have mostly been nothing but only like a 'mirage-chasing' over many thousands of years again, of his continuing living-history. Hence in the perspectives of the human-civilisational advancements over many centuries; it is indeed not that man through his various kinds of Organisations and Institutions all over the world, has not been making any worthwhile multi-dimensional efforts for enduringly establishing any kind of really effective 'global order of peace' but alas, it has always been nothing but rather an embarrassing 'game of hide and seek'. In any case, these are irrefutable facts of the human-history and at the same time these historic-aspects are equally true during our present modern-era as well.

We may now desirably prefer to move away from certain paradoxically philosophical-aspects of the man's age-old living history as interestingly touched upon in the immediately above paragraph, since those subjects by themselves being vast and hence, we are obviously not at all equipped to delve into those areas of human-learning as well as they do not form part the scope of our continuing discussions. Now coming back to principally geostrategic and politico-military perspectives of the multiple and also highly complex aspects of the matters-military, along with advancements of various human-civilisations around the world since their beginning; we learn that otherwise too, the **dominant aspects of the military-leadership are indeed most crucial, concerning the all aspects of soldiering**. Hence, it won't be stretching the matter too far, if in addition we may also say that the aspects of military-leadership are indeed pivotal to the warfare along with its all possible manifestations, since the very beginning.

Aspects of **military-leadership**, both in their concepts and practices, are indeed very vast in both their depth and dimensions and certainly not at all simplistic, as they may appear at the first glance. Precisely in this context, there is one very appropriate, crisp and focused book titled 'The Art of Military Leadership' authored by a retired Indian Army Officer **Major Udaya Chandar**. Major Chandar's book is short, highly focused and indeed very thought-provoking. In few hundred pages only, this book has been able to bring in informatively, the valuable historic and contemplative perceptions of many famous scholars, statesmen and military-commanders the world-over; since many centuries. Such as valuable thoughts of **Machiavelli (1469-1527) of Italy, Harry Truman, Jomini (1779-1869), Field Marshals Slim, Earl Wavell, Montgomery, Rommel, Man stein, Guderion, Generals of the U.S. Army Douglas McArthur, Eisenhower; Marshal Foch and Lawrence of Arabia** and so on, were perceptively discussed.

In addition to the above, well-known ancient thinkers **Sun Tzu of China and Chanakya of India** too, have discussed age-old art of military-leadership aspects in their well-known treatises as well. Further the 8th century another great **Chinese thinker Tu Mu** too, has given his own thoughtful reflections on the very ancient art of military-leadership. Besides the present modern era, many great military commanders belonging to both the ancient as well as medieval ages; who have brilliantly practiced those golden concepts and art of military-leadership in their respective battles and thereby in this respect, they have drawn historically enduring lines on the pages of living military-history in their own right; by means of extra-ordinary saga of their shining dedication of soldiering and valour. History rightfully shows many unforgettable examples, such as **Macedonian Alexander, Greek Hannibal, Roman Julius Caesar, Mark Antony, Egyptian Ramses II and later Queen Cleopatra** (*in her own feminine manner.*) and so on. In this respect, **Joan of Arc of France** has her own history of valour and fearless sacrifice as well.

Among the Indians too, there are many historic personalities in this regard, such as **Chandragupta Maurya, Ashoka the Great, Samudragupta and then Rana Sanga, Babur (*Though initially to begin with his life as the founder of Mughal-empire in the medieval-India; Babur came from foreign lands, but after his decisive victory over the then Delhi's emperor Ibrahim Lodi he permanently settled down in India, and the rest as they say is history.*), Akbar the Great, Maharana Pratap, Shivaji, Queen of Jhansi Laxmibai, Chandbibi and many others. From Far-East too, we have come to know many, like General Giap and so on. Among the Japanese during the second World-War; we may note Generals Tojo as Japan's Prime Minister, Yamashita, and Rear Admiral Raizo Tanaka** and so on. Further during the present modern-era, the independent India too has produced a number of noted military commanders at all levels of the military-leadership, both in peace and war.

In the 'Foreword' to the earlier mentioned Major Udaya Chandar's book, **India's Field Marshal K.M. Cariappa** has very thoughtfully written that, "The author (*Major Udaya Chandar*) has emphasised among many factors in connection with this subject, the following four: (1) Leadership, (2) Man-management, (3) Morale, and (4) Discipline. A good leader has to bear these four factors in his mind at all times". In the much larger perspective of leadership in general terms in fact; leadership-aspects are equally important and applicable in all fields of human-endeavour, and also much more so in the present fast-advancing modern-era; such as challenges of political leadership in a modern democracy, civil-services, technical and medical services, Corporate-sector and then, independently practicing various professionals and businessmen as well. **But in this precise context, it is historically true that initial concepts and the art of leadership and then in all its gradual manifestations later first and foremost fructified in the multiple domain of highly complex matters-military, in the very long and living human-history, in the making.** Then later-on as concepts and practices of the art of leadership slowly commenced to be identified, understood and imbibed in the above mentioned numerous other developing areas of the multi dimensional human-endeavour.

In fact in the specific context of our ongoing discussions at this point, while briefly taking-on these historic facts the principal aim in this regard, has been to put these interesting aspects only, in a desired perspective. Interestingly, the modern absolutely apolitical and highly professional Indian Army has precisely defined the multiple aspects of the military leadership as the fine-tuned art of influencing and directing men to an assigned goal in such a way as to obtain their obedience, confidence and also their loyal co-operation. In fact, the Australian Army too, holds the similar perceptions in this regard. Famous British commander **Field Marshal Viscount Slim** has said in a focused manner that 'military-leadership is the spirit, compounded of the commander's personality and vision.... In fact, it a mixture of the example, persuasion and compulsion as well. Hence in our context, it has been rightly said that leadership in fact, is situational. Further it is also the art of influencing human-behaviour as well as innovative ability to successfully handle them, to achieve the desired goal and successfully completing the given task in quickest of time, and along with minimum of losses in any warlike situations.

It may appear to be somewhat relevant at this point; to mention few interestingly informative words regarding words concerning the famous as well as legendary personality, **'The Lawrence of Arabia'**. Major Udaya Chandar mentions that, "Thomas Edward Lawrence is a scholar turned soldier. Due to religious upbringing, he never even mixed with the neighbours' children. As outstanding scholar, he won many prizes and scholarship. Pursuing study of history and archeology, he visited Syria in 1909. Till the War (*i.e. the first World-War*), he spent his time with the Arabs in the Middle-East. As the War broke out, Lawrence joined the Military and the Political-Intelligence department in Egypt and very soon became popular as 'Lawrence of Arabia' and 'Prince of Mecca". (*Interestingly, some military-historians have even gone to the extent of recording that Colonel Lawrence was able do such jobs alone, by wisely handling the situations as well as Arabs around him that, for doing the similar jobs, minimum an Army-Division would have been required. Anyway this perception too, has its own critically well-considered logic as well; since historically it cannot be refuted that 'Lawrence of Arabia' did in fact, achieve exceptionally extra-ordinary feats; in the then prevailing extremely violent war-like situations in the Middle-East. Moreover, he was not only a brilliant scholar, but a great thinker and visionary of the highly complex matters-military as well.*)

What precisely is meant by the **'art, craft, theory and sciences' of the fundamental aspects of military-leadership**? In this specific perspective, first and foremost we may again emphasise that the fundamental aspects of the military-leadership in some way or other is historically as old as the history of warfare itself. We may ask as to what we precisely mean by the **'military-leadership theories'**? Interestingly, it is true that all sorts of theories on definitely the highly engaging aspects of the military-leadership have been put forward by many scholars and thinkers engaged in different disciplines of human-knowledge and pursuits of learning. Many of them are learned psychologists, behavioural scientists, environmentalists and not the least, learned scholars belonging to the numerous subjects concerning the areas of 'Humanities'. Hence naturally, such scholarly initiatives since taken by those learned experts and thinkers have very interestingly thrown-up engaging debates in this regard. Obviously, these highly educative intellectual activities do throw-up many new ideas and in the process; scholars too are persuaded to begin thinking 'out of the box' and thereby, fruitfully more new as well as more vigorous churning of ideas commences as well. **According to Major Udaya Chandar's earlier referred book two specific theories in this regard, have been put forward: (a) Traits theory and (b) Leader-Group dynamics related situation theory.** But another naturally expected point has come-up that no theory has completely been accepted and, that is understandable. But reassuringly we do learn that fundamentally, the thematic-aspects behind almost all conceptual frame-work in this regard; do principally remain the same. In fact, these are just put forward in different languages and styles since as they may suit the thought processes of different scholars and experienced experts, from the different backgrounds and countries and also, belonging to the different situational timeframe.

In any case, these are indeed understandable but at the same time; they are by themselves interesting as well as engaging too, since these do constitute multiply explorative studies. But interestingly, based on his vast experiences in the actual practices on the multiple aspects of military-leadership for years together at different levels in war, his consistent learning and very high I.Q.; **Field Marshal Earl Wavell** has very thoughtfully said that, "No amount of study and learning make a man a leader unless he has the natural qualities of one. He must have 'character'; which simply means that he knows what he wants and has the courage and determination to get it.... He lists the following qualities of a successful general:

> (1) First and foremost, we may say courage, health, and youth. (2) Character. (3) Knowledge and interest in men. (4) The fighting-spirit and will to win. (5) 'Common sense' among mental qualities (or knowledge of what is, and what is not possible). (6) Really sound knowledge of the 'machines of war' i.e. of topography, moves, supplies and the aspects of administration."

Since the 'art of military-leadership, both in peace and war' are as old as the history of warfare itself so naturally as both the mechanism for conducting a war as well as the 'technologies of war' gradually commenced advancing and developing, along with the much larger civilisational-advancements of man the multiple aspects of the 'art of military-leadership' too, simultaneously got gradually more and more advanced, sharpened and fine-tuned as well. In the specific context these phenomena are naturally inter-connected and also at the same time naturally, do form part of the highly complex and continuously advancing and relatively larger aspects of the 'matters-military'. In this respect, we may prefer to also refer to another somewhat older book titled *Psychology and the Soldier* and authored by **Norman Copland** where very captivatingly find a unique 'caption' as to **"A good Cause puts life and courage into men's hearts"**: The Soldiers' Catechisms, 1944.

The above referred book also informs us many well-known **valuable aspects of 'morale'**; principally concerning **multiple facets of the 'profession of arms'**. While elaborating a little on this vital but otherwise quite hidden factor; we may contextually emphasise here that aspects of morale regarding everything related to the soldiering, is equally weighty as well as extremely important during both peace and war since both situations are inseparably interconnected. **No wonder that since ages, many military-historians and learned thinkers as well as historically well-known military-commanders have very knowledgably said in their own distinct words that concerning the soldiers and multiple-aspects of soldiering it would always be unpardonable foolishness on the part of anybody to neglect vital aspects of 'morale', since responsible for the overall management of vast and also highly complex 'matters-military'** In addition, at this point we may also very relevantly stress-upon sadly that, since the very beginning India has been regressively lacking the multiply "essential strategic-culture and thoughts'; Indians (*at some points of time, certain exceptional personalities apart.*), have seldom been able to cultivate the requisite awareness and thereby, visionary-sensitivities of the paramount importance of precisely concerning again the multiply effective management of the natters-military; in India's highest national interests.

There are some schools of thought, which maintain that 'morale' is a kind of secret-weapon. Well, morale is certainly not a type of conventional offensive or defensive 'weapon' but, morale being invisible in its many kinds of manifestations it is in many ways definitely in all sorts of warlike situations, a very far-reaching 'force-multiplier'. Hence, it is quite understandable that since very ancient-times, all historically well-known military-commanders, such as Alexander the Great, Rameses-II, Chandragupta Maurya, Ashoka the Great, Akbar, Raja Man Singh, Maharana Pratap, Shivaji, Napoleon Bonaparte, to Generals Mac Arthur, Patton, Rommel, Eisenhower, Montgomery, Lord Wavell, Charles D' Gaulle, to Cariappa, Thimayya Marshal of the Air Force Arjan Singh and Sam Manekshaw and also many others the world over have been very conscious about the multiple-aspects of morale within all ranks in their Armed Forces and thereby, always ensured its maximum possible high maintenance during both peace and war. Therefore, **Field Marshal Chet Wood** has very appropriately said for all active as well as newly commissioned military-officers that, **"The safety, well-being and welfare of your Nation comes first always and every time; the safety, well-being and welfare of the men under your command comes next and; your own safety, well-being and welfare comes last, always and every time"**. These soldierly golden-words have become a kind of military-catechisms, for the world-famous Indian Military Academy since the very beginning of this magnificent Institution in the year 1932.

Norman Copeland has very befittingly written that, "The ultimate object of every battle that is fought is psychological rather than physical.... More than ever today, Armies are in need of morale.... There can be no proper morale where there is no self-confidence.... Training, equipment, physical condition, leadership, discipline, spirit de corps—these must do everything they can do to build-up morale, and admittedly they can do a great deal. But yet a great deal more can be done. (*Yes, how true: even in the modern-age today, all types of cooked-up or real frustration, dissatisfaction, disgruntlement, 'withdrawal-symptoms' among the military-men, suicides or even fragging are mainly the results of failure on the part of established military 'chain of command'; besides many other kinds of situation-specific based causes, since thousands of years of the man' living 'military-history' in the making; and obviously also, within the continuing multiple-advancements of the various human-civilisations around the world.*).... Every leader must beware of the temptation to belittle the enemy's worth. In 'Mein Kampf', Hitler confesses by the rude shock the men of his regiment received in the Great War (*i.e. World War-I*) when they found themselves opposed to British troops. (*Like Germans during World War-I; later the Pakistanis did the same thing with respect to the Indians, consequently to their own rude shock.*).... Another powerful antidote of fear is the craving to be important. Sir Archibald Wavell points out that, 'Napoleon did not gain the position he did so much by study of rules and strategy as by a profound knowledge of human-nature in war.—His outstanding knowledge of psychology..... Needless to say that officers and non-commissioned officers must set an example in their own conduct.... Women of loose character haunt environs of every camp. And saddest feature of all, irresponsible girls many of who have scarcely reached their teens, are caught-up in the vortex and brought to a state of moral ruin.... Hall who as a disinterested investigator has no axe to grind, goes so far as to say, 'War of all the occupations of men, because it is the most exciting and the most strenuous, makes not only possible but imperative for its supreme success, the highest degree of chastity." (Copeland, pages 13, 24, 25, 48 & 49)

It is indeed logical that while discussing psychological-aspects of the soldier and his profession of soldiering, during both peace as well as war-like situations; we could not in any way have 'ignored' the **subject-matter of 'fear'**. Fear is a natural instinct being an inherent part of man's nature. In fact, as per the unending evolutionary-process of the mother-Nature as far as we know, so is the case with all the living-creatures on this Earth. **Therefore, if fear is an inherent segment of man's basic nature, then it is further logical that same is the case with soldiers as well**. But then the point, as broadly prevailing in the larger civil population, naturally comes out that if the soldiers too are prone to fear then how can they fight the war? Well interestingly the point is that in case of the soldiers, it is only partially true. **In fact, in order to enable the soldier to overcome the 'feeling and grip of fear' in a war-like scenario, with desired military effectiveness the successful role of vital as well as highly complex aspects of the military-leadership at its various levels, essentially come in.**

The above mentioned important point is one of the crucial reasons, as to why we have earlier been emphasizing naturally the essential and highest need for impeccably effective aspects of the military-leadership during both peace as well as warlike situations. Anybody who is little meaningfully familiar with the profession of arms and soldiering and also, those who are really knowledgeable about the vitally important and much larger geostrategic and politico-militarily complex-essentials, commensurate to one's highest national-interests would always be in a position to really understand the already above talked about aspects of the complex phenomena of military-leadership in action, both during peace as well as war. A soldier when he joins the Armed Forces as a recruit is immediately sent to his concerned basic and vigorous military-training Establishment for gradually refining and shaping him up, through a highly rigorous and well-designed comprehensive training-programme; into a first class and well-chiselled fighting soldier, sailor or air-warrior. Therefore, one of the many highly complex manifestations of the military-leadership in action comes-in right from the selection-process of the recruit to his above mentioned training-programme onward and, naturally gets desirably activated to oversee, motivate, influence as well as guide him; in order to enable him to successfully go through it all, for the intended purpose. And then subsequently, in stages other aspects of the military-leadership in action and with its total comprehensiveness come into play, as per the arising needs.

In fact we further learn that during the presently ongoing fast-changing, high-tech oriented modern age one very engagingly interesting point has emerged as to that these are now no more confined to the functional areas of the all levels of military-brass, since these have very crucially spilled-over into the highest national-level arena of any Nation's political-leadership on both sides of the political-fence, i.e. ruling as well as oppositions' benches. Additionally in fact, these aspects of a Nation's political-leadership get all the more pronounced as well as crucial in a vibrant and functioning liberal-democracy. Are the politicians anywhere, more so those prone to practicing possibly any kind of 'jingoism' without any worthwhile substance into it, and at the same time much more so in today's sovereign democratic India, are paying any attention and sincerely caring to listen at all? Regarding India in this specific and crucial perspective, we would be discussing them in more details, in the subsequent chapters to follow.

Now again coming back to the aspects of one of the inherent segments of the basic human-nature and i.e. 'fear', we may prefer to again look into the already earlier referred **Norman Copeland's** book where in the author says that, "It is the part of every leader to supplement in his subordinates that invincible courage, which is the most miraculous attribute of human-nature.... **Fear may be briefly defined as the emotion caused by an impending evil. The natural reaction to it is concealment of freight.... If the morale of an army is to be consistently maintained it is imperative that the troops should have knowledge of all the facts relating to their situation, even when these are painful and calamitous. There are times indeed, when facts which might prove useful to the enemy have to be concealed; but apart from those men should be told the worst as well as the best.**" (Copeland, pp. 20-21)

In the above quoted paragraph, Norman Copeland very reflectively talks about certain crucial aspects of the military-leadership and the morale-factor, and also their paramount role in **effectively**

dispelling fear from the hearts and minds of the troops in any warlike situations and understandably much more so, when they find themselves in the midst of extremely adverse circumstances. In the man's long and living military-history, since very ancient of times successfully facing such adverse situations and also motivating the troops as well, to cheerfully do the same have always been a great challenge for the military leaders/commanders at all levels in the battlefield or war-zone. In fact, well-known commanders since ages have been emphasizing that in such adverse situations it is most important to keep cool, never to panic, be always cheerfully optimistic in both thoughts and action and at the same time; keep motivating and leading the troops to encourage them for doing the same and never even remotely think of giving-up. Having soldierly pride in self, keeping some inspiring ideals in mind and having unshakable faith in one's regiment, arms and equipment and then no less; always motivating and inspiring the troops to do the same, by means of personal examples and positive talks; are indeed pivotal towards magnificently conquering success itself.

We further notice Major Udaya Chandar in his already earlier referred book on *The Art of Military Leadership* (pages 373, 391 and 392) has written that, "The true leadership comes on the battlefield, in times of stress; when morale is sagging and an atmosphere of glooms sets-in. Here the leader must be prepared to make any sacrifice for the success and must be prepared to draw on all his resources. Sir Ian Hamilton has indeed brought the essence of military-leadership in the battlefield, in most illuminating way: **there is no surest test of military leadership than service on the battlefield**."

In addition then in our specific context at this point of our ongoing discussions, we do also notice certain important factorial-parameters, important to the military-leadership in action with all its possible manifestations, such as: 1. Combat leadership, 2. Espirit de Corps, 3. Battlefield Condition, 4. Fatigue related to fight the battle and also, consequent stress and strain and visible war-casualties around, 5. Combat isolation, 6. Situational awareness concerning, lawlessness, 7. Combat power, 8. Destruction of the Enemy as per own plan, 9. Restoration of Combat-Power, as far as possible, 10. Realistic awareness of the combat-leaders in the field: about fear and panic, 11. Action reduces fear, 12. Men to be made objectively aware: of the actual Situation of the battle, 13. Companionship: among the troops, 14. State of statics: among the troops, 15. State of discipline: among troops, 16. Calm behaviour, 17. Religious faith: of the troops, 18. State of loyalty: of the troops, 19. Physical fitness: of the troops and the officers as well, 20. Personal problems: among the ranks and files, 21. Removal from the battlefield of any combatant as deemed necessary, 22. Initial engagement to battle: by the troops under their leaders, 23. Adjustment to the developing combat: under the direction of the concerned commanders.

By now we have since discussed that **as the various ancient and then subsequently medieval-era human-civilisations have been unendingly advancing quite fascinatingly the complex and parallel progress of the man's another age-old aspects of matters-military too, have been advancing as well**. We do perforce also notice one very far-reaching and powerful aspect of the same phenomena. And interestingly, that aspect is that man's both continuing larger and multi-dimensional civilisational advancements as well as the simultaneous multiply-developing highly complex aspects of the matters-military; have not only been inseparably inter-linked but at the same time, mutually influencing as well. Now in our precise context in this regard, the paramount necessity, *much more so for the Indians* is not to review these highly complex phenomena of mutually influencing as well as simultaneously advancing man's civilisational-aspects together with those of the matters-military, through a kind of *Nehruvian 'myopic-glasses'; since perforce showing Nehru's* *topsy-turvy and illogically-fixated and in addition, fictitiously 'disarmament-oriented world-view and then to our dismay, totally devoid of any realism-based strategic-criticality*'; but essentially to review and comprehensively study them with the desired vision and strategic-prudence.

Nehru was born sometime during end decades of the nineteenth century and then died in the year 1964, as then still serving as the sovereign democratic India's first Prime Minister. No doubt he was in his own right a great leader, scholar, thinker, writer and visionary. But most unlike Sardar Patel, he had perhaps a lingering inclination to get detached from the realities of world as well as the then prevailing ground realities of poverty-stricken India and also displaying a kind of unstoppable tendency to keep living in his own largely fictitious 'make-believe world'. **Sadly in fact, after the passing away of India's first politically-cum-administratively, exceptionally strong Deputy Prime Minister-cum-Home Minister, Sardar Patel, on 15th December 1950 there was no political personality towering and also strong enough left, then on Indian political-scene to effectively maintain checks on at times, Nehru's highly avoidable *'flights of imagination'*. In fact, Sardar Patel has also been very befittingly called as 'The Bismarck of India'**. But Nehru was indeed a very popular charismatic political-leader and hence alas, if he won't have had a nature to remain a kind of 'lonely-person', unable to openly take along his team of political-colleagues for working with the desired democratic effectiveness he would surely have done miracles for India and at the same time, he won't have committed certain 'blunders'; which he rather so naively did, causing great embarrassments and harm to the Indian-Nation and thereby, even today India is unfortunately paying all sorts of 'damaging-prices' and suffering as well.

Principally and historically the most unexpected-cum-disappointing point concerning Nehru, as independent India's first Prime Minister; has been that he was no routine-type ordinary as well as quite forgive and forget type of Indian political leader of his times. He was highly learned and visionary, great thinker, personally honest, patriot and accomplished writer as well, since his writings as well as his well-received books, so reassuringly have shown us. So far as the knowledgeable aspects of history, including some brief dwelling upon the subjects of man's living military-history, pertaining to both India and the world are concerned no doubt that he indeed had a very vast and in-depth grip over them. These aspects of Nehru's remarkable and vast scholarship have indeed been very lucidly displayed in two of his books namely, *The Discovery of India* as well as *Glimpses of World History* for those sections of generations to read and enjoyably learn, who may have an inclination and urge to further enrich their knowledge, to do so in this regard.

In this specific respect, we may even quite interestingly notice that in his earlier already referred book *Glimpses of World History*, on page 211 onward, he has been in a highly scholarly manner, cared to briefly delve-upon largely the **not so well-known aspects of both ancient civilisational as well as military segments of the 'Maya Civilisation', the then belonging to the geographical areas of both central and also northern part of the South America**. Hence very saddening and most frustratingly disappointing point for the Indians and their totally apolitical Armed-Forces in this regard, right in the very recent beginning years of the politically sovereign and independent India's continuing living-history, have been that how could Nehru, with indeed his such exceptionally reassuring background, as the first Indian Prime Minister, became so callous about the vitally important and highly complex 'matters-military' and thereby at the same time, so naively ignored his own nationally crucial 'geostrategic and politico-military aspects' of India's both the internal as well as external security needs? Another no less important point in this regard remains and more so for the Indian-Nation that both internal and external security matters are never separable at the higher strategic-levels, since they are always interlinked and overlapping as well.

Now again coming back to the **geostrategic and matters-military related aspects of the medieval-era** it may appear relevant to look into as to how noted historians have in their own rather different manners, have done a kind of broadly historic date-line oriented demarcations—among the different civilisational-ages in a sequence, since very ancient of times and as reflected in the living human-history as being made available to us. In this specific perspective interestingly, S.R. Sharma has written that, "The transitions from the Ancient World to the Modern is difficult to express in definite chronological terms. But the line however arbitrary, must be drawn somewhere. In the history of Europe the capture of Constantinople by the Turks (1453) is taken as a clear turning point. In the history of India, the commencement of the Mughal rule (1526) is considered by some as a suitable stage from which to begin our *(i.e. India's)* 'modern' period. However, both these happenings in the history of the world had their beginnings in the rise of Islam, which therefore may be taken, for all practical purposes, as the 'watershed' which divides the two stream in World History." (Sharma, 1948, p. 259)

Well, the above quoted lines are indeed very informative and interesting too, but surprisingly even a learned professor; while defining such classifications as given above and even if according to him, it happens to rather arbitrary has come from ancient-era directly to the modern-age and thereby totally skipping the medieval-age, so prominent as well as historically existing in between them. But certainly that is not the total 'fact' here and hence quite befitting to a professor, he quite reassuringly goes on to further very knowledgably write on page 275 in his same book that, "The outstanding feature of this period of transition from the ancient to the modern world are summed up in the word **'Feudalism'**. There is greater agreement regarding its characteristic features than its chronological limits. But roughly we might consider the millennium from the fifth century A.D. to the fifteenth century A.D. as comprising the Middle Ages, of which the earlier half constituted the Dark Age of barbarian invasions and the latter half that of medieval feudalism and chivalry." *Well that is indeed very interesting and precise too; but another relevant point in this regard, emerges that* ***in the broader geostrategic and politico-military perspectives, the ancient as well as medieval histories of India, then China and also subsequently Europe respectively, are at the same time, if similar in some ways such as principally a saga of feudalism, chivalry and war-ridden then; they also happen to be totally different to each other as well.***

In the precise Indian contexts, we may say that as the 10th century was coming to an end very unsettling historic developments in the then Indian Subcontinent were happening, and hence these may be taken as a kind of watershed dividing the age-old Indian ancient-era and also, gradual commencement of Indian medieval-age. Politico-militarily, that was indeed the turmoil-ridden period in India, when old ancient-era powerful and magnificent Indian empires were already decaying and disintegrating and from among their historic debris, many new and some of them hopelessly insignificant too were coming-up and irresponsibly began to fight among themselves for their own geographical space as well as their own kind of chivalry-oriented military recognition as well. And broadly during those human-blood stained years, Sun over India's historic ancient-era was gradually setting and consequently, again extremely war-ridden medieval-age was fast sprouting. Again during the end of 10th century **Mahmud Ghazni** had begun invading into the Indian territories. Indian history of that period clearly shows indeed with great irony that, a series of highly destructive and violent foreign invasions commenced taking place from the north-western sides; when India had become politico-militarily quite weak and consequently, very vulnerable.

In these ironically tragic Indian historic perspectives, logically a broader question arises as to why the living Indian-history has been so embarrassingly shaping-up in a way, it has indeed been?

Thematically in this regard, the principal reason for this has already been discussed and analysed earlier a number of times in different contexts. And obviously that has glaringly been that **India, since bygone ages, has regressively not succeeded in evolving, cultivating and imbibing any meaningful and vibrantly enduring 'essential strategic-culture and thoughts'; with due strategic prudence, and since as per her own distinct socio-cultural and politico-military ethos, matters concerning governance as well as multiple thought-processes**. Tragically this has indeed been humiliatingly a great historic-negation in the context of India's both civilisational as well as matters-military areas. Once we begin reviewing these precise issues in our continuing contexts, with desired depth as well as strategic-criticality, we may also notice that historically these aspects are both vast and multi-dimensional as well.

Hence in addition, we are also made aware that, when we are perforce made confined to our own since planned in advance, the precisely defined principal 'subject-matter' oriented parameters along with our own limited knowledge-horizon; and in turn, these are indeed also able to remind us that these issues of man's learning and continuing research, are very vast and then, both inter-disciplinary and multiply complex too. Going on to the historically geostrategic and politico-military aspects pertaining to **China** we do notice that the Chinese-civilisation with its age-old multiple manifestations is no less magnificently ancient in its own totally distinct ways than those aspects pertaining to India. And at the same, geographically China is much bigger than India, but age-old Indian Societies have historically been more complex and amazingly much more socio-culturally diverse in many ways. In the much larger socio-cultural perspectives, historically such amazing Indian diversities are in the highly complex areas of culture, religion, languages, food, traditions of clothing and dresses, social-customs and so on. Then China too has been culturally no less rich if not so diverse and at the same time, definitely more invention-oriented in the areas of science and technology during the bygone ages, since many centuries prior to the coming of the now historically magnificent Industrial-Revolution. Prominent among them in this precise area may be cited as discoveries of ink, manufacturing of fine porcelain-wares, primitive but indeed the then revolutionary printing-presses, gun-powder and so on. **One very unique and interesting aspect in the field of writing has been that the Chinese since their civilisational beginning itself do not have had any alphabets at all. In the place of alphabets, they have developed a well-evolved ancient-system of using symbols and figures.**

But in the very vast and highly complex areas of the matters-military, historically since the very ancient of times China has decisively been excelling India in many ways. In the complex and very challenging areas of the warfare historically China has never remained confused, so unlike India and thereby they never did allow many kinds definitely mindless and confusing philosophical thoughts and also, geostrategic and politico-militarily crippling mental-fixations, so regressively triggered by typically damaging socio-cultural as well as religious traditions and perhaps irrational ethos. We have since discussed these civilisational-cum-matters-military multiple-aspects concerning both China and India, in different contexts in the earlier chapters as well. In addition then, in this chapter too earlier, while discussing the legendary thoughts as well as timeless treatises of both Sun Tzu and Chanakya we have done the same in the specific contexts of the now ongoing geostrategic and politico-military oriented our discussions, briefly covering again contextually, the vast and highly complex areas of the matters-military. In our ongoing perspectives in fact, in course of our continuing discussions as well as allied contextual reviews we do objectively notice many historically lingering reasons for such state of affairs in this regard and they too, have earlier been covered briefly in various contexts, pertaining to both China and India as well. And historically we do find that due to their different value-systems and socio-cultural as well as politico-militarily otherwise rigid fixations both nations have suffered and have been paying heavy prices in many ways.

Historically with respect to China since very ancient of times Lionel Giles (2008) has very informatively written that, "**Accustomed as we are to think of China as the greatest peace-loving nation on earth, we are in some danger of forgetting that her experience in war in all its phases has been such as no modern nation can parallel**. Her long military annals stretch back to a point at which they are lost in the mists of time. She had built a great wall and was maintaining a huge standing army along her frontier centuries before the first Roman legionary was seen on the Danube. What with the perpetual collisions of the ancient feudal states, the grim conflicts with Huns, Turks and other invaders after the centralization of the government, the terrific upheavals which accompanied the overthrow of so many dynasties, besides the countless rebellions and minor disturbances that have flamed up and flickered out again one by one, it is hardly too much to say that the clash of arms have never ceased to resound in one portion or another of the empire. No less remarkable is the succession of illustrious captains to whom China can point with pride. As in all countries, the greatest are fond of emerging at the most fateful crises of her history."

The above quoted perceptions have historically, socio-culturally, politico-militarily as well as socio-psychologically been irrefutable and prophetic too in this regard, since very ancient of times. With respect to the Chinese Nation-State, in the larger global geostrategic and politico-military perspectives; singularly the most relevant as well as of very far-reaching importance during the present modern-age for any nation (*and much more so for India.*), obviously gets noticed is that this aspect can only naively be neglected by any nation to its own multiple-disadvantage or may even be to its 'peril' at some stage or other. It doesn't mean at all that China has been a nation of only war-mongers and barbarians. China, since very ancient-age, has been a nation of very old, truly magnificent and multiply rich civilisations inhabited by great, thinkers, philosophers, artists belonging different fields of art and culture as well as highly creative populace. Hence, it is no wonder that China's history since many thousands of years have been so colourful and enchanting. Further, it cannot be denied that since many centuries, China's contributions towards many other nations' civilisational advancements as well as cultural enrichments have truly been valuable and indeed magnificent.

But in the larger geostrategic and politico-military global perspectives and then in the highly complex domain of matters-military; since ancient-era, China's postures towards other nations have hardly been peacefully reassuring. In addition, since geographically China is one of the world's few largest countries, with vast natural resources and assets as well as having world's biggest national population, her far from peaceful living military-history and also an inbuilt socio-political tradition for coercive-diplomacy in turn, naturally appear to be internationally threatening. These perforce trigger a kind of lingering fear-psychosis among other nations and hence, all the more so for her many relatively smaller and also few bigger neighbours and therefore, understandably including modern-India as well. When we may look into the earlier referred book concerning Chinese-history, titled ***China* by John Keay** we do notice some very revealing comments on the first page after the cover itself by some of the internationally well-known newspapers/magazines, regarding this book. Interestingly, we find that the *Independent* on Sunday' has written that, "Ambitions....this book has many virtues, not least its refusal to adopt a Eurocentric perspective. It also reminds us that to talk of one China, or one Chinese history, is absurd.... China's past was contested and fragmented: Intoxicatingly interesting." Further the *Traveller* magazine has commented that "An epic history of China.... As China begins to dominate the contemporary world order a comprehensive record of this vast nation, its extended ancestry and distinctive culture is particularly timely. **There's no way of understanding China's stirring future without a sense of its awe-inspiring past."**

Approximately we may historically say that China's medieval-era began sometime around 750 A.D. or so; according to historian John Keay, when during the centuries between 755 and 1005 A.D. reconfiguring of the Chinese empire was undertaken. This in fact, began during relatively the turbulent rule of the earlier Tang dynasty. In this context, John Keay has written that, "Unlike the former to later Han succession, the Tang unbroken was unbroken, comparatively orderly and rarely challenged. In terms of scholars, poets and artists, the second half of the dynasty was as illustrious as the first. Jikaku Daishi, the Japanese pilgrim better known an Ennin, who travelled widely in China in the early 840s, found the roads secure, government effective and Buddhist devotion ubiquitous. The level of economic activity was especially impressive, with barge convoys laden with grain, salt, charcoal, coal and timber plying the rivers and canals. Commercial and cultural contacts with central Asia were maintained via the Uighurs, whose fluctuating domains fingered west from Mongolia to Xinjiang and Ferghana.... With a fine sense of history, these military rulers in the north-east harked back to the pre-imperial 'Warring States' period. They favoured the loose vassalage that had been on the offer from the later (Eastern) Zhou kings in Luoyang and as of old, they vied for the conjectural status of the sovereign power's *ba* ('hegemon' or 'protector'). The north-east's then name of Yan had been revived by An Lushan; in a similar spirit, his successors would compare Tang demands for their submission with those of the Qin First Emperor when conducting China's first unification." (Keay, 2008, pp. 279, 280 & 283)

In fact, Chinese history clearly informs us that as during China's ancient-era the medieval China was not all that violence and war ridden and there were peaceful periods as well. Further during those medieval-era centuries we find that in China, there happened to be good developments in the areas of transport, through land, rivers and sea as well; and then there were far-reaching progress in the fields of trade, commerce as well as industries too. In addition, after the end of the Song dynasty in medieval China during a period of nearly two centuries from 1235 to 1405 AD., there were three more imperial dynasties, namely Mongol Yuan, the Ming and Manchu Qing dynasties respectively for relatively very large periods and spilling-over well into the present modern-age as well, i.e. nearly up to as recent as 1911 A.D. In a way historically, this prolonged period spanning well over a number of centuries from nearly 1225 to 1911 A.D. or so in the Chinese living-history may be called relatively, as a period of multiple-progress and development if not absolutely that of military peace and social tranquility. In fact, during critical review of China's chequered living-history, in our contextual geostrategic and politico-military perspectives, we do also perforce notice that her history's pages are largely a saga of the clashes of arms and soaked in human-blood and hence obviously, what we do universally mean by substantial military-peace and social-tranquility; have seldom been any segment of the Chinese socio-cultural as well as socio-political 'psyche'.

Well, human-history since its inception and always in the making has many sides or perhaps dimensions. In this precise perspective, we may interestingly recollect that in this chapter itself earlier while discussing as to what we do specifically mean by history and what history really is, we have since briefly delved upon these aspects along with a kind of listing of history's many 'types or varieties' as well and thereby, covering man's multiply different activities, since the advent of various human-civilisations. Now coming back to China's medieval as well as earlier period of the present modern-era we do also notice that during the just above mentioned periods of the Chinese-history, if there were multiple progress then at the same time specially during ironically the Ming dynasty there has been a kind of decline in the international standing and reputation of China as well, with respect to the contemporary European Powers. And then, those European Powers' naval-ships as well as mercantile vessels too had begun frequenting China's eastern ports. **All such mercantile marine-traffic were certainly not evil-oriented but during those times, there was indeed a big surge in the multiply-damaging for China; the 'opium-trade' by those European Powers.**

And ironically for this India and specially her present-day States of Bihar and the Eastern U.P., turned out to be as **main source of 'opium-supply'** and the older Indian city of Patna having river-traffic infra-structure, turned out to be the main port of the embarkation of opium-cargo, on way to the Chinese ports, since mentioned above. In this respect, rather surprisingly another historic fact is that for furthering their business interests at the Chinese ports those European traders and shipping-barons had then resorted to craftily bribing the Chinese guards and the mandarins, put on the spot-duties by their Emperors. **Interestingly, this episode of a sort of rather 'organised' and then later a kind of 'institutionalised' corruption-culture became the birthplace of extensive as well as perhaps all pervasive corruption among the civilian-bureaucracy in the world**. And from here the expanding and socio-cultural as well as socio-political and highly corroding bureaucratic corruption culture and bribing began overwhelmingly global. Hence no wonder that today, almost all nations in the present modern-era world, have been infected, may be to varying degrees; by both this socio-cultural as well as politico-economically crippling virus. Needless to say this stinking phenomenon of corruption and bribe, indeed perforce had several damaging effects on the overall multi-dimensional aspects of the vast and highly complex matters-military. And thereby, the again very corroding negation of any nation's fighting-machine just cannot be overestimated. We would naturally be discussing these aspects in more relevant details in the subsequent chapters.

John Keay in his earlier referred book concerning the Chinese-history, on pages 341 and 343 has written that, "Talent and industry, along with fugitives and colonists, had gravitated south with every incursion into the north. The demand for ink and pigments was never as high as among the artists and calligraphers, poets, philosophers, playwrights and diarists who plied their brushes in the benign climate of Southern Song. Block-printed books of reasonable price now circulated widely, connoisseurs and collectors drooled over landscape paintings of exquisite delicacy, and the production of fine lacquerware reached industrial proportions. Above all, it was the classic period of Chinese ceramics. Technical mastery combined with a revival of interest in archaic forms to produce stoneware of elegance and restraint for a rich discerning domestic market. Meanwhile the booming export demand was increasingly supplied from southern kilns within waterborne reach of the coast, most notably those at Jingdezhen in Jiangxi province. Here high quality deposits of clay enriched with kaolin made possible the production of a shell-like ware whose transparent glaze gave a blue-white (*qingbai*) tinge."

In the centuries before the advent of historically unique and momentous 'Industrial Revolution' if we may critically review the above narrated marvels of man's praiseworthy achievements, in our precise geostrategic and politico-military perspectives whether concerning China, India or any other the then civilisational Nation-State, then in any case, essentials of the governance-oriented strategic prudence on part of the concerned King/Emperor or any ruler as well as their top military-commanders; demand that along with such multiple cultural, industrial and economic progress aspects of the matters-military too, like suitably training and arming the armed-forces as part of effectively preparing deterrence-oriented defence and so on, not neglected. **In this specific context since ages, ironically India has mostly been found wanting but, in case of China that has not been so and, the Chinese have since been displaying a somewhat better sense of history**. General George Washington, the founder as well as the first President of the United States of America with exceptionally great historic-vision, is understood to have said that, '**If you want enduring peace, always be prepared for war.**' But in this truly sensible as well as strategic perspective too; China's lines of strategic-thinking and matters-military-oriented foreign policies as well as geo-political ethos have been totally different.

The precise point in this regard is that, as per George Washington's thoughts; remaining prepared for war with requisite strategic prudence, for ensuring peace is one thing and then; mostly believing in a kind of rather threateningly coercive-diplomacy as well as using hot-war at the slightest opportunity, as a pro-active instrument of the 'State-Policy' is quite another. It is reasonably hoped that the points there in, have been made reasonably clear as well as put in the desired perspective in this regard. We many now recollect that we have earlier, in a context, talked about that in a rather precise geostrategic and politico-militarily comprehensive historic-perspectives; both India and China have been found to be avoidably bleeding and suffering in their own distinctly paradoxical ways. Well in fact, it is broadly true that during both ancient as well as medieval historic-ages; meaningfully enduring peace have been largely elusive for most the Nation-States and, perhaps may be few exceptions apart. But the prolonged saga of the historic-tales; constituting China's living military history, have altogether been distinctively different. Interestingly in a way, the above mentioned perceptions, do reasonably substantiate the comments, in a context, by the British newspaper magazine, 'Independent on Sunday' on John Keay's already earlier referred book on 'China' and has since quoted earlier; that 'China's past was contested and fragmented', do come out to be reasonably acceptable.

It is also historically true that since both India and China have been exceptionally ancient civilisations; we may find in our contextually continuing 'strategically critical-journey' at this point that, some kind of the civilisational-origins of both these countries do mysteriously melt-away into the fog and mist of perhaps some pre-historic antiquity. Further, it is a fact that both ancient nations have been regressively ridden with all sorts of feuds and internal wars and then, certain external aggressions as well, in their own different ways. **But any kind of perhaps parallel similarities largely end here and then, the Chinese have seldom betrayed their own cause in the hands of the aliens, the way the Indians have been doing for the narrow and selfish greed of some amongst them, as well as due to their some kind of dirty jealousies as well.** Historically, the Chinese didn't have the kind of tendencies to so easily give-up, like the Indians, before their foes. But in this specific respect and at this precise point, this is indeed partially true only and the complete truth in this regard; is totally different. In fact, even foreign historians and globally famous military commanders with vast war-experience have acknowledged that since ages in chivalry, Indians have been second to none, but have repeatedly suffered due to the lingering lack of the earlier already talked about, the same 'essential strategic-culture and thoughts' on the part of their ruling elites, as well as highly damaging betrayals from time to time, by certain traitors lurking among the members of those Indian-elites only.

In the larger context of the above mentioned certain aspects **General Sandhu** says that, "The best example of the superior fighting qualities of the Hindus (*i.e. then the Indians only.*), in conventional warfare was at Tarain. In the first battle of Tarain **Muizz-ud-din** tried to defeat **Prithviraja** by fighting a conventional, static battle and was thoroughly beaten. In the second battle of Tarain he used the mobility of his cavalry and had no difficulty in beating Prithviraja's static force. The Indians refused to learn the lessons that an army which makes full use of the mobility its cavalry would defeat a force which employed it as a mere as adjunct to other arms. Indeed they chose to unlearn the lesson after they had successfully employed it under the Guptas and for a few centuries thereafter. The result was inevitable—they did not have a chance against the Turks. Indeed the efforts of static Hindu armies to hold up the mobile Turks may be compared to the Allies trying to hold up a German blitzkrieg during the early years of World War II. No amount of courage, determination or bravery would avail them. Defeat under such conditions was certain. Under the circumstances with self-created handicaps perhaps the Rajputs did well. However the Turks never achieved moral ascendancy over the Hindus who continued to resist. The Muslims could only control towns, rural areas remained largely unaffected. Local revolts continued to occur and kept up Hindu pride in their own culture, their heritage and in their in themselves. Rajasthan was under nominal Turk control for only a few years; some of the kingdoms regained independence within a few years others took somewhat longer." (Sandhu, 2003, pp. 138 & 139)

Well in a limited or may be perhaps in a specific context, the above quoted perceptively historic reflections are definitely quite true and then at the same time; while even just hurriedly glancing over the India's past, these may appear to be somewhat encouragingly reassuring as well. And hence in that perspective, the points are indeed well-taken. But in any case in the precise context of India where she has now already so noticeably emerged as a liberal, democratic and sovereign Nation-State, that is not the end of the whole story at this point of our multiply-continuing discussions, and substantially that too in the overall perspectives of our main subject-matter in this regard.

We may always need to remember that the man's authentic living-history, since its inception, just does never fear or favour at any stage of its own continuous making of any historic personality, a remarkably ancient civilisation, even a multi-religiously secular as well as enchantingly colourful society or any big sovereign Nation-State. In other words, this aspect does essentially constitute a principal segment of the requisite 'strategically-critical politico-military sense of history'. Historically as well as with respect to the critically broader geostrategic and politico-military perspectives we have since spelled-out, in the immediately above paragraph, certain avoidably embarrassing 'failings' since last many historic-centuries, on the part of India and the Indians which do also remain authentically recorded in the pages of living Indian history. These since need to be taken objectively in their proper historic perspectives so that necessarily crucial and comprehensive course-corrections may always pro-actively be resorted to, by the Indians in the conduct of their multi-dimensional statecraft as well as processes of governance; with due strategic-prudence and vision, towards the **'games and their rules', which nations have been playing in another vast and highly complex arena of the international power-politics**.

As since already discussed, the principal reasons for their failure to do so; have regressively been on their part, the total lack of the **'essential strategic-culture and thoughts'**. Interestingly in fact, these aspects too, and especially more so for India and the Indians do additionally constitute other essential segments of the earlier talked about 'sense of history'. Hence, the points which **General G.S. Sandhu** has rightly talked about have certain positive attributes regarding commendable aspects of the soldiering as well as soldierly tenacity, on the part of the culturally distinct sections of the historic Indian populace and are definitely true and historically reassuring as well. But since man's long and highly complex living military-history very emphatically has been teaching us that these by themselves are certainly not enough and at the same time they are certainly of no avail in the absence of the crucially **'essential strategic-culture and thoughts'**. These aspects in fact, are a part of the crux of our ongoing discussions, at this point and of course within the overall parameters of the main subject-matters in this regard.

Further interestingly, **history of the patterns of life and societies in the medieval-era European-continent** is in fact no less chequered and war-ridden. Well, with respect to the presently ongoing super high-tech oriented and extra-ordinarily fast-moving—the entire gamut of today's overall human-life pattern, the larger societies' ethos as well as the value-systems are by themselves becoming so overwhelming that they in turn, have even begun to influence the thought-processes of the modern-era's historians, thinkers and social-scientists rather than more appropriately, these to be the other way round. In this specific as well as in fact, global contexts, from Professor S.R. Sharma's already earlier referred book we learn, Eileen Power as having said, *"Today the historian is interested in the social life of the past and not only in the wars and intrigues of the princes"*. This is indeed an interesting comment, and may perhaps be limitedly true and that too, in a specific context only. Well in any case, these perceptive comments are largely subjective only and hence, these aspects may also be taken as debatable.

In Europe's historic perspectives looking back a little, we do notice that when once the **mighty Roman-empire** began to gradually decay and finally collapsing almost the whole of Europe turned towards the times of a chaotic transition. As the saying goes, when a very big, decaying and mighty tree falls on the ground earth begins to somewhat shake for a few moments. Hence looking into the Europe's extremely violent and war-ridden ancient-past its gradual and prolonged transition into the post-Roman imperial-era, perhaps had to be far from smooth and peaceful. During those bygone historic times, Europe due to many reasons, politico-militarily perforce suffered from many weaknesses. In geostrategic and politico-military perspectives, these multiple weaknesses got manifested into a number of invasions into the European region, both by the Mongoloid and other barbarians and Huns, broadly from the east, and then later on by the Islamic herds of Turks and Arabs from the south and across the Mediterranean. This period of socio-political and military turmoil, during the beginning of medieval Europe, lasted for a number of centuries. But human history in fact pertaining to any historic-age as well as also concerning any geo-political region of the world do interestingly have many facets and sociologically, many distinct complexities as well.

Hence historically therefore, the above mentioned period of multiple turmoil during the beginning of medieval-era in Europe and, after the final eclipse of the once politico-militarily mighty Roman-empire many kinds of historic internal transformations too, began to take place in Europe. One both socio-religiously and socio-culturally important development that took place was that, Pope as a formal as well as Institutionalised Entity, got further established and rather strongly firmed-up in Rome and thereby, the ancient-city of Rome and including its some nearby territories; then got shaped into an independent and sovereign City-State, where Pope himself has been reigning supreme. Interestingly, besides Pope's own ceremonial palace and body-guard contingents the main responsibilities for the territorial defence and effective military-security of the Pope's Rome began to rest on the larger Italian Nation-State. In addition to all this, as the invading barbarian and Huns got settled into their new European territories and then, gradually got themselves civilised and integrated into the main-stream societies there; both socio-politically as well as politico-militarily the times in the medieval-Europe itself, relatively turned somewhat more stable and settled, **but certainly not peaceful at all**. Historically as the times were changing, these transformations too, brought in many kinds of new developments such as, **a number of royal and ruling dynasties,**

broadly all over in continental Europe; got sprouted and established along with the then, their new territorial Kingdoms.

What we have since discussed in the above paragraph in fact, historically points towards a rather new kind of turmoil as well as relentless clashes of arms in the medieval-Europe's societies and then triggered by a new kind of militarily-cultural soldierly-ethos and value-systems. Historically, these kinds of medieval as well as distinct 'matters-military' oriented socio-cultural value-systems were not at all unknown to the much ancient Oriental-societies and Nation-States in the East. But in the East, more prominently in China and India their much older and socio-psychologically more complex-cum-intriguing manifesting forms as well as socio-cultural and politico-military textures were quite different from their medieval-European counterparts. Then as further few centuries slowly passed during the medieval-era, these aspects were getting quite pro-actively displayed, largely in terms of relentless 'clashes of arm' based on sometime real and sometime fictitiously irrational causative-factors.

Then in fact, world-history further shows that these phenomena were relentlessly rampant in both parts of the world, i.e. medieval-West as well as the East, in their own socio-culturally-cum-militarily totally distinct ways. These phenomenal developments in turn then, made the men and their societies, during those times, prisoners of new kinds of overwhelmingly dominating and expanding socio-cultural values and soldierly-ethos. Subsequently these in turn, ensured that the **medieval-era man's further global living military-history in the making in fact; became no less violently war-ridden and soaked in human grits, tears and blood**. Perhaps, all these may sound somewhat ironical but in any case, these are indeed the hard historic-facts. And learned historians and social-scientists have interestingly identified the main 'fulcrum' of these phenomenally momentous developments as multiply-complex and powerful manifestations of man's age-old **"Feudalism"**. Historically in fact, this lingering and highly complex phenomenon of the multifaceted age-old feudalism, so integral to the man's inherent nature, is as old as the history of the gradual advent of 'civilised-man' himself in this world.

Interestingly, if we begin to critically review the multi-dimensional as well as perpetually continuing passages of time, specifically during the presently on-going super high-tech modern-era, contextually in our socio-psychological, geostrategic and politico-military perspectives, then we are very clearly able to notice the existence of this feudalism in distinctly different manifestations, in almost all the present civilisational-societies thriving today in the different geographical regions of the world. In fact, since initially the gradual advent of various very ancient human-civilisations in geographically different and far-off areas the world and then simultaneously, as man himself began to get more and more 'civilised' the lingering and socio-psychologically, highly complex 'phenomenon of this feudalism' too, silently got germinated and overwhelmingly manifested as an inseparably integral segment of the man's nature as well as his own psychological make-up. So far perhaps both anthropological-scientists as well as the learned historians have not emphatically said that whether this 'phenomenon of feudalism', as we may understand it today really did exist among the remotely pre-historic and cave-dwelling man.

Now, an interesting question arises as to what in fact, is meant by this feudalism? Socio-psychologically in a way, we may say that feudalism is a powerfully lingering and highly regressive mental-fixation and also, so inherent in the basic psyche of man. At the same time in addition, we may also say in other words that feudalism is a greatly unbecoming as well as malignant civilisational-negation of the man himself. At this point, we may find it rather interesting to recollect that in fact, it is no wonder that since ages and in different contexts, various accomplished saints and religious-leaders belonging to all major religions of the world have been preaching that if a person sincerely wishes to advance and uplift himself or herself spiritually to a higher-plain then first and foremost, a relentless effort has to be made to eliminate all traces of ego altogether from within.

This brings us to a natural fact concerning the man's psychological make-up and his inherent nature that both 'feudalism' and man's 'ego' are, fundamentally as well as complementarily, 'two sides of the same coin'. On the multiple aspects of feudalism as well as in specific context of the socio-political organisations and also, along with the politico-militarily laid-down policy-guided demands from comparatively higher or certain kinds of privileged classes of citizens in the then ancient and medieval European societies, with respect to the prevailing systems of land-holdings **Professor S.R. Sharma** has explained,

> "Feudalism was a very complex organisation of society based upon the land-holding with specified obligations of service. In the words of **Bishop Stubbs**, 'It may be described as a complete organisation of society through the medium of land-tenure, in which from the king to the lowest land-owner, all are bound together by obligation of service and defence: the lord to protect his vassal, the vassal to do service to his lord; the defence and service being based on and regulated by the nature and extent of the land held by the one of the other. In those states in which have reached the territorial stage of development, the rights of defence and service are supplemented by the right of jurisdiction. The lord judges as well as defends his vassal; the vassal does suit as well as service to his lord. In states in which feudal government has reached its utmost growth, the political, financial, judicial, every branch of public administration is regulated by the same conditions. The central authority is a mere shadow of a name.... In the absence of any effective central government, justice was administered by the feudal (from feud = fief) lord in his manorial court. The manor was his estate. On it stood his great castle or fortified residence. The surrounding lands were held by his tenants, or subtenants. The former held from him directly; the later from their immediate superiors. Below all were the serfs and slaves who, indeed, formed the major portion of the population." (Sharma, 1948, pp. 276 & 277)

Greatly revealing and indeed historic contents of the above quoted lines are in addition also historically, of significant relevance to the title of this fifth chapter, i.e. 'Military Aspects and Their Parallel Progress', with respect to the certain important and relatively larger aspects of the highly complex 'matters-military'. It is also quite true that during the present modern-era, when now the democracy and mankind's multiply democratic-thoughts are indeed, the call of this age and more so in the precise perspectives of multi-dimensional matters of the entire gamut of governance. Anything concerning the aspects of 'feudalism' is naturally a matter of great affront and, it is totally accepted and understandable as well. In other words, we may also say that democracy with all its connotations is indeed a total antidote to feudalism. But in the specifically multiple perspectives of our continuing strategically critical-review, precisely in the geostrategic and politico-military contexts; of the multi-dimensionally complex living-history of man, since the antiquity of various human-civilisations and more so at this point specially concerning the highly chequered medieval-age aspects of the feudalism have indeed, added some additionally complex-dimensions to the man's no less complex and also imposingly living military-history. **And historically we do also notice that politico-militarily, these phenomenal aspects concerning feudalism are certainly not only Europe-centric, but indeed global as well.**

If we may look back into both ancient as well as the medieval-era Indian military-history concerning certain elite Indian warrior-clans, specially the Rajputs, we do notice that feudalism with all sorts of its regressively mindless manifestations, were indeed breathing very heavy on almost all aspects of their rather no less brave but, somewhat misguided soldiering and socio-culturally clan-driven obsessive chivalry. Well, similarly when we look into the age-old indeed living Chinese military-history, we do come across broadly the similar story-lines; but of course, historically different and distinctly Chinese both in socio-cultural as well as politico-military perspectives in this regard. But again historically, the multiply intoxicating heavy doses of the highly complex aspects of man's same age-old and still lingering feudalism have prominently been found to be indeed substantial, in almost all civilisational human-societies and their Nation-States in all corners of the world. And logically therefore, how could the highly complex and vast aspects of the matters-military as well as the allied aspects of soldiering with all its fascinating dimensions could have been historically 'spared' by the same feudalism, of its above mentioned highly 'intoxicating-doses' in this regard? Hence, these aspects too, give a glimpse of the overwhelming reach of such aspects of that lingering feudalism.

Now coming back to the medieval-era Europe historically we do notice that those complex aspects of the same kinds of feudalism, in their both socio-culturally as well as politico-military European 'models and display' are indeed decisively visible on human-blood soaked pages of their military-history as well. And interestingly triggered by this feudalism and their many regressive aspects, we find that misguidedly **typical medieval-era chivalry-motivated 'knight-errantry'** was largely the order of the day and then in turn; it also became militarily-fixated obsessions with warriors and their professionally single-tracked military-commanders at various levels. They in turn got regressively motivated towards getting embroiled in **various kinds of highly damaging 'private-warfare'**, with many violent fall-outs and naturally they all together resulted in great-negation of both, the human-civilisational aspects as well as the magnificence of the truly selfless and impeccably higher-aspects of the profession of arms. And further, they resulted, both socio-politically as well as politico-militarily in a peculiar situation where in some leading and relatively more powerful vassals became even militarily stronger than the King himself. Thereby, these disturbing developments altogether that those largely out of turn rising vassals, naturally got overwhelmingly driven by their own rather uncontrollable over-ambitions.

Interestingly in our contextually historic-perspectives, we do also observe that all these together ensured **other kinds of multiply complex and far-reaching 'recipe as complete', with many prolonged and momentous consequences**. Those rather rebellious and rising vassals began displaying many kinds of politico-military postures, challenging the king and his authority over his entire kingdom and at the same time; they began clashing their arms against each other as well. All these historically ensured many kinds of comparatively smaller as well as bigger wars and any of them rather smarter and also gaining more successes could even succeed in overthrowing the king and then, declaring himself the king. But in the bargain, any kind of peace was even far from getting even remotely visible on any of such historically beleaguered kingdom's horizon. Then interestingly, history shows another dimension to it, and, i.e. all the rebellious vassals didn't readily accept the new king on the throne and hence some of them even militarily perforce carved-out maximum possible territories from the original kingdom itself, as their own independent kingdoms which were relatively smaller, but no less independent of the central throne.

Well, the above narrated turmoil-ridden phenomena in turn, perforce triggered another round of similar violence and clashes and consequently at times historically similar cycles of wars were repeated. These were no doubt, typical to medieval-age but it was certainly not only specific to that era's European region alone and, since it was largely common to the medieval-societies in other continental-regions too and historically in their own distinctly different ways. Hence at this point of our ongoing discussions, we find that since man's medieval-age global history is generally known to be wars and violent clashes-ridden and hence, its historic-pages are soaked in human tears and blood. In other words analogically, we may even say that in a way, it was similar to today's tremendously fast-moving modern times' in the global corporate-sector many **(M&A), i.e. 'mergers and acquisitions'** among the companies, operating and doing business in both their respective national and then at times international markets as well. In addition, another quite interesting point concerning the present modern-era in this regard, is obviously visible in the both national as well as international markets is that these mergers and acquisitions among the various companies were sometime mutually negotiated and sometime hostile too, depending upon the circumstances.

Well, the above mentioned, a kind of analogy, is certainly quite limited in its perhaps similarity or so and also, it basically remains confined to its own limited context and hence, may appear to be even debatable therefore, it may be considered in that perspective alone. But at the same time man's history since its commencement in the remote antiquity of time, has many facets and also many dimensions as well and hence in the much larger perspectives; war and its all accompanying consequences, has been one of many such facets and the multiple-dimensions of the man's same age-old history. But at the same time, war and all its historic aspects, have always been dominatingly decisive and no wonder that the warfare or more appropriately, we may say that the aspects of vast and highly complex 'matters-military' since ages, have principally been occupying the 'driving-seat' of a type of rather 'vehicular' complex-phenomena of the man' living-history in its own way, being continuously in the making.

Now, regarding the equally fascinating facets and dimensions of man's history and then, in the specific contexts of medieval-era Europe, it may appear to be quite appropriate to again refer to Professor S.R. Sharma's book as that, "**The growth of towns which gave a death blow to feudalism was the dominating character of the later Middle Ages. While feudalism throve in the rural parts, urban centres developed a different kind of life. Here commerce and industry rather than agriculture were the decisive factors**. Guilds and corporations of merchants, and manufacturers not only brought prosperity to the towns, but also imparted to them a spirit of independence. Under their aegis too, as under the fostering industry of the monasteries, culture was preserved and developed.... In medieval Europe there three institutions of such civilizing importance: they were the monasteries, the feudal manors, and the organised towns.... The monasteries planted often in out-of-the-way places and far off wildernesses acted as far flung outposts of civilisation. The fortified castles and their surrounding self-contained manors served to preserve society from the forces of anarchy.... They were also the arteries through which flow the commerce of the times. Though they were units linked up as fiefs in the feudal system, they were too powerful to be swallowed by the encircling forces. They either bought out or valiantly fought for their liberties and thereby earned an important place among the 'estates of the realm. **Their citizens, the burghers or burgesses, were the creators of the commons of modern democracies.... Florence, the city of Dante, Petrarch, Boccaccio, Machiavelli, Michael Angelo, Leonardo da Vinci, Galileo, Amerigo Vespucci, and Medici, was 'the most illustrious and fortunate Italian republics**.' Despite the handicaps of her inland situation, Florence still became, 'through the skill, industry, enterprise, and genius of her citizens, the great manufacturing, financial, literary, and art centre of the later medieval countries....indeed, as respects the number of her great men, Florence is perhaps unrivaled by any city of the ancient or modern world save Athens.' (*Quoted from Myers, General History, p.436, and hence, the above lines indeed, reveal the magnificence of many enchanting facts in this regard*)." (Sharma, 1948, pp. 279-281) These above quoted lines clearly show us that many Italian city-republics were indeed passing through their 'golden-age', during the medieval-era Europe in spite of the overall situation then prevailing in perhaps the whole of Europe, largely very turbulent since as per the broader civilisational considerations.

Man's living-history very prominently shows us, perhaps that the medieval-era Europe, when reviewed with geostrategic and politico-military criticality, to be quite turbulent and, indeed war-ridden. In a perspective, our above continuing discussions precisely on matters concerning the **feudalism** now, with its further prominently powerful and somewhat socio-culturally as well as politico-militarily enlarged manifestations; historically indicate that feudalism-oriented mind-set and socio-psychological fixations in fact, have had a substantially far-reaching influence on the medieval-age European societies, along with the governing-cultures of their Nation-States and many aspects of both the soldiering and chivalry as well. At this specific point of our ongoing discussions, while remaining very much focused to the multi-dimensional remotely age-old profession of soldiering and chivalry as well as geostrategic and politico-military subject-matters of the **arms, aims and aspects**: *(well quite interestingly; the widely-known terminology, once often used earlier, by an Indian Army Chief of Staff,* ***Late General J.N. Chaudhuri.****)*, the above mentioned **matters-military related aspects of feudalism during now the bygone medieval era, were again certainly not only Europe-centric but no less influencing and although at times somewhat hidden, but present in various other substantially more ancient Asian human-civilisations; notably such as China and India, and also in few other civilisations in different geographical regions** as well.

Specifically in the multiple contexts of the matters-military, precisely with respect to the medieval-age Europe **Late Field Marshal Montgomery** writes that: "When the Western Roman Empire disintegrated, its eastern counterpart at Byzantium lived on, and in struggle for survival, with the Arabs and then with the Turks and Bulgars, is a fascinating story. **In 622 Mohammed led his followers from Mecca to Medina, and began the expansion of Arabianism and Islam**. The Prophet won the first military victory, but the great leaders of Islam were to be **Khalid ibn-al Walid** and **Amr ibn-al As**. Within a hundred years the Islamic empire extended from the Aral Sea to the upper Nile, and from the confines of China to the Bay of Biscay. Only one power in that century was able to resist the Arabs, Byzantium— and even she lost the south-eastern part of her empire. Then as the Arab drive exhausted itself in southern France the Franks came back into prominence." (Montgomery, 1983) (*All these indeed are quite chequered and violently war-ridden episodes of the medieval-era European-history*).

Interestingly in China, during the whole of medieval-age and further into the present modern-era, right up to the very recently past mid-twentieth century, there had largely been relentless 'change of guards', from one ruling imperial dynasty to another in historic succession, and thereby at times some expansion and then, even some shrinkage of the empire; and then much later in the year 1949, communists have come to power; indeed through a series of bloody clashes of arms. Earlier in the varying contexts, and also in the last chapter as well we have since briefly discussed these subject-matters, while identifying a series of the historically chronological successions of the many Chinese ruling imperial-dynasties. In this precise context we do also notice that historically each such Chinese-dynasty had, both socio-culturally as well as politico-militarily distinctly different historic-features, unforgettably unique qualities along with again peculiarly different, historic-eccentricities as well. Historically moreover, encompassing all these numerously bygone-era centuries the intriguing living-history of China, including her corresponding highly complex matters-military and a series of war-records—substantially establish that during all these prolonged historic time-frame, China has indeed seldom, both internally as well as externally, experienced any relatively long periods of peace.

It might not have historically been so intense but in medieval-India too, ironically periods of both peace and war along with actually a series of clashes of arms have been playing a sort of the 'game of hide and seek', since as may be noticed in the pages of again a highly chequered living Indian-history. Paradoxically, similar ongoing situations are even now somewhat prevailing, though now in substantially different situations; in the presently continuing modern-era's sovereign democratic India as well. We have since dealt with these principally matters-military related historic aspects concerning the medieval-India, in different contexts and in some detail earlier; in this as well as in the last chapters.

Now, so far as the historic-aspects regarding the medieval-era subject-matters of armaments and allied war-like military-assets are contextually concerned, at this specific point of our ongoing discussions certain multi-dimensionally far-reaching scientific and technological breakthrough had since been achieved by man during the ancient-era and subsequently in the medieval-age too. And simultaneously along with them, the momentously continuing advancements of the various prominent human-civilisations have also made unforgettably a world of difference, in the vast-arena of the highly complex matters-military. In addition then, in our precise contexts historically, we may be able to suitably enlist them in the date-line based chronological sequence, as below:

1. Discovery of **Copper**—nearly 6000 B.C., 2. Invention of **Wheel**—nearly 4000 B.C., 3. Discovery of **Bronze**—nearly 3000 B.C., 4. Discovery of **Iron**—1500 to 1600 B.C., 5. Discovery of the indeed revolutionary **Gunpowder**—around 1260 A.D., 6. Approximately after a century or so, some relatively rather primitively designed **Cannons** historically appeared on the scene and finally, 6. From 1415 A.D. onward, **Artillery** as a distinct military-arm historically began to make decisively far-reaching difference in the conduct of warfare since based on the professional ingenuity of the concerned military-commanders, present in the battlefield.

With the revolutionary discovery of gunpowder and then subsequently, with the gradual advent of artillery as a distinctly lethal military arm we may also say that highly complex gamut of the conduct of warfare was never going to be the same again. In broader contexts here, we may also recollect that with the coming of iron historically and then subsequently, by means of relentless efforts and through both trial and errors as well as human-ingenuity driven innovative-processes, various arms and accessories, such as swords and shields, daggers, pikes, arrows and many other allied age-old armaments along with many valuable supportive-equipment for the infantry-arm, cavalry and its horses and artillery too were able to change their characteristics and systems of making as well.

In fact, path-breaking discovery of iron was indeed instrumental in bringing-in many changes in the above mentioned military-related activities. Obviously, the important aspect of durability and reliability too, of the above listed armaments and their many valuable accessories, were substantially enhanced and in turn, leading to their more military-dependability in the battlefields. Further, both logistically as well as tactically in the bargain the essential support-tails for the fighting forces too, were able to be somewhat trimmed, resulting in some welcome advantages for the commanders on the move and in the fields, with their fighting-columns. In these respects, iron in fact, brought-in many such advantages, which remain otherwise naturally hidden as matters of routine, unless in a context, viewed and analysed little critically. **Then little later historically, we do notice the momentously revolutionary dawn of the multi-faceted 'Industrial-Revolution'.** This unprecedented historic-phenomenon in fact, has changed the whole world, and things in all fields of human endeavour were never the same again. It is indeed a hard historic fact and since as discussed in different contexts earlier, that warfare of almost all sorts of the humanly conceivable types has always been an inseparable and no less decisive as well as dominating aspect of the man's continuing living-history, since the gradual dawn of various human-civilisations in far-off different regions of the world. Now coming to the no less engaging subject-matters concerning arms, aims and aspects during the medieval-era globally, while critically studying and analyzing them, we do notice that discovery of the **gunpowder** during 1260 A.D., and gradual triggering of the **artillery** sometime in 1415 A.D. onward, as a distinct and highly valuable military-arm of the then fighting forces were indeed great turning-points in this regard. And hence, it has rightly been observed that the entire gamut of warfare were never the same again.

In these indeed far-reaching perspectives, **Field Marshal Montgomery** has written that, "**Possibly the most momentous invention in the history of warfare has been gunpowder**. The first man in the western world to record the formula for gunpowder was Roger Bacon, an English monk, in 1260. His formula was seven parts of saltpetre, five of charcoal, and five of sulphur. 'With such a mixture you will produce a bright flash and a thundering noise, if you know the trick....' Fifty years then passed before the first gun appeared. The earliest type of cannon was known as a *pot-de-fer*, because it was like a vase. In an English M.S. (*i.e. Military History.*) it is shown mounted on a bench, and loaded with a crossbow-type arrow. The gunner fired it by thrusting a red-hot rod into the touch hole.... **From 1415 artillery began to affect the character of the warfare**. Cannon by now were of cylindrical shape, and guns were able to hurl projectiles of 200 pounds weight. **In 1415 Henry V laid seize to Harfleur**. When mining failed he resorted to artillery. He had ten cannon, and the chief engineer, **Master Giles**, organised a steady bombardment, night and day, particularly concentrated against the walls flanking one gate. After twenty-seven days the gate and barbican were in ruins. Master Guiles then set alight the woodwork with an incendiary shell, and as the breach was stormed the town surrendered. **As Henry V's artillery battered down the walls of Harfleur in 1415, the era of impregnable fortifications passed**." (Montgomery, 1983, pp. 106 & 114)

In a historical perspective at the precisely at this point, the above quoted lines from a well-known book do inform us so as to, how the gunpowder and also along with the advent of the artillery in geostrategic and politico-military considerations were able to revolutionize both the military-concepts and conduct of the medieval-era European warfare; indeed both at tactical as well as strategic levels of the military-command in the field. Historically this happened during 14th and 15th centuries and their impact was in fact, so momentous that both medieval-age Kings and Queens and their military commanders were compelled to afresh design and develop their fortresses and defensive-fortifications, along with their war-oriented military tactics and strategies; for effectively coping-up with the above mentioned devastating matters-military technologies.

Well, there is another very interestingly new aspect then, having come-up in this regard. That being the otherwise indeed so pivotal fighting component universally for any Army: the infantry having been rather forced to play a kind of relatively a secondary role to cavalry-arm, for a number of centuries were now perhaps befittingly elevated again to the primary-role, quite decisively in the actual battlefield. In fact in other words, perhaps it has been quite correctly observed that artillery along with the musket amusingly played the role of a great leveler, among the various fighting and supporting arms of an Army. The recorded living military-history of man further informs us that as further centuries moved on multi-dimensional development of military-technologies too, gathered momentum thereby bringing-in new kinds of advancements. In this respect, we do also notice that by seventeenth and eighteenth centuries the design as well as functional aspects of both the muskets and the artillery-pieces, experienced far-reaching improvements. As a consequence, the new muskets became lighter, less unwieldy while carrying and aiming and in addition then, more accurate while firing as well. Similarly during the later medieval-age, artillery cannon too, became more accurate while shelling and also increased their ranges while bringing-in more intense devastation as well as lethality on to the enemy's warlike assets in the battlefields.

As further times passed-on, musketry-oriented infantry-arm of the Army became more professional and militarily well-drilled and focused. These soldiers were able to fight as a tactically better and more cohesively formed body of men and thereby under their commanders, they were always in a position to quickly get-in a packed series of lines and then, commence intense volley of fire with devastating effect on to the enemy. In fact, the man's living military-history further enlightens us that not only the muskets, but also the artillery-pieces too, gradually underwent many revolutionary changes in designs as well as improvements in their performance in the fields. Slowly, weights of the guns were reduced without compromising their strength and reliability. Ranges of these cannon were definitely increased further along with their accuracy as well. In addition then, many kinds of improvements were gradually brought-in, in the cannon's integral gun-carriages too resulting in both militarily tactical as well as strategically many revolutionary advantages in this regard. In fact in altogether changed contexts, such multiple phenomena are even now, quite relentlessly continuing in the highly complex military-domain of the presently ongoing super high-tech modern-era as well. These fascinating aspects of the matters-military would more appropriately be discussed in somewhat more detail, in the later chapters.

Interestingly a little earlier, we have since mentioned the militarily quite effective seize in 1415 A.D. of the historic fortress of Harfleur, situated within the French territory where cannon indeed played a decisively prominent role. Well, that was not the end of this story, and within few years of this setback tides had changed due to the relentless hard-work and revolutionary progress achieved by the French, towards their proficiencies in the challenging fields of the clashes of arms as well as the demanding aspects of the military-leadership. This is indeed a very fascinating historic story proving the soldierly doggedness as well as totally unwavering and relentless pursuits towards achieving memorable successes, in the both highly complex and ruthless arena of the matters-military as the following paragraph would bear out.

Field Marshal Montgomery writes that, "After **Henry V's death in 1422** the French gradually drove the English off their soil. One of the leading spirits in the French recovery was **Joan of Arc, a peasant girl**. The Duke of Alencon reported that Joan was an expert at the preparation of the artillery. Maybe Joan did have some God-given military ability; certainly she restored the morale of the French soldiery. But the superiority of the French in artillery was perhaps as great a cause of their success. **The great artilleryman was Jean Bureau; in 1449-50**, Bureau and his brother and his brother conducted, it is said, sixty successful seize operations in the reconquest of Normandy. At the battle of Castillon I 1453 the French army under Bureau's command had 250 cannon, and the crossfire and enfilade by the French artillery caused the English heavy casualties. (*Now, coming to* *the Hussites in Eastern Europe and their commander:* ***a unique military-genius, Zizka and his brilliantly original exploits***.) Already some years earlier, field artillery had been used in Eastern Europe as a third formula for the defeat of heavy cavalry by infantry. **John Zizka** was the military leader of the Hussite movement in Bohemia, a religious nationalistic and popular uprising which began in 1419.... In the following wars Zizka developed his highly original military system. He used ordinary peasant wagons of a suitable size, mounted guns on them, and arranged them in a circle—just as the American pioneers were to do four hundred years later. **For defence, the system of wagon leaguers defended by firearms gave him the advantages of a fortress combined with mobility**. (*Well, as the saying goes that everything is fair in both love and war. In the specific contexts of the warfare, in fact pro-actively battlefield-oriented both military-hardware based as well as tactical-compulsion-cum-initiative-oriented innovations, in many ways do constitute a highly valuable segment of the otherwise very vast basis for the successful conduct of any battle or even the entire war, with both tactical and also strategic brilliance on the past of military-commanders at various levels. The quite complex and vast living military-history of the mankind, since ages is full of many such unforgettable historic-examples in this regard.*). The wagons were usually sited on a hilltop. Each wagon carried two or three small guns, and heavier guns were mounted on special carriages. Zizka was the first commander to use field artillery systematically since the experiment of Alexander at the Hydaspes. The rest of the Hussite army consisted of a few cavalry for scouting, and infantry crossbowmen. When Zizka was satisfied with the training of his army he set out for Prague and by the end of the campaign of 1421 the Hussites were in control of most of Bohemia. Zizka himself was blinded by an arrow, but he continued to command, planning his battles on the basis of exact information about the enemy's forces and dispositions given to him in answers to his questions. Zizka himself died of plague in 1424 aged only forty-eight. 'The whole army was overcome by immense sadness', and his soldiers were thereafter called 'orphans'. Undoubtedly they had lost a great leader. Zizka was a man of enormous personal courage and force of character. Unfettered by tradition, he ingeniously exploited limited assets to produce and original and successful tactical formula. He was to exercise a powerful influence on tactics and strategy in Eastern Europe." (Montgomery, 1983, pp. 114 & 116)

The above narrated lines offer important lessons concerning the highly complex aspects of matters-military; are indeed all the more **vitally important for India and the Indians, and perhaps singularly much more so, for the today's modern sovereign democratic Indian Republic, especially in the multiply challenging contexts of the presently prevailing, very fluid and threateningly complex geostrategic and politico-military situations around her. Needless to say that indeed these in turn, perforce heavily influence modern India's overall both intra-national as well as international security-environment**. In the overall above perspectives of, and at this precise point of our continuing discussions and in addition, in the extra-ordinarily relevant contexts, particularly regarding today's modern India, one important point again most sincerely wished to be submitted here that, just because this author happens to be an Indian; does not at all mean that in any case, he happens to be unnecessarily critical or perhaps even 'hostile' about anything concerning India or the Indians, whether in any historic contexts or regarding the more turbulent and ruthlessly competitive present times, the **so vast as well as highly complex aspects of indeed the singularly vital the same matters-military**. In any case it is needless to emphasise again that it is far from the truth in this regard.

In fact, contextually and appropriately highlighted since a number of times earlier, and broadly in similar perspectives the centrally pivotal aspects in this regard, have always universally been the age-old but timeless **"multiple-aspects of the essential strategic-culture and thoughts". But**

ironically for the Indians today, it is indeed very embarrassing as to how the Modern, sovereign and democratic India's politico-bureaucratic combine have been so 'callously' or perhaps even 'willfully' and 'shortsightedly' neglecting these highly complex and yet perhaps supremely vital aspects of the 'essential strategic-culture and thoughts', both internally as well as internationally in the contexts of the all-encompassing as well as many newly emerging security-challenges. We would also be discussing these aspects contextually in more details in the subsequent chapters.

If we may delve upon the vast as well as multiple aspects of military-history, with the requisite strategic and intellectually discerning strategic criticality, then perforce we are made to be substantially conscious of the hard-facts in this regard that, any kind of even the far-reaching increase in our military strengths, sizes of our Armed-Forces and also in addition, sizeable progress in both military-hardware as well as the multiple infra-structural wherewithal just cannot reassuringly guarantee any sovereign Nation-State's security, in the absence of again the above mentioned **multiple aspects of 'essential strategic-culture and thoughts'**. In fact, the living military-history also further does show us that, these timeless aspects of the 'essential strategic-culture and thoughts' do have many additional multiply vast and vital facets. And perhaps, these are all-encompassing as well.

We may quite rightly notice that since almost the initial stage of our continuing "strategically critical journey, through the humanity's past, present and possibly future: in global geostrategic and politico-military perspectives"; we have particularly been talking about the crucially important and timeless multiple-aspects of the 'essential strategic-culture and thoughts', in so many contexts while taking-up various subjects-matters as well as issues and also, as and when these have been cropping-up from time to time in this regard. This in any case, quite clearly shows their far-reaching importance and in addition, their indeed over-riding decisiveness in the entire gamut of any Nation-State's existence, well-being and governance and so on. Now at least at this point of our ongoing discussions, contextually quite an appropriate question may be raised as to **what precisely are meant by these crucially important aspects of the 'essential strategic-culture and thoughts'**? Interestingly, in the very vast as well as multi-dimensional aspects of the entire gamut of a Nation-State's basic phenomena of existence, well-being and multiple processes of governance in any case what is not meant by them and their manifestations in this regard: **The multiple-aspects of the 'essential strategic-culture and thoughts' in fact, decisively affect and then indeed overwhelmingly influence almost every segment of a man's life, peace and tranquility within the all-encompassing total framework of his society and the Nation-State in this world.**

Hence in these perspectives, it would indeed be rather quite a short-sighted view as to if, one may opine that those multiple-aspects of the 'essential strategic-culture and thoughts' principally concern the areas of matters-military and even more specifically to the conduct of warfare. In any case, that would indeed be quite an erroneous opinion in this regard. We may recall that earlier we have already very precisely emphasised in different stages and contexts of our these on-going discussions, that since very ancient of historic times and as per both the geostrategically as well as politico-militarily; the altering compulsions, and also as dictated by the multiply transforming situations and challenges (*i.e. not only military-aspects related.*) during differently fascinating phases of the man's living-history perpetually in the making; it has indeed irrefutably been **vital to fruitfully cultivate, evolve, develop and simultaneously imbibe and adopt the timelessly vital aspects of the 'essential strategic-culture and thoughts'**. The principal points in this regard, are that **multiple-aspects of the 'essential strategic-culture and thoughts' are indeed pivotal to the crucially and timely triggering of the multiply strategic thought-processes on the correct path and**

simultaneously acquiring the requisite vision; on the part of a Nation-State's politico-military leadership, towards both reassuringly and successfully fulfilling their inescapably duty-bound obligations to her and the peoples within her domain. Now, we may even thoughtfully ascertain that these multiple-aspects are indeed all-encompassing and certainly not only 'matters-military-centric'. In other words, fundamentally they are the 'keys' to the all-round well-being and multiple-developments of the man, his society and in turn, his Nation-State.

We have just above mentioned that the present modern-era is highly complex and that is because the overall quality and ways of life of today's man has totally got altered, fast-paced and more demanding. He is now forced not to take things casually and easy in any manner. Well, in certain ways the irony is that he is even forced by his overall environment, to be quite analytically calculative and surprisingly be 'strategically' critical as well as cautious even in his personal life. It may perhaps sound socio-culturally and socio-economically even paradoxical, but it is largely true. Well in today's context in any case, it appears to be broadly understandable but that is a different matter altogether. Today's world is fast-changing because rates of both the multi-dimensional and multi-directional advancements of the so many multiple disciplines of the modern sciences and technologies are being very fast and increasingly accelerating, as well as more and more influencing on the lives of modern men, their societies and then, their Nation-States as well. All these engaging and fascinating phenomena are highly complex, concentric by themselves and then simultaneously enlarging too; with at times bringing-in many kinds of perplexingly multiple-impacts on the overall situations and environment around the man.

Man's living-history teaches us that any society or a Nation-State, which has been able to fruitfully, during her initial civilisational multiple advancement-processes or at the earliest at some time during those processes, evolve, cultivate and proactively imbibe the multiple-aspects of 'essential strategic-culture and thoughts' historically she has been found to much less regret and hardly suffer any humiliating set-back or any kind of socio-cultural and/or politico-military tragedy, caused at some stage by any kind of adversely developing circumstances or even due to some evil-designs or hostile-actions by her adversaries. But most ironically and rather unfortunately, India has been lacking these vital attributes since both the ancient-era and the medieval-age and then paradoxically, these vitally important multiple-values have not been adequately taken cared of as since desired and even now, during the present modern-age and that too, not only in the domain of matters-military alone, but in other important areas of the Nation's management of her affairs and the overall systems of governance by the modern and democratic India's politico-bureaucratic combine. In many ways ironically, this so regressively lingering 'state of affairs' has indeed been very embarrassingly pathetic, for both the Indians as well as the Indian-Nation.

In these India-specific perspectives, whenever we peep into the Medieval-era Indian history we perforce find many 'historic-chinks' in this regard. In the contexts of our discussions at this specific point, we may prefer to take-on few important historic facts/episodes from the military-history of medieval India but before that, we may like to briefly discuss certain larger but very closely related and also quite relevant historic aspects in this regard. We have since discussed that even during the few centuries before historic advent of the Christ-era Alexander from Macedonia of ancient Greece invaded India, through her north-western borders. And then through the same route, barbarian Huns came as invading herds; followed much later by Kushans and so on. Historically, it is a great irony that invasions into India in fact, have hardly ever stopped. **Politico-militarily, another very humiliating point in this regard has been that, all such invasions into**

the Indian territories by the foreign powers from time to time, have largely taken place through India's north-western borders. And the great irony of ironies in this regard, has been that at no point during those momentous times it ever occurred to any of the concerned Indian rulers or monarchs as to why and how it was so and then, both geostrategically and politico-militarily, why, how and what all could possibly be done by them both individually or collectively; in order to ensure that such hostile military invasions into India, were effectively tackled and prevented.

Now with respect to the above, one important point arises as to such embarrassing aspects regarding the crucially important geostrategic and politico-military dimensions, during both the ancient-era and medieval-age of living Indian-history while objectively reviewing, we find that these have indeed been largely the historic-saga in this regard then what all were the principal causes for them. While analytically focusing on these historic aspects concerning the socio-cultural, political and military dimensions of the living Indian-history in our contextual perspectives, we find that the Indian rulers/monarchs along with many of their military commanders during those times, lacked the essentially meaningful sense of history and thereby, the overridingly decisive geostrategic and politico-military vision so essential for the overall governance as well as both internal and external security-aspects of a kingdom or Nation-State, with successful purposefulness and geo-political sense of dignity. These irrefutable historic facts have indeed been India's great undoing, and along with historically repetitive geostrategic and politico-military embarrassments, in spite of her fascinating magnificence of wonderful civilisational advancements and great socio-cultural globally acclaimed heritage, since antiquity of the beginning of ancient-era; as recorded in the pages of man's long history. Now at this stage of our critically analytical discussions concerning India and the Indian-history in our relevant contexts in this regard, since as mistakenly believed commonly in India; these 'chinks' in the armour of historic India just do not in any manner pertain to the domains of India's matters-military but almost to the all aspects of her processes of governance since ages and in addition, in aspects of the Indian-peoples' socio-cultural and socio-psychological ways of living and thinking as well. In fact historically, there would hardly be any other important Nation-State in the world, with such magnificently ancient advancing civilisation, whose historic-saga in the above discussed specific perspectives would be containing such unpalatably embarrassing episodes.

Hence, with respect to the India's such repetitive geostrategic and politico-military failings as well as civilisational negation, the principal question arises as to what all precisely have been the central causes for them? **Well, obviously it is needless to say that the main reasons for all these have been that since ages, India as a Nation-State has been regressively lacking the above discussed multiple-aspects of the "essential strategic-culture and thoughts"**. Historically indeed, there are many reasons with distinctly different dimensions for them, and they have since been discussed in some detail, not only earlier in this chapter but in the preceding chapters as well, and in addition with all the more historic facts and detail in the last chapter, while contextually dealing with the ancient India's multiple civilisational as well as other related aspects in this regard. We may now recall that while discussing the just mentioned multiple-aspects of the ancient Indian-civilisation in the last fourth chapter; we have clearly identified and emphasised another connected overriding historic reason since ages as the vested-interests-driven sinister designs and manipulations, both in the multiple-faceted and regressively class as well as caste-fixated Indian societies and royal-courts, by the powerfully entrenched and overwhelmingly influential **"Brahmin priest-class"**. In these contexts, if we may care to discerningly review the age-old continuing living

Indian-history in the making then, we may clearly notice many kinds of highly damaging manifestations in the India's socio-cultural, socio-political and politico-military spheres of the crafty initiatives, selfish motives and even dangerous conspiracies of the same Indian "Brahmin priest-class'. **In fact in many ways historically since ages, they (*i.e. the same 'Brahmin priest-class.*) have acted as a stumbling road-block in the path of the above discussed: gradually cultivating, evolving, developing and pro-actively imbibing by the Indians and the Indian-State; the multiple-aspects of "essential strategic-culture and thoughts".**

But in these contexts now, in the presently ongoing ultra high-tech modern-age when for India too, times have overwhelmingly changed and at the same time much more so since political-freedom in the year 1947, India has magnificently transformed into a sovereign, secular and fast-developing modern democratic-Republic and, where in the present information-age modern era the above mentioned once very powerful 'Brahmin priest-class' have also largely withered-away and hence, now having without any say whatsoever in the entire gamut of national-governance. And even then as to why and how the today's modern Indian, both geostrategically and politico-militarily, rather very awkwardly are yet to fruitfully come to terms with these crucially important aspects and thereby, still failing to make requisite situational corrections towards befittingly cultivating, evolving, developing and pro-actively imbibing those attributes of the "essential strategic-culture and thoughts". The past is already over and it is so important for us even today in the sense that, it contains many invaluable lessons for the modern Indians both in the contexts of the present and the future. But it may also be said that life in general, largely belongs to the present and hence it becomes imperative to anyone concerned to mend the ways of both thought-processes and action, and make essential course-corrections in this regard. In these precise contexts, the Indians and the modern Indian-State have yet miles to go and that too, very fast with the requisite strategic prudence so that lost times are made-up for good, in the best possible supreme national-interests. It is also said that the 'water flows down the gravity.' and hence, are the India's top-level political-class across the Party-lines and also the high officials, both civil and military, sincerely listening for the larger national good so that, the requisite awareness also gradually but intensely, creeps-in into the middle as well as lower-level both the political-activists, Government functionaries and military-personnel respectively for the needful in this regard.

Now coming back to India's medieval-era living-history, India has historically been subjected to repeated foreign military-invasions. History further shows that in spite of all these since ages, the Indians and the Indian rulers/monarchs as well as the key decision-makers have refused to develop the requisite sense of history and thereby, productively learn from the history. We have since in the last fourth chapter discussed, that **the advent of Islam**, both as a new religion as well as largely a distinctly new civilisation, took place during the very beginning of medieval-age of man's history in the world. As the Islam took roots largely in the middle-east to start with, substantial migration of population began to take place in the central-Asian region, with the dawn of 8th century and this historic phenomenon continued for a number of centuries. In fact, migration of many kinds of tribes since spread and settled in the vast areas covering Iran, Iraq and Afghanistan, into the then in comparison, scarcely populated areas of the north and north-western India. Traditionally in the living Indian-history, these numerous belonging to the middle-east as well as the central Asian region have together been addressed as the Turks, for easy understanding and so on and we then notice that many historians and scholars have recorded their at times quite violent entry, as a period of the medieval-era Muslim-invasions into India. Here again, there is no need to repeat as to why and how, both geostrategically as well as politico-militarily all these historic happenings took place from the north-western routes into the Indian territories, since these have already been delved-upon quite in detail a number of times in various allied contexts earlier. In these contexts, the noted Indian military-history author, **Maj. Gen. Gurcharan Singh Sandhu** has written that, "The Turks were 'a nomadic fighting race of rude horsemen galloping the steppes with boundless energy, not quite barbarians but devoid of human sentiments, grossly materialistic, burdened with an all-pervading racialism and the fierce orthodoxy of the neo-convert. They were, however, intensely practical men of action. The cultured Arab and Persian had nothing of the ferocity of temper and iconoclastic zeal which was predominant characteristic of the Turks who were utterly devoid of those higher qualities which make-up the dignity of man. The passions of these nomadic tribes could easily worked-up to fever heat by a fanatic leader who held out a prospect of plunder or proselytism.' The last wave of the Turks lost ground to the first wave of the Mongols.... **Muizz-ud-din's** (*a member of the Gauri tribe.*) next campaign was aimed at raiding Gujarat through the desert. He was encouraged to adopt this route because it was followed earlier during the century (*i.e. during the 11th century: in 1178*) by several Turkish raiders into Rajasthan. The destruction of a Turkish army by the Chahamana Anahilladeva in 1162 is recorded. His son Kalhana makes similar claims between 1161 1179. An invasion and capture of Nagaur by Bahram's governor in Punjab recorded in the *Tabaqat-i-Nasiri* probably took place during reign of Kalhana. The Rajputs had apparently forgotten the bitter lesson of Mahmud's devastating raids and took no steps to co-ordinate their efforts to deter more serious invasions which were sure to follow." (Sandhu, 2003, pp. 109 & 110: 'A Military History of Medieval India').

Now, if we may go in for the historic review of the **ruthless and highly destructive raids of Mahmud of Ghazni** into the then Indian territories, that would indeed be a long story and at the same time, since some of details are quite well-known in this regard. Interestingly in important military-cum-strategic perspectives here, General Sandhu also in his above referred book, on page 55 writes that, "The first raid into India in September 1000 was limited to the capture of a few fortresses in the vicinity of Peshawar; raid was obviously a well-planned preliminary operation which should have served as a warning to the Shahi ruler. For the next campaigning season Mahmud assembled a force of 15000 select cavalry and a large number mounted and dismounted free-booters near Peshawar where Jayapala arrived with a force of 12000 horse; 30000 infantry and 300 elephants to prevent Mahmud from advancing any further. Jayapala avoided action for some time because he was expecting reinforcements; he had not taken anticipatory action after the previous season's offensive and tactically significant activities of his opponent whom he, apparently, underestimated.... Jayapala was released on a promise to pay a ransom of 2,50,000 dinars and 25 elephants; one son and a grandson were retained by Mahmud as hostages. Jayapala's son Anandapala paid the indemnity and secured the release of the hostages. Jayapala, having suffered three successive defeats, held himself unworthy of the Shahi throne. He crowned his son, Anandapala, and burnt himself to death."

Now, if we may care to look back into the above quoted paragraphs from General Sandhu's book with the desired geostrategic and politico-military criticality in our perspectives then we clearly observe that, there are indeed overriding historic-reasons for all these otherwise quite avoidable, historically factual embarrassments for the medieval-era Indian society and their numerous kingdoms in this regard. In their broader contexts since ages, the essentially comprehensive strategic prudence as well as far-sightedness perforce demand that the historic realities both 'sweet and bitter' in fact, should be taken in proper perspectives along with an imaginative 'sense of history', and a sort of correct positivity and thereby requisite lessons are learnt with far-reaching and all-encompassing futuristic vision. And historically now, it is needless to say that in this regard, India has indeed been mostly found wanting in many ways. Hence in these contexts now, coming-back to the above asked overriding historic-reasons, for all the not so palatable recorded episodes in the pages of living military-history, we may refer to the earlier few paragraphs above and where in, those specific reasons have again been emphasised. **The historically pivotal-point for those overriding reasons in this regard, has always been that not only as per the crucially important dictates as of the geostrategic and Politico-military considerations, but also with respect to the comprehensiveness of the multi-faceted socio-cultural as well as socio-political aspects of the overall governance processes of the Indian societies and Nation-State since ages, India has either failed or callously never cared for, with due historic vision and strategic-prudence to gradually cultivate, evolve, develop and imbibe the multiple-aspects of "essential strategic-culture and thoughts" in their totality**. And again it is needless to say that, the man's living-history has never forgiven and will never forgive at any point any human-society or Nation-State for such repetitively lingering lapses.

Well, there are many such historic-episodes since recorded in the continuing Indian living military-history, substantiating thereby their above mentioned historically irrefutable and overriding reasons in this regard, and in our on-going discussions, contextually we have picked-up only few examples here for our purpose. At the same time and as also briefly mentioned earlier; there has never been any dearth of soldierly chivalry, sense of sacrifice and the allied military-attributes in this regard, on the part of Indians since ancient times. Aspects of these magnificent historic qualities on the part of Indians have indeed been quite unreservedly acknowledged by both Indian as well as Non-Indian military historians, in their well-noted writings. But as already emphasised in the last fourth chapter as well that, only those above mentioned attributes on the part of any human-society or a Nation-State anywhere in the world, are not sufficient at all. The principal point intended to be driven home is that, these indeed so crucial attributes in this regard will have to be perforce comprehensively supplemented and then, both geostrategically as well as politico-militarily reinforced by the since above discussed multiple-aspects of the same "essential strategic-culture and thoughts".

Now so far as the magnificent aspects of soldierly-chivalry as well as other allied military attributes, on the part of Indians as mentioned above are concerned; as indeed a medieval-era example, we may recollect the name of **Hemu or Hem Chandra**, who historically came on the scene during the 16th century. In this respect, **General Sandhu** in his above referred book, on pages 475 and 476 mentions that, "**Adil Khan** appointed a competent and an exceptionally brilliant Hindu, Hem Chandra (Hemu), as his *wazir*. The history of Northern India during the next two years is, indeed, and account of the brilliant yet tragic career of this extraordinarily sagacious and unusual Hindu of the Muslim period who came very close to re-establishing a Hindu rajya in Northern India but by a quirk of fate. Hem Chandra, a *Dhusar Bania* by caste, was the son a petty shopkeeper of Quitbpura near Rewari, then part of Alwar. He obtained employment under Islam Shah, came to his notice and gradually rose to the post of *shahna* (superintendent of the market in Delhi) and *daroga-i-Dak-chauki* (head of the department of Post and Intelligence). He was also entrusted with other important business which gave him experience in military command. Adil Shah recognised the great merit and qualities of Hem Chandra and appointed him as his Prime Minister and 'left the full control of political and military affairs in his hands.... In all, according to Akbar Nama, Hem Chandra fought twenty two battles with the opponents of Adil Shah and won all of them." (Sandhu, 2003, pp. 475-476)

Now in continuity at this point, we do find from the pages of man's overall living military-history that there are indeed numerous historically captivating examples, with respect to the Indians and their numerous kingdoms belonging to the ancient, medieval and modern ages and then to the present India as a single democratic and sovereign Nation-State as well. But in spite of all this, the overall living military-history of India since its very beginning couldn't have been very reassuring as well as desirably motivating for the peoples residing in the Indian subcontinent, due to one overpowering single-lacunae in this regard. And that kind of singular lacunae concerning India, has been analytically spelled-out and emphasised too, in our contextual perspectives in the above preceding paragraphs and perhaps adequately as well. Then, we may also say that in the larger geostrategic and politico-military contexts, living military-history does show that these are causes and thereby, there have perhaps inevitably been certain precise course of events, since having taken place as well. In other words, that's precisely is the way for any particular Nation-State's living military-history, to keep taking shape for the future. Based on all these historic facts as well as events, military-historians in the exploratory manner study and research onto them for writing the relevant pages of the same living military-history as further enriching parts of the multiply magnificent and continuously expanding store-house of the human knowledge.

Now in these specific contexts concerning India, Field Marshal Montgomery writes that: "**The course of warfare in India has been dictated to an exceptional degree by the natural factors of geography, population movement and climate. Between the Himalayas and the Vindhya Range in Central India lies Hindustan, a vast fertile land plain with no natural defences. Before the British assumed control no power took responsibility for guarding the north-west frontier, and from earliest times immigrant peoples crossed the passes. In due course Greeks and then Turks, Huns, Mongols and Persians came the same way**. Between 2400 B.C. and A.D. 1500 the actual inhabitants of India were everywhere defeated by the foreign invaders and pushed southwards, the invading movements being generally hated by the Vindhya Mountains. In southern India large areas are hilly and dry—unsuitable for the movement of considerable bodies of people. This country lends itself to resistance against both invasion and internal government by loose-fighting warriors such as the Marathas.... **War was most prominent in the politics and literature of the ancient Hindus. Yet the documentation available allows us to gain only a very imprecise understanding of this military history**. Kautilya (*i.e. Chanakya*), author of a manual of statecraft called the *Arthasastra* (c. 100 B.C.), is a useful authority for, significantly, he includes a military treatise of some merit within his political theory." (*In our contextually larger perspectives, we have since already delved upon Chanakya's Treatises earlier in this chapter, with multiple and indeed unique magnificence as well as some 'shortcomings' in precise geostrategic and politico-military perspectives and with specific focus on the vast aspects of India's matters-military, since her very ancient past and then at the same time, we may recall that we have also since made a brief comparative studies in the contexts of our presently on-going discussions, both Chanakya's and the Treatises of Sun Tzu of China in this regard.*). (Montgomery, 1983, pp. 127-128)

Now in continuity of the above and on the same pages, Field Marshal Montgomery again writes that, "From the dimmest period of the past to the nineteenth century A.D. the bow was the chief weapon in India. Arrian describes the Indian bow in 326 B.C.: 'The Indian infantry have a bow equal in length to the man who carries it. Nothing can withstand an arrow shot by an Indian archer, neither shield nor breastplate.' Later on the sword rose in esteem, and club and spear, discus and sling were among the other weapons also used. Chariots, with crews of from two to 12 men, were in use for many hundreds of years. Indian horses, however, were of an inferior breed, and the disasters at the hands of the Greeks and Turks were armies having a poor cavalry element by armies consisting essentially of good cavalry—this despite the fact that **Indian armies in the first century B.C. are the first cavalry to have been equipped with the stirrup**. Shields were carried by all except the archers and the very poor. They were made of the skins oxen or tigers and of bamboo or matted creeper, and were decorated with emblems. Only a minority had body armour; a type of chain armour existed, but padded coats of quilted cotton were more often used." (Montgomery, 1983) (*Well, the comments on weapon-systems, in use by the soldiers in India since the remotely ancient-times could even be debatable since at times substantially in variance, as often seen in the writings of different scholars and military historians across the world. In any case, instead of being even remotely judgmental in this regard we may need to critically review them in a perspective, for our purpose of study.*)

While critically reviewing the very long and continuously living Indian military-history, we do notice many kinds of multiply complex age-old 'imperial' historic facts captivatingly surfacing before our eyes. The historic-facts indeed are that India is a very vast geographical entity having a unique and magnificent age-old Indian-civilisation perpetually living and enduringly advancing, since the antiquity of civilisational-time in our world. Prominent scholars, historians, thinkers and writers from all over the globe, have indeed listed it as one of the leading ancient human-civilisations in the world, since the beginning of man's history. Well, these historic-aspects in a context are all very fine but at the same time, it is also historically true that at a later stage of the present modern-era India was more of a multi-dimensionally magnificent civilisational-entity only, and certainly not a Nation-State strictly in the socio-political, geostrategic as well as politico-military parameters. And in addition, she was certainly at times able to even rise for some time as a kind of 'perplexing' Nation-State, though temporarily under some charismatic King-Emperor with historic geo-political and politico-military brilliance and strategic-vision resulting in establishing a highly flourishing, rich as well as politico-militarily mighty historic and vast Indian-empire. And interestingly then even at times, just sentimentally if not territorially, a kind of socio-cultural and/or socio-religiously motivated 'Nation-State'.

But as per well-accepted historic parameters in this regard these were phenomenally somewhat temporary, certainly not having the requisite enduring continuity enabling the Indian societies to take shape as an united and strong 'Politico-military Entity' and thereby, purposefully cultivate, develop, imbibe and proactively adopt the earlier discussed multiple aspects of those far-reaching **'essential strategic-culture and thoughts'**. The hard historic realities remain that just plain sentiments, patriotic-thoughts and literally enchanting but fictional-poetry by themselves do not in any case, constitute a strong, thriving and both geostrategically and politico-militarily enduring Nation-State. In the above contexts and at precisely this point of our continuing discussions if we may critically recall the no less long and certainly much more violent and turmoil-ridden multiple-aspects of the living Chinese military-history and the Chinese-civilisation then with quite different kinds of interesting historic facts, begin surfacing in this regard. Hence we do notice that perhaps at no stage of her great living-history, **Chinese peoples were never almost totally devoid of the multiple aspects of those 'essential strategic-culture and thoughts in the manner Indians were ironically as well as quite unfortunately found wanting in fact, so regressively and repeatedly as well**. In fact further, even in today's modern and internationally prominent democratic and sovereign Indian-Republic the present-day duly elected political-leaders belonging to all sorts of political-shades and other prominent decision-makers as well as influential authorities, do not perhaps have any time or meaningful inclination to render any serious-thoughts towards those overridingly important aspects of the 'essential strategic-culture and thoughts'. And hence, it is needless to repeat again that the **man's continuing and living-history perforce never spares any one human-society, Nation-State or any person in any manner whatsoever**.

Now, returning to these contextual aspects of the living Indian-history; we do find since the very ancient-era and then also continuing in the medieval-age, the Indian subcontinent was having numerous relatively big and strong, medium-sized as well as small Kingdom-States within her geographical territories. Many of them are historically prominent and since well-known as well but, quite a few of them not commonly so well-known. And interestingly, one such princely-state happens to be the **Mithila-Kingdom** located in the northern-part the present-day province of Bihar, bordering the sovereign State of Nepal. It is historically an area of fertile agricultural lands and encircled by India's prominent rivers, with Ganga in the south, Kosi-river in the east and Gandak in the west and hence, geostrategically quite water-locked as well. Interestingly, this happened to be then perhaps the only Hindu-kingdom in almost the whole of vast north and eastern medieval-India when Muslim invaders consisting mainly of newly converted Arabs and Turks were on the territory-conquering spree in India, entering through the same commonly adopted north-western passes and routes.

But Mithila was consistently and fairly successfully resisting those ransacking invaders. When medieval eleventh century was coming to an end, a ruling dynasty was established in Mithila by the **King Nanyadeva** from Karnataka. With respect to these and other magnificent aspects concerning the 'Mithila-kingdom', Maj. Gen. Sandhu further writes that "The audacious act of defying Muslim authority seems to have provoked **Sultan Ghiyas-ud-din Tughlaq** to invade Tirhut on his way back from a campaign in Bengal during 1324-25. **Isami** has left us a detailed account of the campaign. The Sultan advanced from the bank of Kosi into Tirhut whose king sought haven in a thick forest.... For about a week patrols went out to kill every Hindu in the vicinity. Then the Sultan left for Delhi and left **Ahmad bin Tablaga** to carry on with the siege. **Harisimhadeva** successfully resisted all efforts of the Muslim army to capture the fort. After some time he left for Nepal which acknowledged his sovereignty and organised resistance against any further advance by the Muslims who occupied Simraon, the capital in the Nepal Terai. During the confusion and chaos in the later part of Mohammad Tughlaq's reign, the Karnataka family regained control over Tirhut (since in that territory the brave and tenacious kingdom of Mithila was situated.).... An invasion of Tirhut was carried out by Nusrat Shah of Bengal (1519-32) and the principality was finally subjugated.... The importance of Mithila lies in its role as a repository of Sanskrit learning, as a refuse for Hindu scholars and as the preserver of medieval Hindu culture for several centuries after the Delhi Sultanate was established." (Sandhu, 2003, pp. 404-405)

Well, there are a number of other historic examples similar to the one mentioned above. But due to the above repeatedly elaborated in different contexts during our continuing discussions earlier; and also **since so regressively persisting in fact, the pivotal lacuna in the vast and highly complex spheres of the statecraft, governance and the matters-military, concerning those multiple varieties of the bygone era's relatively small and large Indian kingdoms,** which in turn, imposed crippling limitations on them when the inescapable needs arose in both geostrategic as well as politico-military terms, to get united with the essential multiple-vision for victoriously facing the foreign invaders and decisively drive them back from the larger Indian sub-continental territories. **And it is needless to say that, if that would have been the case since so momentously positive in all respects the entire multiple-course of the then overall Indian-history in the making, would have been almost totally different.** But alas that was not to be and obviously, the faults were certainly not in the 'stars' of those Indian-kingdoms but alone in the socio-cultural psyche and rigidly fixated attitudes of the Indian-rulers, their advisers, their **lingering and entrenched Brahmin priest-class**, other influential decision-makers, their armies' top military-commanders and amazingly, even the kingdom's powerful traders. Indeed, what a highly depressive scenario has so overwhelmingly been persisting and of course, with at times few historic-exceptions here and there apart among the elites in the societies within those kingdoms belonging to that bygone age and thereby, the recipe remained regressively complete for the repeated disasters, humiliations and untold sufferings in the face of a number of foreign-invasion into the larger Indian territories. And conclusively, it is again needless to say that the **obvious pivotal-reason in this regard has indeed remained the total absence of the same 'essential strategic-culture and thoughts'**.

In continuity in these broader and comprehensive perspectives, when we render even a critically discerning glance over the age-old Chinese-history, we do find that China has been no less violent war and turmoil-ridden in many ways. But it has not been so depressing and mostly getting subjugated by the foreign-invaders, as indeed that of the then contemporary India. In China for more than one hundred years beginning from 650 A.D., the mighty Tang dynasty ruled over mammoth Chinese-empire. But in China by then, and in their own distinct historic-methodology and system, the **multiple-aspects of 'essential strategic-culture and thoughts'** had already been cultivated, developed and fine-tuned; imbibed and pro-actively adopted up to a commendable extent, in the thought-processes as well as geostrategic and politico-military approach of the then mighty Chinese-emperors. We may have a glimpse of these Chinese socio-political and cultural subject-matters and developments in the already referred book *China* by the Scottish scholar of history, John Keay.

John Keay writes that, "Only the methodology and means for enforcing this ambitious arrangement left something to be desired. **Confucian doctrine**, formulated during 'Warring States' era and partly in reaction to it, was adamant about civilian control over military affairs. It was one of the features that distinguished China's culture from that of its nomadic neighbours. Though expansive and often downright aggressive, at few moments in history could the Chinese empire be characterised as militaristic. Military matters were traditionally as a subordinate function of the bureaucracy; under no circumstances should the bureaucracy become a subordinate function to the military. Theoretically, standing armies were anathema, professional soldiers as parasites. Troops, whether conscripts or *fubing* (territorial militias), should be farmers-on-horseback and peasant-with-crossbows; generals should be, and usually were, bureaucrats in uniform. They were commissioned for a single campaign and, unless appointed, reverted to civilian life after it. With military success offering a short cut to the highest office, a term disparaging an upstart as one who 'went out a general and came in a chancellor' (*Chu fiang ru xiang*) gained wide currency." (*We do find that even during bygone era, Chinese elites were having a very high degree of awareness of the crucially important geostrategic and politico-military perspectives and hence consequently they were pro-actively able to quite judiciously as well as innovatively economize on the empire's military-spending but at the same time, without permitting any chinks to develop in their armour. These achievements on their part, in vast and highly complex areas of the matters-military could have been unthinkable without the decisive and also integral-presence of the 'essential strategic-culture and thoughts', in their entire gamut of the whole process of imperial governance of the colossal Chinese-empires. In any case when viewed in multiply larger perspectives in this regard, we may even emphatically maintain that in the comprehensively global geostrategic and politico-military terms these then were no small matters specially when during those times, the whole of Europe was reeling under their historic 'dark-age' and the ancient Indian civilisational territories were repeatedly being subjected to the all sorts of ruthless foreign-invasion. With respect to the modern sovereign democratic* ***India:***

China's that kind of the dynamic system of bureaucracy, so innovatively then established during the above narrated regime of her historic Tang dynasty, just cannot be compared with the perhaps a kind of official 'file-oriented' bureaucracy, continuing in the today's independent democratic India with any possible stretch of politic-bureaucratic imagination whatsoever). (Keay, 2008, p. 259)

The above quoted scholarly paragraph and its brief critical-analysis in our broader perspectives; clearly indicates to us that China did not undergo certain otherwise avoidable humiliations in the decisive spheres of matters-military and warfare and that too repeated ones; which India had to perforce suffer and also, all largely due to her own making. As historically consequent fallouts in this regard, India both geostrategically and politico-militarily, got repeatedly overpowered by the alien invading forces and then, subjected to different kinds of foreign-rules since the beginning of the medieval-age for many centuries continuously.

Well in historic perspectives, this was certainly not the path then for the Chinese history to undertake for its own in a way, parallel but distinctly different gradual shaping-up, regardless of all these aspects otherwise multiply affecting the mankind in this world. We may further prefer to elaborate that these historically factual aspects in addition, did ensure that China's rulers and elite decision-makers since ancient-times were indeed able to gradually and comparatively, better develop and fine-tune their strategically prudent and comprehensive multiple-vision; in social, cultural, socio-psychological, socio-political and also military perspectives. And hence, these did make further far-reaching difference in almost all spheres of their gamut of the various processes of governance. Concerning the above discussed historic aspects in our contexts at this precise point, we do very informatively find that John Keay narrated a very unique historic phenomenon belonging to the period, when the medieval age was fast coming to an end and the present modern-era had already come in, and when competing European Imperial-Powers were already making deep inroads into almost all the multiply-lucrative geographical territories of both Asian and African continents, since situated in the south and south-east of the Occident.

John Keay covering certain historically important aspects during 1760–1880, has written that, "...Manchu Qing dynasty's long-reigning 'Three Emperors', received tribute-tendering deputations from some of 'All under-Haven's farthest extremities. One came from Kalmuck (Kalmyk) Mongols.... The second deputation supposedly hailed from Pegu, the erstwhile capital of Burma. And the third came from London, George Viscount Macartney of Dervock, ambassador and leader of this first ever embassy of from a British sovereign to a Chinese emperor, was at pains to emphasise that his mission had nothing to do with tribute. Rather was it an exchange between equals, between 'the greatest Sovereign of the West and....the greatest sovereign of the East'.... The formal reception was held in a vast marquee within the imperial park at Jehol, the Manchu's summer retreat north of Beijing and just beyond the Great Wall. Macartney, who had risen at 4 a.m. to dress, chose pink for the occasion. Over a suit of embroidered white velvet, he wrapped himself in the technically 'crimson' mantle of satin, lined with white taffeta, which was proper to a Knight of the Bath.... Two ships, one a sixty-four-gun warship, the other the largest and fastest vessel in the English East India Company's fleet, had taken a year to convey the mission to the northern port of Tianjin. From there, after a short river journey, the whole circus had been hauled to Beijing in 85 wagons and 39 handcarts, requiring 209 horses and 2,495 porters. Macartney had counted them. It was not excessive for the 'tribute mission' that the Qing flags announced it to be; but for an embassy it was enormous. In fact it was possibly the greatest ever dispatched from Britain.... In keeping with the scale of the thing, Macartney's diplomatic wish-list was also ambitious. China and Britain were to establish amicable and reciprocal relations. He was to persuade the emperor to agree to his staying on as a permanent

British ambassador in Beijing and to encourage the dispatch of a contemporary Chinese embassy to London. Three new ports on the China coast were to be opened to British trade.... **Macartney did his best. An experienced diplomat and an appreciative guest who went out of his way to cultivate good relations with the Qing officials, he even negotiated the tricky issue of the kowtow; he says he merely genuflected as he would to his own king. But otherwise he failed dismally.... At enormous expense, most of it borne by the East India Company, the Macartney mission achieved nothing; but it was not therefore insignificant. Allowed to withdraw to Guangzhou via the Grand Canal and the Yangzi before sailing home, Macartney and his colleagues observed something of the Qing Empire at the height of its power.... Macartney anticipated this, but his carful observations on the political, social economic climate being both acute and prescient. Though himself humiliated, his visit would come to be seen as the harbinger of China's humiliation.** Of the three major themes of the Qing period—the successful expansion into inner Asia, a mixed record in managing the empire and its vastly increased population, and the disastrous handling of the insistent Western demand for commercial penetration—it is the last which has usually received the greatest attention." (Keay, 'China', 2008, pp. 446-448)

We get a far-reaching glimpse of the 'relentless power-game: that nations do keep-on playing in the vast, multi-dimensional and very much threateningly challenging arena of the international power-politics, competing trade and commerce and, ruthless struggle for multiple-supremacy; both regionally and globally, in this world. **But here again, the centrally pivotal factor remains the same, i.e. the multiple-aspects of the 'essential strategic-culture and thoughts' supremely driving and guiding all these activities and initiatives which the nations keep indulging-in among themselves and, much more as well.** The overriding fact in these perspectives remains that the central sources of both the multiple-vision and strengths, required for consistently playing the above mentioned 'international power-game rather successfully are in fact, nothing but the same **multiple-aspects of 'essential strategic-culture and thoughts'**.

And at the same time any nation, which just doesn't pay historically the requisite attention to these supremely important aspects and keep regressively refusing to make necessary multi-dimensional efforts to cultivate, develop, fine-tune and imbibe them fruitfully, then in spite of her otherwise many historically magnificent civilisational attributes, repeatedly keep suffering on many fronts as well as getting humiliated among the comity of Nations in this world. In any case, man's long history since the very beginning of human-life remains the best and most irrefutable source of both evidence and authentication in this regard. **Paradoxically even today, India and the Indians are yet to see reasons with essential strategic-prudence and, come to terms with them in spite of history relentlessly coaxing them to wake-up and take the multiple corrective-actions.** Now again coming back to the above narrated historic episodes between then Britain and China it is otherwise clear and historically well-accepted that, both of them were both geostrategically as well as politico-militarily prudent and always quite alert in guarding their strategic naivety and geo-political short sightedness respectively. But as history later borne-out that since the British mostly as usual, being comparatively more smart and multiply-shrewd in this regard, won the game in the long-run and historically in the end. But during those historic episodes, where was India then especially when the Indian territories were being used by Britain as springboard, for launching their shrewdly multiple-initiatives towards the then China during her Qing regime? The sad ironical historic fact remains that India was then a subjugated colony under the imperial British-rule and all the resources as well as money provided for that **Macartney's expedition** was being provided from different Indian resources and wealth then being at the disposal of the imperial

British rulers, through their shrewd as well as always scheming British East-India Company operating and also commanding politico-military powers in India at that time. Howsoever the Indians keep-on glossing over these and many other historic-facts in varying contexts here and there, but the plain-history always remains unalterable and unchanged, regardless of the socio-psychological, socio-cultural and socio-political feelings and any kind of fixations of any society or Nation-State in this regard.

Therefore, it has been rightly said that the man's living-history since the beginning in any case, is friend or enemy to none and thereby spares no society or Nation-State, anywhere or during any historic-era during its perpetual living as well as being in the making. But life keeps on marching ahead and hence since Macartney's expedition into China earlier, as few more centuries came and went, and the 20th century during the fast-track modern age, much after the second World-War came to its full bloom, the equations of the above mentioned international power-game, between both the modern Britain and China had substantially changed in many ways. In addition, as since times had changed, the rules of the same highly complex-game too had perforce undergone major changes and the mighty and almost global British-empire had become a thing of the past and consequently, had shrunk into the pages of man's chequered-history in this regard. We may relevantly recall at this precise point of our continuing discussions that when Britannia was ruling the waves globally, the British had again with her typical shrewdness and strategic-vision had then already acquired the Hong-Kong islands on a long-lease from China. And before the 20th century came to an end, this lease in question had already expired. By then, in a way post-communism China had already become a globally fast developing economy and emerging as one of the prominently leading and imposing military-powers in the world.

Further, we do notice that, indeed India has always been consisting of the quite rigid as well as perhaps incurably regressive caste-oriented and in addition, socio-culturally hierarchy-fixated age-old societies. In our principal and comprehensive geostrategic and politico-military perspectives it is a historic fact that these highly negative aspects pertaining to the Indian societies since very remote ancient-era, have indeed been so damagingly lingering. But that is not the complete story in the pages of the living Indian military-history. At the same time, it is also historically true that since those bygone ages Indian societies and their numerous kingdoms have had many magnificent warrior-classes and their clans, though largely caste-oriented and again socio-culturally rigid or even at times, self-damagingly inflexible, but in spite of all this their soldierly spirit for the principled military-valour and numerous historic saga of their socio-psychologically, cheerful readiness for self-sacrifice in the battlefields, have never been second to others. In this respect just to cite few examples, **we may mention: Rajputs, Marathas, Dogras, Sikhs, Biharis, numerous clans belonging to southern areas of the vast Indian peninsula, India's colourful tribal-classes, Jats, Mahars and later-days' Muslim-Pathans as well as superbly magnificent Gorkhas and, many others as well. Hence, it is no wonder and perfectly understandable as well that with the major socio-cultural multiply positive-reforms and together along with, unquestionable faith as well as total soldierly dedication towards the modern democratic India's pro-active and all-pervasive secular-ethos, her modern and highly professional-cum-apolitical Indian Army's numerous top-class Infantry-Regiments have been constituted and now quite successfully stand organised and also named after the above mentioned famous warrior-classes**. For the brave Rajputs historically, fighting wars has always been a superbly heroic pastime. We find that beginning of this tradition of Rajputs may be traced, along with the commencement of India's ancient epic-age. We may recall that a glimpse of the multiple-facets of India's epic-age have since been given in the last fourth chapter while dealing with the enchanting aspects of the ancient Indian-civilisation. **In this perspective, General Sandhu has quite interestingly mentioned that, "Under such circumstances a mastery of the strategy and tactics of war should have been every Rajput chief's major concern and preoccupation. It was not so because no strategy or tactics which placed the enemy at a disadvantage was considered fair or permissible.** (*Look at this: even no connection at all with perhaps, common-sense based intelligent military-prudence!*) It was deemed cowardly to use deception or to gain surprise. Their battles were standardised and may be compared to knightly tournaments of the Middle Ages in Europe.... Death in battle was an ultimate achievement. There was little generalship and all that a general could do was set a good example during the fight and let matters take their own course. Once battle was joined, he was not expected to interfere with its course. Eventually one side gave up the fight when its soldiers were allowed to leave the battlefield." (*In any case, now in today's modern thinking in the vast and highly complex areas of the 'matters-military' it all appears to be militarily irrational and thereby, according to the internationally well-accepted norms and theories in the multiple-fields of both strategy and tactics almost totally illogical as well. But in addition, it is also a magnificent fact that, a number of medieval-era great Rajput military-hero and chieftains such as* ***Rana Sanga, Maharana Pratap, Raja Man Singh and many others****, have indeed left inedible marks on the pages of India's living military-history.*)

Well, unlike the medieval-age Rajput warrior classes and their clans-based social structures and hierarchies, the magnificent **Marathas** were quite militarily prudent and flexible too; and much of credit in this regard, goes to one of the India's famous military-leaders, strategists and warriors during the medieval-age, and that is the great Maratha, **Shivaji Maharaja**. He was indeed one of great and earliest pioneers, with highly flexible and innovative tactical-thoughts in the complex areas of guerilla-warfare. The central tactical aim in his operational-plans for attack was precisely two-pronged, i.e. as far as possible to achieve total surprise over the enemy and at the same time strive hard to render him a real lethal-blow, for causing harm to his fighting-potential and morale of his troops. And in this particular gamut of warfare, he was truly unparalleled and hence, rightly he ultimately became a historically famous figure. Shivaji as a monarch, as well as brilliant supreme commander of the great Maratha-empire, situated then in the Deccan territories of the medieval-era peninsular India, which he was instrumental in magnificently building, always believed in unexpectedly attacking the enemy first. In order to effectively facilitate the launching of his military-plans, he first acquired and also build a number of formidable Forts cited at tactically very advantageous locations as his nodal strong-points to enable militarily multiple-uses such as, logistical-centres, places for both tactical-retreats and bases for launching his operations and also, for providing reinforcements. It was indeed all his unique-concepts and thereby, he was always successful in making the best uses of the terrain and also, militarily dominate, and observe the areas around as well. All these aspects have well-deservedly made him an immortal hero in the history of medieval-India.

Since earlier times, Marathas have been known for efficient cattle-breeders and raisers. They were fond of horses of good breeds as well, and they were also proficient in taking utmost care of their domesticated animals. **In fact, Shivaji's main-stay of his armies was the first-class cavalry**. General Sandhu has mentioned that, "According to **Colonel Tone** who served with the Maratha Cavalry, 'they had a very thorough knowledge of the care of horses and their breeding.' Shivaji was instrumental in the import of a good basic breed of horse. He preferred to import horses from north India and even arranged to purchase West Asian animals from ports in the Deccan. This resulted in the local breed but horses continued to be imported from Kandahar and Tibet until the end of the 18th century. The mainstay of the Maratha cavalry, however was the indigenous *Dakhni* horse which was brought to a high state of perfection.... **Colonel Broughton** saw these animals in Sindhia's **camp** and describes them as, '...The sum of three and even four thousand rupees is often paid for a Dakhni whose pedigree is well known and so fond are the Marathas of these beautiful and valuable favourites, that when they can afford it, they will feed then on wheaten cakes, boiled rice, sugar, butter and other similar dainties'." (*All these are indeed historic and militarily captivating as well.*) (Sandhu, 2003, p. 761)

Then the Sikhs of India too, are indeed second to none in valour and cold courage in the battlefields. Magnificently for them and indeed in the much larger contexts as well the Tenth Guru of the Sikh religion, **Guru Govind Singhji**, remains the highest and the most glorious source of the martial inspiration. In fact in the similar larger contexts, **Maharana Pratap, Shivaji Maharaja** and many others too remain magnificently iconic and medieval era sources of highly enduring martial inspiration. Similar are the traditions in their own specific manners and highly motivating medieval as well as modern-era military-historic aspects, regarding the examples of chivalry with respects to the Dogras, Madrasis, Biharis, Jats, Gorkhas, Pathans and all other troops. These have been elaborated briefly in the earlier paragraphs and thereby constituting magnificent martial sources of the unique and befitting soldierly inspirations in the battles-cum-wars.

In these segments of warfare concerning the Indian-military, since ages and even now in the mind-boggling presently continuing modern-era it is naturally obvious that the military-commanders at all levels as well as multiple-aspects of the captivating subject-matters of the military-leadership, both in peace and war do continuously play a very complex and crucial role in every possible situation. Well, at this specific point regarding the multiple-aspects of the Indian military-history, in the larger contexts of our continuing discussions and also to maintain rather a mandatory objectivity in this regard; the Indians are always expected to acknowledge and meaningfully accept that **the British did play highly historic fruitful role in multiple-ways for implanting, nursing and furthering these aspects (*though of course for their own compulsive imperial-cum-colonial reasons*) in the Indian-military, on their arrival and subsequently, their own colonialism-oriented multiple geostrategic and politico-military activities and, with indeed remarkable and highly focused far-sightedness in the vast Indian subcontinent**. Along with all others, these are also interesting facts of the man's living as well as expanding military-history since, continuously in the making.

Now as we begin critically studying the no less vast and highly engaging medieval-age European military-history we do come across many unique and revealing historic-aspects and facts, in the context of our ongoing discussions. In this respect, **Field Marshal Montgomery** has mentioned that, "Between the battles of Bouvines (1214) and Morat (1476), warfare was the rule in Europe. The disintegration of the Holy Roman Empire was furthered by the struggle between the Papacy and the Hohenstaufen imperial family, by the Swiss wars of independence, and the wars of the Bohemian Hussites. The power of the English state was consolidated by Edward I's conquest of Wales (1277-95), although the English attempt to conquer Scotland (1296-1328) was a failure. Then came; the clash between England and France, the series of English invasions between 1337 and 1453, together known as the Hundred Years War. From the military point of view these wars are immensely important. The knights, the castle and feudalism gave way to infantryman, to firearms and to professionalism. Strategically it became clear that the defensive with the weapons available could win battles provided the enemy could be induced to attack, but to win campaign offensive action was necessary. Tactically superiority passed to new missile weapons (*i.e. artillery to begin with.*), and these, rather than shock-action, became decisive factor in battle." (Montgomery, 1983, p. 103)

The above quoted paragraph gives us a sharply-focused and, indeed, a critically panoramic view of the militarily transitional phase in the conceptual strategic and tactical aspects in the larger domain of the matters-military in Europe. In addition then, we also do very interestingly come across contextually here another rather universally true aspect of the inherent nature of mankind in this regard, being no less proactively present among the medieval-age European men at arms and perhaps the entire matters-military as well. And man's that very specific inherent aspect happens to be a kind of die-hard 'resistance to any change'. In these perspectives, with historic precision regarding the medieval-age European aspects of warfare Field Marshal Montgomery has provided a number of informative examples in his book substantiating man's natural instinctive resistance to any kind of change in the multiple-aspects of soldiering, weapon-systems and weapon-platforms.

In perhaps the all-encompassing and much larger contexts, we may also say that the same remains equally true in all kinds of highly complex and vast spheres of multi-dimensional as well as relentless human-endeavour. Well, it is also never meant that man in any field, never accepts progressive and path-breaking changes at all. Man since the ancient-ages, does accept changes but the processes of accepting and fruitfully imbibing them are indeed always very slow and at times in certain contexts, quite struggle-oriented and also friction-ridden as well. Further, we do also find that as we move.

Now coming back to the medieval age European aspects of the warfare living military-history indicates to us that during the ancient-era battle of Adrianople in 378 A.D. itself, the then militarily decisive supremacy of the heavy-cavalry was historically established for anyone to professionally take notice as a kind of lesson for the coming years. Sometime during the middle of thirteenth century, when obviously the medieval-era in Europe was historically at its zenith any Western European military-commander was able to militarily learn and imbibe the many professionally valuable lessons then, since already thrown-up earlier, by a number of battles constituting the historic crusades with inescapably compulsive matters-military-oriented acumen for futuristic-vision in this regard. Obviously, we may also say these do always remain part of the much larger and multiple-aspects of inherently complex human-nature itself.

In these specific contexts, historically interesting and also very valuably **crucial lesson happens to be that the militarily tactical sum of both the infantry and cavalry (*i.e. in the present context: the armour.*), whenever imaginatively deployed with battle-specific co-ordination and thereby, always adhering to the all-important principle of the 'unity of command'; the decisive advantages are far-reaching as well as much greater than the linear sum-total of these two fighting arms, whenever operating separately**. Now, in the present high-tech modern-era as we critically begin to go deeper during this particular lesson's explorative studies, we do perforce begin to come across their vast and militarily complex depths and dimensions; especially during their professional applications in the live as well as fast altering real warlike situations. In fact, the above mentioned crucial lesson with all its changing manifestations, since as the multiple-aspects of military-technologies, too, keep perpetually advancing are being equally important and valid in the presently ongoing modern-era as well.

Further as we proceed, we do also notice that the above mentioned aspects concerning man's inherent nature of die-hard fixations with respect to the all sorts of 'resistance to change' have in fact, been equally true in case of all other kinds of medieval-era military weapon-systems as well as strategic and tactical concepts of warfare; on the part of military professionals, thinkers and also the then Europe's military-commanders in the field. Be it artillery and its projectiles, coats of arms and shapes of steel-helmets, otherwise remarkably advancing multiple-processes of the man's innovatively developing conceptualization, design and development of many kinds of weapon-systems, weapon-platforms and so on we do prominently find the identical fixations, concerning 'resistance to change' in this regard. Interestingly, as further time progressed a similar phenomenon remained in place, i.e. the advent of gun-powder and all its multiple fallouts and their military applications as well. Discovery of the gun-powder led to the multiple-processes of the gradual development of concepts and improved designs of the many kinds of small, medium and heavy firearms. In the larger contexts of our on-going discussions, we may since be aware that in an overall totality, the presently continuing super high-tech modern-age has been witnessing in fact, truly mind-boggling and unprecedented advancements in the multi-dimensional fields of now indeed the amazingly numerous disciplines of modern sciences and technologies.

Historically in fact, we may take further notice that irrespective of any kind of fixated resistance or no resistance, changes do get manifested for actual materialisation everywhere, in their own different ways. In addition to the various kinds of weapon-systems and weapon-platforms since discussed above; in the medieval-era Europe, age-old strongly fixated feudalism as well as culture of knighthood gradually began to be things of the past and thus, leading to simultaneous professionalization of the European armies. Interestingly then, we may also notice that increasing dependence on various kinds of the systems of mercenaries too, commenced. And it is needless to say that the military culture of mercenaries in turn, had its own advantages and disadvantages as well. But interestingly again due to the same socio-culturally fixated reasons, always getting activated almost in any conceivable situations the age-old and deep-rooted highly regressive culture of feudalism in any case, was not going to be uprooted that easily. As quite rightly the saying goes that 'old habits die hard' and mostly tends to linger-on regardless. But at the same time, certain relatively smarter European-societies, especially the English, were able be somewhat quicker in adopting changes.

As the culture of feudalism began to gradually wither-away particularly in the military domain and thereby, parallel much advantageous process of the steady professionalization of the armies also commenced, the earlier entrenched and whimsical practices of recruitment also began to give space to the more efficient and military profession-based systems of recruitment. These in turn historically led to the raising of much better trained, more efficient and disciplined professional armies in medieval-era European Nation-States. At the same time, as we critically proceed further in this regard in historic-contexts, we may bear in mind that those European Nation-States then were mostly different kinds of feudalism-based monarchies. Interestingly therefore, from socio-cultural as well as politico-military considerations and then in the larger perspectives, historically we may also observe that these aspects' combined dynamics resulted in different kinds of socio-political complexities, coupled with increased socio-economic urges for looking beyond. Then as more times passed with increase in homeland populations, thereby resulting in different kinds of even competitive manifestations of overseas imperial ambitions among them.

Over the centuries perhaps the seeds for the compulsive urge was sown again among those medieval-era European Monarchies to competitively go overseas for acquisition of their imperial-colonies with all their socio-political and military wherewithal, into the vast continents of Asia, Africa and North as well as South Americas. Well interestingly, we may also say that historically the phenomena of their multi-dimensional and proactive ambition-driven initiatives with all kinds of rather built-in complexities sooner than later resulted in a type of imperialism-based competitive-race among those unchecked and also expanding European politico-military powers. As these prolonged phenomena of relentless European imperialism-driven processes of overseas expansionism saddled on their military-might, were progressing; continuous and unprecedented advancements in the various fields of sciences and technologies were also getting triggered and accelerated as well. In a parallel manner, we do also observe that as centuries roked by, medieval-age was ending too and simultaneously then, the gradual dawn of the modern-era was taking place. Moreover, we may perhaps be interested to know the main reasons for the historic medieval-era hundred years' wars between England and France. In this respect, Field Marshal Montgomery (page 111), has written that, "In 1337 the Hundred Years' War between England and France began. The issues included the throne of France, the recent French support of the Scots, and rivalry over the Flemish wool trade, and long standing border warfare between English and French sailors in the Channel. From a few disjointed invasions by the English, it developed into a war of attrition, really amounting to the repeated devastation of Artois, Normandy, Brittany and Aquitaine by ruffian bands of professional soldiers." (*Well, while naturally accepting the historicity of these and many other such wars we may find them rather as mindless wars with the then lingering regressive fixations, as mostly being naturally quite avoidable. No wonder and as also already quoted little earlier, this famous British Field Marshal has rightly commented that in the medieval-era Europe, warfare was largely nothing but, to rule over the entire continent*).

In our contexts, while critically reviewing the broad geostrategic and politico-military aspects of the medieval-era European Nation-States and their governance, the then military superiority of Spain cannot be overlooked. And in these precise perspectives, Field Marshal Montgomery has mentioned that, "**In the history of European warfare in the sixteenth century the leading nation was Spain**. She took a major part in all the nationalistic-dynastic wars of the century; furthermore, in the new commercial offensive of European imperialism, the ships of Spain with those of England snapping at their heels, were responsible for the tremendous development and expansion of the strategic and diplomatic horizon of Europe to include the Atlantic and the newly colonised continents of America. In the development of military and naval technology and tactics, Spain was again a leader.... The forays of Europeans over the seas of the world in this period were made irresistible by development of the sailing ship armed with cannon—described by Cipolla as essentially a compact device which allowed a small crew to master unparalleled masses of inanimate energy for movement and destruction. Here the most creative individual, in the navigation and tactics of these revolutionary ships was an Englishman—Francis Drake. But the Spanish seamen, though outclassed at the defeat of the Armada in 1588 has been as ready to experiment as any." (Montgomery, 2000, p. 145)

Now, again coming back to the aspects of historically checkered and also multiply progressive matters-military of the medieval-era Europe, quite expectedly a number of historic military-personalities have valuably contributed in shaping-up the man's living military-history. Hence, we find that in the then European continent too, famous strategists and military-commanders with their own distinct historic talents and sharp military-vision have indeed brilliantly conducted their numerous military-campaigns. Befittingly in the process, they have left the noted histories of their campaigns and in addition, saga of their display of the brilliant aspects of military-leadership in war and at times, even some of their war-dairies. In this precise context, we may cite as suitably historic examples the names of **Marlborough, Nelson, Napoleon and Wellington** and many others as well. And we would be discussing about them as well as their geostrategic and politico-military historic achievements briefly little later. But for more historically multiple-details regarding the aspects of their brilliant military-vision and leadership and military-exploits in war and strategic-thoughts and so on, we may suitably refer to various books on the living military-history in this regard. In fact in larger contexts, we may also recall that in the many preceding pages, we have since delved-upon many of the noted historic military-personalities and their professionally brilliant military-attributes being par-excellence in their own distinct manners and also, belonging to the both ancient and modern ages. We would be witnessing the same phenomena in the present modern-era as well, when we come to their many more numerous aspects in the subsequent forthcoming chapters.

During the beginning of fifteenth century or perhaps even little earlier, historically in the Medieval Europe, the military-manoeuvre and so many kinds of initiative-oriented as well as strategically motivated innovations, on the part of military-commanders in the fields had begun playing decisive roles in the then European wars as well. These vast aspects of the **manoeuvre warfare**, in perhaps all kinds of super high-tech-oriented and extremely fast-moving military-operations acquiring three and even more dimensions and in all kinds of fast-changing situations, terrains and weather, in the presently continuing modern-age too, do indeed play important roles in many ways.

Well, quite fascinatingly from the matters-military points of view now in the modern-era, we may further observe that multiple-aspects of the manoeuvre-warfare have also been expanding and acquiring additional dimensions and in the process if these developments are opening perhaps newer avenues, opportunities and space for taking more innovative initiatives by the military-commanders in the fields then, at the same time; they indeed keep throwing-up many unprecedented militarily threatening challenges before them too. **In fact and as earlier noted by us too, many well-known strategic-thinkers, military-historians and experienced military-commanders the world over, in their own respective ways of the wisdom-oriented thoughts and writings have emphasised that even during their active military careers, very few officers and commanders at almost all levels have been able to fruitfully grasp the finer nitty-gritty of the multiple-aspects of manoeuvre-warfare, in spite of their continued training and war-games throughout their service careers in their respective Armed Forces.**

In the context of medieval Europe in these perspectives, we may again refer to Field Marshal Montgomery's words, "In 1495 the victory of Charles VIII at Fornovo against the combined armies of Venice and Milan ended the somewhat absurd era of *condottiere* (a lead of a mercenary band of military adventures) warfare. For the past two hundred years the wars of the Italian cities had been conducted by the band of cavalry under professional soldiers known as *condottiere*. **Many campaigns by this time had become so scientific that battles were no more than manoeuvres, in which armies would surrender as soon as they were technically outflanked or cut off from their base.** The Italians therefore got a rude shock when the French with their Swiss mercenaries crossed the Alps, with the intention of taking the towns by storms and slaughtering prisoners. Charles VIII's irresistible progress up and down the length of Italy in 1494-5 showed what the artillery could do. In recent years the **Jacques de Genouillac had made certain technical improvements in artillery weapons, including the introduction of gun carriages, of trunnions to elevate guns, and the use of metal instead of stone cannon balls**." (Montgomery, 2000, p. 146) (*In this fifth chapter itself and in earlier chapters too, we have earlier discussed in various contexts during our continuing discussions in this regard, certain other important aspects of the phenomena of the ongoing multi-dimensional* *advancements of the weapon-systems and weapon-platforms since the beginning of ancient-age and so on, briefly covering their genesis as well. In the same precisely continuing perspective in this regard, Jacques de Genouillac's contributions as cited above too, are quite battle-oriented results-giving and valuably progressive contributions; in fact as also emphasised earlier, being obviously triggered by the innovative ingenuity of a highly focused and active human-mind).*

Coming back to medieval-era Europe's legendary warriors and outstanding military-commanders, we find **Marlborough's** name and the saga of his soldierly brilliance and exploits, with unique prominence. History tells us that Marlborough was an outstandingly multi-faceted military personality.

In fact, the noted records of historic military saga of the series of Marlborough's great military victories, do indicate that he was a visionary and brilliant tactician of his times. Since he was always able to see through the developing situations in the battlefield, he remained flexible and could timely make highly beneficial changes in his plans and retain both his initiatives as well as surprises against his enemy. A noted example in this regard, was his timely change of plans in the battle at Blenheim against the militarily strong Franco-Bavarian Army, where he realised his error in deciding to attack the enemy's much stronger front and thereby, he corrected it to achieve victory. In the medieval-era Europe, as we have already seen that the wars were calling of the times then, and these in turn perforce got transformed into a kind of relentless struggles of achieving the decisive military-superiority over others and in addition, go in for the contests when needed to at least maintain some 'workable' balance of power as in case of prolonged war of the Spanish succession during the years 1703-13. Interestingly now in the present modern-age, specifically since the mid-twentieth century post-world war II years, we are now familiar with the re-adoption of this older phrase of the 'balance of power', that is being frequently used by almost all the leading nations in different contexts.

With respect to the above and Marlborough as well, **Arthur Birnie** has written that, "As in later wars, the number of the combatants enlarged the field of hostilities. There were fighting fronts in Flanders, Germany, Italy and Spain. There were naval battles in the Mediterranean. There were clashes in Asia and America. In its imperial aspect the wars had its deepest significance (*For facilitating the sprouting of seeds for the birth and expansion of overseas empires by those competing European geostrategic, politico-economic as well as leading military-powers.*). Graver issues were at stake than the Continental diplomatists suspected. In Europe, the balance of power was restored by the humbling of the French monarchy. But further afield, Britain was launched on the first stage of her progress towards the world empire.... The rest of Marlborough's fighting was done in Flanders, a country studded with fortresses and seamed with watercourses, making mobile warfare difficult. Here the French had entrenched themselves behind a defensive system of ditches and earth works—the so called Lines of Brabant. Stretching from the Meuse to the Scheldt, it was a position that could not be turned. In some ways it reminds us of the trench line that ran in 1914 from the sea to the Alps.... Marlborough was a unique combination of the diplomatist and the soldier. As negotiator and general he provided the cement that held the Grand Alliance together. From him came the general direction that allowed the wars to be brought to a successful conclusion. Alone among the allied commanders he realised the connection between the separate fronts and the need for coordinating the efforts of the different armies." (Birnie, 1942, pp. 100, 117, 119)

Further in this specific context, **Field Marshal Montgomery** has since written that, "**John Churchill, Duke of Marlborough from 1650 to 1722**. He was a military genius with great diplomatic skill, and it was he who was responsible for the rise of the British army to become one of the foremost armies in Europe....it was Marlborough who prevented this factor (*i.e. the 'warfare of fortification' as then developed by Vauban*) from paralysing generalship. It was age of cavalry; but the introductions of the bayonet raised the status of the infantry soldier and caused the end of the pikemen.... Marlborough had seen a wide range of professional service, including service as a colonel of infantry under Turenne in 1674-5. As a man he was difficult to get to know, and his private moral record has certain blemishes; but against this can be set the devoted care at all times for the men under his command, and the tireless service of his country. He was an equable man and unfailingly courteous and charming. He had a complete mastery of his profession, and while he saw the problems of the war in perspective as a whole, he also overlooked the essential detail, tactical or administrative. This was the time when the art of command was meeting increased complications. While the rise of armies and the range of strategy continued to increase rapidly, the organisation of the administrative machine did not keep pace. Furthermore, the communications in Europe were slow and politics complex, so that a commander, particularly of an allied army, had many responsibilities as a diplomat. At the same time he had to deal with politicians in his home country." (Montgomery, 1983, pp. 179, 182 & 183)

Importantly in the larger perspectives of this project, the living military history of medieval-era Europe, also shows that during the militarily challenging times of Marlborough many far-reaching and indeed revolutionary reforms both in civil and military spheres too were taking place in Europe and, more particularly in the British Islands. **Historically, the 'Bank of England was established in 1694 and this in turn, brought-in revolutionarily multiple fallouts for the English Nation both in civil and military fields.** This important development in fact, facilitated England to historically achieve a kind of head start and to begin taking pioneering steps on the paths of accelerated economic advancement and at the same time, they were also able to make the then systems of military command and control relatively better equipped in the complex fields of both the military campaigns and the management of logistics as well, in comparison to the other medieval European nations. Obviously then due to this important development in England at that time, it decisively put the English and the Dutch in militarily quite an advantageous position with respect to the French military forces. Similarly in many areas of the then matters-military, professionally facing **Marlborough** while always rising to any such demanding challenges, he was also able to brilliantly siege them as opportunities for the unprecedented operational tactical as well as strategic visionary creativities.

During the 17th and early 18th centuries, Louis XIV as then an aggressive king of France was rather fruitfully able to grasp the aspects of certain path-breaking and highly beneficial processes of multiple military-reforms as already begun earlier, during the times of Gustavus. In these overall processes of multiple military-reforms and much more, historically, Marlborough in fact was able to observe them in his own professional military-vision as well as both strategic and tactical highly creative thought-processes and thus visibly contributed brilliant ideas by proactively applying them in his own campaigns in advancing art and craft of warfare. One historic aspect in this regard, has been then in the areas of military-recruitment systems and logistical organisations, wherein the lingering feudalism-based fixations which in any case was naturally full of rotten corruption and nepotism, so strongly entrenched since the ancient-ages.

These were largely abolished and a path-breaking system of professional recruitments of troops, largely based on merit being the criteria of selection was substantially introduced. Raw-troops were strongly drilled and made to undergo the well-planned systems of rigorous military training. Results were indeed revolutionary: troops came out of those military-training establishments as militarily well-chiselled, toughened and disciplined soldiers and thus under highly professional and competent military-leadership, to able to make any European medieval-army proud and also to be trusted even in any tough and exceptionally demanding situation of any battle-cum-war. Marlborough was able to practice these concepts brilliantly, and also further trained his armies for forced long-marches with speed and grit so the troops' morale was always maintained high. **Essential tactical elements both speed and surprises were his key factors in his campaigns since he was a military stickler for prudently managing all important details in his planning and management of multiple logistical aspects**. In this respect, Marlborough followed the appropriately designed systems of magazines, for advantageously positioning and managing for provisions so that his campaigns never suffered in these aspects at crucial moments.

Further, the above mentioned aspects ensured that Marlborough, with his brilliant military and visionary-leadership as a commander never permitted the most valuable aspect of military-initiatives at any stage to be passed-on to his adversaries in the field. **On the armament side during his times the slow and cumbersome matchlock gun was replaced and flint-lock system, ensuring relatively much faster rate of fire, was adopted as the main infantry-weapon. In addition then, a separate Corps of Grenadiers was formed and the entire army was professionally re-organised around the infantry, since infantry was made the main teeth-arm. And in continuity, professional position of the Engineers was also suitably upgraded and also Artillery was militarily integrated better with the army as a whole.** During Marlborough's times a number of historic military personalities with multiple-talents came on the scene and in turn they too valuably contributed in the fields of military thought-processes and innovations as well.

Well, during our continuing medieval European-era specific discussions in these precise perspectives, we perhaps cannot overlook another historically noted figure, **Fredrick the Great**, a famous General of those days' Prussia. He was also a great soldier and military-commander. Interestingly, he happened to have some surprising similarities with Napoleon in many ways. Fredrick was also a man with a relatively shorter height and also not having very strong and impressive physical attributes. But he too was a gifted military leader in many ways. Similar to Napoleon, he was also a talented intellectual soldier, in spite of his not so impressive physical sides of the personality. No wonder that, Fredrick too during his times, was able to decisively influence many by his vision and thoughts and thereby during his times, could even be instrumental in changing the then course of history. When he came to his accession, history indicates that then the Prussian infantry was perhaps decisively best in the medieval Europe. And in addition, also similar to Marlborough, Fredrick was another brilliant general to not only believe in the multiple military-values of the artillery in war but was quite imaginatively capable of successfully exploiting its capabilities in war. In fact during Fredrick's times, due to the visionary leadership of his father, the Prussian infantry was so much trained and drilled with skill; that during their advancing deployment, they could come back into columns and then, fast wheeled back into the lines with perfect timing and precision; for crucially putting the enemy in a kind of confusion and even off-guard. Further in terms of the tactical operations, the 'oblique-order' is particularly associated with his name wherein his well-trained infantry could be wheeled-back and then obliquely charge on to a relatively weaker flank of the enemy, depending upon any emerging situation and the given choice of time.

Just to name a few, we may take note of brilliant generalship of **Conde** and **Turenne,** and then **Vauban** was a brilliant military-engineer with great strategic-vision for imaginative fortress designs and also historically, he was highly skilled and expert in offensive siege craft. Another historic military personality in Europe during that later part of medieval-age in Europe was a **French Marshal Villars** who too, was able to achieve a number of victories in his campaigns. Interesting historic aspect during those turbulent times in Europe, **when medieval-era was at last coming to an end and thereby, dawn of multi-dimensional 'Industrial-Revolution' was indeed taking place;** when after losing a number of wars, with the decisive factors of Vauban's brilliant fortification-warfare and along with outstanding generalship of Marshal Villars, France was in a position to secure relatively advantageous terms in the treaty of Utrecht in the year 1713. These highly informative facts from medieval-era European military history are indeed able to show in the larger perspectives of our ongoing geostrategic and politico-militarily critical discussions, as to how decisive roles have always been played by the prominent historic military-leaders in shaping-up the living military-history, since in their own extraordinary distinct ways.

Now, it appears quite appropriate in our present specific contexts to delve upon other famous historic military personalities such as, **Nelson, Napoleon and Wellington** respectively. Though we may also wish to remain aware that during those slowly closing years of medieval-era in Europe, besides these three very famous military personalities, there indeed were a number of other outstanding military-commanders who too contributed in their own distinct ways in the making of living military-history during those momentous years. But no less important historic facts in this regard are, that the persona of Nelson, Napoleon and Wellington with brilliant professionalism and outstanding attributes for military-leadership as well as humane in temperaments both in peace and war, were also able to substantially transform them into historically immortal and indeed charismatic visionary thinkers, military-strategists and commanders in their own distinctly unique ways. During those times, Europe was able to witness the **advent of mobile artillery** with the unprecedented rapid-rate of fire though hardly any substantial improvements were seen in the designs and actual performance abilities of any European Navy's warships. But historically it is also true that from naval perspectives, one encouraging development had been that externally steel reinforced with some kinds of stiffeners attached coppers-plates were put at the bottom of the ships for improving the speed and endurance in the sea.

The above militarily developmental aspects, with respect to the weapon-systems and weapon-platforms along with their mobility-potential, now may technologically appear to be quite simple and primitive from the present superb high-tech age parameters, but during those medieval times these were certainly remarkable in many ways. Contextually during those prolonged momentous times, we may also again be reminded that then, in fact, the medieval-era was setting-in and simultaneously modern-age was on its way, with all its unprecedented and far-reaching manifestations to begin with and then becoming fast-paced, along with multi-dimensional as well as multi-disciplinary scientific and technological advancements that changed things forever.

In this respect, we may also say that many kinds of powerful revolutions were coming-in and since they in fact themselves were being triggered by the **fast-paced dawn of overwhelmingly powerful and multiply influential aspects of the historic 'Industrial-Revolution'**. All these resulted in multiple kinds and almost total transformation of the man's ways of life with new kinds of commitment and challenges, his overall environmental situations all-around him and thereby; his place in this world. And then came, the mind-boggling and a kind of even 'Post-Industrial-Revolution-Age' with all its multiple fallouts. In other words, things for the man and his overall quality of life; were definitely going to be never the same again. We would be discussing these aspects in more appropriately and in somewhat substantial details, in the forthcoming seventh and eighth chapters.

Admiral Nelson's full name was **Horatio Nelson**, and his was born in the year 1758. From his childhood he was talented in his own way and then he joined the Navy very early in his life and by the time, he entered the British Navy he happened to be hardly thirteen years of age. Nelson was indeed gifted with many outstanding qualities of an outstanding military-commander both in times of peace and war and at the same time, as a Statesman as well. In addition, he was dedicated to his profession of soldiering and thereby focused, both intellectually as well as militarily too. **Hence in case of Nelson both as a person and a naval sailor, the recipe was complete to historically shape him up as a brilliant Admiral. From the very beginning Nelson with his multiple attributes, was able to futuristically visualise that the Britain being and island nation, her formidable military-assets and strengths lay in the sea and her decisive supremacy over there.** Now in the overall geostrategic and politico-military perspectives, we would find that Britain was rightly to effectively take multiple initiatives in the sea, for realising her multi-dimensional short as well as long-terms objectives. And in these larger dimensions; Nelson with his multiple talents and unquestioned loyalty, during those closing years of the medieval-era in Europe, made historic contributions in the service of his Nation.

Living military history shows us that in comparison to France during those years, Britain had a relatively more capable and stronger Navy, with around sixty naval warships with multiple military capabilities. In addition to this, she had well-trained and much better disciplined adequate naval manpower, manning these warships as well as her shore-based naval and other allied strategic establishments. Further to crown all these highly valuable British naval assets the British-Nation had a well-educated, trained and professionally highly groomed hierarchy of the naval-officers of all ranks and military expertise. **During those historically checkered years, when momentous developments were taking place in the whole of the European continent no other country or military power in Europe could display such vision and make their naval strengths equal to that of Britain.** In the coming centuries, these momentous developments were indeed forever going to decisively change the entire subsequent course of the larger and multiple living-histories in the making; of the whole of Asia, Africa, Australia and also the entire Americas. We would be discussing these aspects in more detail and dimensions, in the subsequent forthcoming chapters.

But in our contexts of the ongoing discussions, luck-factors in certain respects, specifically regarding the above mentioned points, there certainly was no sort of any windfall for then the British-Nation. Historically in fact, it was all due to the decisive strategic-shrewdness, multiple-vision, focused thought-processes, discipline and consistent hard work of the British leadership at all levels and her peoples, both in the military and the civil-sectors. Hence consequent to all these, during those momentous times in Europe; Britain did finally emerged as a decisively lead player in this regard. But as we continue to critically review all these in our contextual perspectives, we certainly find it was not at all any cakewalk for then the British-Nation. Interestingly in fact, the centuries' old rivalries between France and Britain did come again to the forefront and hence naturally, France was not going to sit idle and let this pass that easily over her rich multiple-heritage and destiny. Moreover, France too then did not have just a push-over Navy, since they were also very much having their own naval wherewithal to give a really tough time to the British sailing as well as her naval warlike initiatives in the sea.

Precisely with respect to the above, **Field Marshal Montgomery** has mentioned that, "The

French used their Navy to damage British shipping and to threaten invasion. British naval strategy in war had various aspects. Besides the usual functions of her sea power—the protection of the British shipping and damage to her enemy's, and protection of her shores from invasion—Britain also attempted to use sea power to confer mobility on her army. Several overseas operations were thus conducted, in Flanders, in the West Indies, and on the coast of France. But since they were unsuccessful this policy was abandoned for the time being and the British concentrated on blockading the coast of France, in order to stifle French trade and to force into battle and French fleet which dared to leave harbor. Between 194 and 1805 the British Navy won six major victories.... The old warship remained technically much the same, now at last British sailors discovered the most effective way to use it. One reason for this was the outstanding ability of some of the British seamen, notably Howe and Nelson. A second reason was introduction of a comprehensive signaling system, devised by Kempenfelt and Howe; this enabled commanders to progress from preset and rigid formations to more elaborate and flexible tactics.... Howe or Nelson had a more difficult task than their successors; if the plans of approach had been faulty they could not remedy them by signal; we, in the days of oil, fuel and wireless communications, are apt to forget how different conditions were in the days of sail." (Montgomery, 1983, pp. 211 & 212)

The historically famous and tragic Battle of Trafalgar on 21st October 1805, was the last battle so successfully fought by Nelson against the powerful French Toulon fleet of the French Admiral Villeneuve. The French commander did his best to breakout but, Nelson gave a hard chase to the French fleet and forced it into battle on his terms. It was indeed a naval engagement between the two leading rival powers of Europe of those days, which did historically displayed militarily skillful and highly innovative maneuver of the fleets of warships by two top rival naval commanders, in order to encircle and threaten each other for decisive victory in a historic naval-battle during the end of medieval-era in the English channel separating the two fighting nations. But the famous British Admiral, Nelson was decisively able to both outclass and outwit his rival French Admiral Villeneuve and in the process fierce naval battle ensued resulting in the capture and destruction of a large number of French warships. Ultimately, Nelson did achieve an unchallenged victory over the French fleet.

But the French too, gave a fierce fight in the sea then and hence they in fact extracted a heavy price from the British, in terms of the highly valuable life of Nelson during the prime of his career. **Militarily, those were the days of sharp shooters and snipers and Nelson was tragically hit with a fatal bullet injury killing him on the top deck of his flag ship itself**. That was indeed the gloriously victorious end of one of the all-time great naval personalities of the world, who always believed in leading and maneuvering his fleet in any naval battle, by positioning his flag-ship in the commanding front of his fleet. In this respect historically, another fact also emerged and wherein after this Battle of Trafalgar, Britain achieved the position of being an unrivaled naval power over the seas. This aspect subsequently did bring far-reaching and long-term fallouts wherein, Britain was able to also achieve, both geostrategically and politico-militarily in fact, multiple head start; while building her overseas historic colonial-empires all around the world and thereby, historically it is well-known that peoples then began to even call that, 'Britannia was ruling the waves'. Now coming to **Napoleon** in our specific perspectives, he was another historically multi-faceted personality and famously unique military-commander. There is an internationally famous biography of Napoleon giving many historic and interesting aspects of his personality, psyche and thinking of this legendary figure. Napoleon's strategic concepts of warfare were based on a kind of four-cornered conceptual-framework. These were: 1. Numerical strength of his battalions constituting the field formations of his armies. 2. Concentration of force at tactically well thought-out plans. 3. Always adhering to the vital principle of the unity of command and 4. Speed of action in battle for giving a kind of knock-out blow to the enemy and in addition, seizing and always decisively maintaining initiative in his own hands. Interestingly in our larger perspectives in this regard, in order to understand as to why and how Napoleon was in fact, what he had been and why and how he thought and did all that, we may have to first give a critical glance over the multiple-situations prevailing then in France along with the threats and challenges confronting the French Nation as since perceived by the French people and their leaders.

Napoleon was born in the year 1769 in Corsica, situated on the mid-northern side of the Mediterranean. At the very young age of nearly 10 years, he was admitted and studied as well as received valuable military training with full dedication, in the military Institutions of France for ten years. These were indeed most educative and valuable formative years of Napoleon's early life, which were able to lay the essential foundation for subsequently grooming him into a historic military-genius, a remarkable soldier and a great commander of his armies in war. In addition, from the early years of his education and training itself he became a dedicated student of the living military-history and also, writings of many great commanders and strategic thinkers on which he could lay his hands on. With these learnings and deep studies of such writings, he realised the importance of patriotic feelings for one's nation and zeal to serve one's country with the best of abilities and unshakable dedication. He began his military career by joining the French regiment of artillery and that further added to his knowledge and experience, also enabled him to grasp the real military values and decisive punch rendered by the artillery, if employed with precise sense of timings as per the well chalked out plans of any battle and in co-ordination with the larger strategy of a military campaign.

Interestingly, Napoleon's life and psychological make-up has also shown that since his youth itself he was very ambitious, always thinking high and big and at the same time, he was politically ambitious as well. Hence, when he got the opportunity to stay in Paris he never went for any relatively unrewarding job and, displayed his keenness to cultivate friendship with right kind of people and at right places. But his military brilliance and series of successes in a number of wars with France's adversaries, came first and became quite instrumental in turning him into a magnetic personality and charismatic leader, which in turn made him famous and also took him to places in the highest level socio-political and military circuit in France. At the young age of nearly thirty only, he was promoted to high military ranks and given the field-command of French army stationed in Italy, where he was charged with fighting against the much larger and well-trained and better equipped force of the Austrians and Sardinians. But to his dismay he found that his troops were ill-trained, poorly equipped, ill-fed and hence plagued with low morale and very bad soldierly discipline. In fact, very few commanders would have really taken over such a command and much less, would have really been able to sustain the same and that too with such historic success. But indeed, while looking into both by any militarily professional and strategically critical-yardstick, we learn that within less than a fortnight itself, Napoleon was historically able to achieve a real miracle.

With his charismatic leadership, professional brilliance and unwavering dedication to his cause and undoubted loyalty to his troops; he was able to totally transform his army into a disciplined fighting force with high morale. And no wonder when the actual battle commenced, the outcome was decisively in favour of Napoleon. This was his first remarkable success, which was indeed a brilliant display of the unique synergy achieved by Napoleon through a wonderful combination of his charismatic leadership, his skill to motivate and unfold the zeal to fight among his troops and officers and then, be determined to achieve victory against all odds. It has been appropriately said that, in such challenging situation for the French army in Italy then; Napoleon was in fact, remarkably successful in triggering the soul of the army situated there, so essentially first coming back to life. After achieving a series of victories in Europe then, in the year 1798 Napoleon was launched on another long-term and historic politico-military expedition in Egypt.

Here much wider aspects of Napoleon's leadership qualities and vision, regarding the much larger and other than exclusively military, dimensions of the man's living-history with integrated intellectual awareness concerning the future demands too come into light. While proceeding towards Egypt, he also very thoughtfully included a team of learned scholars of the man's ancient-history, archeology and even Egyptology for gathering the valuable data, multiple-information and knowledge in this regard and then, also for writing and compiling scholarly treatises in French, for posterity. But, with respect to the larger essentialities of the strategic-aspects concerning such costly expedition in Egypt during that prevailing situation in France, there are differences in opinions and, some experts have even questioned the overall geostrategic and politico-military sense behind his this historic venture. Well, these are weighty questions of history and hence, meant for in-depth debates among the historians, strategic thinkers and other experts concerned. Interestingly then, another side of Napoleon's multi-faceted personality emerged and came into play; and that being that in the year 1799 itself, while leaving the command of his army into another general's hands, he hurriedly departed back to France with his own politically ambitious agenda.

On reaching Paris and with the help of a kind of political-coup in France, he was successful in securing the position of the First Consul. With this far-reaching historic development, it became quite established that Napoleon had occupied a sort of unchallenged position as a Military-Dictator of France. Hence, it is needless to say that subsequently this indeed brought many fateful and also at times, tragic consequences for both the French Nation and Napoleon himself. But in the beginning as the First Consul of France, Napoleon proved to be very promising with outstanding abilities and naturally; his uncommon attributes enabled him to be a highly successful ruler. But destiny in some cases on its own seldom turns out to be rosy and more so; everything in this regard, is far from being under total control of man. Historically, with his military victory over the Austrians in the year 1800 ended the war in Europe for the time being, he was so relentlessly fighting one after the other and at the same time, displaying his visionary statesmanship; he even manage to secure a sort of peace-treaty with Britain.

But ironically, the threatening military legacy of France and also as being continuously perceived by the other European-Powers in fact, totally preceding the Napoleonic-era seldom allowed them to believe such French peace-intended initiatives and political gestures, otherwise obviously in the larger interests of everyone concerned in this regard. And in the larger perspectives at this precise point of our ongoing discussions, as we proceed to critically look into the man's living-history since continuously in the making always, we do clearly notice that whenever trust among all concerned is at the premium and mutual suspicion is calling the shots, peace and socio-political tranquility seldom prevail. In fact contextually as per the geostrategic and politico-military parameters in this regard, the living-history also shows us that in any such prevailing situations, ironically any well-intentioned move or even politico-diplomatic initiative by any one party is seldom taken or understood as intended for; and hence it is taken as a kind of alarming threat. History is full of such episodes, in fact both internationally and even intra-nationally as well. Similarly in case of the then France too paradoxically, due to all sorts of omissions and commissions on all concerned and including France as well and, this war-torn nation was again sucked into a series of violently destructive wars. And consequently, there was seldom any peace in the Napoleon's life and career

On Napoleon, Field Marshal Montgomery has written that: "**Napoleon was a man of outstanding intellect, energy and willpower. He dominated all around him, and was totally self-centered. Caulaincourt, for ten years a close companion, said of him: 'He always applied all his means, all his faculties, and all his attention to the action or discussions of the moment into everything he put passion.' Napoleon was a master of strategy: the range, speed and co-ordination of his operations were unique**. Since his armies lived off the country and roads were becoming good, he could move fast. He himself was a great builder of roads. He formulated his plans on the basis of the information supplied by his staff, headed by Berthier. Information was kept up to date and was immediately accessible on every relevant subject. Minute research preceded the organization of a campaign, and Napoleon himself issued the final orders on everything.... His strategy was always offensive.... Although continually outnumbered in the theatre of war, he seldom fought battle without local superiority at the point of conflict. He always kept his eye on developments in neighbouring areas, and he planned his campaigns to unfold in such a way that the maximum political advantage could be seized immediately the actual fighting ended successfully—although his foreign minister Talleyrand felt that he exploited his victories too thoroughly to be diplomatic. (*This aspect of his mental-makeup too, perhaps might have been another dominating reason as to why any meaningful peace always eluded Napoleon, in every possible conceivable manner.*) **After 1805 Napoleon developed a new strategic technique suitable for armies.... The army corps, a self-contained formation consisting of two or three divisions, had been introduced**. Part of the secret of his success remained rapid and accurate movement.... Napoleon's tactics were also offensive and long prepared.... As he said, 'there is a moment in engagements when the least manoeuvre is decisive and gives victory: it is one drop of water which makes the vessel run over'. His eye for the ground was outstanding; Caulaincourt wrote that 'he seemed to extract men, horses and guns from the very bowels of the earth'. Infantry was the chief of Napoleon's armies...and the psychological of massed troops in column was considerable. Columns had already proved their worth in the revolutionary wars." (Montgomery, 1983, pp. 220 & 221)

Interestingly in the above historic context itself **Arthur Birnie** has written that, "By 1800 the lines of Napoleonic strategy had definitely become fixed. **Jomini**, the great contemporary military theorist, had discovered its secret and expounded its principles in his *Traite des Grande Operations Militaires*, published in 1804. The later Napoleonic campaigns fall into one of two classes. When Napoleon was weaker than the enemy he operated on interior lines, as in Italy in 1796. When he was stronger he attempted an enveloping movement, as in Marengo.... On the battle field Napoleon's chief aim was to get on his opponent's flank or rear.... A blow to enemy's flank was napoleon's recipe for victory. In his own words, 'it is by turning the enemy, by attacking his flank, that battles are won'. For his infantry Napoleon retained the formation in columns, but slightly modified in order to secure a combination of fire power and shock. The half-deep order (*ordre demi-plein*) was a kind of mixture of line and column, and was sometimes called the *ordre mixte*." (Birnie, 1942, pp. 238, 240 & 241)

But interestingly, man's much wider living-history since the beginning also emphatically indicates to anybody who cares to study or even review it with not only the requisite geostrategic and politico-military criticality, but in addition within the conceptual frame-work concerning encompassing of the much wider socio-cultural, socio-economic and socio-political dimensions in this regard; and more so when in the concerned country, where any really meaningful and vibrant liberal democracy is to take birth and get developed with the desired strength and healthily pulsating life; any leader there, historically displaying a really meteoric rise, is most likely destined to meet a kind of nose-dive downfall as well. Paradoxically even with rather a tragic historicity in a way, that exactly turned out to be fateful destiny in case of the historically Great Napoleon Bonaparte. Perhaps a series of outstanding successes since achieved by Napoleon, both in war as well as in the multiple-domain of the highest level national governance later, made him rather obstinate in approach and somewhat self-damagingly rigid even in his crucial strategic thinking as well. At the risk of going rather tangential in one's own analytical perceptions in this regard one may perhaps say that a sum-total of all these proceeded fast to make Napoleon overconfident that he could never go wrong or even much less ever get defeated in a battle by his any enemy. **And in this specific context interestingly, the living-history of man also shows us that, any such recipe is rather sure to bring any similar earlier mentioned kind of crashing nose-dive downfall**. And then, history also naturally indicates to us that ultimately the '**Waterloo**' was waiting with tragically similar kind of consequences for Napoleon.

The historic battle of Waterloo was fought between the two all-time great Europe's military-leaders on 18th of June 1815, the man's living history with all its multiple-dimensions; had already entered the present modern-era. Interestingly, this battle was not fought either in the islands of Britain or even within the territorial limits of France. The battlefield happened to be at the ridge of Mont St Jean, located near the village of Waterloo in Belgium and situated nearly at 16 km. south of the city of Brussels. In fact, much more momentous development, triggering the commencement of presently on-going modern-era; with its mind-boggling and immeasurably multiple fallouts and also, globally influencing the entire humanity in so many ways; had already begun much earlier. And obviously that singularly unprecedented momentous development was historically the far-reaching **'Industrial-Revolution'**.

We have earmarked a separate chapter for this singularly most influential historic development here. Now so far as the historic battle of Waterloo is concerned, this too did change the entire future course of subsequent history of the whole of Europe, and thereby in fact with globally geostrategic and politico-military multiple-connotations. If we may here pause a little, in order to critically read in our larger specific contexts, between the lines of the living-history of a couple of the highly turbulent centuries, in the lives of mankind in this world interestingly, we may even say that the post-Waterloo years in the European continent with its somewhat changed socio-political and militarily-strategic environment to begin with did also facilitate in its own ways, the subsequently accelerated multi-dimensional advancements of the above talked about, **that historically unprecedented Industrial-Revolution.**

About the **Battle of Waterloo** has been mentioned that, "The final battle of Napoleonic wars saw the French Emperor's ambitions crushed once and for all. Despite flashes of his former brilliance in the campaign, Napoleon was unable to break up the Allied armies, and his fate was sealed by the Prussians marching to Wellington's aid on 18 June rather than falling back after their reverse at Ligny.... The Emperor (1769-1821) with 71,000 men of the French Armee du Nord attacked an Anglo-Netherlandish army of 60,000 men under the Duke of Wellington (1769-1852), who was joined by Prince Gebhard von Blucher's Prussian army that evening.... In a superb defensive battle, Wellington's army was able, with great difficulties, to hold off Napoleon's disjointed attacks until the Prussian army arrived.... Napoleon's escape from Elba and restoration of the Empire could not be tolerated by the Allies, who sought to crush this threat to European peace.... Defeat at Waterloo forced napoleon's second abdication, after which he was finally exiled to St Helena in the south

Atlantic.... As the battlefields go, that of Waterloo, as compared to Borodino (1812) in Russia, was a very compact and dense one where there was to be a lot of action in a compressed space in the course of a single day, 18 June, that was to change the course of European history forever." (Christer Jorgensen, 2011, pp. 165 & 167)

Interestingly, the year of the birth of two charismatic soldiers, in the man's military-history's, outstanding and inspiring commanders of men in the battle fields in their own historically distinct ways and also at the same time great rivals were the same, i.e. 1769. **Wellington** came from an aristocratic family of Ireland and, from childhood, he was noticed as a promising boy with special talents in both mathematics and music. At the young age of 18 years, he joined the then British Army and also, definitely appeared to have an attraction and inclination towards military service; and as history has subsequently shown that in the army he was destined to make it great and turn out to be one of the all-time greatest military personalities in both peace and war and also in fact, for all time to come. Destiny brought him to **India** in the year 1796 and, India then being under the colonial rule of the British East India Company. Naturally in the vast and both historically as well as socio-culturally colourful and also, socio-religiously tradition-based human-societies with great legacies and in addition, gloriously ancient civilisational past, Wellington found much wider fields open to him with many kinds of new opportunities and challenges as well.

Hence luckily during his long stay in India, Wellington was able to indulge in deep studies of the art and craft of soldiering, the vast fields of military sciences and technologies as well as finer nuances of the art of military-leadership, as applicable in both peace and war situations. In fact, while in India and whenever he found time, he delved deeply into the works of great military-historians and thinkers. Here he was able to militarily train himself more extensively and also, he got opportunities to put his knowledge in practice in fact, in real warlike circumstances and naturally, the range of his original thinking and approach in operational planning; did not go unnoticed by his senior commanders and got well-deserved recognition. Historically for the British military-hierarchy stationed in India then, singularly most valuable contribution of Wellington Napoleon was to find a tactically effective answer to the powerfully galloping and charging well-trained and competently commanded units of the **Maratha cavalry: since as glorious legacy of the Great Shivaji**.

Whenever we may care to study the living military-history of the times of Wellington, contextually with some desired strategic-criticality we are indeed able to notice **that during his long nearly nine years' stay and dedicated service in India in fact, turned-out to be professionally very fortunate for Wellington**. His militarily active years in the vast Indian subcontinent turned out to be extremely valuable on-the-job education-cum-training assignments; for his future career and accelerated growth in the somewhat rigid hierarchy of the military-command in the then British Army. Perhaps even more important professional aspects in this regard, have been his zeal for multiple studies concerning not only the matters-military but his consistent interests in the subject-matters of many other disciplines of the vast storehouse of human-knowledge. And hence during his long and very fruitful years in India, he got relatively lots of time and opportunities as well to consistently indulge in his valuable pursuits in life and consequently, satisfy his hunger for knowledge to the maximum possible extent.

Ironically at a little later stage, it is rather a different matter altogether to begin within his career that on his return to Britain in the year 1805 the people concerned with the areas of decision-making as well as exercising authority, hardly cared for Wellington's attributes and achievements more so during his stay and highly chequered service in India. Well quite interestingly anywhere in this world, that's precisely the way the entrenched bureaucracy keeps functioning; and even more so during the present fast-moving and commonly believed perhaps superb high-tech era as well. And it is needless to say that at times, even otherwise considered quite objective, precise and rather so-called efficiently result-oriented military-bureaucracy too fall in the same line and only keep following regressively, the relatively same old beaten-track. But how long can any such system anywhere, whether intentionally or otherwise, continue to keep any exceptionally bright star in the shadow, and more so who in any case when he or she is destined to make exceptionally own glorious history in the further course of time? (*Globally, a kind of some interesting comments on—civil and military—entrenched bureaucracy, anywhere!*)

Hence whenever, we may look into the annals of man's living-history in our larger geostrategic and politico-military perspectives in this regard or even otherwise we clearly find that since the very ancient of ages a truly expanding galaxy of such numerous historically prominent personalities, have indelibly been drawing glorious lines on those pages of history by making their distinctly own exceptional mark in the so many different areas of the relentless human-endeavour. In fact in other words, that's precisely the way man's multi-dimensional and vast living-history is continuously being made since the very beginning. Now coming back to the life and times of **Wellington**, quite informatively we find that Field Marshal Montgomery has written that, "In India he gained a sound military training and built up a considerable local reputation. No British commander had previously worked out a tactical technique for dealing with the hordes of Maratha horsemen, but **Wellington** found the answer with a line formation. His supply lines were well organised to give him a long line into the enemy country and mobility once he got there. His most notable engagements were the storming of Ahmednagar, one of the strangest fortresses in India, and his victory in the early of Assaye in 1803. In 1805, after nine years in India, he returned to England. Although the reputation he had gained in India counted for nothing at home, he had accumulated very valuable experience—in pitched battles against heavy numerical odds, the organization of supply, forced marches, sieges, and coaxing of difficult allies. Between 1805 and 1808, he occupied himself with politics, becoming Chief Secretary for Ireland. It was his political interest rather than his position in the army which secured him his next appointments. Following his success in the Peninsula in 1808, Wellington returned there in 1809 to command 21,000 men. The British army of this period differed in many ways from the French. Most notably it was not a national army but an old-style small professional army, of the type which the military theorists were then writing off as obsolete. The contrast between the careers of Wellington and Napoleon is characteristic. Officers' commissions in the British army were obtained by purchase and interest, and were normally available only to men of gentle birth." (Montgomery, 1983, p. 229)

The above quoted paragraph, from a noted historic book of a famous late top-level British military commander of the second World-War era, indeed do mirror some interesting but strange aspects concerning the functioning as well as few organisational-systems and policies there in the British army, working during even the end of the medieval-era Britain as a prevailing practice. But in fact, another astonishing aspect historically emerging here is that, the above narrated certain facts then prevailing in the British army appeared to be even somewhat satisfactorily working. Some may even feel temptingly call it, typically British as well. Anyway, that is another matter but in our wider contexts at this precise point, it perhaps appears to be historically exclusive and also no less strangely typical to the larger British societies and culture. While critically reviewing certain historic aspects concerning quite chequered times of the Wellington-era in this regard, we may notice that in spite of now in retrospect, looking very strangely weird system of granting commissions as officers in the British Army on payments then mostly to eligible young men belonging to upper-classes of the British societies, the prevailing health and morale of the existing military-command system among hierarchy of the officer-corps and subsequently in turn, down to the ranks and files appeared to be comparatively better, than those in the rival but no less also historic and battle-worthy French Army.

One glaring difference in this regard, has been then that, in the French army the system of command was mostly autocratic and hence tending to be at times even undesirably dictatorial as well as somewhat not so soldierly as well. This in a way may be called not so healthy a culture of the military command and control system and this in turn, did create certain irritating problems at times, even to a truly charismatic French military-leader and warrior, like Napoleon Bonaparte as well. And consequently, the French Marshals suffering from somewhat frustratingly low-morale and perhaps, forced to be active in the distant fronts, even tended to twist the war-orders of Napoleon too as it happened nearer home, when the Monarch was then militarily active and under heavy pressure, in the far-off Russian snow-bound territories in the winter. But with respect to the British, it certainly was not the case up to any such extent and also, the British officers tended to be more soldierly participative in the military command-system down the line and hence professionalism too, then prevailed in the British army up to a much greater degree. If we look into these aspects little more deeply in our broader perspectives at this point, then we may find that historically since already many centuries gone by the much larger socio-cultural aspects, geostrategic and politico-military traditions as well as the integrally structural framework of the British societies and their perhaps inherent psyche in this regard, have been distinctly different from those of the French societies.

Further while critically peeping through the multiple-windows of man's vast living-history, into these otherwise always both geostrategic and politico-militarily quite important and in addition, very far-reaching aspects with their due comprehensiveness, we would be coming across historically hidden but then also critically revealing factual aspects in this regard. Hence, as we further proceed along, in our this continuing strategically critical journey then, we would interestingly be able to perhaps find the principal reasons so as to why the British mostly prevailed over their main rival and certainly no less strong and resourceful the French, during their mutually competitive and prolonged struggles for building their respective overseas colonial-empires later in almost all the continents across the Globe. We would find these interesting historic aspects coming on the surface, during our further continuing discussions; in more relevant contexts, in the next chapter. In addition, if we may care to review these historically multiple-developments in their much larger perspectives of the man's living-history, then the broader outcome of the competitive rivalries between both the Britain and France in Europe, tended to further repeat during the multi-dimensionally competitive imperialism, in securing their own respective colonial-empires in the overseas continents as well. During this overseas empire-building phase of the living-history in addition to France and Britain, other medieval-age European imperial-powers, such as the Dutch, Portuguese and Spanish too, jumped in this fray. Further rather amusingly, with respect to the present modern-era France and Britain their otherwise age-old perhaps irreconcilable rivalries, tended to continue in their new incarnations even right up to the end of the very recently past cold-war era as well.

If we continue to critically review as per above then, we may notice that these historic aspects with respect to those two European arch-rivals, especially during those then unprecedentedly momentous times perforce took many mutually hostile dimensions as well. Interestingly in our larger perspectives in this regard, we may further clearly notice that those momentous years were in fact overlapping in multi-dimensionally extraordinary ways since both the ending medieval-era, along with the simultaneously beginning of present modern-era with their many consequent complexities. Then these now recently passed historic complexities, even tended to manifest their own distinct kinds of dimensions as well; since at the same time, another unique as well as unprecedented 'Industrial-Revolution' was also beginning with its so many previously unimaginable fall-outs as well. We may perhaps get inclined to say that even the man's multi-dimensional vast aspects of the living-history since continuously in the making, have in the process; its own strangely peculiar ways. Now coming back to the times of Wellington, that period's violently turmoil-ridden military-history does also inform us that many far-reaching positive developments too were then being triggered by few historically eminent military personalities including Willington himself, in both the organizational structuring as well as in the areas of military command and functional-reforms in the then British Army, which were no less weighty, beneficial and with enduring fall-outs too in many ways, for the entire British Nation as well. Military-history informs us that **General Moore** on taking over his command, was very imaginatively able to thoroughly overall and stream-line the multiple-aspects of military discipline through the army's ranks and files and in addition, the system of their comprehensively consistent training as well.

Another revolutionary change in the areas of military command as well as administration took place, when the **Duke of York** was made administrative head of the then British Army. Interestingly at this point, we may feel some necessity to emphasise that the multiple-aspects of military administration are largely different from those of the civil-administration, with some of their distinctly separate dimensions. Administrative aspects anywhere are always complex and these are no simple matters as some may tend to believe but in case of the military, the chief of administration is directly responsible for many additional areas such as streamlined management and control of the multiple aspects of logistics and many other things as well. **Duke of York was visionary enough to undertake many pioneering steps with focused aim, for establishing both military academy as well as staff-college for then the British Army.** It is now needless to say that both these Institutions are indeed the very foundational alma-mater, for looking after multiple aspects of training of officer-cadets and the budding commanders in any army respectively.

It is historically interesting to learn that like Napoleon, Wellington too, always wanted to go into all the important details concerning almost all the areas of military-command. Wellington with his inherent attributes along with his hard training and grooming as an outstanding military-commander, always remained fully aware of both the importance and complexities of focused training of his troops, aspects of the management of finance, supply and transport and many other dimensions of the logistics as well as other additional areas of military administration. In fact, Wellington knew that if any commander remains negligent with respect to the above aspects, is sure to invite disaster and hence destined to come to grief. Wellington was not only a talented military commander in both peace and war but, a great visionary as well. **He was also historically instrumental in first introducing and organising the corps of both sappers and miners and military-police, in the British army of his times.** The singularly turning point in Wellington's military-career was his historic success in the prolonged Peninsular war, then fought in the Spanish territories. Then his success in the battle of Waterloo, interestingly came as an' icing on the cake'. But, whenever we may look into the pages of military-history concerning Wellington, we do find that his good luck too played no small part at the most crucial of times. Hence, it has been rightly said and also proved repeatedly ahead that in addition to the multi-dimensional professional competence, a general has to be lucky as well for successfully facing the many imponderables of war.

All these aspects of unforeseen human-progress are no doubt true and in our overall perspectives here, these would befittingly be discussed in the forthcoming later chapters. But at this precise point of our continuing discussions concerning the bygone medieval-era and also in addition, as per our specific perspectives both geostrategically as well as politico-militarily, we do notice that this era's engaging history, was indeed no less chequered, in both socio-cultural and socio-political considerations as well. **Hence historically, with the gradual end of ancient-era and simultaneously the dawn of medieval-age; we may indeed see again a gradual spurt of the many kinds of historic kingdoms and the empire, all over the world.** This by itself is another unique historic-aspect of now the bygone medieval age. These very interesting phenomena have historically been quite prolonged and at the same time, in certain regions of the world they have continued right up to the time-frame when in fact medieval-era was ending and modern-age was gradually coming-in, along with its own many kinds of amazing dimensions and facets as well. In fact in our contexts at this precise point, the larger history additionally does show to us that historically, the establishment of the many kingdoms and empires in the medieval-era Europe did face many kinds of problems other than socio-cultural and politico-military in this regard.

In this respect, **Jawaharlal Nehru** has very informatively mentioned that, "**Kublai Khan** (*A great Mongol King of medieval China*) had sent a message to the Pope asking him to send 100 learned men to China. But Pope did no such thing. He was in a bad way at the time....**this was the period, after the death of Emperor Fredrick II, when there was no Emperor from 1250 to 1273. Central Europe was in a frightful condition then, and there was disorder, and robber knights plundering everywhere. Rudolph of Hapsburg became Emperor in 1273, but this did not improve matters much**. Italy was lost to the Empire. Not only there was political disorder, but there were the beginnings of what might be called religious disorder, from the point of view of the Roman Church. People were no longer so docile and obedient to the orders of the Church. They had begun to doubt, and **doubt is a dangerous thing in matters religious**.... More terrible than war was and massacre even, there came the **great plague to Europe about 1348**. It spread all over Europe from Russia and Asia Minor to England; it went to Egypt, northern Africa, Central Asia, and then spread westward. **It was called the Black Death** and it killed off people by the million. After a third of the population of England died, and in China and elsewhere the death-toll was stupendous. It is surprising that it did not come to India. This awful calamity reduced the population greatly and often there were not enough people to till the land. Owing to the lack of men, the wages of workers tended to rise from their miserable level. But the landlords and property owners the parliaments, and they passed laws to force the people to work at the old miserable wage and not to ask for more (*In fact, these and much more of different kinds indeed have been the saga of the ruthlessly cruel exploitation of the helplessly miserable peoples by their fellow human-beings since the advent of different human-civilisations in the world, and these as usual in turn, perpetually advancing in so many multiple ways. Even today in the fast transforming super high-tech modern age, we find different manifestations of same barbaric exploitation regardless.*). All over western Europe these peasant revolts took place one after the other." (Nehru, 1934, pp. 265 & 272)

While trying to review the historic coming-up of the above mentioned kingdoms and empires in various regions of the world, we may perhaps have to look back in to the certain historic events in a perspective preceding these imperial phenomena since beginning of the medieval-age. We may recall that in Europe, with the gradual demise of the once mighty Roman-empire almost whole of European-region indeed had a very trying time, facing lot of turmoil and ruthlessly bloody clashes of arms, indeed triggered by many kinds of barbarian invasions. And historically by the time, these upheavals were getting gradually settled, the medieval-era had already begun experiencing its multiply historic dawn. Invading barbarians were already carving-out their own territories and at the same time, they were settling-down in their new European homes and thereby they themselves were gradually getting assimilated in the mainstream societies and also as a consequential socio-cultural fall-out, got civilised as well. But indeed in the obviously concerned contexts at this historic point in the medieval Europe, in the geostrategic, socio-political as well as politico-military perspectives as a result, no power-vacuum and absence of any kind of the then systemic-governance could in any case remain for long. Hence, the historic phenomena of the process of establishment of many monarchies in Europe and thereby, establishment of their various kingdoms and empires in the different territories in that continent in any case had to get materialised there. In larger considerations, these aspects are in fact, universal in nature in similar socio-political and politico-military circumstances.

In fact, even before coming to Europe's medieval-era historic socio-political and politico-military complex phenomena, concerning the gradual establishment of various kingdoms and empires there, while further critically reviewing the medieval Europe's living-history, we notice that a couple of historically decisive wars too were fought between the rival Christian and aggressively invading and also territorially expanding Islamic forces. So as earlier too mentioned during our contextually ongoing discussions; we do find that these wars, besides their other inherent socio-cultural and politico-military causative-factors; were also ironically perforce made to acquire fanatically religious dimensions as well. This is paradoxically again another inherent socio-psychological aspect of the man's nature since the very beginning, and also we may recall that these too since already discussed in our different contexts earlier, and hence thereby man's multi-dimensional vast living-history have naturally been also acquiring similarly allied aspects as well. During those medieval-age historic centuries and as years were passing by, we may again recall that due to man's inherently relentless ingenuity and exploring multiple efforts, many new inventions and innovations too, were taking place. Interestingly as we have already seen that the multiple phenomena of these many kinds of new inventions and innovations, are again in fact integral segments of the much larger phenomenon of man's evolutionary process on this planet Earth since the very beginning.

In this specific context, in *Great Battles*, p. 99, it has been mentioned that, "Gunpowder and artillery revolutionized European warfare after its introduction during the fourteenth and fifteenth centuries and forced military commanders to develop fortresses and military tactics and strategies to deal with this new, devastatingly effective technology.... During the seventeenth and eighteenth centuries, muskets became less unwieldy, lighter and more accurate. By that time massed infantry armies made up of professional, regular troops fought each other in huge, packed series of lines, firing at each other with musket and cannon shot at close range. In this deadly, close-range fire-fights just a handful of musket volleys could decide the outcome of a battle."

The above quoted lines here now may appear to be rather primitive and also mindlessly devoid of even basic tenets of the tactical concepts regarding both the deployment of the bodies of troops as well as imaginative applications of armaments in the battlefields. But any such line of thinking and approach perhaps had their own rationalities in the past, in the applications in war. With the passing time and also with the advent of the unprecedented **'Industrial-Revolution'** did further revolutionize the entire multiply-complex aspects of the matters-military as well. These in fact, multi-dimensional revolutionary changes overwhelmingly included the areas of the armaments, military communications, mobility of military-forces and many other allied factors too, and consequently the aspects of warfare were never the same again. Among the above talked about historically decisive medieval-era battles having taken place in Europe, with far-reaching implications and also which did change the subsequent course of the living-history then in the making there **first the battle of Constantinople (5 April to 29 May 1453) and then the battle of Lepanto (07 October 1571)**, do come in our minds. In fact, the living military-history do clearly indicate that in this context, there are definitely few other battles too but for some due illustration at this precise point, we may look into these two historic battles to further our continuing discussions.

With respect to the Battle of Constantinople in *Great Battles* (page 101) it has been mentioned that, "The fall of Constantinople in 1453, its once impregnable walls beached by Ottoman siege guns, brought to an end the Byzantine Empire and established the Ottoman Empire, as the main threat to Christendom in the Mediterranean and Eastern Europe for the next two centuries. The Byzantine, once the bulwark of the Christian West against the Arabs and Turks, had been defeated by the Seljuk Turks at Manzikert in 1071 and in 1389 their descendants the Ottoman Turks defeated a Serb army at Kosovo Polje. By 1400 most of Asia Minor was in the hands of the Ottomans and it seemed only a matter of time before the once flourishing Imperial Capital of a million inhabitants would fall to them.... Turks used for the first time, massive siege artillery to blow a hole in the Theodosian Wall protecting the city. The city of Constantinople was the Imperial Byzantine Capital (303-1453) located on a peninsula by the Bosporus, the Sea of Marmara and facing the Golden Horn in what is now Turkey (*The main geostrategic and politico-military reason for this battle being*).... **The Ottoman Turks had expanded their empire in every direction and wished to make Constantinople their capital and eliminate the Christian stronghold behind their lines. The city was taken with great bloodshed, and became the capital of the Ottoman Empire."**

But the Christian warriors and their able commanders, to their great credit, were certainly not the type to sit idle and just sulk over their earlier above mentioned bashing in the hands of the Ottoman Turks. Very soon they were able to militarily prepare with great comprehension and were in a position, to give the Turks, a befitting reply with the desired military-decisiveness. And in this precise context, the living military-history shows that the D-day, for this historic military-decision through an imaginatively planned offensive action in and around the Mediterranean and, in favour of the Western Christian Powers; soon arrived. And this resulted in the already earlier mentioned **historic battle of Lepanto**, which was successfully fought by the Christian armies as well as well-trained naval fleets, with great operational co-ordination and wartime military vision. In the above cited and many more historic battles, having been fought from time to time both in the vast continental Europe as well as in Asia, **the principal geostrategic and politico-military reasons-cum-driving-forces behind these extremely violent and repeated clashes of arms again and again, in various historic battlefields have been the irresistible geo-political urges for the establishment of expanding empires, by the multiply competing Mohammedan and then European Powers, then so active in the world**.

And now coming back to the **historic battle of Lepanto, fought on 7th October 1571**, it is written that, "In the last great Mediterranean galley battle, the combined fleet of the Western Christian powers decisively defeated the Turkish fleet thanks to superior gunnery and better armed fighting men. The Turks' losses were so great that their naval power never really recovered.... **Philip II** of Spain and **Pope Pius V** founded the Holy Empire to fight the Ottoman Turks under **Sultan Salim II**. The League's fleet was commanded by John Juan of Austria (1545-76) while the Turks were led by Ali Pasha (d. 1571). Lepanto was the greatest Christian naval victory against the Muslims and the last sea battle fought exclusively with oared galleys. *It was fought* in the gulf of

Patras on the western coast of Greece, taking its name from the Port of Lepanto (Naupaktos). The Ottoman Turkish Empire posed a deadly threat to Italy and the rest of Europe. Christian gunfire smashed the Ottoman fleet, breaking Turkish naval power in the Mediterranean." (Christer Jorgensen, 2011, p. 109)

Hence like the above mentioned a couple of battles for the imperial supremacy of one Power against the other and then, for the expansion of one's empire, into the new territories and also as mentioned earlier; many more battles were consequently fought between the warring Powers. As it is commonly said that establishment of one's empire anywhere and, it is so even now is never easy and peaceful at all. We may interestingly recall and as already delved upon in some detail in different but certainly in quite related contexts, in the first three chapters that as man was progressively getting evolved and also, began getting civilised over various periods of time, during both the pre-historic as well as remotely bygone ancient historic ages, he had since begun to slowly live in a kind of primitive settlements and, by clustering in loosely formed different groups and then, later even as since simultaneously evolving early and different tribes as well. If we may care to critically look much deeper into these aspects, particularly in case of the man then we notice that as man got gradually more civilised and also organised in different groups as well as historically evolving tribes and naturally, as very primitive to begin with they also commenced to acquire their own distinct group-identities and group-dynamics being guided by their own group-interests and urges to fulfill their many newer desires/wishes and so on. Further, we may well even ask as to what all are the causative-factors towards in fact, multiply giving births to different types of these so-called necessities?

Well in fact, the causative-factors in this regard, are also phenomenally linked to the same highly complex processes of continuing evolution of the Supreme Mother-nature, precisely with respect to the man, in our context over here. Hence presently at this point it appears that as man got progressively evolved and both consequently and simultaneously, different ancient human-civilisations too got gradually evolved and began multi-dimensionally advancing as well, then at the same time many historically far-reaching developments also, commenced in the lives of men.

As we move further with our continuing 'strategically critical journey' within our pre-defined contextually specific perspectives, especially with respect to the medieval Europe at this precise point, we do notice that during the fourteenth century there indeed were very violent conflicts involving all sorts of players-cum-stake-holders, within the highly complex historic processes leading towards formation or further consolidation of the leading medieval-era European States. In these historically complex processes, we further observe that the culture of powerfully entrenched **'feudalism'** happened to be the principal driving-force in all sorts of the typically intriguing and also sometimes counter-productive battle line-ups; and then even at times, with irrationally fixated politico-strategic calculations or the operational-planning by various fiercely contending Parties.

In the above precise context with substantial relevance, **Richard Overy** has written that, "The 14th century was a tumultuous period of state building in which rulers sought to define frontiers and develop centralized bureaucracies. They were challenged by parliaments of nobles, knights and townsmen, seeking concessions in return for taxation. Furthermore, the cost of frontier wars forced rulers back into the arms of their subjects. War added to the turmoil caused to 14th century Europe by famine and plague.... As regional despots gained power over the cities, conflicts within their walls died, culminating in the conquests of large areas of Lombardy and Tuscany by Giangaleazzo Visconti, duke of Milan. Few Italian republics stood firm against the trend, though

Florence made much capital of its self-proclaimed dedication to the cause of republican liberty. The Pope himself decamped in 1305 to Avignon in southern France, a more peaceful city than Rome. Papal centralization was visible was visible in the tighter control over Church appointments and taxation, successes that generated bitter criticism from reformers such as Wyclif in Oxford and Hus in Prague. The return of the Papacy to Rome in 1376 led to deep divisions among the cardinals and the outbreak of a schism that lasted nearly 40 years, The conflicts between France and England for control of France, above all Gascony, erupted again in 1294 and then in 1337 (the Hundred Years' War). Initial English successes coupled with internal strife gravely weakened France and great princes such as **Valois dukes of Burgundy** exploited this royal debility to create powerful state-lets of their own." (Overy, pp. 140 & 141)

Now coming back to the aspects of establishing kingdoms and building empires in the medieval-age Europe with human sweat, grit and blood we find that ironically, the entire history of these happenings over there has been extremely violence-ridden and that too, in already the then war-ravaged Europe. **Well, historically in fact medieval England took the lead in Europe, as we may see that nearly during 870 A.D. Alfred the Great first established the first British kingdom there.** He was a visionary in his own right and he succeeded in giving England, a sort of codified-laws for the royal governance of his kingdom. **In all these historic processes, whether in Europe or anywhere then the highly complex matters-military along with their arms, aims and aspects too, decisively played no small-role, and that too multiply in a parallel manner in this regard**. It has always been historically true that establishment of Monarchies and thereby, simultaneous conquests of as much of territories as deemed politico-militarily feasible for setting-up the requisite kingdoms with their demarcated and also both strategically and militarily secured borders; perforce meant conflicts of competing claims-cum-interests and hence, nothing but wars between the military forces of rival Monarchs. Indeed very relevantly in this regard; **James I. Stuart** has said that, "It is atheism and blasphemy to dispute what God can do; ...so it is presumption and high contempt in a subject to dispute what a king can do, or say that a king cannot do this or that." Well, quite interesting, isn't it? But in the perspectives of historic objectivity, we may even maintain that it remains always true.

In fact, the above mentioned aspects of conflicts of various kinds of claims-cum-interests and hence, the consequent wars between the rivals have never been limited or confined to any specific age or to any one particular continent. The various causative-factors in their roots, along with their textures and different kinds of manifestations may have remained mostly varied both in their forms and contents, but these varied conflicts and wars are continuing during even the present high-tech modern-era as well and that too, due to different real or perceived reasons. Interestingly, in today's multiply accelerating and fast-moving high-tech era modern world and along with in the present scenario of the relentless games of international power-politics and with their own mind-boggling complexities, coupled with the all-round prevailing insincerities as well as lack of trusts among nations and then the emerging and altering global geostrategic and politico-militarily complex situations as well, political-leaders belonging to different nations and socio-political ideologies on any opportunity before the media, even begin to talk about need and desire to solve various international tension-generating problems. Though at times as few exceptions, such efforts have somewhat succeeded too, but those are different matters and mainly due to different kinds of compulsive reasons as well. We would obviously be discussing these aspects in more details in later chapters.

But so far as the bygone medieval-era is concerned, as in Europe and so in the then other continents as well, man's living-history shows that, any kind of solution to any disputes among various kingdoms/empires; were mostly solved by means of resorting to the arms. Then during some points of that living-history perpetually in the making, there have been very welcome periods of peace and tranquility, though sometime short or long as well, in some such kingdoms/empires; and this in turn, resulted in the periods of very encouraging as well as beneficial times of all round civilisational progress, which naturally did enrich the then fortunate (though mostly temporary too.) peoples concerned of the societies, with those kingdom/empires. Any such desirable and historical phenomena in this regard happen to be much part of the same living-history. In our ongoing discussion certain dominating aspects of the man's inherent nature, in various ways get manifested rather collectively in both behavioural and functional domains of man's socio-political and also imperial entities as well, such as in so many kingdoms/empires mushroomed all over, since the very ancient era onward, in the whole of Asia, Europe and all other continents of the world.

Hence in general, since the man mostly comes to peaceful terms regarding any kind of dispute with any other person and in the similar manner, history further very clearly shows to us that, any particular kingdom/empire too, mostly agrees to any otherwise very much desirable and highly valuable peaceful resolution of any specific dispute with any other rival kingdom/empire mostly, if only compelled to do so by certain unavoidable circumstances. Certain kinds of wishful thinking apart, the undeniable ground-level hard facts remain that if the above mentioned any such dominating compulsions are not present then ironically, man mostly tends to fight in some way or other to settle any dispute. In today's rather snobbishly so called 'civilised-world' a person may not otherwise agree to these perhaps strangely sounding perceptions but historically, certain irrefutable realities concerning the mankind; just cannot be wished away. And historically again, we do also find largely the identical some kind of irrefutable scenario with respect to the historic kingdoms/empires as well in any particular societies and in any part of the world. In fact, interestingly this is one principal reason as to why the man's vast living-history since the very beginning in this regard has been so chequered in the making, and at the same time always soaked in human sweat, grit and blood.

Now again coming back to the aspects of medieval-age Europe's monarchies and to begin with, concerning England, Professor S.R. Sharma has written that, "England on account of her insular position, developed faster and outgrew the need for monarchial rule earlier that most other countries. Her earliest king to whom we need refer here was **Alfred the Great (871-901 A.D.)**. His memory is still cherished among the greatest of that country. He is rightly regarded as the creator and saviour of England and figures well in history as well as literature. An inscription on his statue at Wantage beautifully sums up his great qualities and services thus: **'Alfred found learning dead, and restored it. Education neglected, and he revived it. The laws powerless and he gave them force. The Church debased, and he raised it. Alfred's name will live as long as mankind will repeat the past.'**.... Next after Alfred the most memorable monarch of England was William the Conqueror (1066-87). Though a Norman, and ruler as much of Normandy as of England (after his victory over Harold at Senlac or Hastings, 1066), he left a permanent mark over English history or institutions." (Sharma, 1948, p. 332)

Hence those were the phenomenal and rather violently turmoil-ridden methods, through which the various medieval-era European Monarchies and then historically to begin with in case of England, were born. In this interesting context as we further continue with our critical review, the above highlighted historically very far-reaching civilisational phenomena were mostly materializing through relentlessly resorting to a series of the clashes of arms. We may also recall at this precise point that, in the initial chapters only we have noted that the man's living-history since continuously in the making, has mostly been the highly chequered 'history of warfare'. We would be further visualizing that as man's living-history interestingly, marches-on and enters into the present modern-era identically the similar historic phenomena in various altered contexts keep on repeating all over the world.

Perhaps no segment of the humanity in any corner of the world; has ever remained unaffected by these all-encompassing and human blood-ridden sweeping civilisational phenomena in fact, at any point of the man's living-history. In other words and in the larger perspectives, we may also say that the same **living-history of the making and in its overall process perhaps always keeps on ruthlessly extracting, a kind of its own rather consequential prices; in terms of sweat, grit, human-blood, sufferings and thereby, all sorts of violent wars and human miseries as well.** These appear to be very weighty and ironical aspects of the man's living-history and continue to decisively influence the principal courses of that living-history. Now coming back to the historic aspects of **medieval-era Monarchy in France**, **Professor S.R. Sharma** has written that, "French history, as distinct from that of the Western Franks, began only with the accession of **Hugh Capet** (968 A.D.) whose dynasty continued to rule France until it was replaced by the **House of Valois** in the first quarter of fourteenth century. Even during that period the Capetians shared their dominion with the English, as the Angevin Empire in France lasted from 1154-1204. Besides, the country was rent with feudal strife until the accession of **Louis VI** (1108-37), the fifth monarch of the line, who at least partially got the situation in hand. **Henry II** of England has possessed more of French territory than the French monarch, but under **John** (of England) and **Philippe II** (of France) the position was completely changed. England lost all except Guienne and Channel islands; and, 1216, **Louis, the son of Philippe**, landed on English soil by invitation of the English barons, to displace John (*Interesting, isn't it? Well historically, those were the times between the rivalry-ridden France and England in the medieval-era Europe. Though quite fascinatingly, today these may perhaps sound like a ping-pong match.*). Philippe was the main founder of the French monarchy.... **Louis XIV** was indeed, the proto-type of Grand Monarchy. He supplied the model which was copied by all later rulers, each according to his genius and capacity.... Louis XIV also decorated his court with poets, playwrights, philosophers and scientific men. **Boileau** laid down the canons of style; **Corneille** gave French drama its rhetorical and classical form; and **Racine**, its final perfection and polish. The popular **Moliere** (1622-73) wrote his incomparable comedies and **La Fontaine** his simple and satirical fables on the foibles of society. **Voltaire** called the age Louis XIV 'the most enlightened age the world has ever seen;'....it gave to French culture a stamp and prestige which was to survive the loss of French political ascendancy, and even the downfall of Grand Monarchy itself. But there was another side to this picture. Louis XIV revived religious intolerance in France by his revocation of the edict of Nantes.... 'Under his rule,' writes **Wells**, 'were carried out the dragonnades,' a peculiarly malignant and effectual form of persecution. Rough soldiers were quartered in the houses of the protestants and were free to disorder the life of their hosts and insult their woman-kind as they thought fit (*Too bad and, it was definitely demonic indeed.*). The worst legacy of Louis XIV was however, a legacy of wars; ruinous to France and ruinous to Europe and the world, though initially it looked like success. His reign opened with the French acquisition of Alsace, as a result of the ***Treaty of Westphalia (1648)*** which ended the thirty years war." (Sharma, 1948, pp. 336-339)

In the above pages, we have briefly since delved upon the historically engaging aspects of the struggles for the establishment, consolidation and any possible expansion by the various Monarchies; principally concerning their respective empires precisely, in the western part of the medieval-era

Europe where in fact, the leading historic countries have been England and France. European living-history concerning those chequered times in this regard; clearly indicates to us that **principally it were England and France, mainly driven by their ruthlessly competitive ambitions, sometime so-called perhaps understandable weightage of their necessities but, largely by their own respective greed and then also historically, no less coupled with their rather unchecked and arrogant imperial desires to territorially expand their respective kingdoms/empires**.

And in addition thereby, while continuing with our critical review of the medieval-age European living-history; we do also clearly observe that the same history was also rather ironically being written with human-blood, depicting a saga of a series of the clashes of arms in the various battlefields in Europe then and hence consequently; no holds barred stories of the blood-shed, soldierly chivalries, glories of the concerned clashing empires, obsessively imperial leadership of their rulers and, along with the epoch-making heroic-deeds as well as remarkable military-vision of their respective commanders both in the times of peace and war. But at the same time in these larger perspectives interestingly; we do also notice that historically, a rather distinctly magnificent saga of the medieval-age Europe's remarkably creative socio-cultural-cum-intellectual civilisational advancements too were being written and in the process; multiply enriching the entire humanity as well. Hence, in the above quoted paragraph from Professor S.R. Sharma's book, giving some historically interesting details of certain medieval-age French Kings, we do also find some aspects of the man's civilisational advancements in the then French empire. Man's living-history in this regard, also shows similar aspects with respect to the England, Germany and other European nations as well.

Such living-history of the medieval-age kingdoms/empires, their royal dynasties and so on were not confined to the Western Europe only but in addition, the similar living-history also got manifested in the Eastern Europe and Scandinavia. Historically in fact, the different forms, external and socio-cultural-cum-politico-military textures naturally, might have been somewhat different from those of the Western European kingdoms/empires, but the inherent medieval-era monarchial ambitions, greed and various kinds of imperial-cum-strategic contents in this regard, were largely identical and thereby in these perspectives the contextual processes had to remain the same and hence, geostrategically predictable. In other words, in the contexts of our ongoing specific reviews with the necessary strategic-criticality in this regard, we may also say that during the medieval-age of military chivalry as well as the armed-clashes largely remaining as the pivotal-policy for statecraft the consequential living-history of the then Eastern Europe too, had to be an equally human blood-soaked saga of a series of historically feudalism-oriented clashes of arms between the rival monarchs and their armies.

Now in this context, where 9th and 10th century Scandinavia is concerned, Richard Overy (page 100) has mentioned that, "In Scandinavia itself, powerful kingdoms began to emerge out of the disarray of the Viking-age. Most notable was the emergence of Denmark under three kings, **Gorm, Harold and Sven**. By the 10th century Denmark had become a powerful kingdom, and, under **Cnut** (1014-15), the centre of an Anglo-Scandinavian empire. Much of the southern Norway was united under the rule of **Harold Finehair** after the battle of Hafrsfjord in the 890s, but after his death in the 930s Norway occasionally came under Danish lordship. Sweden meanwhile was united only at the end of the 11th century under the kings of Uppland. On the fringes of the Scandinavian world Iceland was, by 930, and independent commonwealth (without a king) while Greenland was colonized by Norwegians from about 985."

In these interesting contexts of the medieval-era rather whole of Europe, as we proceed further, we do also historically notice that the 12th and 13th centuries respectively, have been perhaps extraordinarily epoch-making. We may recall that little earlier, we have since briefly delved upon these aspects with respect to the Western Europe where then the centralized bureaucracies, parliaments of nobles, chivalrous knights, Papacy and townsmen and so on, played their own epoch-making historic roles. These then had many kinds of fallouts, concerning both fighting of wars and some initiatives having been taken for peace-making as well. Similarly in these historic contexts, we do interesting notice that somewhat identical living-history then, also got shaped during those centuries in Eastern Europe as well. Historically in this regard, any similarities/dissimilarities apart but it may be somewhat erroneous on our part if at this precise point, in the course of these continuing critical reviews in our larger perspectives; we tend to view these aspects with respect to both the Western and Eastern areas of medieval-era Europe respectively, in a kind of water-tight separate compartments since, both had mostly been interrelated as well as integral.

In addition then in the above perspectives, we do also historically observe that during those epoch-making medieval times in the Eastern Europe relatively much bigger and well-established as well as well-consolidated princely States got fructified. Here too, in the larger geostrategic and politico-military contexts, systems of governance and then along with this, the methodologies for policy-formulation as well as the practices of the matters-military were both imperial-cum-feudal in contents and character. At the same times then, as per the prevailing medieval-era oriented traditions in this regard, **certain powerful politico-religious factors** too, had naturally come in the above mentioned aspects of overall multiple systems of governance in those Eastern European States. In these perspectives we have since mentioned above that both Western and Eastern European States/Empires were not functioning in a sort of distinct 'water-tight' compartment and hence, they were integral and historically overlapping and, mutually influencing as well and accordingly the same have been somewhat substantiated here. As already put earlier in different contexts and also we are already aware of the facts that, these precisely are the ways for the mankind's 'living-history' continuously in the making onto its 'expanding civilisational canvas', since the very beginning itself.

Precisely in somewhat larger contexts with respect to the above; we cite a historic battle of Tannenberg, having been fought on 15 July 1410. Regarding this battle with far-reaching historic fallouts in the medieval-era Eastern Europe in *Great battles*, it has been mentioned that, "The decisive battle of Tannenberg, between the army of the Teutonic Knights and the Royal Army of Poland-Lithuania, brought to a halt the eastward expansion of the hitherto invincible military order and secured the independence of Poland-Lithuania.... Founded in Acre in 1190 to defend the holy Land, the Teutonic Order was a military order of warrior-monks whose military and religious reputation could be equated with the Knights of **St. John** or the **Templars**. But unlike their rivals the Teutonic Order moved the seat of their activities and crusades to the southern and eastern coasts of the Baltic.... The army of the Teutonic Order under their Grand Master **Ulrich von Jungingen** (d. 1410) fought the Royal Army of Poland-Lithuania under **King Wladislaw II Jageillo** (1350-1414). The Royal Army of Poland-Lithuania with Tartar (Mongol) and Russian auxiliaries annihilated the more experienced and heavily armed Teutonic army. (*The place of the battle was*) on a shallow grass plain between the East Prussian (Mazurian) villages of Tannenberg and Grunwald. The Royal Army was caught by surprise when the whole Teutonic army bore down on them before they were prepared but the Teutons squandered a great opportunity to secure victory by not attacking first. The Polish-Lithuanian victory halted the Teutonic Order's eastward expansion and effectively broke its power." (Christer Jorgensen, 2011, p. 83)

In fact during those momentous times, Hungary became a dominant player around her considerable territories including Poland as well. But with politico-military intentions to conclusively get hold of the Polish throne, she was not completely successful though the then Lithuania had succeeded in taking almost full control of the territories between the Black sea and then, right up to the Baltic. As in the medieval Western Europe and so in the eastern kingdoms too, persisting religious dissent had become an influencing factor in these imperial power-games; among the main contending players. And then in such prevailing situations interestingly; perhaps the multiple-aspects of almost the all-powerful **'feudalism'** were historically, always active in many highly visible as well as in hidden manner; in deciding the subsequent course of man's living-history over there.

Hence, those aspects of the feudalism and its multiple manifestations were the same and also, similarly powerful in the chequered living-history of the Eastern Europe as well. Further the man's living-history shows us that, during the bygone medieval-era feudalism happened to be at its peak, in its concepts regarding socio-cultural ethos and thereby many different kinds of regressive customs and traditions as well as practices were followed all over the world. **If one may critically look into the medieval-age living-history of Russia, South-East Asia, China, India and even in countries of the Far-East one would find the identical phenomena concerning feudalism with its all aspects, having got manifested in all these adjacent and far flung areas, though in apparently different forms and in addition then also in bizarre socio-cultural systemic fixations. Interestingly in these precise contexts, we do also find that with the passing of historic medieval-age it was not that those multiple-aspects of the age-old feudalism became things of the past but in fact, feudalism in all its no doubt, altering manifestations happened to be no less powerfully lingering and hence, passed on into the incoming and presently ongoing modern-era as well**. In our larger contexts, we would more appropriately be discussing these aspects in subsequent chapters as well.

Now in our these precise contexts, coming back to the Eastern Europe as we proceed further, we do notice that both geostrategically and politico-militarily during the medieval-age, the countries over there got spared from certain very unsettling economic situations. In these respects they happened to be more fortunate and perhaps one reason could be that, though a number of wars were fought in those East-European countries but to their relatively good fortune, they most likely did not land-up in situations of getting devastatingly war-ridden or so. Hence, the rulers of these countries were able to draw their firmed-up international boundaries and interestingly these relatively more secured borders did endure right into the recently passed 20th century. Now, coming to the aspects of some of the multiply vast subjects of the matters-military, we do notice that not much time had in fact passed, since the advent of fire-arms as important weapons-systems for fighting wars, that these had already reached Eastern Europe as well. And subsequently cannons too, came on the scene. But during those times in the Eastern Europe too, cavalry remained the main arm in the battlefields and interestingly as in the West, it continued to remain so, even in the face of those initial fire-arms as well. In fact, in front of the pike-carrying charging cavalry columns, the infantry continued to face difficult times till now may be the simple-looking but during those times, certainly revolutionary discovery of the bayonet and thereby the commanding dominance of the cavalry-arm certainly began to decline.

As we proceed further and then, we do also notice that during the much earlier medieval-era times, **Mongols** indulged in the invasions of Russian territories as well. And then these invasions of the 8th-9th centuries, did in fact, have far-reaching impacts on the making of subsequent history of medieval Russia along with also formation of the first historic State in the Kievan Russia. Those were historically the highly momentous times in the early medieval-period Russia as such and interestingly orthodox Christianity too, had begun playing active roles in these geostrategic-cum-politico-military games of since entrenched culture of feudalism-based struggles for power as well as territorial supremacy. In fact, Christianity came to Russia earlier itself and, through interactions with the Byzantium Empire. During those years, Russia was yet to firmly come-up secured with its own powerful and vast Russian-empire with, both imperially as well as politico-militarily strong, viable and enduring foundations. In the larger geostrategic contexts, we find that a number of river-systems in Russia between the Baltic and the Black sea provided the Nature's valuable gifts for using them as waterways, both for the trade-cum-transit as well as for the military movements. In politico-military contexts, these water-transport facilities acted for the Russians as double-edged weapon. When rivers were not frozen, it could act as some obstacles for any advancing enemy columns but when the rivers were frozen hard consequent to the proverbial Russian winter, any accustomed enemy could walk or ride over it without much difficulty for conquering additional adjacent territories.

And hence, as per the consequential ongoing processes of man's living-history in the making; Russian territories were subjected to few foreign invasions, such as that of the galloping and conquering medieval-era Mongolian cavalry. Savage Mongols captured large chunk of the Russian territory while advancing and in turn, they were able to sort of also balkanize Russia into a number of weaker, battered and subjugated Princely States. Obviously in our larger perspectives, we do also observe that these developments in the medieval Russia were consequent to the blood-stained clashes of arms and hence, were never peaceful. Then as we move we do notice that during the latter half of the seventeenth century, **Peter the Great** was in fact able to lay a relatively much stronger and secured foundations for the modern Russian-Empire. We would be discussing these historic aspects concerning Russia in our precise contexts, in the next chapter. During the early medieval-era in the 8th and 9th centuries a large part of the Russian territories specially the Kievan Russia, were subjugated by the **Vikings**, i.e. historically known as the **Rus** here and the man's living-history further indicates that under those Rus in fact, the first Russian State was established and somewhat politico-militarily firmed-up for quite some time.

This historic Viking supremacy over these areas, resulted in some expansion of the territories of their Kievan State and also Rus were able to develop many kinds of multi-purpose communication-systems. They could develop a very advantageous river-based water-transport system connecting the Gulf of Finland, through the river Neva and the river Volkhov, and thereby, across through the Dnieper and then, further across the Black Sea and then on to the Byzantium. All these enabled the Viking Rus to historically expand towards southern region and hence right during the year 882 A.D. itself; **Kiev was made the Capital**. And consequently all these combined together, accelerated the fast and multiple development of Kiev as strategically, a very important city there right during those early times itself. But after some centuries, when the challenging time came and during 1237-8 A.D. tragically this early State of Kievan Russia couldn't stand the onslaughts of invading Mongols.

These **attacking Mongolian columns** chose such period for their invasion, when the rivers happened to be frozen and this tactical advantage substantially helped them to conveniently cross over the Volga and drive deep. On the way ruthless Mongols led by their commander **Khan Batu** did not even pause while destroying a number of relatively wealthy towns, including the famous Capital city of Kiev. Tragically, all these were great trying times for the Rus since the clock was turned back for them, for quite a considerable period of time. Now at this point of our ongoing discussions, it may appear to bring-in few thoughts concerning the persistent and repeated Mongolian historic invasions in various parts of the vast continents of Asia and Europe and so on and in addition, their consequent devastating after-effects. In our geostrategic and politico-military perspectives at this precise point, we may say that during those medieval-age times, these masquerading Mongolian herds were quite savage, since they did not ever believe in militarily clean and soldierly conquest but, in devastating the areas with ruthless destruction wherever they were able to step on.

In fact, in the above narrated historically ironical, civilisational negation but at the same time, with their many remarkable military-conquests of the then **Mongols; John Keay** has written that, "Seldom can more mismatched adversaries have squared up to one another. The **Song** (*a ruling Dynasty, earlier in China*), with a massive army but a wretched military record, had been retreating, on and off, for nearly three hundred years; the Mongols, though fewer and occasionally repelled, had been advancing for fifty years, had never lost a war, and had conquered most of the known world.... According to both Christian and Muslim sources, bleached bones and charred timbers marked their trail; the slaughter of those who offered resistance was generally as comprehensive as steel and muscle could manage. Urbanisation of much of Asia was halted in its tracks, fragile agricultural systems never recovered, and population figures are thought to have nosedived. **Like the plague, which they may unintentionally have spread, the Mongols came from nowhere and were suddenly everywhere**. The unity of the peoples of Mongolian steppe, no less than the authority of their new ruler, depended on maintaining the momentum of success. Only by more raids, more proof of invincibility and more redistributed plunder—livestock, gold, furs, textiles, foodstuffs, craftsmen, presentable women and enslaveable children—could loyalty be rewarded and allegiance of others attracted." (Keay, 2008, pp. 350 & 351) (*Well, while looking into these perhaps demonic sides of Mongolian history of the medieval-era, we may find them as great civilisational negations as well as bizarre, but historically in our contexts facts remain facts. Need we say more in this regard?*)

But historically in this context, when we critically delve deeper in these aspects then to our surprise; we do also find rather some positive/mixed sides too in this regard. **Richard Overy** has very informatively mentioned that, "**The Mongol pastoral nomads from the depths of Asia, achieved conquests of unrivalled range, from the eastern frontiers of Germany to Korea and from the Arctic Ocean to Turkey and the Persian Gulf, though their invasion of Japan and Java failed. Whole peoples were uprooted and dispersed, permanently changing the entire character of many regions.... The Mongol conquerors came into contact with three main religions: Buddhism, Islam and Christianity. The Mongol elites were shamanist, although many Turco-Mongols adhered nominally to other faiths. However, they were attracted by the high culture associated with the great religions. Many of their administrators were Muslims, Buddhists and Christians, and the Mongol courts had become truly multi-cultural, multi-ethnic and multi-religious. The Mongols universally exercised religious tolerance, although many have assumed that they were merely hedging their bets with the Almighty**. In 1258 the Islamic capital, Baghdad, were captured and plundered, the caliph slain and massacres perpetrated, but the operation was overseen by Muslim advisors, carried out with the assistance of local Muslim rulers and their soldiers and supported by the Caliph's Muslim subjects, Kurds and Shi'ites. Even though the

Mongols had sometimes promoted Nestorian Christianity and Buddhism in Persia, 1295 the Mongol ruler, Ghazan Khan, tactically declared his state nominally Muslim. In China meanwhile, the Yuan Mongols had endorsed Buddhism, thereby strengthening the position of that faith in the Far East.... The Mongol conquests were achieved through a mixture of bold strategy, iron discipline, brilliant propaganda, charismatic leadership and the disorganization and ineptitude of their adversaries. **The Mongol armies were directed by able generals and proficient administrators, who eventually began to believe their own propaganda: that they had a mandate from heaven to rule the world. Within a couple of generations the Mongol elite had transformed themselves from rude and coarse nomadic steppe warriors to cultured and sophisticated statesmen commanding the respect and obedience of their cultured subjects. Where once the appearance of their tents and armies stabbed fear and horror into whole cities and provinces, their courts later became magnets for seekers of adventure, glory, culture and wealth from around the world. Genghis Khan terrified the world with his cruel and plundering horsemen, but he gave birth to a legacy of splendour**." (Overy, 2008, pp. 121-124)

Now, with the above historically informative quotations in this regard, leaving the Mongols' historic and global military exploits and also their so-called sweet and sour multiple-legacies, we may prefer to again have some more look into the multiple-aspects of medieval-era some additional European countries and empires. We may also recall that earlier in these perspectives, we have since discussed some historic details concerning the British and the French; and their turmoil-ridden history of establishing their kingdoms/empires, the leadership aspects of their monarchs, statesmen and the military-commanders in their respective challenging domain; as well as their some noted civilisational glory and accomplishments as well. We may again note that at this stage, we have since touched-upon these historically interesting aspects: with respect to the British, French, Dutch and Spanish and so on, in our larger perspectives in fact and prior to our coming to their own geostrategic and politico-military colonial adventures overseas, and in the process along with their inevitable rivalries among themselves. Hence, while looking into the aspects of the medieval-era German empire interestingly we may notice that a kind of prominently parallel institution and legacy of Papacy, along with the Throne of Frankish kingdoms too playing then no less significant a role. Pope and his Papal institutions, in their own perspectives and manner in these historic kingdoms in the medieval German territories were strategically quite focused regarding their prerogatives and authorities vis-à-vis the royal thrones. In fact historically while looking into the concerned socio-cultural aspects with respect to the medieval-era Europe we may also notice that the Institution of Pope by itself was very important, since the Papacy had many imperially prominent functions to perform. And hence naturally, the Papal institutions were quite authoritatively conscious of this fact and their roles as well; since they also were not at all willing to entertain any sort of interference in this regard. Earlier with respect to England, France and the Italy, and in different contexts; we have since touched upon these aspects.

The German emperors of those times were also conscious of their own royal obligations and priorities regarding ruling over their large territories in the medieval-age Europe. If we may critically look into these historic aspects even deeper, we may also clearly notice that in the medieval-Europe; there was a well-established socio-religious tradition that Pope or the concerned Bishop duly selected and appointed by him in fact; had the Church-sponsored rightful authority to award the imperial title and also, to ritually consecrate the kings as well. It may perhaps appear important to bear in mind that various Frankish kingdoms were established and consolidated historically, from 840 to 843 A.D., and subsequently among these kingdoms fortunately the history has not been of so many violent clashes of arms and even at times, rather irrational battles as in the further west. Obviously, there were differences among those Frankish Monarchies but these were seldom driven to the extreme, and at the same time these ruling monarchies were also bound together by numerous socio-cultural, religious as well as kinship connections.

But at the same time, aspects of feudalism as since already brought-in in the different contexts earlier were powerfully present and thereby, the royal governance in these Frankish kingdoms were personalized and highly centralized as well. In this precise context interestingly; one distinctly far-reaching rather historically positive aspect happened to be that, the Frankish kings and their royal-families were themselves dedicatedly faithful and devoted to the Church. This naturally had many reassuring fallouts for the Papacy, their monastic houses and consistent missionary zeal and initiatives in far-flung regions. Now in our larger perspectives at this precise point, so far as the **vast areas of the matters-military** as well as other imperial dimensions regarding the **Far East** are concerned and especially pertaining to **Japan** and **Korea**; we would perhaps more befittingly be discussing in the next chapter.

*

CHAPTER 6

Medieval Ages, Birth of Empires and Dawn of Democracies in the West

Before we come to the historically epoch-making phenomena of since both geostrategically and politico-militarily rivalry-driven and leading imperial European Powers' birth of overseas Empires; as well as the dawn of democracies in the West in fact, we have to briefly visit certain important and prolonged events preceding them. In other words, during the much earlier centuries of the medieval-era; it may appear to be quite befitting to briefly delve upon the earlier geneses of these subsequent and highly engaging processes regarding fruition of those European overseas empires, as consequent to the man's living-history and the politico-military multiple-initiatives having been taken by those leading European Powers, for establishing their respective colonial-empires in almost all the continents of this world.

Hence interestingly in these precise contexts, we may first prefer to look into the **historic aspects of much earlier vast and intercontinental empire-making ruthless zeal on the part of the Mongolian monarchs/military-chiefs and their 'spine-chilling' armies during the 12th/13th centuries onwards**. To begin with at the first step itself, this specific suggestion may understandably appear to be rather strange but in our precise perspectives, if we may critically look deeper into these aspects and then, we may quite historically see some connectivity in this regard. And in the process as we begin to proceed further, we may also notice that in certain ways; **the age-old feudalism** too, at times got manifested quite differently among their rulers and the peoples, in comparison to the others. **The facts remain that in those earlier times, since naturally the Mongols lacked skills for sea-faring and so on; and in spite of it, they could establish their vast intercontinental empire, mainly through land-routes.**

We may recall that during our earlier discussions and also in different contexts, we have since talked about the great but highly ruthless military conqueror and Mongolian empire-builder named, **Changiz Khan**. It was during the first half of the 13th century, that Changiz Khan had died and **his son named Oghotai** came to the throne to rule over the vast Mongol-empire. It may perhaps be needed to clarify that **these Mongol Khans were no followers of Islam** as many people among the masses tend to believe and in addition these Mongols in the beginning, were traditionally not at all devoutly religious with respect to any specific religion. Their subsequent history also shows that during later periods and at different times and places some of the Mongols got converted to the Christianity, some to Islam and then interestingly, some to even Buddhism. Well in any case, that is a different matter; but historically during those earlier medieval-era centuries, these religious aspects too, mattered significantly in the imperial politico-military multiple-processes of the building of the then unprecedentedly vast **Mongolian-empires**.

Changiz Khan's son Oghotaiin fact, was most unlike his father since he was psychologically a peace-loving ruler and even, did not believe in relentless wars, and ruthless blood-shed for the sake of conquering more and more new territories and then thereby, also cruelly persecuting the subjugated peoples. Further in fact, he was a different kind of feudal lord and quite humane as well; since he was sincerely desirous of giving a maximum possible spell of peace in the lives of Mongol populace and, along with undisturbed social tranquility as well. Considering the different kinds of historic manifestations of the already powerfully entrenched as well as socio-psychologically lingering culture of the age-old feudalism during those times and, since almost among all other ruling dynasties and their subjects anywhere this indeed was altogether culturally some enhanced imperial thought on the part of the Mongol King Oghotai. But alas, under the circumstances and as the destiny was to have it; Oghotai's such encouraging thoughts were historically not going to be enduring. In fact, with some even not very long spell of peace **Mongols had since recouped and were again in good shape, along with their renewed military wherewithal under the command of their highly competent General, named Sabutai** in Baghdad for their second devastating invasion of Europe.

In precisely the above contexts, **Jawaharlal Nehru** (*Glimpses of World History*,page 256) has since written that, "The armies and generals of Europe were no match for **Sabutai**. Carefully preparing his ground by sending spies and advance agents to the enemy countries to bring information, he knew well what the political and military situation of these countries was before he advanced. On the field of battle he was the master of the art of war, and European generals seemed just beginners at it in comparison to him. Sabutai marched straight to Russia, leaving Baghdad and the Seljuqs in peace on the south-west. For six years he marched on and on, plundering and destroying Moscow, Kiev, Poland, Hungary and Cracow. In 1241 a Polish and German army was annihilated at Liebnitz in Lower Silesia in Central Europe. The whole of Europe seemed to be doomed. There was nobody to stop the Mongols. **Frederick II**, wonder of the world though he was called, must have paled before this real wonder which had come out of Mongolia. The kings and rulers of Europe gasped, when suddenly unexpected relief came. Oghotai had died, and there was some trouble about succession. So the Mongol armies in Europe, undefeated though they were, turned back and marched east to their homelands in 1242. Europe breathed again.... Meanwhile (*the Mongols*) had spread in China, and finished off completely the Kins in the north and even the Sungs in South China. **Mangu Khan** became the Great Khan in 1252, and he appointed **Kublai Khan** the Governor of China. To Mangu's court at Karakorum came great concourse of people from Asia and Europe."

Earlier in the above contexts historically, we have since mentioned that certain aspects of different religions too have probably always played, even socio-politically quite important roles in the geostrategic and politico-military arena of such military conquests, for settling various kinds of scores and expansion of empires as well, in different continents. And then interestingly, there have also been some kinds of competitive rivalries among the religious leaders and holy personalities for prominence in the various royal courts in both Europe and Asia. In fact, these religious leaders were even competitively quite keen to persuade to bring-in the respective kings/ monarchs into the folds of their respective religion if possible, for obvious socio-political as well as cultural reasons. And hence, historically during the 13th century, we find same being the case in the court of **Kublai Khan** as the Mongol governor of China as well. In this regard, we have noticed that the Pope from Rome had taken special initiatives by sending an emissary to the court of Kublai Khan in China to convey his personal message to the Governor and we may also recall that we have since recorded the episode concerning this Papal initiative, in some related contexts in the last chapter itself. And interestingly in this respect, the Papal initiative was then also able to trigger historically far-reaching fallouts in a context other than religion altogether.

Along with the historic phenomena of the exchange of envoys and messages as well as to and fro convoys, naturally there were merchants and business men too, travelling in the company of those convoys. History further shows us that in particular, the prominent and initiative-driven keen merchants from Venice; took interests in travelling all the way to the royal court of Kublai Khan in Peking and there they were especially welcomed as well. Similarly **two brothers from Venice, named Nicolo Polo and Maffeo Polo** too had arrived in the court of Kublai Khan and they were well received over there, since they had earlier travelled right up to Bokhara in search of their desired business. Then subsequently during their second journey again from the West to East, all the way through the land route and along with the Christian monks; they also took another young man named **Marco Polo, who also happened to be Nicolo's son**. And then as subsequent events kept unfolding, the living history in fact in its womb was destined to have, in some historic contexts; an especially memorable place as well as prominence for him. Since at the start of their journey through the highly demanding land-route all the way to China, many years had already passed and many unsettling changes had occurred in Venice before they could come back to their homeland. At the same time, we also learn that true to his inborn genius and hence along with his prolonged and tough journey Marco Polo could fruitfully utilise his time in certain remarkably creative activities, which have their own historic significance and at the same time no less valuable in specifically the present contexts of our continuing discussions in this regard.

In these specific contexts interestingly, Jawaharlal Nehru has written a very short but highly informative chapter as '**Marco Polo: The Great Traveller**' in earlier mentioned book in which he has provided historically certain significant details regarding **Marco Polo and his memorable and detailed travelogue** as well. In our larger geostrategic and politico-military perspectives and specifically, at the present point of our continuing discussions the multiple historic aspects of the Mongols and the establishment of their vast empires, in fact pre-dating the later times' advents of the overseas empires of a number of rival European Imperial Powers and, covering almost all continents of the world have certain important connectivity with respect to the multi-dimensional manifestations of these European imperialistic-culture and 'military-cum-naval power projections-based' as well as strategically very shrewd fruition of their unprecedented transcontinental colonialism. And in these precise contexts, if we may care to delve deeper into the historic details regarding the Mongols and their unchallenged military-prowess and thereby their zeal for ruthless conquests of vast territories and merging them into their expanding Mongolian empires and along with the highly valuable information given in the famous Marco Polo's travel-accounts too, have played their own motivational as well as indirectly contributory roles in the much later times' fruition of those imperial European Powers' urge for different kinds of global colonialism. In the contexts of the medieval-era living history of India interestingly, we are already aware that conquer of the Indian sub-continent and also founder of the historic Mughal-Empire in India, **Babar,** was himself also historically a famous descendant of the Mongols.

Now coming back to the **persona of Marco Polo**, during his initial prolonged journey in the company of his father and uncle through the historic old caravan route towards China; and in course of his journey spanning over few years, Marco did have enough time and hence true to a kind of his inborn genius was able to learn a number of new and useful languages such as, the Chinese and Mongol as well. His lingual attributes did become valuable assets for him and hence in Peking then, he became a great favourite of the Mongolian Governor of China, Kublai Khan. Marco Polo spent many years in China in the dedicated and distinguished service under Kublai Khan, in important assignments and also with high ranks. **In fact, Kublai Khan was so fond of Marco Polo that he was very reluctant to let the three Polos leave his royal court and return back to their home in Venice**. But at last he permitted them to leave, especially Marco Polo; and that too under great pressure to send a most valuable 'living-cargo' to his another Mongolian clan named **Argon**, who happened to be the Mongol ruler in their empire in Persia. Interestingly, that highly valuable 'cargo' happened to be a young Mongolian princess, along with her escorts as provided by Kublai Khan himself as the would be bride of Argon himself. And that's how the three Polos were able to leave Peking accompanied by the young princess, who was to be first delivered to Argon himself in Persia; on their way to Venice. That is another matter and as the fate was destined to have it, that by the time the three Polos were able to reach to Persia, aged Argon was already dead, and hence that young Mongolian princess was married to the son of Argon from his first wife, who too then was no more alive.

Marco Polo and his companions then, left China via the sea-route on way to Sumatra (now, part of the modern-era Indonesia), and located in the South-East Asia. They all stayed there for quite some time, for the necessary rests and replenishments of various essential supplies for their onward journey. He found there another quite different kind of historically **famous Buddhist Empire of Shri Vijaya**; which interestingly, was totally of Indian origin. In this precise context, we may recall that we have since already noted in the earlier chapters that, how and when the Buddhism as a religion; was first brought in and firmly implanted in the South-East Asian islands from India by the then prosperous ruling Indian Monarchs and more so, without any violence or war. **Marco Polo was naturally quite impressed and he then, came to southern India across the Bay of Bengal**. And here, he found another powerful and prosperous kingdom of the Pandyan dynasty; along with a flourishing sea-port of Kayal on the eastern coast of this Indian kingdom. Marco Polo and his party were quite impressed by these multi-dimensional medieval-era civilisational developments in the southern India and hence naturally, they were motivated to have a long stay over there. From there they began their further journey toward West for the Middle-East, with their destination in Persia. And on reaching Persia after two years of their journey, then under the Mongolian rule; and as already narrated in the above paragraph, he delivered his highly precious 'living-cargo' to the Mongol prince and son of late Argon, for their happy royal-wedding.

After some stay in Persia, the Polos then further commenced their journey further somewhat north-west and across the Mediterranean Sea to their final home-bound destination in Venice. **And finally they were able to reach Venice in the year 1295. In fact in between, many years had already passed and many changes too had taken place in their homeland, since Marco Polo had left Venice nearly twenty four years earlier. Naturally there were hardly anybody in the city of Polos, who could place them correctly and recognise them. At the same time even the bigger ironies were that no one really believed the authenticities of their travel-stories all the way to China and then their return journey along with the visits to the kingdoms/places in South-East Asia and further on to India, Persia and so on.** Those were the days and such was the overall environment of ignorance prevailing then, during a sort of zenith of the medieval-age. But only this was not the end of their problems on their return to Venice, since otherwise too, they had arrived back in their home when their glorious city of Venice had already landed in politico-militarily highly depressing and unsettling situation. In this precise context, we may recall that during those times in certain areas of southern Europe such as Greece and Italy historically; there was a kind of prevailing system of the independently administered and quite prosperous 'City-States'. And medieval-era European living-history further informs us that, these independent City-States were having fixations regarding all sorts of real or imaginary and then, mostly irreconcilable rivalries/ jealousies among themselves, including with military dimensions.

Hence for **Marco Polo** on the eve of his return to home then, another even bigger difficulty came as the rival City-States of Venice and Genoa got entangled in a kind of all-out war. They had their own different kinds of grievances against each other, with their certain opposing geostrategic and politico-military dimensions since those were the times of the medieval-age European culture of knighthood in shining armour and military chivalry as well. Both the independent City-States of Venice and Genoa were powerfully rival Naval-Powers as well and naturally then as a consequence, great naval-battles too commenced between them in the Mediterranean. But unfortunately for the Venetians, Venice militarily got subjugated and as an obvious result, thousands of the citizens of Venice were taken prisoners and as the fate would have it; among them was the great traveller Marco Polo as well. In course of his historically epoch-making and exceptionally long journey from and then back to the Venice, visionary Marco Polo had already maintained valuable detailed notes of his observations and highly fascinating information as gathered by him; during his travels for so many years. Hence during his years in prison in Genoa, the legendary Marco Polo was able to systematically rewrite his famous travelogue as ***Travels of Marco Polo*** for posterity.

In this historically very fascinating context, **Jawaharlal Nehru** has beautifully written that, "In these travels Marco describes China especially, and the many journeys he made through it; he also describes to some extent Siam, Java, Sumatra, Ceylon and South India. He tells us of the great Chinese seaports crowded with ships from all parts of the Orient, some as large as to carry crews of 400 or 400 men. He describes China as a smiling and prosperous country with many cities and boroughs; and manufactures of 'cloth of silk and gold and many fine taffetas'; and 'fine vineyards and fields and gardens'; and 'excellent hostelries for travellers' all along the routes. He tells us of the special messenger service for imperial messages. These messages travelled at the rate of 400 miles in twenty four hours by relays of horses—which are very going indeed. **We are informed that the people of China used black stones, which they dug out of the ground, in place of firewood. This obviously means that they worked coalmines and used coal as fuel. Kublai Khan issued paper money—that is, he issued paper notes with a promise to pay in gold, as is done today.** This is most interesting as showing that a modern method of creating credit was used by him. Marco mentioned much to the excitement and amazement of people in Europe, that a Christian colony under a ruler Prester John lived in China. Probably these were some old Nestorians in Mongolia.... **Marco's story was and still is, a wonderful story of travel. To the people of Europe in their tight little countries with their petty jealousies it was an eye opener. It brought home to them the greatness and wealth and marvels of the larger world. It excited their imaginations, and called their sense of adventure, and tickled their cupidity. It induced them to take to the sea more**. Europe was growing. Its young civilisation was finding its feet and struggling against the restrictions of the middle Ages. It was full of energy like youth on the verge of manhood. **This**

urge to the sea and the quest of wealth and adventure carried the Europeans later to America, round the Cape of Good Hope, to the Pacific, to India, to China and Japan. The sea became the highway of the world and the great caravan routes across continents lessened in importance." (Nehru, 1934, pp. 263 & 264)

We may here also recall that little earlier, we have observed that when contextually reviewing these aspects, in the larger geostrategic and politico-military perspectives we may also find that the then unprecedented multi-dimensional aspects of the Mongols' far-reaching military conquests and firmly establishing the vast medieval-era Mongolian empires, covering most of the continental territories of both Europe and Asia and along with the valuably historic information contained in the fascinating travelogue of Marco Polo, that these too, were having their own distinctly motivational roles in triggering the subsequently huge historic phenomena of establishing even much bigger overseas empires, by the ruthlessly rival and highly ambitious European imperial Powers of those medieval-age times. At this precise point of our on-going discussions, if we may delve deeper into the above phenomenal historic aspects in our geostrategic and politico-military perspectives we may also contextually visualise **that the ignorance about the multiple-aspects of larger world, leads to a kind of mental darkness and thereby, perforce breeds all sorts of amusing and at times even some harmful naivety in the highly complex areas of "essential strategic-culture and thoughts." While just opposed to this, we may also say that valuably multiple-information in this regard, is both enlightening as well as enriching to the knowledge and then it is also rightly said that knowledge by itself, in all its dimensions as a great asset, is power**. For meaningful perceptions of these concepts, **multiple-aspects of the 'essential strategic-culture and thoughts'**, remain the essential prerequisites.

In this precise context, we may now clearly see that the very interesting historic information contained in the just above quoted paragraph from Nehru's well-known book does also point towards the same aspects of our thoughts in this regard. If we may care to mentally travel back into the 11th to 14th centuries' medieval-era situations and overall socio-cultural as well as politico-military environment, the then prevailing in the European societies and their various rather extremely feudalism-fixated Nation-States; then in our precise contexts at this point of our continuing discussions, we may in addition, vividly visualise the kinds of ignorance and consequent mental-darkness those various European ruling-elites and their subjects were living into; so far as the aspects of larger world were concerned and especially more so, with respect to the historically much more ancient and remarkably better enlightened and civilised societies and, also the systems of governance in practice in different kinds of the Nation-States, since situated in the much larger Asian continent. At this point, it appears somewhat relevant for us to recall that earlier in the last chapter in certain contexts we have since noted that the pivotal value-systems then existing in those medieval-era European societies and their socio-political entities were principally based on the die-hard feudalistic military-culture of chivalry and knighthood in shining-armour. When in our larger perspectives, further critically reviewing the various dimensions of these extremely rigidity-oriented socio-cultural and politico-military aspects, so overwhelmingly present in the medieval-era Europe; then we are able to paradoxically visualise as to in what sort of dark and narrow mental lanes, the Europeans were living then, with hardly any windows opening towards the outside larger world.

Hence historically in our contexts of continuing discussions and then at this precise point, while putting these medieval-era European aspects in broad analytical perspectives, we find that no wonder as to why those European Nation-States were almost perpetually in a state of war among themselves. We may also admit that in historic perspectives, such opinions may be matters of conjectures but at the same time in a context they are undeniably apparent as well. Interestingly we have since earlier too noted that history spares no one and at the same time, it is never influenced or dictated by any one, howsoever, mighty and powerful in any manner. In other words, man's living-history, with all its multiple-manifestations, keeps on marching ahead regardless and expanding in its multiple contents and dimensions, on to the unending 'canvas of time', and hence on the way, keeps on recording and interpreting, with maximum possible objectivity the events as well as, the so many aspects of their consequent fallouts, as they keep on occurring and unfolding before the living history's discerning eyes.

We may even appropriately say **that history can never objectively be considered dead in any manner since it is the history alone, which is able to keep alive the humanity's past with all its dimensions and connotations and in addition as always, being aided and abated by the advancing marvels of the various disciplines of sciences and technologies; since the advents of various ancient human-civilisations in different ages of the same history and, in different geographical areas of the world**. We have since also noted earlier that, the man's living-history is multi-dimensional and hence naturally, it has many segments. Some of these segments of history and along with some other disciplines of human-knowledge, as deemed relevantly useful and helpful; in the contexts of our continuing strategically critical journey through humanity's past, present and possibly future, in global geostrategic and politico-military perspectives and as would now clearly be seen that, the same won't have been possible at all, without the availability of the man's same living-history. Hence it appears rather imperative that these aspects too, may essentially be seen in their own proper perspectives. **Now coming to the closing-aspects of the once unchallenged and unprecedentedly vast Mongolian Empire, since in various contexts, and, along with its some other prominent geostrategic and politico-military aspects as already discussed earlier in this as well as in the last chapters, we notice the same phenomena of the imperial rise and fall in case of this Mongolian Empire as well**. In the process, we have also touched upon the various very fascinating aspects of the life and memorable creative-activities, the great traveller Marco Polo and then in addition, briefly highlighted the multiply contributory values of his endeavour too and the interesting historic glimpses of the same, we would be seeing as we proceed further in this chapter itself. As it was destined, the 13th century Mongolian ruler of China, Kublai Khan had already passed away soon after Marco Polo had left him for his home in Venice. We do notice that after the passing away of a great Mongol Kublai both imperially as well as politico-militarily, the situations in China did not remain the same and hence many kinds of upheaval and unsettling socio-political scenario got cropped-up. When Mongols first made their inroads in China through their devastatingly ruthless military campaigns, they had founded their own ruling dynasty over there, known as the Yuan-dynasty. This dynasty, with their successive Mongolian princes coming to the Chinese Throne, ruled only little less than a century from 1279 to 1368. As part of the now commonly known universal sociological phenomena of **periodically historic rise of nationalism** then got powerfully manifested during those unsettling times in medieval-era China too, and thereby bringing down basically the foreigners' Yuan dynasty and consequently, the Mongols were chased away from the Chinese territory in fact, much beyond the historically unique and great Wall of China.

Now, if we further look critically deeper into the above discussed historic aspects concerning founding of empires by any alien power thorough the flash of swords and/or crafty politico-diplomatic initiatives with shrewd strategic-plans, its ruling-dynasties as well as consequently with certain allied decisively powerful geostrategic and politico-military dimensions, we do notice that such empires with some discriminating imperial mindset and arrogance, do not survive very long since they happen to be totally non-indigenous to begin with. History further indicates that in course of time, two factors get germinated and develop into powerfully challenging factors and hence, multiply-challenging to the rule of any such alien imperial-power. In this case we do also find the same being equally true. Hence, the above perception appears to be largely true and universally applicable to all kinds of such empires anywhere as interestingly, we would be seeing the same phenomena being manifested in these precise perspectives and in course of our discussions, as we cover the last phase of the passing away of medieval-era and then simultaneously enter into the beginning phase of the present modern-age as well.

We further notice that the two prominently decisive factors getting germinated in course of time, in any imperially subjugated country with her restless native population and hence, they happen to be the; **(1) birth of some kinds of the feelings for nationalism and, (2) highly charged and overwhelmingly sentimental growth of patriotism**. Factually it is a different matter that the contents, textures as well as socio-cultural and politico-military dynamics of the above mentioned two decisively driving factors are largely situational and may historically vary in case of different subjugated peoples and their Nation-States and also in addition, as pertaining to the socio-cultural and politico-militarily fixated historic value-systems of the concerned ruling imperial powers. And interestingly further at this precise point, we may also obviously notice that the socio-political dynamics of the above highlighted two factors as well as their various inseparably connected and as, hinted above aspects do all together project highly complex processes and they are so, understandable as well. Now as we proceed further with the engaging historicity of the above mentioned factors and their various aspects in this regard, we may also notice that at the roots of all these are the **historically period-based manifestations of the multi-dimensional clashes (not necessarily violent.) of both the differing civilisations as well as newly emerging powerful ideas.** Then with respect to the subjugated peoples and their territories we find that in our contexts, as we begin conceptualising them and their so many allied aspects all taken together; we also do begin to visualise their enormous complexities as well.

Now again coming back to the historic aspects, the end of Mongolian Empire in the Medieval-age China after the death of Kublai Khan and thereby, demise of the ruling Yuan dynasty over there, we notice here too the same phenomena, with their distinctly Chinese contents and textures, getting manifested and consequently triggering the politico-military eclipse of that Mongolian Empire. And as the situations then kept on developing over there, the entire revolution in this regard, was indigenously Chinese. Hence, after the Mongols were forced to leave China, an indigenous Chinese, named **Zhu Yuanzhang** who was prominently leading this revolution then, declared himself Emperor and also, again established a ruling Chinese dynasty historically known as the **Tai-Ming Dynasty**. The Ming dynasty in China ruled for nearly 300 years, from nearly 1368 to 1644. In the history, the medieval-era Ming-period in China is known as an age of glory and substantial prosperity. But all segments of the historic Ming-period of China are not equally rosy and when objectively reviewed with analytical criticality in our larger perspectives in this regard, the otherwise shining-armour of Ming too, has some chinks as well. The broadly true and understandably larger point is that, any segment of the man's living-history, as precisely this one here is never a simply linear record of one dynasty passing by and a new ruling dynasty coming to the throne. Any such phenomenon, at any point of history and in any country, has multiple facets with some being very glorious and shining and some not so palatable for the people concerned and hence, not so reassuring. In fact that's precisely the way that the man's living-history in any particular civilisation or period anywhere, is continuously being made with all its complexities and dimensions, since the beginning of the history itself.

And hence, the same remains equally true in case of the medieval-era Chinese Ming Dynasty as well. In this context we may again refer to **John Keay** and he writes that, "Nor were the Ming quick to learn the misadventures of the Mongols while ruling China. Though piously claiming to be restoring the institutions and rituals of the Tang and Song, in practice **Zhu Yuanzhang** seemed bent on reconstituting the empire he had just toppled. **Khubilai's** division of the country into provinces was retained, with modifications; and both in the provinces and Nanjing **'the formal structure of governmental institutions....began by imitating the Yuan almost exactly'**. It was the same with the military. Zhu's genuine determination to improve the lot of the cultivator, a legacy of his dreadful childhood, meant holding down taxation and giving a high priority to agricultural development, including military colonies. But with this proviso, the challenge of demobilizing and resettling his armies was not dissimilar to that faced by the Yuan. Thus like the Yuan, 'the emperor made his army into a distinct occupational caste within the population, created a hereditary officer class to govern it, (and) gave the military officers a clearly superior status compared to their civil equivalents'. Edward Dreyer actually calls them a new conquering horde, but this time Chinese in origin. Generals became nobles with hereditary titles and fiefs; and imperial offspring became princes with supervisory responsibilities at the apex of this great military structure.... But compared to the Mongol Yuan, Zhu Yuanzhang was even worse. Courtiers and ministers were beaten in his presence, sometimes to death, and as of the great purge in of 1380 and accounted for those 30,000 to 40,000 lives, scarcely a day passed without mass executions. No reign of terror in Chinese history can compare with it. Obsessed with controlling every aspect of government and deeply suspicious of any who might ridicule his deficiencies of birth and education, the emperor dispensed with the office of chancellor and himself took on the executive role in government. The examination system was eventually reinstated but only with a view to improving the supply of bureaucrats, and not to restoring their influence. In the 1390s it was he military hierarchy, the emperor's own creation, which became the target of his suspicion. A word of complaint from a long serving general in 1393 brought his immediate execution, followed by those of his supposed associates to the tune of four marquises, an earl, a minister, ten other nobles, sixteen chief commissioners and an unspecified number of junior officials, plus the extended families of all of them, giving a grand total of over fifteen thousand. When 1398 the emperor himself died, the empire heaved an almighty sigh of relief." (Keay, 2008, pp. 373 & 274)

In a context in the above quoted paragraph, we indeed find some amazing facts then prevailing in the medieval-era China under the Mongolian Yuan-dynasty's rule. Another historic fact with respect to the almost all societies the world over since ancient times has been that, the socio-racial, socio-cultural as well as socio-political mosaic and their multi-dimensional dynamics have always distinctly remained varied and highly complex as well. In this respect, **so far as the display and practices of the aspects of imperialism, and especially in case of the alien imperial-rule, in any particular subjugated land, the imperial considerations, based on the racialism-based fixations and prejudices; have always remained overriding**. It may even perhaps be considered as typical socio-psychological paradoxes concerning the man and his psyche **but, facts do always remain facts on the pages of man's living-history**. And the same aspects have been found to be no less true and pronounced, during the Mongolian Yuan dynasty's rule in the medieval-era China. History clearly informs us that during that 14th century China too, the ruling Mongols had formulated their own kinds of imperialism-based grading of the Chinese population for deciding their preferences; for employability in the service of empire at different levels in the hierarchies in both the military as well as civil sector services. The principal aims in this regard have always been that in all kinds of employments in the empire; the indigenous Chinese people were always discriminated against, in all possible ways, in favour of the Mongols.

In fact, we may also historically say that the **pivotal compulsions with respect to the imperialism-based theories and practices of any alien imperial-power's rule in a subjugated country have to be based on the concerned conquered peoples' prevailing socio-cultural, socio-psychological and politico-military—historic and situations-based guided-designs of the conceptual-framework for successfully enforcing the imperial-rule over the overpowered territories and their indigenous peoples, with all its manifestations**. While further critically reviewing these interesting historic aspects, precisely in our larger perspectives, we do very clearly see that, as the above mentioned overridingly important aspects have been true in case of the Mongolian Yuan-dynasty's imperial rule in the medieval-era China but certainly with altogether different textures and contents, **the same have equally been so; in the later-date Rival European imperial powers' overseas colonial rules in the numerous countries of the continents of Asia, Africa and the Americas. At this point, it may also be equally appropriate to mention that same has also been no less the realities (*at times, even very ruthless and extremely nauseating too.*), during nearly two and half centuries' British imperial rule in the Indian subcontinent**. In fact with the desired historic objectivity, we may also appropriately say that the beginning of man's living-history itself and though with different textures, contents and connotations; these precisely are the ways, since the multiple-aspects of the imperialism have compulsively been practiced the world over.

Another historic point, which may also appear to be quite influentially related in this regard, happens to be that the societies in almost all the Nation-States, and more so in case of relatively the larger Nations; the populations inhabiting them, are seldom racially as well as socio-culturally homogeneous at all. Though in certain countries by their physical appearance, colour of their skin and other such features they may largely appear to be identical at a first glance to any foreigner, but similarity mostly ends there itself. Interestingly in some ways, and more so contextually at the present point of our continuing discussions, these aspects too, with respect to the peoples concerned anywhere; also assume some decisiveness in the multiple process of the above discussed imposition of imperial-rule in any subjugated country by the concerned conquering power. Similarly in China as well, the large Chinese population too, has a number of sociological divisions such as; Han (*mainly northern Chinese*), Khitans, Jurchen, Koreans and so on. Hence any such peoples-oriented varieties historically prevailing in China too, the basic sociological homogeneity couldn't have survived. Now, since we have already mentioned that aspects of the imposition of imperial-rule anywhere by any alien dominating power; perforce have to be based on the implementation of some kinds of socio-political discriminatory policies and hence, Mongolian Yuan-dynasty too during its rule in China, followed similar policies. In this respect, man's living-history shows us that during Mongols' imperial-rule in the medieval-era China, with such discrimination-oriented socio-political hierarchies; on the top came the Mongols themselves, next mainly Uighurs, then central Asian Muslims followed by some other coming foreigners and at lowest end, came the various kinds of the above described largely indigenous Chinese peoples.

As we proceed along further in this regard, we would interestingly be noticing the even more colourful and broadly similar rather ruthlessly imperialism-oriented phenomena; as we begin dealing with the multiple aspects of the establishment of overseas empires, by the leading and rival European imperial-powers in the different continents of the world. Before we come to them, it appears quite befitting to first briefly but no less critically, delve upon similar aspects of, and even more ruthless in some ways; concerning some other medieval-era imperial powers and their exploits in the Far-East.

Quoting Nehru again with some very precise relevance, he historically mentions and also along with in the larger geostrategic and politico-military perspectives in this regard that, **"Civilisations, like empires, fall, not so much because of the strength of the enemy outside, as through the weakness and decay within**. Rome fell not because of the barbarians; they merely knocked down something that was already dead. The heart of Rome had ceased beating when the arms and legs were cut off. We see something of this process in India and China and in the case of Arabs. The collapse of Arabian civilisation was sudden, even as their rise had been. In India and China the process is long-drawn-out and it is not easy to spot it." (*How appropriately precise and relevant the above quoted lines are in this regard; both historically as well as, in the contexts of presently on-going modern times too. Isn't it? Well, at this point, we may also say in our larger geostrategic and politico-military perspectives; since remotely very ancient of times; that, in case of the both rise and subsequent fall of the civilisations as well as empires; the positive visionary awareness of the* ***multiple aspects of 'essential strategic-culture and thoughts'*** *and then ironically, their self-destructive negligence as well; are principally at the pivotal centres of both such rise and fall respectively. We may recall that, we have since delved upon these far-reaching aspects, in some details and at the different points of our continuing discussions, and also in different contexts here. We would also quite appropriately further, be coming to this one of the most important and influential factors; again and again and, as deemed necessary later in this regard.* ***The overriding point here apparently visible remains that the multiply-decisive aspects of this 'essential strategic-culture and thoughts'; can neither be overemphasized, nor would be prudent at all to ever neglect the same at any stage and in any manner, while exercising the multi-dimensional aspects of the statecraft by the governing Institutions/Monarchs, under any kind of political-system for the governance anywhere; in its all possible manifesting totality. This paramount compulsive aspect in this regard, remains always timeless and, universal as well***). (Nehru, 1934, p. 209)

Man's living-history informs us that **Japan**, as a somewhat ancient and distinctly different civilisational-society as well as a typically feudalistic Nation-State to begin with, has first been found to be mentioned, in the ancient Chinese records of the third century AD. Then **Korea** too, though again distinctly different in many ways is equally parallel and ancient as well. Interestingly here, a largely common, far-reaching and powerfully influential threads of epoch-making civilisational-history are seen to be passing through the highly chequered past of both Japan and Korea. And these historically common 'thread' are principally constituted of the substantial influences of the numerous aspects of the world's one of the most ancient and glorious civilisations: and here it is needless to say that, this obviously happens to be the **China**. At this point, it is important to bear in mind that when we are dealing with these highly chequered, interrelated, interactive and again, historic civilisational-aspects of these three countries, in our larger perspectives, then we are not dealing with the Aryans here, but distinctly quite different the Mongoloid populations. Now coming back to the three nations of **China, Korea and Japan**, we find that the latter two nations are situated in the Far-East, whereas China is a very big and almost a kind of the continental-size nation by itself and located in and as part of the northern Asia. Again out of the three, China is the most ancient civilisation and thereby her history is believed to be originating interestingly, in the fog of the remotely very ancient of times. In addition, we are also required to bear in mind that both the very ancient antiquity as well as multiply vast complexities of any very large country and her colourful societies substantially bring upon their peoples, simultaneously both the numerous aspects of again the multiple civilisational assets as well as also some kinds of heavy burden as well. And like India, China too, has historically been no exception. But interestingly one rather praiseworthy historic truth in this regard, remains that both China and India have substantially maintained with their own and in distinctly different historic manners the gloriously ancient as well as continuously surviving and also healthy civilisational entities, since the very beginning. And needless to say that very few ancient civilisations, since the inception of man's living-history itself, have been able to historically display such kinds of multi-dimensional resilience as well as truly vibrant survivability.

But the history also indicates that ironically in China, the gloriously and unique Chinese Confucian civilisational aspects with their own remarkable originality, have since got withered-away with the passage of time because of many kinds of unsettling and turmoil-ridden developments and in addition also, during the much later modern-era sprouting of the none so peaceful and also not at all enriching fruition of the communism, as a socio-political philosophy, and also certainly this being a very devastating political ideology. If we may pause a little, and care to delve even deeper into the now still lingering and prevailing communism-oriented overall scenario in the present-day China and at the same time also, keeping in mind the larger but very precise parameters concerning the geostrategic and politico-military perspectives, then we may be able to have the far-reaching glimpses of the extents of multiple-devastations perforce inflicted upon the age-old and also once so gloriously magnificent Chinese societies and their remarkably unique civilisation. With respect to the above, it is found to be obvious that very ruthlessly, the first and foremost negation in this regard, always happens to be the destructive subjugation of rather very sensitive principles of the human-rights aspects.

Hence, howsoever, one may try to gloss-over any such civilisational negation concerning any particular society or Nation-State, but the truth cannot be wished away. In spite of the very impressive scientific, technological, industrial and economic progress now taking place in the present modern-era China, the above mentioned multi-dimensional damages inflicted upon the Chinese societies and also along with, the very tragic negation of the ancient Chinese civilisation by again the turmoil-ridden and extensively violent, revolutionary and forceful imposition of rather a very extreme form of communist thoughts and, their socio-culturally devastating political-system for her governance cannot be brushed-aside from the discerning eyes of the larger humanity abroad. Such is the extent of undeniable devastations caused by the any kind of totalitarian-system or ruthless dictatorship of politico-military governance. In any case and also, whatever may be one's belief in any particular political ideology, such hard historic facts can never be erased from the pages of the universally acceptable man's living-history, in any manner. We would further be discussing these overriding aspects of the man's living-history, in some more details, in the subsequent chapters, concerning not only China, but some other nations as well.

We may also recall that in previous chapters we have briefly discussed certain thoughts, in different contexts, of **Professor Samuel Huntington** (*The Clash of Civilisations and the Remaking of World Order*), wherein in course of his arguments, he also says that religion happens to be the central defining characteristic of civilisations; which is obviously not appearing to be widely agreeable at all. Interestingly, we have since also quoted earlier in this regard, the other no less weighty thoughts of different well-known scholars as well, countering Huntington's perceptions.

In this interesting context, if we look into the religion-based precise aspects of China, Korea and Japan we find that during different centuries in the ancient-era itself, the societies in these three countries were greatly influenced in their multiple socio-cultural dimensions, by the teachings of one common religion known as **Buddhism**. And we know that Buddhism first took birth in India and then afterward, it travelled to China and then subsequently to Korea and Japan. We further, in addition, also find that religion has been playing some kinds of influential roles in the overall very fascinating and multi-dimensional advancements of the civilisations; but religion perhaps cannot be considered as the sole defining characteristic in this regard. We have in different contexts, also discussed that multiply decisive aspects of the geostrategic and politico-military considerations have always been compulsively overriding since ages, in both the advancements of civilisations as well as promotion of any particular Power's urge for promotion of imperialism and thereby, building of their empires.

Hence, it is the one's overwhelming motivation towards establishing geostrategic and politico-military domination over as much additional territories and also, geographically as far as this being situation-wise feasible, which in turn, leads to the promotion of one's own imperialism-fixated urge to build empires or expand areas of influence. But in this precise context, a couple of important points do also clearly emerge and hence, the first being that the above mentioned motivation is one thing and then; second being the aspects of the feasibility in this regard. At this point, we may also mention that just matters concerning one's 'flights of imagination' only, doesn't carry any such ambition anywhere since first, the necessary wherewithal are always first needed to be put in place. Further at the same time, man's living-history also clearly shows that since ages, at the root of any kind of substantial success-story in this regard, has always been first, the **compulsively cultivating, evolving and imbibing for proactive applications of the mandatory features of the 'essential strategic-culture and thoughts'**, towards the desired fruition of the aspects of above mentioned motivation. Hence, while critically reviewing these weighty points since being discussed here we may clearly notice that in these contexts, the above mentioned mandatory features of the **'essential strategic-culture and thoughts'** do assume far greater decisive importance than certain aspects of the religion, since as found to be perceived by Professor Huntington. Moreover, we may again recollect that at a number of times earlier, and also in our larger geostrategic and politico-military perspectives, we have since already emphasised the **singularly most important as well as unsurpassed decisiveness of the multiple-aspects of 'essential strategic-culture and thoughts'**.

In this specific context, Late **J.N. Dixit** has since mentioned in his already earlier referred USI (India) National Security Paper 1997 that, "The validity of his argument (*i.e. Huntington's*) that the fault-lines between civilisations are becoming the central lines of conflict in global politics and that ideology and power equations are no longer relevant. This is intellectual misrepresentation and factually erroneous perception. International politics, as the events since the end of Cold War have proved repeatedly, is still governed by, motivations of geostrategic dominance, desire to have access to and control of natural resources, to have exclusive capabilities for advanced technologies and to prevent conflict situations which do not serve the specific interests of the more powerful countries. The effectiveness with which these objectives can be met still depends on political stability, economic strength and technological capabilities.... One can do no better than to quote Will and Ariel Durant in evaluating Huntington's work, in concluding volume of their *Story of Civilisations, the Lessons of History*. They say: **'Obviously historiography cannot be a science. It can only be an industry, an art and a philosophy. An industry by ferreting out facts, an art by establishing a**

meaningful order in the chaos of materials, a philosophy by seeking perspective and enlightenment, the present is the past, rolled up for action, past is the present unrolled for understanding. But total perspective is an optical illusion. History smiles at all attempts to force its flow into theoretical patterns or logical grooves. It plays havoc with our generalisations, breaks all our rules. History is baroque." (*It is great, so precise and thought-provoking indeed. Isn't it? Here we do also see that the above quoted analytical thoughts in the larger contexts and also at this point of our continuing discussions, quite befittingly become overlapping in these respects onto both, the aspects of medieval-era and the present modern-age as well. So far as the multiple-aspects of the presently on-going modern-age are concerned, we would be covering them, in some more details, in the later chapters*).

Now in these contexts, the man's living-history perhaps quite clearly and conclusively shows that so far as the advent of some remotely ancient and hence the earliest, human-civilisations in different continents of the world are concerned; that the historically interesting fruition of multiple-aspects of the phenomena of building of empires, during different historic-ages the world over; have always first needed the properly effective placement of the **four principal prerequisites**. The **first** and foremost, of these four timeless prerequisites, happens to be the supremely desirable and vibrant **'essential strategic-culture and thoughts'**; the **second** being, **strategically the imaginative development, well-rounded and the dedicated nursing of all aspects of the matters-military**; and then **third**, being in the possession of **strategic skills for the statecraft at home**, and then **fourth** but no less important, is **clearly the simultaneously successful conduct of the multi-dimensional art of the highly complex international games of the rivalry-infested geo-politics and diplomacy in the highly dicey and slippery international arena of power-politics** as well. Hence, while keeping in mind the above mentioned four prerequisites in this regard, and then critically reviewing the overall aspects of the building of empires in different geographical continents and, in different historic-ages, in the larger geostrategic and politico-military perspectives, man's living-history emphatically shows certain undeniable hard-facts, which may further be seen in the following paragraphs below in this regard.

With respect to the above mentioned four principal prerequisites, man's living-history also shows that these too, have their own historic complexities, substantially in the multiple-processes of these prerequisites having been seen, meaningfully understood, imbibed and then proactively included in the entire gamut of the governance, by the concerned Monarch and/or both the ruling civil and military elites of any particular Nation-State. In these relatively somewhat hidden, but no less decisive as well, aspects of the man's same continuing history in the making, in addition we also find that with respects to the none of the Monarchs, Kingdoms or Nation-States and, at any point of history concerning them had ever been able to fully or optimally do so, in any strategic-cum-fruitful manner. Moreover at the same time, as we further delve deeper into these no less important and interesting dimensions of the living-history, we do also find that contextually, the relative degree with respect to each of the above explained four prerequisites, does always act as changing variable too. Hence since beginning of the very ancient-era itself, in any such historic Kingdoms or Nation-States and their ruling-elites; different kinds of permutations/combinations concerning the degrees of those four prerequisites in this regard, have been found to be historically getting materialised and thereby substantially influencing the same living-history in the making. And therefore in these contexts, these are precisely the multiple-aspects of the above mentioned historic complexities, with respect to those four principal prerequisites.

In these perspectives, we find that the one overriding historic-fact emerging and, that being during any historic period and anywhere, only those geo-politically ambitious Monarchs or Nation-

States have been able to politico-militarily establish any historically noted and enduring Empire, who have been able to learn, understand, imbibe and then, adequately practice the above mentioned four principal prerequisites, with comprehensively desired strategic-vision or at least have decisively been able to successfully do the same, more than their concerned adversaries. We may recall and that may be in different contexts and earlier also, we have mentioned that since the advent of various human-civilisations in this world, living-history has been teaching us that, in the rather 'ruthless' arena of the competitive game of relentless power-politics; there is no permanent friend or enemy, but only one's permanent national-interests. If we delve deeper into these history-making aspects; then very interestingly we find that at any given point of the living-history, the contents and priorities of one's certain national interests may vary, but irrespective of that, the fundamental permanency of the well-identified and also well-defined given sets of one's supreme national-interests; always remain paramount and unassailable.

History also clearly shows that any Monarch or Nation-State who has callously neglected or compromised the same has rather mercilessly been compelled and made to pay a very heavy price, by the tenets of same living-history. Interestingly, we since also find that when any concerned Monarch or Nation-State is able to adequately take care of the multiple-aspects of **'essential strategic-culture and thoughts'** and these in turn, do fruitfully enable him or her to pro-actively see through the above mentioned timeless historic realities, while there is still enough time, and prepare with multi-dimensional strategic prudence for successfully facing any kind of lurking threats. And it is needless to say that, historically encompassing almost all the continents in this world, man's living-history is full of various kinds of similar saga, including both positive and/or negative episodes concerning different Monarchs or Nation-States since the beginning of history itself.

Now, first specifically coming to the **Korean Nation-State** in this regard we do find her certain very revealing and engaging historic aspects, pertaining to the later periods of ancient-era and also in continuity into the earlier periods of medieval-age and at the same time historically, these are also found to be having close links in some manners, **with both China and Japan**; and interestingly to begin with, **concerning the one particular ancient religion-based aspects with India as well**. Historically in these perspectives, while referring again to Nehru we find that, he has very informatively written that, "In Korea, after a period of division and conflict, a united independent kingdom was established in 935 AC and this lasted for a long time—about 450 years. Korea takes much of her civilisation and art and methods of government from China. Religion and something of art go to her as well as Japan, from India, through China. Japan situated far to the east, almost like a sentinel of Asia, carries on her existence, more or less cut off from the rest of the world." (Nehru, 1934, p. 203) (*Historically with respect to Japan, these aspects were true broadly during the 18th/19th and an early year of the 20th centuries only; but this is understandable here since, Nehru had written his this book during the early years of 1030s*). The Fujiwara family is supreme (*it was then so.*), and the emperor, who has recently become something more than a clan chief, is kept in the shade. Later on, came the Shogun. (*We would be discussing the multiple-aspects of Shogun and Shogunate, when in our contexts here, we take-up Japan little later*).

Not so numerous as in China, but during the ancient and medieval ages in Korea too, we find a couple of historic Korean dynasties, with their own distinct kinds of imperial ambitions; gradually surfacing and establishing their own kingdoms over there. As shown in the above quoted paragraph, and mentioned that, enduring rule over the kingdom in unified Korea, since the beginning of early 8th century A.D. itself, and lasting for nearly 450 years, belonged to the **Koryo imperial clan of Korea**, during those enriching years. Then subsequently, **another ruling clan, known as Choson,** got subsequently emerged in Korea; but the historic change-over from the Koryo to Choson dynastic-rule, was not so violent at all. This is historically rather of some exceptional nature, which the world was not accustomed to see at all. If we pause a little at this point here, and go deeper into such concerned historic aspects of the Korean societies and their politico-militarily oriented kind of feudalistic Nation-State of those times then rather somewhat encouragingly, we find that left to themselves the Koreans were not so much imperially violence-prone, like their other neighbours. But alas, the Korean living-history of those historic period and further on also no less so during the many bygone years of the present modern-era as well was tragically not destined to be peaceful at all. It is certainly a great tragic irony of the living-history of Korea. And as we have since already emphasised earlier that the man's living-history favours none at all, anywhere and at any time and hence here too we find that historically, the main the then highly feudalism-oriented and ambitious Nation-State was, principally that of Japan which has repeatedly invaded Korea in the past.

While unambiguously maintaining the necessary historic-objectivity in the larger geostrategic and politico-military perspectives at this precise point of our continuing strategically critical discussions, we may recall that since the beginning of man's living-history itself, the dynastic-rule anywhere and based on the feudalistic imperialism and also unchecked ambitions to expand one's empire, has perforce resorted to the blatant military invasions of her neighbouring kingdoms or Nation-States. Viewed in the larger historic perspectives, we find these to be irrefutable, timeless and rather universal facts of history. And the same historic aspects too and as have been tragically seen always, were then getting manifested with respect to the repeated Japanese invasions of the medieval-era Korea as well. And further on, in our larger geostrategic and politico-military perspectives interestingly, we do notice that the multiply strategic location of Korea in the Far-East of world's geographical-map too (*which the Koreans couldn't have helped or altered in any way.*), gives her unique geo-political importance. At the same time then, we in addition also are required to historically bear in mind that at no time of the man's living-history, and hence all the more during the bygone ancient as well as medieval ages, no politico-militarily strong and highly ambitious imperial-power, whether operating earlier regionally, or subsequently later globally and then even more intensely in different manners during the present modern-era as well, would have ever been able to ignore these Korean geographical parameters. Hence, we may also say that this too, is one of the numerous ways and methods of the perpetually continued making of the man's living-history, since the very beginning itself.

We have since noted that the age-old numerous magnificent aspects of the ancient Chinese civilisation, had substantially influenced the historic Korean societies and their Nation-State in the past. Hence, we may further quite appropriately say that since the ancient times, China has been the multiply historic mentor of the Korean civilisation itself. In these contexts, when delved deeper; we may interestingly find that this is perhaps the first recorded history of the marvels of civilisational influences and multiple interactions between two societies in the Far-East. In addition, we have also seen that the **historic Choson imperial dynasty** took over the rein of Korea from the **earlier Koryo dynasty**, sometime during the end of 14th or very beginning of the 15th century. We find historically that, during the Choson imperial rule in Korea, the multi-dimensional magnificence of influences of the much more ancient and already highly developed Chinese civilisation and the value-systems; over the Korean societies and systems of governance of the Korean Nation-State, were at their zenith. And during the Choson rule, Korea was able to develop her own politico-military wherewithal and thereby, was remarkably able to face and survive the Japanese invasion sometime during the end of the 16th century.

In these contexts precisely, we may historically notice another magnificent point that, such multiply vast and deeper impact of the Chinese civilisation over Korea had gradually in fact taken place, without any kind of her the then imperial politico-military subjugation by China. Further viewed in the civilisational interactive perspectives, we find these phenomena of the Chinese multiple influences over Korea, getting manifested since the ancient times more through the historic socio-cultural multiple-paths, than any kind of direct imperial-rule by any dynastic-Monarch, then occupying the Chinese throne. **Hence, the man's living-history also shows us that if ever historically allowed to get freely flowered in certain circumstances the socio-cultural aspects and their multiple-influences can even be much more far-reaching and enduring, than threatening might of the sword.** In fact, while objectively viewing these in our larger geostrategic and politico-military perspectives; all we may say that, these aspects are historically situational and hence, as permitted by the various emerging circumstances, possibly anywhere and at any point of history, since rather as permitted by the 'advancing waves' of the living-history itself. But even during those later years of ancient-era and then, further continuing into the entire medieval age as well; all was not that golden in Korea in every respect. While in these contexts, when discerningly looking into the vast and highly complex Chinese living-history; we find that China too, had never been known as a very peace-loving Nation-State at all. Moreover, few exceptions, if at all apart the extensively far-reaching and multiply perhaps all-encompassing civilisational influences, from a very powerful and large country into the neighbourhood; always have their own kinds of sweeping repercussions. And since as borne by the man's living-history too, that in course of time, when any particular society or Nation-State's principal religion also gradually but in a fast-track, gets politicized; the recipe becomes highly complex resulting into multiply unsettling fallouts. And unfortunately in these numerous historic contexts since put together, Korea too couldn't escape such rather inevitable repercussions. Hence as usual, history never spares anyone in any manner.

In these multiple contexts concerning Korea then **Richard Overy** has mentioned that, "**Choson Korea was a model Confucian State**. Dress was Ming; law was Ming; government organization and recruitment was Ming; and Korea established its own international order by accepting tribute from Manchurian tribes and Japanese, while the Ming took tribute from Korea. Social ideals derived from *The Family Rites of Zhu-xi* (Chu Hsi): reject Buddhism; practice ancestor rites; establish primogeniture; and enhance clan cohesion. Everyday life was metaphysical, and philosophy became politicized. Choson was not plagued by powerful eunuchs, but court factionalism emerged to produce political gridlock. In 1590, the Japanese threatened to invade China through Korea. Court opinion split along factional lines and nothing was done. The Japanese came in 1592, and did not withdraw until 1598. East Asia fell into war, with Japanese fighting Chinese in Korea using European arquebuses. Even a Portuguese priest accompanied Christian samurai to Korea. Korea was devastated. Japanese atrocities and pillage left lasting scars, **and the war confirmed Korea as the geopolitical pivot of East Asia**.... From 1800 (*i.e. during the present modern-era itself*), a succession of Boy kings resulted in weak leadership and strong in-laws. Corruption expanded and government became impotent from the 1860s. Systemic decline and climatic disturbance produced vulnerability to Japanese imperialism, which used gunboat diplomacy to obtain an unequal treaty in 1876. In 1882, Korea signed its first treaty with a Western power (USA), opening the country to Christianity and further trade. Economic dislocation and a new religion produced a peasant revolt in 1894. Japanese and Chinese troops intervened and stayed to fight each other in Korea for the first time since 1592. With victory over China, Japan turned to eliminate a Russian threat and was victorious in the Russo-Japanese War (1904-5). Japan annexed Korea in 1910." (Overy, 2008, pp. 174-175)

The above quoted paragraphs indeed contain a very brief, but clearly a very informative glimpse of the highly turmoil-ridden and chequered history of Korea, since the ending phase of the medieval-age and then, continuing into the present modern-era and even right up to the beginning of the First World War. Hence, since forced by the numerous historic circumstances and, certainly not at all at her free initiative or choosing; Korea remained cursed by many clashes of alien military forces on her soil, and also at the same time; had to suffer ruthless politico-military subjugation and exploitation, for a number of years. And such an adversely depressing situation for the unfortunate Koreans, in fact, lasted right up to the **end of the Second World War in the year 1945 and thereby; annihilation of the menacing Axis Powers**. But even then as a great paradoxical irony of the man's living-history during the present modern-era's last 20th century, the above mentioned important geopolitical typicality of Korea was not going to leave the Koreans alone and in peace. Coming to the East and Far-Eastern Asia, we now find there too a totally new and unprecedented geostrategic and politico-military situation prevailing, which are again historically, not of the Koreans' own making. What an irony of fate, and also a series of geopolitical circumstantial developments as well as many emerging complex situations and in addition that the world may never be same again but, the Koreans are still shackled by those very numerous factorial circumstances and fresh emerging situations and in turn, being made to suffer as well. Since, the northern borders of Korea are linked to China and with the end of Second World War and subsequent beginning of the Cold-War; in the 1950s only, another war between the USA and the Communist Bloc began in the Far East and with some kinds of circumstantially compulsive and geographical irony again the battlefields for this war too here, happened to be nowhere else but in Korea that this war came to be known as the Korean War of 1950s. Well, we may even say that one doesn't have to minutely scan through the pages of man's very vast living-history in order to know as to what all happens to any geographical area or a Nation-State, when it is perforce made to act as battlefields; where again alien military-forces had to fight a devastating war. **The end result for the time being for Korea was that, she got artificially divided with a kind of ruthless geopolitical 'surgical-knife' by the rival Foreign Powers; and for which the Korean suffering masses had given no free and 'sovereign' consent whatsoever.**

And hence for the helpless victims, that's the way in this world, the power-rivalry based geopolitics functions regardless and ironically as a result with no faults of theirs, the Korean peoples are still suffering. **Now, Korea remains divided in two parts: the North and South Koreas; with millions of Koreans having already died in such wars. North Korea is suffocating under an extremely communism-oriented ruthless dictatorship, where even fresh-air is not permitted to blow through the masses, perhaps to get them any little relief. South Korea is altogether a different ball-game and politico-militarily remains an U.S. Protectorate. Situation remains tense always, with both North and South Korean Armed Forces facing each other round the clock, in a kind of eyeball to eyeball situation. But geopolitically, the prevailing situations in the Far-East and along with the U.S. geostrategic and politico-military dominance over there; South Korea first got systematically industrialized and subsequently, has moved far ahead and today, she is even justifiably known as an economic as well as industrial 'Asian-tiger'.** We would be coming to these aspects more appropriately in some more detail in the subsequent chapters.

Now in our larger perspectives, we may prefer to come to, in a way rather parallel **Japanese Nation-State**; which has quite different, but no less engaging and turmoil-ridden, historic contours and dimensions in this regard. If historically and along with an intended criticality and also in our larger perspectives, we make efforts to peep into a sort of the remote past of Japan as well as her distinctly different societies along with her almost totally different socio-cultural aspects, socio-religious value-systems and ways of life, then, we may come to some very unique and altogether different reasons for all these, since the ancient-era itself then prevailing for more than millennia in the Japanese Islands. Geo-politically in this regard, we may possibly identify the main reasons for all these, concerning Japan, to be again largely geographical and that interestingly being the situational location of the **'Yellow Sea'** in between the Chinese mainland and Japan and, then the **'East China Sea'** itself. As per the above and as historically pertaining to the very ancient-era in this regard, the first of the two obvious facts emerging during our ongoing critical review at this precise point, happens to be that the above mentioned seas have separated the Japanese Islands from the mainland China and Korea. And the second reason being that, the ruling Mongols in China, and also heading the then unprecedentedly vast Mongolian Empire were no sea-faring peoples, otherwise waters of the above mentioned seas were creating no major marine gaps over there.

Hence, contextually the above mentioned and now obviously appearing to be so simple a geographical fact, but historically during those ancient times being so consequential is one of the very many momentous conjectures of man's age-old, multi-dimensionally very vast and also continuing living-history. In other words, we may also visualise that the living-history's age-old and revolutionarily very far-reaching man's professional marine acumen, along with the aspects of ship-building technologies during those ancient times, were since yet to travel to the Far-East and gradually get embedded over there. In these contexts, if we may delve little deeper and then, we observe in the pages of man's very vast living-history, as to why Japan could remain aloof as well as politico-militarily independent all along for so long during the many centuries of the earlier ancient-era. Viewed in these interesting perspectives in a context here, it is historically no wonder as to why even **Mongol Kublai Khan**, then ruling over China and also, **the ancient-era's great traveller Marco Polo** couldn't cross over into Japan. Interestingly in this regard, we find this as a typically unique and historic example, as to why sometimes and also circumstantially how a particularly specific kind of geographical location and its situational setting may historically turn out to be both, due to geostrategic and politico-military reasons, accruing perhaps certain very far-reaching and special kinds of historic advantages, or even some disadvantages as well.

But at the same time, man's living-history also shows that any such situation-specific conditional advantages do carry certain kinds of unavoidable prices too, ironically as integrative riders. And contextually during those bygone historic times we do notice here that in case of Japan then, this specific price historically happened to be the consequentially accompanying 'aloofness'. But these kinds of accompanying prices, in any case and in any historic sense; cannot universally be generalised. Hence, during our ongoing multiple discussions again we do observe that, fascinatingly strange are the ways of man's multi-dimensional living-history, since continuously in the making. Since under the mother-Nature highly complex and immeasurably larger, unending and infinite scheme of things, man's behavioural responses in different situations or even in an identical situation in two different locations and/or at two different historic points, are never the same and also unpredictable too. And hence similarly those strange ways of the continuous making of man's living-history too, are seldom predictable and obviously therefore, cannot in any manner be charted by the man in advance; since in this regard there are always so many unpredictably complex variables as well.

Now at this precise point and also, critically in our larger geostrategic and politico-military perspectives, historically may we be permitted **to perhaps draw an interesting parallel, concerning the ancient as well as medieval ages' Great Britain and of course at the same time, also keeping**

in mind certain decisively major differences too in this regard? We are since aware that the historic British Nation-State too, is territorially a cluster of Islands and hence having been geographically kept separated from France by the English Channel and thereby, also from the continental Europe. In addition, we are also aware that this geographically situational-setting and its own multiple-aspects for the British Nation have had decisively major-roles to play in the making of her living-history, since the beginning of those ancient times. Hence, it is also no wonder that since the very beginning of the British Nation as a civilisational as well as geopolitically, a distinct entity she has continuously remained independent and sovereign, but certainly not aloof at all. Since the advent of man's both sea-faring acumen and naval-powers, the British people have also been one of the leading pioneers, with respect to the relentless sea-bound adventures. In addition, since the beginning of the European living-history of both ancient as well as the medieval ages, the British had a well-developed tradition of building both big and different types of ships and also in addition, raising a world's very powerful, well-trained, well-equipped and professionally well-led Naval Force. And these in turn, effectively and always prevented other European naval powers from setting foot on the British Islands. We would appropriately again be coming into some later periods of medieval-age, and also then, continuing into the modern-era's prolonged historic phases of the building of overseas and truly global British as well as other historically rival European Powers' Colonial-Empire too, little later in this chapter itself.

Now, contextually coming back to the precise point where we just a little while ago, have left the historically multi-dimensional and larger geostrategic and politico-military aspects, again critically concerning the Nation-State of Japan in this regard we notice that, as the much earlier periods' history of Japan is exclusively unique so is her modern-era's unprecedented fast-track awakening too. But at the same time, while befittingly looking into these fascinating historic aspects in our larger geostrategic and politico-military perspectives; we also find that contextually such interesting historic-attributes are not Japan specific only, but during different historic time-frames and also, with differing textures and contours; the same are also found to be true in case of ancient Greece, medieval British Nation and then, modern-era Russia as well. But any historic similarity if at all in this regard, is also found to be ending there only since historically highly chequered path in each case, has clearly been very different. **Hence in case of Japan, while looking into these aspects contextually with a critical objectivity we do find that her historically exclusive uniqueness in this regard, have been her multi-faceted socio-cultural, socio-political and socio-psychological all round transformation-cum-modernisation and which in turn, have been unprecedentedly quick-paced, highly focused and comparatively peaceful as well.** And the same is historically not at all completely true, in case of the other nations cited above in this regard, though no doubt in certain respects they too, may be taken as being very appropriate examples here.

Professor S.R. Sharma has very informatively mentioned that, "The awakening of Japan is unique and unparalleled in human history. Such all-round transformation as we find in modern Japan has, no doubt, been accomplished by many another country, but only after a long process of natural evolution as in England, or by violent revolution as in Soviet Russia. In the ancient world, Greece displayed a sudden and surprising gush of energy, after the overthrow of Persia, and created a wonderful culture but Greece could never be united. The feeble imperialism of Athens proved abortive in the face of the irrepressible centrifugalism of the Greeks. The Napoleon of Greece (Alexander) was a foreigner, and his work was even less effective than that the little Corsican. But we have in modern Japan the rare combination of the creative energy of the ancient Greeks, the revolutionary fervour of the modern Russian, and the industrial and technical efficiency of the

English. And all these characteristics have come to forefront within less than a century (*Since with respect to the year 2012, Prof. Sharma had published his book 1938; this specific time-frame may appropriately be altered as around one and half centuries*). Indeed in the *Awakening of the East*, though chronologically the last, has been the most significant and portentous....(*Quoting from Will Durant's book on 'The Story of Civilisation' II, p. 829, Prof. Sharma says that*) A recent writer has divided Japanese history into three periods: **(i) classical Buddhist Japan (1522-1603)** 'suddenly civilised by China and Korea, refined and softened by religion, and creating the historic masterpieces of Japanese literature and art;' **(ii) feudal Japan of the Tokugawa Shogunate (1603-1868),** 'peaceful...isolated and self-contained, seeking no alien territory and no external trade, content with agriculture and wedded to art and philosophy;' **(iii) modern Japan (since 1853)** 1868, 'seeking foreign materials and markets, fighting wars and irrepressible expansion, imitating the imperialistic ordour and methods of the West, and threatening both ascendancy of the white race and peace of the world." (Sharma, 1948, pp. 402 & 403)

As per the above mentioned perceptions and, contextually when these are honed into the multiples-aspects of the Japanese living-history we may also visualise that simultaneously in the process, and down those lanes of history in the making, the dynamic peoples of Japan were remarkably able to gradually cultivate, evolve and fruitfully as well as proactively imbibe the **multiple-aspects of 'essential strategic-culture and thoughts'**. Interestingly in this precise context, man's living-history also shows that due to different historic reasons as well as also, due to the different kinds of perhaps historically exclusive socio-cultural, socio-psychological and socio-political fault-lines existing in the societies of the different Nation-States, every nation has not been able to substantially and equally do the same, and in equal degree as well. Now, concerning the medieval-era Japan, we do notice the phenomenal rise of one historic Japanese personality named **Hideyoshi**, who during the 16th century; made great efforts to socio-culturally and politico-militarily transform Japan from Buddhist-Japan to action-oriented and somewhat outward looking, but feudal-Japan. We may recall as since also discussed in different contexts in the earlier chapters that, **universally the various aspects of age-old feudalism, have numerous textures, dimensions and socio-culturally as well as socio-politically, many kinds of fixated manifestations as well**.

Hence, in Japan too then, the multiple-aspects and again the display of feudalism as it gradually got germinated and strongly firmed-up in the medieval-era Japanese societies, had its distinctly own exclusive kinds of features and fixations. In medieval Japan, during **Hideyoshi's** times, though such aspects of feudalism were order of the day; largely based on them only, he was able to pro-actively transform and organise Japan politico-militarily and, also pro-actively up to a great extent. But ironically, these phenomena had their own inevitable methods of extracting prices from the concerned human societies as well. Hence, man's living-history in a context, has also addressed him as **the Clive of Japan**. Interestingly down the ages, in case of the every historically noted personality in every conceivable area of human endeavour, there are different and consequently highly decisive reasons/occurrences in that person's life earlier, which are instrumental in triggering the gradual transformation of his/her total personality so that in course of the life, he or she is able to do and achieve what all are perhaps destined for him, in the wombs of history to be subsequently made and also draw his/her own indelible lines in the concerned pages of that living-history itself, for posterity. **Now in these interesting perspectives, again coming back to the education and personality-development aspects of Hideyoshi, we notice that he was perhaps found to be rather an uncorrectable boy by his family elders and hence, abandoned to fend for himself.** This may appear to be even amazing, but as his destiny would have it—this proved to be a great turning-point in his life and also consequently thereby, momentously far-reaching. **Hideyoshi grew, got educated and trained by the historically famous 'institution' of Samurai**, which in turn marvelously shaped him for being capable to move-on, on his epoch-making path in the Japanese living-history. **It is well known that the centuries-old legacy of Samurai happens to be the famous historic tradition of the swordsmen in Japan.** After finishing his typical kind of education and severely demanding training, under the strict guardianship of Samurai, Hideyoshi began with his sudden military-moves into Korea in around 1592, but that proved to be a misadventure. Anyway, he didn't lose his heart, since he as a person, full of energy and remarkable sense of history, was no shortsighted man. He had many extraordinary 'geostrategic and politico-militarily oriented' instinctive qualities and vision and in addition he had also substantially possessed the already earlier listed four prerequisites, as essential attributes for any empire-building venture with the decisive might of both the sword and his multi-dimensional strategic-vision.

In this precise context, Professor S.R. Sharma has written that, "**Hideyoshi had the foresight of Sir Josiah Child (who in 1685 dreamed of 'the foundation of a large, well-grounded, sure English dominion in India for all time to come')**: 'With Korean troops', he assured his Emperor, 'aided by your illustrious influence, I intend to bring the whole of China under my sway. When that is affected, the three countries (Korea, China, and Japan) will be one. I shall do it as easily as a man rolls up a piece of matting and carries it away under his arm." (Sharma, 1948, p. 403) (*So in our larger perspectives, interestingly, the above quoted highly reflective few-lines* ***very vividly show as to who precisely was Hideyoshi*** *and in addition, what all he was capable of doing, or he then though so, for Japan; in the momentous process of drawing distinctly his own indelible lines during his times, on the pages of man's living-history.*).

As man's living-history further moved-on, after Hideyoshi during the medieval-era Japan, the history making reins of that country subsequently moved into the hands of another Japanese legend named, **Iyeyasu** who commanded unchallenged sway over the destiny of Japan in the period from 1603 to 1616, i.e. for nearly 13 to 14 years. He belonged to the **historic culture of Shogunate in the medieval Japan and since Shogun in Japanese, means a military commander or general**. And that precisely was the historic system, as since got cropped-up in the medieval-era Japan; which historically in fact, was able to transform Japan into a dominantly military-culture-oriented and highly feudal Nation-State and in addition also, along with a very ambitious kind of an uncontrollable urge to geostrategic and politico-militarily, build and expand her empire. Hence in our larger perspectives and at this precise point, if we may critically delve deeper, we interestingly notice that in fact, it was principally the Japan who began calling the shots then; and perforce made herself largely the lead player, in the overall history of the distant medieval-era Far-East region. And historically all this continued for a long time in fact till the arrival of no less kinds of geostrategic and politico-military as well as extremely imperial-cum-colonial culture-oriented and also, more powerful and rivalry-ridden European seafaring Powers, in the Far-East. First and foremost result was that **with the arrival of both geostrategically and politico-militarily much more crafty, shrewd and with a multi-dimensional vision to patiently become long-term and more capable overseas players; the era of a kind of the then historically unchallenged Japanese supremacy over the region, became a thing of the past**.

Coming to the highly momentous historic aspects of the medieval-era Japan, then under her Shogunate multiple-sway history further tells us that, as Shogun rose over the Japanese horizon they in fact began to exercise more commanding powers than the Monarch himself, over both her medieval societies as well as the military. Consequently these realities then in Japan had their own multiply far-reaching fallouts. During those medieval years in Japan for a very long time; Shogun had remained quite understandably, highly feudalistic as well as strictly clannish too since they kept coming from the very conservative Japanese Minamoto family. And the clan they came from was called **Tokugawa** and hence historically, this medieval Shogun dynastic-cum-feudalistic regime in Japan, came to be known as **Tokugawa Shogunate**. In many respects, the medieval-age Japanese Shogun-era, though feudalistic as well as highly militaristic; was remarkable, and the first builder of Japan as one of the foremost dynamic Nation-States of the world with momentously futuristic fallouts, affecting the entire humanity. We would be becoming to these aspects, as we continue to proceed further.

Professor S.R. Sharma while quoting Lafcadio Hearn as well as Will Durant, and then himself contextually elaborating further, has mentioned that, "**According to Lafcadio Hearn, 'the Tokugawa was the happiest in the long life of the nation.' Professor Will Durant writes that: 'Iyeyasu organised peace as ably and as ruthlessly as he had organised war, and administered Japan so well that it was content to be ruled by his posterity and his principles for eight generations.' The principles of Iyeyasu were summed up by himself thus: 'Take care of the people, Strive to be virtuous, Never neglect to protect the country**.' Internally Japan suffered from the evils of feudalism, but externally she appears to have been always united in her attitude towards foreigners. The patriotism of the Japanese is unique and ancient: it has been almost their true national religion. 'The great Yamato (i.e. Japan),' wrote in 1334, 'is a divine country. It is only our land whose foundations were first laid by the Divine Ancestor. It alone has been transmitted by the Sun Goddess to a long line of her descendants. There is nothing of this kind in foreign countries. Therefore it is called the Divine Land.' This has been the faith of the Japanese people ever since. As a corollary to it they have ever looked upon all foreigners with suspicion if not hatred. Particularly this has been their attitude towards the white races—the Europeans." (Sharma, 1948, pp. 403 & 404)

The historic and highly revealingly momentous contents in the above quoted lines from internationally known learned authors, bring many relatively hidden or historically weighty as well as far-reaching facts of the Japan's past. And at the same time, we do also notice that those historic facts, concerning the unique Japanese societies and a prominent Nation-State indeed did have multiple kinds of positive as well as negative, momentous consequences for herself. In this interesting context, man's living-history also shows that, it is itself a dynamically as well as multiply-reflective, chronological living-records of again the multiple consequences of man's own positive and negative deeds in the different societies and/or Nation-States anywhere, in any given situation and at any consequential point there-by, of that living-history itself.

Now as we critically proceed further, in our exploratory review in the defined larger geostrategic and politico-military perspectives, we notice in the relevant pages of the multi-dimensionally voluminous living history of man, from the very beginning itself that historically no country, has ever been able to exercise even a reasonable amount of always the most crucial multi-dimensional strategic-prudence in her multiple conduct of the 'tenets of statecraft', neither in the areas of internal governance nor even in the highly complex and challenging domain of the cool, sensitive, visionary and measured conduct of the art and craft of both diplomacy and the matters-military with respect to any actively threatening, or any potentially possible adversary in any kind of the given geostrategic and politico-militarily emerging situation. Ironically, we do also find that, since ages and in their larger perspectives, the above mentioned historic facts are perhaps the greatest paradoxes of the man's living-history. As we desirably proceed in a way, to dissect these aspects microscopically in our contexts, then we do also clearly see that these historic occurrences anywhere, by themselves have never been uniformly similar and consistent, in any manner at all. Historically somewhere and in a specifically given situation, if some socio-political historic entity or an empire/nation has been able to successfully do the same then it has seldom been found to be healthily reassuring and enduring. And interestingly in these contexts and at this precise point the historic Nation-State of Japan too, has been no exception at all. Historically, perhaps the principal reason in this regard mainly appears to be at different historic periods/times the unpredictably varying degrees of the complex processes of the most crucial cultivation, evolution and then, proactively imbibing; the since earlier already discussed, the multiple-aspects of **'essential strategic-culture and thoughts'**.

Now in the above discussed engaging contexts, while coming back to the above noted historic aspects of the **Tokugawa Shogunate** in the medieval-era Japan, the above quoted remarkable principles of the first Shogun Iyeyasu then, for the desirable politico-military governance of his Nation-State, were the first seeds sown in the medieval-era Japan triggering the multiple-process for substantially successful cultivation, evolution and imbibing those multiple-aspects of the 'essential strategic-culture and thoughts' by both the Japanese ruling-elites and the peoples. **At this precise point, if we may recall the very distant ancient-era's, universally true multiple-aspects of the 'Art of War, since historically first enunciated by Sun Tzu of China in his timeless Treatises and along with subsequently, indeed the multi-dimensional historic Treatises on the entire gamut of the politico-military governance, as spelled-out by Chanakya during the pre-Christ era of ancient India, then man's living-history also shows the above quoted short and crisp principles for governance in Japan; as well as the conduct of geo-politics as deemed fit then, by the medieval-era Japan's great historic personality Iyeyasu, perhaps interestingly, may be considered as the third concept, in the line in this regard**. At a first glance, these views as since derived here, may perhaps appear to be matters of conjecture only and debatable, but these too may appear to be having some substance as well.

Looking deeper into the historic aspects of the Tokugawa Shogunate in the medieval-era Japan, we find that by the end of 17th century the Shoguns, one after the other in their successions, were able to successfully subjugate and control the powerful sway of the Japanese feudal barons in their respective territories since known as daimyo in history. This was no small achievement over there then, since the powerful and warlike feudal-lords happened to be in a kind of unchallenged control of the Buddhist institutions then extensively flourishing in the medieval Japan. The extent of powerfully entrenched feudalism since socio-culturally as well as socio-politically then, existing in that medieval-era Japan, may in these contexts be perceived by the historic facts that during the prolonged periods of Tokugawa Shogunate those warlike, rich and landed feudal-barons along with their herds of samurai, though being only even lesser than ten percent of the population, happened to be the most privileged classes remaining busy in nursing all kinds of their own entrenched vested-interests. In addition to these, the merchant-class, forming a sizable portion of the population and also accorded a relatively lower status but under the practices of Confucianism and since as adopted by the Tokugawa Shogunate, as the multiple-tenets of their statecrafts those merchant-classes were substantially given free hands to go about for conducting their business and earn profits.

Obviously in cities in the medieval Japan then, some of those business barons too, became fabulously rich. Ironically as a far-reaching consequence then, the Japanese masses, in such a far-eastern cluster of islands as a medieval-era Nation-State since in a state of relatively deprived as well as toiling majority population were largely left, to fend for themselves. These socio-cultural and political realties then existing in the medieval-era Japan and, though not so visible under many kinds of shining initiatives, deeds and accomplishments achieved under the centuries of Tokugawa Shogunate reign certainly could not be wished away. We may recall that during our earlier discussions in different contexts we have also noted that as a perhaps inevitably a very great historic paradox, in the man's continuing living-history since ages, the seeds of both rise and fall of any kingdom/empire, are perforce sown by the concerned ruling-elites themselves, in their own overall systemic conducts misconducts of the statecrafts. Hence we may say that since ancient times, this historic reality in particular, also appears to be phenomenally universal as well.

At the same time, these some special kinds of socio-political historic phenomena, but largely triggered by other internal or external power-centres and through the various kinds of geostrategic and politico-military means, have their own multiple dimensions. In addition, as and since repeatedly being seen in the man's continuing living-history indeed have their own internally germinating causative-factors, for both the rise and fall of the concerned political entity or regime; remain built-in, in that particular regime's socio-cultural, socio-psychological, socio-political, and/or then socio-religious segments as well. At this precise point, man's living-history also shows that the combination and degrees of such causative-factors, since pertaining to the above mentioned different segments in this regard may consequentially, with their own complexities vary from regime to regime and hence, may never be identical at all. In the above discussed contexts we find that under the Tokugawa Shogunate medieval Japan too, certainly was no exception. But at the same time, the historicity of the remarkably unprecedented politico-military sway of Tokugawa, over the then whole of medieval Japan cannot be minimized. Besides these, Tokugawa had uncommon geostrategic and politico-military vision for consolidating the Shogunate national regime and, make it unchallenged by their potential rivals. They were shrewd enough to view even the Emperor as their possible rival and hence, were also visionary enough to strategically prevail over and dominate the imperial court as well. In addition to these, the Tokugawa Shogunate became brilliantly successful in taking control of the then entire gamut of the Japanese diplomacy and foreign trade too. Historically, it appears that largely **many kinds of negative-fallouts of the powerfully entrenched and socio-culturally as well as socio-politically all pervasive aspects of feudalism**, during those times in the medieval Japan and at the same time, coupled with the prolonged and rather complete isolation of Japan from the external world were perhaps the principal reasons for the Tokugawa's inevitable fall, as would also be seen in the succeeding paragraphs.

Historically later, when the Europeans landed on the shores of Japan, they brought in their own kinds of missionaries-oriented religious aspects of Christianity too, as a new kind of religion, since then totally unknown to the Japanese populace. But in the overall historic perspectives in this regard, we may not forget that Christian-missionaries, besides their built-in programme for spreading Christianity, they also bring-in well-conceived programme for multi-dimensional social-welfare, education and health improvements in substantial manners; in any lands they have stepped-in. But further, due to the all kinds of fault-lines creeping-up on both sides and thereby, the inevitable interplay of the clashing vested-interests, do create socio-cultural and socio-political complexities and tensions as well. These remarkably initiative-driven and sea-faring European Powers, and with the baggage of their own historic reasons in their own continent were also divided among themselves, in their own numerous and mutually rivalry-ridden nationalities and at the same time; during their medieval-era overseas competitive initiatives, were also having their strategically very shrewd and different agendas. Their agendas pertained to many areas, such as geostrategic and politico-military, socio-religious and then, having their own different kinds of programme concerning trade and commerce as well. Hence, when seen in their own perspectives, then related to Japan we may even say that the recipe for the inevitable confrontations, were rather complete and, rest is again no less chequered historic chapter, of the medieval-era, including during the beginning of present modern-age Japan in this regard.

In these specific contexts, **Richard Overy** has written that, "Nonetheless, by the end of the 18th century financial crisis and rural unrest prompted attempts to reform by the Tokugawa. All ended in failure. In 1854 and 1855 the Americans, British, Dutch and Russians extracted 'friendship' treaties from the Tokugawa. These aggressive foreign intrusions coupled with the weak Japanese response provoked crisis. For once unsure of itself Tokugawa solicited advice from the *daimyo (i.e. feudal barons)* and authorization from the imperial court, thus drawing both groups on to the political stage. In 1858, the Tokugawa signed trade treaties with the Americans, then with the British, but without first seeking imperial approval. A wave of radicalism swept Japan which led to the fall of the Tokugawa." (Overy, 2008, pp. 177-178)

By the time Tokugawa regime in Japan had fallen, the 17th century was ending and even sometime much earlier itself; present modern-era had already substantially entered in the European continent. This had its own multiple influences and changes too, since **the unprecedented Industrial-Revolution, with all its far-reaching dimensions and yet with quite unknown potential and capacities was already sweeping fast almost all over the Europe**. These were the momentous times, throwing unprecedented kinds of new opportunities as well as challenges before those rival European Powers, with their overseas ambitions. We may recall that much before they reached the Japanese shores, they were already present in China too, with own ruthless geostrategic and politico-military ambitions and hence, they were in the thick of their competitive multiple-activities, with their **'hands of steel in velvet gloves'** and hence, outwitting and out-maneuvering the Chinese power-centres, by all sorts of means at their disposal.

This is one relatively recent historic example as already mentioned earlier in some contexts; of the **all kinds of games which nations play, in no-holds-barred global-arena of the international power-politics and interestingly, which have their own each nation-specific, multiply compulsive dimensions as well which, in turn, keep driving the nations concerned while playing these games, since regardless of the 'victim' nation's survival, peace and thereby, which is perforce made the playing fields. In these precise contexts we may also say that, what indeed is a momentous irony of the man's living-history herself**. But then, this was since destined, as before the Europeans moved first towards the eastern Chinese shores in the Far-East; they had already both geostrategically and politico-militarily, swept over almost whole of the Middle-East, the entire Indian subcontinent and South-East Asia too. **In these ruthlessly competitive games of geo-politics and in addition also, to multiply subjugate, dominate and colonise as many new global overseas territories as feasible the British, with their relatively superior and visionary geopolitical farsightedness as well as better groomed traditional skills—concerning the multiple-aspects of 'essential strategic-culture and thoughts'; the British Nation-State was ultimately able to occupy the centrestage and, since historically remaining far ahead of their other European rivals**. Hence, with their Naval and military might, it was a matter of time in a short span of one or two centuries only, before those rival European Powers began knocking at the openings of Japanese shores.

In our larger perspectives, we have since very briefly discussed earlier; as to what all they did, while entrenching themselves in the Chinese soil. Moreover, during the times of those European inroads in China and much before their arrival on the Japanese shores, history also informs us that

Japan, thanks to the prolonged regime for more than 700 years of Tokugawa Shogunate and in the process, converting Japan into a viable imperial Asian military-power, Japan was having her own imperially ambitious strategic-designs in China and hence, she had made her own substantial inroads in China. Consequently in the Chinese lands, all these developments then did create a very complex situation, in geostrategic and politico-military perspectives being full of the clashes of arms and imperial rivalries among the multiply competitive rival Powers, with their own kinds of irreconcilable ambitious agendas. All these historic factors in totality did make a highly complex scenario for further more of theirs, intensely competitive geo-political power-games, since already being played by these rival Powers in the Far-East. In the process, they perhaps did never display any genuine concern for overall well-being and the sufferings of the native peoples anywhere.

Hence while looking into these historic aspects deeper, we do also visualise that these rather extremely ruthless phenomena; concerning across the international borders and/or overseas geostrategic and politico-militarily multiple imperial initiatives, have many historic dimensions. Man's living-history also factually informs us that since ages, **any rising and highly ambitious Imperial-Power anywhere and at any given point of history, has always taken such multiple-initiatives for the maximum possible promotion, expansion and consolidation of her imperialistic empire-building historic process, and in addition possibly towards perhaps her own kinds of self-created, unchecked and unjustified 'compulsions' for the any kind of colonialism are also irrefutable global historic facts**. We would contextually be discussing the births of overseas empires and at the same time also historically coupled phenomenon of the dawn of democracies in the West, as we proceed further in this chapter.

We have also briefly discussed earlier, the many kinds of devastating fallouts of the ruthlessly uncontrolled and even very unscrupulous geo-political power-games of the various rival imperial Powers since historically, already having been played in Korea too. But before coming to this, we may also interestingly recall that by that time; **the British, as particularly a dominating leading European Power, had already strongly consolidated their geostrategic and politico-military imperial hold over India and hence with a typical British strategic shrewdness and vision, they had made the Indian eastern coasts as the staging-points for their historic adventures into the Far-East**. Hence, while critically reviewing these historic aspects, specifically pertaining to the Nation-States located in the Far-East, and also in our larger perspectives we do notice that the various streams of man's living-history since travelling through the different historic ages and time-frames and also at the same time, through numerous geographical continents in the world, are all mutually influencing in the processes of their continuous history-making and thereby, interactively integrative as well.

Now, while coming back to the highly checkered saga of the medieval as well as the beginning of modern-era's living-history of the Far-East, we do find that when Japan remained isolated from the outside world, for few centuries since the middle of 17th century itself, ironically time itself had become rather static for her in many ways. But at the same time, far-reaching changes were already taking place in whole of Europe, America, even Africa and then, in the remaining parts of Asia and these in turn, were fast going to further make multiple influences in the whole of world; and interestingly also later-on, in Japan too. And hence during those years, we do also notice that the first thing in Japan happened: that her isolation was perforce immediately made a thing of the past and thereby; time for Japan, was never going to be the same again. The most influential and far-reaching development during these times was that, **the United States of America** had already achieved political independence from the British imperial yoke in the year 1789 and hence, had fast begun the process of consolidation of the U.S. national territories and also at the same time had quickly moved ahead on the multiple paths of recovery as well as accelerated political, economic and industrial developments of the newly emerged and indeed, that of a continental-size nation. Simultaneously as a great fortune democracy had also dawned in USA and consequently for her, soon indigenous germination of the many kinds of far-reaching positive benefits and traditions too, had begun in that large nation of the multinational immigrants; and since geographically located across the Atlantic for the Europeans and at the same time, across the Pacific for the South-East as well as Far-East regions.

We may now contextually visualise as to what kinds of truly momentous and great times then, those recently past centuries had been when the multiple advancements of the **Industrial-Revolution**, had already historically begun gathering the sweeping and overwhelmingly far-reaching momentum in the whole of Europe and, subsequently in the North America as well. At this point, we may prefer to bear in mind that, the most influential and historically path-breaking civilisational-birth of the Industrial-Revolution itself, for the entire humanity had already taken place in Europe, more than a century earlier. We would be discussing these and many more multiple historic-aspects of the Industrial-Revolution in the next seventh chapter. Now, so far as the Japan during those times was concerned; she was already a civilised nation in her exclusively indigenous socio-cultural, socio-political and politico-military dimensions and though Japan was then somewhat fixatedly feudalistic but at the same time, she was internally commanding reasonable state of peace as well. Though Japan then remained isolated, but she had already received far-reaching Chinese civilisational influences and thereby Buddhism flourished there and subsequently Shinto too, which happens to be an exclusive kind of the Japanese religious worship of their ancestors that was revived. Confucius, being perhaps the greatest and most well-known sage, philosopher and visionary thinker of the ancient-era China, had already been adopted by the Japanese peoples, as their civilisational-cum-socio-cultural idol and in addition also, as a source of enlightenment in many ways, during those times. Historically this, too, is a classical example of the **unique and multiple civilisational interactions between not only the two, but three nations, i.e. China, Korea and Japan respectively**.

Hence, we do also interestingly note that the flourishing Japanese nobility, their imperial courts and the aspects of their ways of life in many ways had some resemblance with those of the medieval-era Europe. But as noted earlier that, during those times of the Tokugawa Shogunate in Japan, these were visibly on surface only and the down-trodden Japanese toiling masses were getting restless too with socio-economic discontent, and things were not going to be easy for Japan anymore. But we may also recall that, in the then medieval Europe to both socio-culturally as well as socio-economically, things were far from being smooth and easy in any manner. Simultaneously, in other geopolitical and strategic dimensions; European and American overseas Powers had relentlessly begun knocking at the doors, seeking obviously many kinds of openings in Japan to further their since already earlier mentioned multiple-agenda. Initially these foreign attempts towards Japan were not initially bringing the desired success for them, but these seasoned foreign geostrategic and politico-military Powers were not the kinds to easily give-up and withdraw, since they knew that the duly planned and focused multi-dimensional perseverance with powerful naval and military forces and also along with other resources to back-up, were bound to succeed in the end. Hence, these overseas Powers finally, prevailed in fact, even over the legendary self-assured Japanese socio-cultural and politico-military tenacity.

Man's living-history in this regard, also interestingly shows that though not in the beginning to start with, but by the mid-nineteenth century onward USA from across the Pacific too, became particularly interested in these rather historically recent geopolitical games; already being played by the rival European Powers in the Far-East. Since by that time, the Americans had already sufficiently developed and consolidated their multiply necessary and viably strong geostrategic and politico-military wherewithal, for such US overseas initiatives and sustained adventures. We further learn that sometime during 1853, an American Naval Squadron, after a long journey across the Pacific; arrived in Japan with a detailed diplomatic letter from the US President and addressed to the Japanese Emperor, containing many kinds of 'sugar-coated' attractive proposals from the USA to Japan. The conservative Shoguns took some time for a number of months, but then finally relented and a couple of ports were opened to the Americans, for the purpose. Historically, indeed it was a far-reaching development in the Far-East.

And historically since, this was the first any such Japanese 'gesture' to any foreign Power and to begin with the Americans were in fact, the first to substantially step-on to the Japanese soil, to further their own ultimately the long-term multiple-agenda in the strategic Far-East. **Interestingly, with the arrival of the US Naval-Squadron there for the first time the Japanese were to see then any revolutionary steam-ship**. In addition for the first time therefore the Japanese had the mind-boggling glimpse of a path-breaking fruit of the fast advancing aspects in Europe, of the historically unprecedented Industrial-Revolution. It was obvious in this regard, that it would have been quite naïve to imagine that on learning this momentous development the other leading European Powers, i.e. the British, Dutch and the Russians then would have kept quiet and remained aloof any longer. Hence, they too subsequently moved-in soon and executed their similar multiple-treaties with the Japanese Shogunate, for their own purposes. These were since momentously far-reaching and fast developments, then taking place in Japan and historically after nearly three or two and half centuries, Japan was again substantially opening-up to the outside world and ultimately Japan was gradually but quite fast, going to almost change forever.

Moreover by the time these foreigners, with their strong Naval, military and economic resources, appeared in Japan with their own long-term geopolitical ambitions; the once mighty Shoguns had lost much of the shines of their armour and regime and thereby, their real strengths had already become things of the past. And hence, their once extraordinarily powerful gestures and displays had already become hollow from inside and obviously then; these otherwise unsettling realities in the then Japan, could not remain hidden for long, from the shrewdly observing eyes of those foreign Powers, and therefore consequently; a kind of rather devastating power-game of 'chess' had inevitably started among those foreign Powers, with Japan herself becoming the 'chessboard'. What a far-reaching turning-point was this for the future of Japan with many kinds of further-on, both positive and perhaps few negative fallouts. **Interestingly in these multiple historic processes the aspects of certain inherent attributes of the Japanese peoples and their focused culture in fact then, contributed a lot towards their unique nation's fast modernisation as well as industrialization.**

Now contextually coming back again to the Japanese history, we do also notice that while Japan then consistently marched ahead on her multiple fast-tracks with her inherent socio-cultural attributive aspects, towards both her substantial modernisation and industrialization and also while copying many aspects of the Western ways of life, ironically Japan then still lingered with now, perhaps the irreconcilable **old social cult of the 'Emperor-worship'**. We may also now say here that, this in fact remained indeed amazing. Historically, these aspects happen to be quite strange civilisational sides of the Japanese peoples. Well at the same time in this specific context, while reviewing the larger canvas of man's living-history in this regard; we do also notice that there have been numerous civilisations and Nation-States since the very ancient ages, including China, India, Egypt, Greece, the old Roman and other historic empires and then, along with many modern nations too, where many kinds of unexplainable strange cults and other similar aspects, have been surviving. In any case, that is that and hence, these may objectively be taken in one's strides, since as they are.

But at the same time, we cannot ignore another simultaneous dimension of the dynamic civilisational aspects of the unique and truly focused Japanese peoples and that irrefutably happens to be that, the Japanese Nation multiply forged ahead regardless and indeed multi-dimensionally so far and so fast that, even the highly self-assured Western Nations were amazingly taken aback and got overwhelmed. In these captivating perspectives we may very appropriately, but in altogether almost with totally different geopolitical situations and challenges and at the same time with totally different kinds of the historic-canvas and thereby; at this precise point, we may encouragingly cite a parallel and definitely very encouraging example of the modern Nation-State of Israel as well. We have since briefly discussed certain very engaging historic aspects of the Jewish peoples residing all over the world, population-wise not so big, but no less significant. With respect to the brief comparative studies of the multi-dimensional civilisational and in addition historic geo-political as well as both geostrategic and politico-military aspects of both Japan and China, **Jawaharlal Nehru** (pages 527, 528, 529, 530, 531 and 532) has very informatively written that, "The **old feudal Daimios** were abolished. The capital of the Emperor was taken from Kyoto to Yedo, which was now renamed Tokyo. A new constitution was announced with two houses of Parliament, of which the lower house was elected, the upper nominated. There were changes in education, law, industry, and almost everything. Factories grew up, and a modern army and navy were formed. Experts were sent for from foreign countries, and Japanese students were sent to Europe and America, not to become barristers and the like as Indians have done in the past, but to become scientists and technical experts.... And at the same time as they pushed ahead these reforms, they **spread the cult of emperor-worship**. It was a strange combination: factories and modern industry and a semblance of parliamentary government on the one side, and medieval worship of the divine Emperor on the other.... The contrast between China and Japan during the last half of the nineteenth century is remarkable. Japan rapidly westernised: China...got involved in most extraordinary difficulties. Why did this happen? The very vastness of China, her great population and area, made change difficult. India also suffers from this seeming source of strength—huge area and population.... Then again China's great civilisation had grown up in thousands of years and too closely interwoven with her life to be easily discarded. Again we can compare India to China. Japan, on the other hand, had borrowed Chinese civilisation and could more easily replace it. Another reason for China's difficulties was the continual interference of European Powers. China was a great continental country. She could not shut herself up, as the islands of Japan had done.... So Japan shot ahead, while China was still blindly struggling on and trying, with little success, to adapt herself to the new conditions. **And yet there is another strange fact worth noticing. Japan took to Western machinery and industry and, with a modern army and navy, put on the grab of an advanced industrialized power.... The passionate and self-sacrificing patriotism of the Japanese was closely allied to this loyalty to the Emperor. Nationalism and the cult of the divine Emperor went side-by-side.** China on the other hand, did not take readily to big machinery and industry; but the Chinese, or at any rate modern China, welcomed western thought and ideas and the scientific outlook.... Thus we see that although modern China entered more into the spirit of

Western civilisation, Japan outstripped her because she put on the armour of it, ignoring the spirit. And all Europe praised Japan because she was strong in the armour, and they made her one of their fellowship. But China was weak and did not have Maxim guns and the like. So they insulted her and preached to her and exploited her, caring little for her thought and ideas.... Meanwhile Japan, by her victory over China, had established her position as the strongest nation of the Far East. China had appeared in all her weakness, and all fear of her vanished from the Western Powers. They swooped down on her like vultures on a dead or dying body, and tried to as much as possible for themselves... France, Russia, England and Germany—all scrambled for seaports on the China coast and for privileges. There was an unholy and a most unseemly battle for concessions. Every little thing was made an excuse for claiming additional privileges or concessions.... It was extraordinary—this shameless scramble. Of course, China did not enjoy paring with territory or granting concessions. She was forced to agree on every occasion by displays of naval force and threats of bombardment. What shall we call this scandalous behaviour? Highway robbery? Brigandage? It is the way of imperialism. Sometimes it works on secret; sometimes it covers its evil deeds under a cloak of pious sentiment and hypocritical pretense of the doing good to others."

In the above quoted lines, written with a scholarly precision and conveying certain far-reaching historic facts, principally concerning Japan and China in a context, Nehru has been able to sharply penetrate through the 'real core of truth' of the aspects of western-imperialism, since at times very craftily glossed-over and also exposed the stinking-muck therein. But at the same time, it has also been found and especially with historic manifestations of certain Western-Powers' colonialism in many regions of the world and more so particularly concerning the British and, more specifically in India that sometimes interestingly even some good too, has perhaps unintentionally come out of that 'sinking muck'; and of course at the time, with a price-tag attached to it. And hence we are again able to see that indeed strange are the ways of man's living-history in its making, since the very beginning.

But historically these very significant aspects in a context apart now coming from the author in particular, of those above quoted lines, i.e. Late Jawaharlal Nehru; who himself happened to be a very historic personality as the first Prime Minister of Independent democratic India and, his well-known marvels in some areas of scholarship cannot be denied; but at the same time, his full of naivety, the saga of failures/blunders in the paramount and sensitive domains of India's geopolitics, geostrategic and politico-military aspects concerning her national security too are still difficult to be forgotten by the Indians. Paradoxically all the more so when **Nehru is found to be writing, with certainly his commendable scholarly awareness about the decisive importance and historically path-breaking reach of relatively, quite modern the Japanese Army and Navy, during the recently gone-by momentous centuries in the Far East. Historically therefore, the most weighty and unforgettable question for the Indians arises here as to why and how, he could afford to almost perhaps so callously forget the decisive geostrategic and politico-military importance of sovereign India's Armed Forces when the crucial era of his 'tryst with destiny', during his own Premiership came after 1947 and thereby, making it humiliatingly so costly for the Indian-Nation?**

We have since at times in earlier chapters too and in different contexts emphatically hinted upon the above mentioned, not at all palatable facts, concerning the very beginning years of now the continuous making of independent India's present living-history. **This is in fact, one malignantly lingering and pivotal reason as to why the modern democratic and also multiply fast developing Indian-Nation, even after more than six decades of her political independence**

so ironically has not substantially succeeded in fruitfully cultivating, evolving, nursing, and then proactively imbibing the multiple-aspects of 'essential strategic-culture and thoughts'. In our larger perspectives, we would be discussing these aspects in somewhat more detail, as we proceed further in this chapter itself.

As times moved-on, during the turmoil-ridden 18th and 19th centuries in the history of the Far-East we interestingly find that in course of her imperialistically very ambitious adventure in China, Japan came in direct military clashes with another giant European imperial Power, i.e. the Tsarist Russia as well. In some historic contexts, this perhaps may even interestingly be viewed as a kind of confrontation between the David and the Goliath. Well, in spite of her smaller geographical size certainly Japan was militarily no push-over and at the same time, their degree of socio-cultural, geostrategic and politico-military confidence as well as the unshakable tenacity since already discussed earlier, cannot be wished away. And when the hour of this clash of arms came, Japan was decisively able to achieve a military victory over the then Tsarist Russia. The central strategic reason for this violent clash between the military might of both Russia and Japan then, happened to be for the possession and strategic supremacy over the Port of Arthur in the region.

With this sensationally surprising victory of Japan over both China and Russia, the other Self-confident European Powers, since already proactively present in the Far East, were then taken aback and also both, geopolitically as well as militarily shaken and hence naturally all these historic developments, began to foment troubles for Japan and her subsequent clashes with those overseas Powers. In a perspective here, if we may critically look into these multiply historic developments taking place in the Far East, then we may also interestingly find them, as interrelated manifestations, since as already earlier discussed in different contexts of one pivotal reason that since the very beginning of man's living-history, **the root-causes of the all kinds of military-clashes between two or more Nation-States or Power-Centres, are the clashes of various kinds of the geopolitical, geostrategic, politico-military and also certain specific kinds of weighty economic interests**. Hence at the same time, the above mentioned pivotal-reason in this regard may be a combination of the all of them or any two, or even more than two of the above mentioned national/imperial interests of any particular one or a group of more than one Powers.

Historically in these specific contexts while coming back to China, we find that China was already deep into a multiple series of crises and numerous imperialistic powers had already got entrenched then, in the Chinese soil, with their own ruthlessly rival and exploitative agendas. Japan too had entered China and, was having her own imperialistic designs in Manchuria. Frustrated Chinese had risen as a rebel, since being **exploited and impoverished and in addition also, was even made opium-addicts, by the British; with the supply of opium, ironically as a British 'trading-commodity', from India. Then Boxer-rebellion too, had surfaced in China in fact, actually being indigenously Chinese in true patriotic-sense, but very craftily, the European imperial Powers over there, had termed it as a brigand criminal uprising**. Moreover, China due to many historic reasons as noted earlier were lacking the multiple capabilities like Japan, as a society and Nation-State to be able to effectively reform while there was still time, since necessary for meeting numerous threats and challenges facing her. It was not today's modern China, since now is presently seen mostly to be aggressively flexing her economic and military muscles. **Lots of water has since flown down the Hwang-Ho, and times had now perhaps totally changed for China.**

As noted earlier in China then, USA had already moved in and the otherwise **Chinese patriotic Boxer rebellion** was already very sinisterly termed as unlawful and brigand-like by the Europeans and hence, they wished to ruthlessly crush it by all sorts of diplomatic as well as politico-military means at their command. All kinds of clashes commenced; consequently China was further humiliated and on top of it, another humiliating and totally unequal treaty, since now known as the **'Peking Protocol'**, was forcibly thrust upon China then in the year 1901. Historically as quoted by Nehru from the diary of an English war reporter, accompanying then the European allied forces invading China, that, ***"There are things that I must not write, and may not be printed in England, which would seem to show that this Western civilisation of ours is merely a veneer over savagery. The actual truth has never been written about any war, and this will be no exception."*** (*What indeed is this paradoxical irony of the man's living-history!*).

As quoted in Professor S.R. Sharma's book we find Tennyson as very reflectively saying that, ***"The old order changeth, yielding place to new, And God fulfills Himself in many ways."*** (Sharma, 1948, p. 349) These few words are indeed timeless and universal. Now coming back to perhaps somewhat long awaited and historically **multiple-aspects of the building of overseas empires, by the rivalry-ridden medieval-era European Powers and also, highly checkered historic past of the phenomenally gradual dawn of the democracies in the West**, we find them to be simultaneously progressing with their many kinds of heat and dust, in a number of far-off continents of the world and at the same time, mutually influencing and inter-connected too. In addition further, these historic phenomena have many kinds of different historic reasons; since separately surfacing in different manners, both in Europe and then in different overseas and far-off continents as well.

Further history also informs us that, when sea-venturing and mutually competing European Powers began moving out and landing on to the virgin shores of Asia, Africa and Americas; with their own different kinds of rival and even indigenously perhaps compulsive imperial agendas, back-home in their own countries all was not well and consequently overall socio-economic, socio-political and politico-military conditions were far from being anything but normal. So those were the multiply turmoil-ridden and highly disturbed times and hence, Europe itself was in state of flux. It would indeed be perhaps humorously naïve, for peoples of the bygone era colonies since acquired by the dominating medieval European Powers, in the above mentioned far-off continents of the world that back home the Europeans were very rich and powerful peoples, who could decisively subjugate them and rule over their countries. It was all far from the plain historic truth in this regard and in fact, it was due to the figments of those conquered peoples' imagination and also perhaps, obviously bred due to their all-round then prevailing ignorance as well.

It was an altogether different story and a later historic development in Europe, including the British-Nation, that they subsequently began being rich and prosperous, after they could consolidate their imperial hold over their extensive overseas empires; and commenced transporting the so many kinds of vast natural resources, wealth and money from their colonies to their home countries, through multiple-means of their own craftily designed trade and commerce as well as ruthlessly exploitative and imperially coercive politico-military means. These are again very weighty and vast subjects of the colonial-era world-history by themselves; but we would be having some very interesting glimpses of these historic aspects as we proceed further here, in our larger perspectives. At the same time in these contexts, we do also notice that many kinds of momentous historic developments too, were then simultaneously taking place all over Europe and they were all mutually influencing and overlapping as well. Such as the **unprecedented advent of the Industrial-**

Revolution, socio-cultural and socio-political multi-dimensional renaissance of Europe and at the same time, populations too of the many countries in Europe, including the U.K., had begun expanding rather fast.

If we may care to visualise all these mutually affecting developments since simultaneously taking place, as an overall historic scenario in Europe in a precise perspective, then we also see the situation as very complex and hence, **almost the whole of Europe was then definitely in a state of flux**. In fact, the mind-boggling complexities of the situation then in Europe, along with the various kinds of rather inherent compulsions too then surfacing there can be gauged from the historic facts that, those rival European imperial Powers were no naïve in the aspects of empires-building; since they were almost relentlessly fighting among themselves to grab as much new territories as feasible, and expand their respective empires in Europe first, since the ancient-era itself. Each European Monarch and ruling-elite had his or her own imperially ambitious and mutually irreconcilable agenda as well as various kinds of geostrategic and politico-military compulsions which were perforce driving them to the battlefields all over the Europe, regardless of the their consequences. On top of it and as we may also recall here that during the earlier medieval-era Europe; the age of a powerfully prominent military-culture of chivalry and coupled with its strict codes of conduct too, had dawned on Europe. In all these relentless clashes-of-arms-oriented times then in Europe, peace and social tranquility were a far-cry for the Europeans. If we may review these highly checkered aspects of the Europe's living-history, in a broadly appropriate perspective at this precise point of our ongoing discussions, then we may clearly notice certain kinds of rather uncommon attributes too, getting sprouted among both the different European ruling-elites and the masses as well, in varying degrees.

Besides the above mentioned compulsive developments then taking place in European nations, such as the commencement of Industrial-Revolution, multi-dimensional renaissance and the increase in population; in addition we do also notice that earlier elaborated four essential prerequisites, concerning the aspects of any viable and politico-militarily enduring overseas empire-building processes too, had since already substantially got gelled in varying degrees, among the European societies and their Nation-States. We may need to recall and since specified earlier that, those four essential prerequisites in this regard, are (i) Essential strategic-culture and thoughts, (ii) Strategically, the imaginative development, well-rounded and dedicated nursing of the aspects of matters-military, (iii) Possession of the essential politico-strategic multiple-skills for the statecraft and aspects of governance at home and finally, (iv) The requisite skills for the multi-dimensional conduct of the art of highly complex international game of the rivalry-infested geopolitics and diplomacy. In continuity it may appear to be perhaps appropriate to also mention here that, the above listed third point concerning the possession of essential multi-dimensional skills for the conduct of highly complex tasks concerning the statecraft at home and the internal areas of governance could only substantially and gradually fructify, through again not so peaceful multiple processes of the far-reaching socio-political and imperial reforms and thereby both socio-culturally and socio-politically replacing the age-old fixated aspects of the old-order by the new ones, in the leading European countries and, along with the flowering of renaissance in Europe.

In these multiple contexts, concerning certain exclusive kinds of the historically highly complex and restive aspects of the medieval-era European societies and the Nation-States, we also interestingly notice that, those very European nations were subjugated and callously exploited by their ruthless despotic Monarchs/Ruling-elites and at the same time, certain other nations in Europe too, were subjected to the similar exploitative treatments and hardships, by other neighbouring nations. This was indeed a typically strange but historically true as well, and those were indeed the broadly prevailing socio-economic and socio-political situations, during the medieval-era centuries in Europe. At this precise point, we may also in addition say that, the creation of warlike situations between the nations anywhere and then; leading to the highly consequential and even tragic clashes of arms, are also other kinds of self-corrective mechanisms in this regard.

In these contexts, man's living-history further informs that as the Naval Fleets and other military assets of the certain medieval-era leading sea-faring European Powers, had begun making both geostrategically and politico-militarily well-crafted in-roads into the continents of Asia, Africa and the Americas and simultaneously therefore back home too, all sorts of no more controllable agitations and upheavals had also begun taking place in the continental Europe and even in the British Islands as well. Historically since the very ancient times, as anywhere in Asia or elsewhere; in Europe too, the Monarchy had gradually managed to acquire a kind of highly regressive, feudalism-oriented and socio-culturally fixated hallow of the ruling Monarch's divine right to rule over his or her subjects; based on the personal callousness, whims, fancies and all sorts of the irrational eccentricities and, even irrespective of serious consequences to the masses, in the kingdom concerned.

Professor S.R. Sharma has mentioned that, "The 'Divine Right' on which the Grand Monarchy was based had become so debased that it proved to be a right to exploit (claim) the people for selfish autocracy and luxuries of the kings. But this claim could not be sustained for long in the wake of the progress the mankind was making. Just as the autocracy and corruption of the Church had given rise to reformation in religious matters, so also in the political field there was soon to be a *re-formation*. **The divine right of kings was to give place to the 'Divine right of Peoples':** ***vox populi vox Dei*****, 'the voice of the people is the voice of God' was to be the new slogan**." (Sharma, 1948, p. 349)

In the years to come in numerous ways, the above mentioned perhaps most far-reaching and indeed mentally illuminating thought-process from then-on in the medieval Europe and subsequently the world-over, and also since the perpetually ongoing times entered into the present modern-era too, has indeed become revolutionarily path-breaking. And subsequently then, this thought-process has become the pivotal keywords for the all kinds of revolution, socio-cultural renaissance and then finally, even for the quick-paced germination as well as historically the most momentous phenomenon of the humanity's civilisational dawn of democracies in the world. Hence in our larger perspectives and at this precise point we may even say that **the advent of Industrial-Revolution and in addition, the dawn of Democracies can perhaps also very befittingly be considered as historically the two most prominent and unprecedentedly influential, humanity's civilisational 'milestones'**; during the ending phases of medieval-age, in the pages of man's continuing living-history. The multiple aspects of man's overall qualities of life as well as again, the fast enhancing levels of man's all-round awareness are indeed the continuously enriching, thriving and also living, by-products of those two above identified civilisational milestones in this regard.

Now, coming back to the period of revolutionarily momentous times in Europe here, in our larger geostrategic and politico-military perspectives specifically with respect to the Nation-State of **Holland**, we find that she was then already geopolitically and militarily subjugated by another Imperial-Power in the medieval-era Europe, i.e. **Spain**. It was indeed a great medieval-era irony in

Europe that Holland or her second name also being The Netherlands, was relatively more developed and certainly much more advanced than Spain but was politico-militarily subjugated only because militarily, **Spain** was then much stronger, and with a traditional culture of mostly being battle-hardy for the purpose. Whereas **Holland** was historically even rival to **Italy** in art, literature and culture and, excelled in the skills for manufacturing technologies as well as was then phenomenally successful in the competitive areas of overseas international trade and commerce with their ships reaching to the ports, almost all over the world and hence naturally, their manufactures were highly preferred goods anywhere they could export.

And at the same time, the Dutch people were also very patriotic, having great amount of national pride and self-esteem, and hence they could never allow the Spanish Monarch Charles I and then his successor Phillips II to be peaceful, with their ruthlessly exploitative sort of colonial rule over Holland. Moreover, Spain was still very superstitious and violently fanatic with their Catholic Christianity too, and hence these also added fuel to the fire of politico-religious struggles for freedom of the Dutch masses. Interestingly, these struggles were fundamentally not against the Spaniards, but much more against the tyrannical rule of despotic Spanish Monarchs. In this precise context, the man's living-history has since recorded Holland's these relentless politico-religious struggles as indeed a praiseworthy **Dutch War of Independence**. Historically some learned historians in a context, have very befittingly termed this example of ultimately the successful Dutch War of Independence, as a historic episode of indeed also a civilisational tussle between the above discussed Old-Order and the New-Order in continental Europe. This successful Dutch War of Independence lasted for thirty years and ended with the historic medieval-era (*during its ending years and the simultaneous dawn of the present modern-age.*); **'Treaty of Westphalia' of 1648** in fact, to full satisfaction of the Dutch peoples. This was definitely a great historic event in many ways.

In these perspectives and to the amusement of many of us Europe's medieval living-history also informs us that there was indeed a different kind similar old verses new orders' tussle took place in England then too, with far-reaching and historic epoch-making fallouts in Europe. In Europe of those years then, **one particular coincidence had its own interesting historicity in the sense that during the time of Dutch War of Independence, the English people too had begun intensive struggle to overthrow the Stuart Monarchy in England and, the principal reason being that the earlier grand Tudor Monarchy had socio-politically decayed and 'degenerated' into the highly callous and exploitative Stuart Monarchy. And in addition the time too, was fast and significantly changing in Europe with many kinds of historic fallouts and further in the sense that the multiple-control of the State's as well as the peoples' future, was no more going to be tolerated into the hands of exploitative Monarchs and consequently therefore, the people were restless to take almost total control of their wellbeing and destiny**. Interestingly in this context, we may also again visualise at this precise point, the complexities as well as interconnectivities of the many kinds of multi-dimensionally phenomenal processes of the unending making of man's living-history in the sense that the initial seeds of the many socio-political and cultural reforms being then triggered in Europe, was perhaps first sown in the **unique historic episode of the famous Magna Carta of the year 1215, in England itself**.

In this precise context, **Professor S.R. Sharma** has recorded that, "Nevertheless, the work of Henry II (1154-89) culminated in making his fifth son John (1199-1216), so powerful that his tyranny initiated a new trend in English history. Its first fruit was the famous Magna Carta (1215), the first of a series of great charter on which the edifice of English liberty rests. Before this was wrung from John by the English barons, the English king was an autocrat unparalleled in his authority in

Christendom for six hundred years since Charlemagne. **'The Magna Carta,' says Professor Adams, 'closes one epoch of English constitutional history and begins another.' The principal interest of British history since the Great Charter lies in the growth of Parliament**.... Greenidge has said that soul of Greek history is its constitutionalism. The same may be asserted about England of all modern countries. As the reformation movement culminated in Netherlands in the political liberation of its people, so too, in England it was to get merged in its constitutional struggle. This last was England's supreme gift to the world. 'While Germany boast her reformation and France her revolution,' says Trevelyan, 'England can point to her dealings with the House of Stuart.... During the seventeenth century a despotic scheme of society and government was so firmly established in Europe, that but for the course of events in England it would have been the sole successor of the medieval system.... **Magna Carta (1215) which may be considered as the first great landmark, though it has always ranked as the sheet-anchor of English liberty. Other Charters which followed in succeeding centuries only sought to secure and extend what had already been laid down in that basic document. The barons who fought against King John for their feudal rights and privileges were really the unconscious parents of the English parliamentary system. The committee they set up to safeguard those rights and privileges developed into the 'Mother of Parliaments'**.... They (*i.e. The Stuart*) interfered alike with civil and religious liberty. Meanwhile the nation—particularly the middle classes had become prosperous enough to get restive and intolerant. As **Macaulay** said, 'During two hundred years, all the sovereigns who had ruled England—with the single exception of the unfortunate **Henry VI**—had been strong-minded, high spirited, courageous, and of princely bearing. Almost all had possessed abilities above the ordinary level. It was no light thing that, on the very eve of the decisive struggle between our Kings and their Parliaments, royalty should be exhibited to the stammering, slobbering, shedding unmanly tears, trembling at a drawn sword, and talking in the style of a buffoon and a pedagogue.' **James I** nevertheless insisted: 'As to dispute what God may do is blasphemy, so it is sedition in a subject to dispute what a King may do. I will not be content that my power be disputed on.' So he, and his son Charles I levied taxes, appointed and dismissed ministers, followed policies, and summoned or dissolved Parliaments, as it suited their arbitrary wills. **When their needs compelled them to go to the Parliaments for grants of money, the later bargained for their rights and liberties which had been trampled upon by their reckless sovereigns.... Hence to cut the long story short, he preferred the scaffold to the fate of the House of Windsor. Charles I was executed in 1649 as the climax of Civil War, and England became a republic**." (Sharma, 1948, pp. 333, 350-352)

What a turmoil-ridden medieval-era living-history of the Great Britain; which herself was evolving then, but not so peacefully at all, and also at the same time momentously struggling to get rid of the since already outlived and rather extremely autocratic the **'Old-Order'** and then, imbibe the **'New-Order'** system, as quickly as possible. And the entire quite unsettling process, lasting over around a century or so; was not going to be so easy and peaceful in any case. Nevertheless, these unique historic socio-cultural and socio-political phenomena, consisting of a series of historic events, in the medieval history of the British people as since also so precisely highlighted along with their own significant historicity in the above quoted scholarly lines in fact do emphasise their path-breaking achievements and also at the same time, being enduring trend-setters for the entire humanity at large and its subsequent future, from there on. And here lies the unique historic 'genetic-seeds' for the subsequent dawn of democracies in the West and then gradually in many countries of the world and interestingly further, this path-breaking phenomenon is still going-on in still more countries in Asia, Africa, Americas and, even further on in Europe as well. Now again coming back to the historically specific aspects of the 'Old-Order' withering away and, giving the vacant space to the 'New-Order' we also notice that England was not going to survive for long as a Republic, as since quoted above; after the end of **English Civil-War** culminating into the execution of the British Monarch Charles I himself, in the year 1649. Ironically, England was yet to settle-down finally and peacefully once of all, in her multiple socio-cultural and socio-political domain. We find that the execution of Charles I and declaring England as a Republic; subsequently brought-in to power, no less autocrat and tyrant **Cromwell (1649-58)**, ultimately as a dictator. But Cromwell's some historic achievements too, towards politico-militarily elevating England into an unrivaled position then in the whole of medieval Europe, cannot be ignored. It is a fact that, Cromwell levied heavy tax burdens on the British people then; but with that money during his time, he was able to raise the then unchallenged British Army and Navy, which did lay historically an essential foundation to substantially serve the future multiple British geostrategic and politico-military national-interests during the history-making centuries ahead, since destined to follow.

But at the same time and in a way ironically, the English people perhaps were forced to land **'from frying-pan to the fire'**. What a history all this has been then, in the whole of medieval Europe and wherein, England in fact, became the lead-player in this epoch-making turn of events. And at the same time perhaps, quite unconsciously but also momentously, England had begun 'blazing the trail' in this regard for substantially, the forthcoming 'New-Order-oriented' path-breaking future; to perform the role of a 'beacon-light', for the entire mankind's forward march towards the further civilisational enlightenment and humanity's acquisition of the democratic rights and freedom. We do also notice that these historic civilisational phenomena were never easy and smooth for the man, and interestingly in our contexts not the least then even for the medieval British nation too since historically, with lot many trials and tribulations as well.

From Cromwell's autocratic rule for nearly the nine years and adventurously experimenting with a kind of Commonwealth system to England again reverting to the Monarchy's altogether newly crafted reincarnation into a refreshing and principally supreme but as largely a ceremonial Institution only has been a short and crisp, but no less checkered period of the medieval British history. In our larger perspectives, while critically looking into these aspects of the English-history we do also observe that, in order to serve as an epoch-making 'beacon-light' for in fact the whole world, and for its **gradual forward historic march towards democracy and supremacy of the Constitutionalism**; England had to first substantially reform herself, for the flowering of her vibrant democracy and unique Constitutionalism wherein; all her Constitutional Institutions had their distinctly defined, mandate, tenets of discipline and unassailable place and also, with least possible room for any kind of constitutional confrontation. These historic aspects are in fact, precisely the unique marvels of the functioning of British democracy and the constitutional fragrance of the 'mother of Parliaments', i.e. again the British Parliament.

It may be desired that in an appropriate perspective, the non-British may not have any kind of even little reservation in bestowing the befitting praise, where it is very much due and hence in the above discussed aspects of the momentous years of the medieval British history, that unreserved praise becomes rather duly mandatory too. In these specific contexts, we do further notice that during the mid-seventeenth century itself **as England began to switch-over to the constitutionally ceremonial Monarchy, the real executive powers for the aspects of national governance and also for the law-making authority were finally passed on to the peoples' elected Parliaments and through them to the various governmental Institutions, and together with her Monarchy's**

all dignity, ceremonial pomp, glorious display and stature remaining intact but also thereby, the King could only reign but does not rule or govern. A number of nations in the world, in their own ways and with varying degrees of success, have remodeled themselves on this unique and indeed praiseworthy British-system.

We then find that with respect to **France**, it was the historic **French Revolution**, which was indeed instrumental in triggering this path-breaking phenomenon on the French soil. But due to various historic socio-psychological, socio-cultural, socio-political, and then the distinctly different politico-military reasons, as well as the altogether different culture of geopolitical and strategic fixations the historic methodologies of France in this regard, happened to be quite different from those of the British Nation, and that is understandable as well. We may recall that in some different but allied contexts, we have since briefly touched upon the historic aspects of the no less unique **French-Revolution**. Learned historians are of the opinion that in fact, **the fall of Bastille on the 14th of July 1789, is historically the triggering event for the momentous beginning of the French-Revolution**, and this is logical too. We may also note here that, this was no less epoch-making historic event in its own ways for the entire mankind. The historic-slogan of French origin at the point of the fall of Bastille is still powerfully echoing in many parts of world, with different kinds of socio-cultural and revolutionary fervour. And that slogan happens to be, since now so commonly well-known, as the **inspiring resonance of the three words, i.e. Liberty, Equality and Fraternity**.

In our larger perspectives in this regard, we do also notice that as in the case of England, during her own struggles for liberty and justice, the British Monarch Charles I was executed and similarly then, later on during the French-Revolution, the French Monarch Louis XVI too, was done to a violent death during the year 1792. France was declared a Republic and ironically, a much more violent and turmoil-ridden history was then to follow for the French people; perhaps culminating into the **sudden and sensational rise of Napoleon the Great, to finally his equally sensational fall**, and geopolitically as well as politico-militarily with that a historic era had ended in the French history. But France survived as a republic, obviously minus any kind of an institution of the Monarchy too. In our larger perspectives and at this precise point, we may even say with perhaps substantial historicity in this regard that, the French are French and the British are British.

In these contexts, as we begin to review certain relevant historic aspects of the **two Asian giants, i.e. China and India** we find them living through different kinds of very violent and extremely unsettling phases of their own altogether different kinds of living-histories, and with their distinct and strange turn of interesting events too. So far as **China** is concerned, no doubt that the Chinese peoples were subjected to many kinds of humiliations, first in the hands of her neighbour, Japan to perforce obey many kinds of her autocratic dictates and simultaneously to further Japanese imperial interests as well. Then as the various European Imperial Powers began establishing themselves on the Chinese coastal territories China again perforce was ruthlessly exploited to further their unjust commercial interests and even naked greed. In these multiple processes, those overseas Imperial Powers geopolitically adopted many kinds of shrewd games of coercive diplomacy and along with that if they felt any compelling necessity then they didn't hesitate to trigger any possible form of violence among the Chinese masses too. It was altogether a very harsh phase of history for the Chinese peoples.

But unlike India, China was still largely united with some kind of a central authority, i.e. the Chinese Emperor. In addition, China's socio-cultural and religious historic past and the present then, as well also quite different from India and at the same time, her geographical vastness and complexities were much more challenging too. And so, both geopolitically as well as politico-militarily it was perhaps not feasible for those European Imperial Powers to colonise China, as since the British could already do with respect to India. Hence with their typical geostrategic and politico-military shrewdness these rival European Powers very shrewdly and proactively concentrated on altogether different kinds of imperial strategic game, i.e. instead of promotion of or through their imperialism in China, they concentrated more on promoting their unscrupulous trade and other commercial interests, through coercive diplomacy duly backed by the requisite military, naval and financial might at their disposal.

In these contexts we may recall that we have since earlier discussed briefly the subjects of **boxer-rebellion** as well as the **crafty exports of opium to China**, which was soon to play havoc with the Chinese peoples, their societies and very ancient civilisation as well. Ironically for the Chinese, this opium, then being extensively cultivated in India, under the British rule and direction was soon to replace even the bullion and silver, surprisingly as the principal currency for the international trade and commerce in China. It turned into a pitiable condition for the Chinese, with very far and damaging consequences for the Chinese Nation-State and the rest is history. Pursuits of highly ambitious, ruthless and greedy imperialism, of any Imperial-Power whether European or Asian or from anywhere; have many dimensions and hence as per the concerned Imperial Powers' situational strategies with respect to any particular targeted nation, they get manifested accordingly over there.

Precisely in the above contexts, the Scottish scholar **John Keay** has written that, "The tea, mostly black bohea not from Guangdong but from farther north, was purchased with silver. Barter would have been preferred by the company, but other British exports enjoyed little demand in China; woolens were generally unwelcome in steamy Guangdong, and as the emperor had told Macartney (*since already quoted earlier.*), China has as yet 'not in the slightest need of British manufactures'. The balance of trade throughout the eighteenth century was thus overwhelmingly in China's favour. The Celestial Empire came to rely on this abundant source of silver, while British manufacturers and economists deplored the disregard of the national interest and what they called the 'drain of specie'. **Indian produce** sometimes supplemented silver, notably raw cotton shipped direct from Bombay (Mumbai) to Guangzhou; so did a roundabout trade involving the shipment of **Indian textiles** to south-east Asian markets, then of south-east Asian spices and culinary exotica to China. It was in this three-way trade, conducted mainly by non-company adventurers and agencies, and involving the mercantile communities of the Malay and Indonesian ports, among whom Chinese emigrants were now prominent, that the potential of **Indian-grown opium** first became apparent. In China home-grown opium, mainly for medicinal use, had been consumed, as it had in Europe, for at least a thousand years. **But smoking it—rather than infusing or eating it—for the greater 'hit' nowadays classed as 'recreational' seems to have originated in Taiwan in the seventeenth century.... The habit spread to the mainland**, where the Qing Yongzheng emperor made opium dens and opium dealing illegal but did not actually proscribe the drug. **It also spread to south-east Asia, where Indian grown opium was already circulating and from where both Chinese and Europeans began shipping some on to China from the 1720s**.... Echoing the exponential growth of Britain's tea imports, China's imports of Indian opium topped 13,000 chests in 1828 and had doubled again by 1836. By then, 'total imports came to $18 million, making it the world's most valuable single commodity trade of the nineteenth century.... As of the late 1820s, therefore, **foreign tea purchases no longer required an outlay of bullion; opium credits more than sufficed. Silver no longer flowed into China; it flowed out**. By the 1830s it was flowing out at the rate of 9 million teas a year. The balance of trade had reversed. It was now imperial officials in Beijing who worried about the national interest and complained about 'the drain of specie'. **Within China, and especially in the south, the demand for silver pushed up its value against copper cash, causing the same distress as in Ming times among the cash-earning but silver-taxed classes. Reduced growth, unemployment and urban unrest (another of the cluster of problems associated with dynastic decline) are directly attributable...to the sudden impact of this dramatic and disastrous shift in the balance of payments**." (Keay, 2008, pp. 457 & 458)

It is clear from the above lines that some of the multiple varieties of highly unscrupulous and ruthless games of imperialism have been in practice and also been played by any concerned imperial power over the victim nation and her peoples and, with totally callous disregard of even any disastrous consequences. But specifically here in the above quoted historically factual lines, we find that the victim nation ultimately perforce made to suffer extremely disastrous consequences was China. Further in these historic contexts; as we proceed in our continuing discussions here, we would also be seeing soon as to how India even little earlier, was victimized and exploited though in different fashion and through different kinds of ruthless games of imperialism, by the same overseas Imperial Powers.

Man's living-history since continuously in the making, does have many simultaneous dimensions and facets and at the same time also universally encompassing all continents of the world, are all interconnected and integrative as well. As a specific example, we find that since India was first colonised by the then most prominent and leading European Imperial Power, i.e. the British, and hence under the colonial rule of the British East India company, India's eastern coast's ports were readily available as strategic base and spring-board for launching the imperial mission towards China. Then as further centuries passed all these may be termed as few linked segments of the interrelated chain of the multi-dimensional imperial games, then since so craftily being played by the rival European imperial powers with their different geostrategic and politico-military agenda and 'compulsions', in different parts of the world.

Now with respect to China precisely then; the ruthless games of 'opium-trade' and even the earlier talked about 'opium-war' had perhaps their own typical historic reasons. And hence, while critically reviewing them, interestingly we do also notice that sometimes, even stranger are the ways of since continuous making of the man's living-history. In this precise context, we may recall our brief discussions along with few earlier quoted paragraphs from **the same book on *China* by John Keay** concerning the well-planned and with great splendor, **the geopolitical and diplomatic visit of George Viscount Macartney to China,** duly financed by the East India Company and as per royal orders of the British Crown during the ending years 1793, which finally turned into a rather embarrassing failure for mighty British Power. But in spite of all this, while returning back, shrewd Macartney with futuristic strategic vision, took this unique opportunity to watch China, her geographical topography and her peoples and culture and compiled lots of data, for planning any future similar mission duly backed by a mighty military and naval forces for impact and thus if the need be then, to any possible initiatives even towards coercive diplomacy to succeed, and thereby as a kind of revenge to humiliate China. And thus historically, the forced British opium-export to China and even the opium-war were the consequential results. Hence, needless to say that at times strange are the ways of history and perhaps even stranger are the games of imperialism, which any imperial power plays against the concerned victim nation.

But unlike India, China could not interact so closely and multi-dimensionally with the European

Imperialists and hence, as further few centuries passed she could not acquire historically then perhaps the most valuable socio-political awareness and concepts regarding the modern treatises concerning the **civilisational constitutionalism for ushering in democracy**, as the most humane and enriching system of governance, **and inevitably then China was later-on overwhelmingly 'chained' by the ideologies of communism, which ironically during the latter half of the 20th century, proved to be rather multiply untenable, inhumanly cruel and socio-psychologically suffocating** too. Hence, we find that both China and India interestingly began their further march on altogether two totally different geopolitical and politico-cultural paths, on the ongoing canvas of man's living-history since continuously in the making. Even in the larger geostrategic and politico-military perspectives, we find that these two Asian Giants being also the world's two most ancient, though two very different but historically interactive too, and equally glorious and captivating civilisations, were subsequently subjected to the entirely two different historical experiences, along with their different multiple-fallouts.

During all these historic medieval-era as well as during beginning of the present modern-age, the multi-dimensional games of imperialism and the colonization processes since then in progress in different continents of the world and simultaneously many different but no less important and influential historic phenomena too, were fast progressing during those times. The three historic phenomena were especially notable in this regard—the first one being the **path-breaking gradual dawn of democracies in Europe**, the second one being **advent of the unprecedented Industrial-Revolution** and third but no less far-reaching: the **colonization of the Americas since as building of the first overseas historic empires** by the rival European Powers, with a highly complex and ruthlessly competitive game of geopolitics, on the global strategic chess-board. In our larger perspectives, we do notice that since as the destiny of mankind would have had it surprisingly in all the above mentioned three very significant historic phenomena, the British-Nation ultimately turned out to be most shrewd and successful player in these highly competitive games of geopolitics, among those rival European Imperial Powers. And this too in fact, was no less significant as a far-reaching contributory factor, towards further shaping-up of the man's living-history in this world. We would be discussing in some more detail these historic developments in our larger geostrategic and politico-military perspectives after first briefly touching upon the beginning phases of the historic imperial colonization of India, and aspects of the initial phase of rivalry between the competitive European Imperial Powers on the Indian soil.

Now in these specific contexts, while coming to the South-Asian Giant, we find that from the 15th to the present 21st century, like China the parallel history and the multiple historic experiences of **India** too, have been no less checkered but totally different as well. As the history shows, relatively rather a rapid colonization of the Indian Subcontinent by the tenacious and geopolitically as well as geostrategically shrewd British Imperialists, ultimately resulted in making India as the **'Jewel in the Crown'**, then of the truly global British-Empire. World's history has also very appropriately recorded this indeed a path-breaking historic achievement of the British Imperialism as building of the **second overseas Empire**, as since the earlier colonization of the Americas, has sometimes been called building of the **first overseas Empire**. As since already briefly delved upon, both historic situational reasons as well as the then prevailing scenario in India, were quite different from those in China. Interestingly, we may recall that during the later period of the bygone ancient-era and the very early phase of the medieval-age in a way the imperially decisive colonization of the whole of huge Chinese land-mass had been done by the historic Mongols alone, and which lasted for a number of centuries. During those historic times, the Mongols were both geostrategically as well as politico-militarily most skilled, focused and remained militarily unchallenged for a long period then, in the whole of Asia, Europe and Africa. Perhaps it also appears that, they could so brilliantly succeed in China, because historically both the Chinese populace and the Mongolians belong to the same racial stock. But historically so far as India is concerned, she stood already a colonised territory since long and by the time, really decisive initiatives were being undertaken by the British and other European Powers in India; the geographical as well as politico-military requisite space was already being made readily available, by the fast decaying and collapsing once mighty and glorious Mughal Empire. These were indeed the great moments then in the Indian subcontinent, giving births to the spacious territorial and politico-military vacuum and thereby the British with their long-term geopolitical-cum-strategic vision and a great sense of history, were comprehensively quick to readily grab this huge opportunity to substantially fulfill their far-sighted imperialistic ambitions. As the sixteenth century had already begun, the fast disintegrating and decaying Mughal Empire in India and especially after the passing away of **Emperor Aurangzeb**, the strong and secured imperial lock of the once mighty empire stood already broken with its borders no longer secured and hence, the 'huge imperial gates' were wide open and swinging to and fro as a kind of rather open invitation to other imperial powers.

Man's history in India was repeating itself: first for **Babar** the founder of Mughal Empire in India in 1526, after the slaying of **Delhi's Sultan Ibrahim Lodi** in the Battle of Panipat, those 'huge imperial gates' began swinging similarly for Babar, inviting him to march-on to Delhi and now nearly after two and half centuries; those gates again began swinging for the arrival of European Imperial Powers and, to begin their multiple-struggles of for the decisive imperial superiority to colonise India. In differently larger contexts and as already discussed in earlier chapters, we may recall that the medieval-age historic phenomena of a series of the colonization processes in the Indian subcontinent by the successive foreign powers had already begun much earlier since with the fall of **Maharajah Prithviraja** in the hand of **Mohammed Ghori**. And since then, India was hardly able to substantially recover on her own and thereby, both geostrategically and politico-militarily ascertain in the arena of historically age-based and perpetually changing global geopolitics.

But with the earlier conquerors, before the arrival of British, there was one big difference in the sense, they all permanently began settling down here and got absorbed in the Indian mainstream, unlike the European imperialists. But that is a different story altogether, with its own distinct historicity. Moreover, the historic process of the gradual arrival of Europeans of different nationalities, into India was altogether totally different from the earlier Muslim invaders. Europeans began trickling into India via the sea-routes and arrived one after the other, on both the western and southern coasts, principally as traders for economic gains through import and export of goods, since India happened to be a substantially big consumer market and also, a large source for many valuable goods as well as numerous attractive items, for exports to the European countries. Opportunities for building of empires to them came little later, with historically the inevitable collapse of Mughal Empire in India. Moreover, as since already touched upon earlier in a context, the long processes of movement of the rival European Imperial Powers and sizeable sections their population towards Americas, Asia and Africa during the medieval-era had their own compulsive reasons. First, the process of fanning-out towards different corners of the world, and thereby in some places like in Africa, the Americas and even subsequently in Australia, huge territories of virgin lands were almost readily available to imperially move and settle the growing European population, as a part of their socio-economic and strategic compulsions. Hence these phenomena are by themselves, multi-dimensional as well as interconnected too in their own the then historic and specific situations-based, geostrategic and politico-military contexts.

In the above perspectives and precisely concerning India, **John Keay** in the 15th chapter interestingly on, **'From Taj to Raj'**; has written that, "On the coast (*i.e. Indian coasts, during the years from 1682 to 1750*), the Europeans came and went. From their port of Bassein the Portuguese had acquired an adjacent trickle of islands which afforded good shelter for their shipping. Among the coconut groves on one of the islands they had built a small fort. They called it Bon Bahia or Bombay (*thus the name of that city: Bombay.*). In the 1660s, following an Anglo-Portuguese alliance against their Dutch rivals, the place was transferred to **Charles II** as part of the Portuguese wife's dowry. Although Bombay itself was as yet of no commercial value, the English thus acquired a territorial toehold adjacent the busy shipping lanes of the west coast. To the south, Goa remained in Portuguese hands while Cochin, an important entrepot for the spice trade, had been wrested from them by the Dutch, also in the 1660s. North of Bombay the Mughal port of Surat, superseding the now mud-silted Cambay as the main maritime outlet of northern India, hosted much the busiest Dutch and English trading establishments. From Surat European purchasing agents fanned out into the cities and weaving centres of Gujarat and beyond to place their orders and oversee dispatch. And down to Surat from Ahmadabad, Burhanpur, Broach and Baroda, came the bundled cottons and silks and the barrelled indigo (in great demand for dyeing uniforms) which constituted the main items of export. On the other side of the Peninsula all three European powers, plus the newly arrived French, retained similar toeholds on the Coromandel and Andhra coasts. Textiles were again the main item of trade, but there was a tendency here for the weavers to gravitate towards the European settlements which thus became zones of export-dependent prosperity. None of these settlements was yet of much political importance **but the security offered by the heavy guns and well-built forts were proving an attraction. Additionally their stocks of powder, guns and gunners were eagerly sought by the contending powers in the hinterland**." (Keay, 2000, pp. 348 & 350)

In our larger perspectives as we proceed further, to critically review these epoch-making chapters of the medieval-era vast Indian living-history, we interestingly find many somewhat hidden dimensions, unfolding before us in this regard. At the end of the above quoted paragraph, we find a mention of the politico-militarily competing powers in the Indian hinterland. And they were historically in fact, some relatively big, a number of medium-sized and then also numerous comparatively smaller autonomous-cum-sovereign ruling Princely States in the vast Indian subcontinent. In this precise context, the Indian living-history also clearly shows that since the ancient of times, socio-culturally unique, highly colourful and richly gifted by the mother-Nature this geographically the huge territorial expanse of the Indian landmass, along with the advent and soon subsequently, the multi-dimensional rapid advancements of the ancient Indian civilisation as well as with other parallel human-civilisations; and then coupled with the **equally momentous and simultaneous multiple-developments of the matters-military** with its numerous historic fallouts, also perforce had many historic consequences for the whole of Indian subcontinent and its colourful peoples.

It is indeed a great irony that, the Indian political-class is yet to learn meaningfully, from the man's vast living-history and thereby keeping in their minds, the sovereign India's highest national-interests as a sort of path-finder compass, for playing her multi-dimensional roles with long-term strategic-vision, among the comity of nations since as expected of her. In these contexts, what a sad historic picturisation all this leads us to even now, for the modern, sovereign, democratic and otherwise fast developing Indian Nation-State. It has indeed been so that paradoxically even now, the entrenched Indian Politico-Bureaucratic nexus with its many lingering negative dimensions, seems to be proverbially bent upon behaving like an ostrich, instead of sincerely being ready to develop essential attributes and capabilities for effectively facing the looming as well as multiply menacing the numerous geostrategic and politico-military challenges and, convincingly reassure the nation in this regard. We have earlier too in different contexts, discussed these matters and quite necessarily; we would again be constructively doing so in different contexts, as we proceed further in our ongoing discussions.

Now coming back to the fascinating historicity of the adventurous voyages by the medieval-era's relentlessly exploring Europeans; first we may need to bear in mind that, they perhaps have always displayed in-built ambitions to explore and conquer the sea and oceans first, in order to be able to launch their multiple-missions, into the separate and vast continents situated far-away from indeed their European continent itself. To fulfill these necessities, the leading European nations began with expanding their geographical horizon, improve upon their mapping techniques and also simultaneously, advance their ship-building as well as sea-faring technologies. From the human civilisational considerations in this precise context, we may also notice that perhaps historically the most fascinating aspect in this regard, has been that all these developments on the European shores; since then being triggered by the marvels of human-spirit and relentless ingenuity, were in fact taking place prior to the advent of the little later and unprecedented Industrial-Revolution itself. Hence it has since also very appropriately been said that, **the man's inherent spirit and the urge for ingenuity can never be suppressed**.

Historically in this precise context as we move-on, we do also notice that during the ending years of the fifteenth century two far-reaching and independent developments had indeed taken place which were soon going to very momentously influence indeed, the subsequent making of man's living-history the world over. The first had been when **Christopher Columbus historically set sail on his unprecedented sea-voyage across the Atlantic in the year 1492**; in the most adventurous search of the new world, i.e. huge and largely 'virgin' American continents. Columbus's this indeed history-making successful sea-voyage, thus opened-up the Americas for subsequent European conquests. The second epoch-making historic episode, in a kind of the identical context; has been **discovery of sea-route to the Indian Peninsula, in the year 1498 by another European Vasco da Gama**. And consequently the man's living-history has never been the same again. At this specific point, since we may also say that after these two above mentioned historic developments; the **momentously most influential and path-breaking advent of the Industrial-Revolution** came and hence, a kind of the unprecedented combination of these three indeed unique developments changed the face of the whole world and, also the entire subsequent history-making processes for the all times to come. We would since be delving upon them more in various contexts, as we proceed further in our continuing discussions.

All the above mentioned momentous developments were taking place; when **the historic golden era of the imperial rule of Mughal Emperor Akbar the Great** was over, but the medieval-age Mughal Empire in the Indian subcontinent, was already gloriously enduring. And hence interestingly, the **Mughal Emperors, Jahangir followed by his son Shahjahan since seated on the Royal Throne of huge and powerful empire were indeed large-hearted enough to grant the Europeans belonging to the different nationalities of Europe, their requests for the trading tights in India**. And as a result, the trade between India and the concerned European nations then had already begun to flourish. In these contexts and particularly at this precise point we may like to also bear in mind that during those medieval-era years, India under the Mughal rule; and especially as a result of indeed the praiseworthy and visionary rule of **Akbar**, was already much more resourceful and comparatively richer, than the European nations during those times. And naturally, those shrewd Europeans were indeed visionary enough to meaningfully see through these and many other facts concerning the Indian subcontinent, her riches and other vast natural resources and in addition also, her age-old civilisational assets and legacies as well as other socio-cultural-cum-feudal fixations. Hence, as the British later decisively triumphed in India; to finally take her on as both geostrategically and politico-militarily, most valuable segment of their Global Empire, she was then befittingly graded as the **'Jewel in the British Crown'**.

And even little earlier to the above mentioned historic development; after the Europeans landing on the Indian shores, their trades in India did expand soon in many directions. Those European settlers as traders in India then were able to export many kinds of valuable goods to Europe and in return, they brought in some **European produce and specially the superior breed horses as well as other much needed military hardware like heavy guns, military-related technologies and experts for training the Indian forces of the Mughal empire and also those of other Indian Princely States**. Such military training included rigorous military-drill, European field-tactics related concepts while taking on the opposing forces and also, the various aspects of military leadership at different levels of military command and control, as per broadly then prevailing in the medieval-era European Armies. We may recall that in the last fourth chapter, we have since noted as to how the **great Maratha ruler and brilliant military strategist, Shivaji was so particular in importing the European horses for his powerful cavalry, along with other allied military-inputs**. Ultimately the British decisively triumphed in India, in the multi-dimensional struggles among the rival European Powers. In these substantial areas of the history making, with **decisive inputs of the matters-military during the medieval-era Indian subcontinent** then, we find that all these were the unprecedented developments historically, with far-reaching influences for the all-time to come and, also subsequently indeed in the whole of South Asia. We would soon be discussing these historically momentous as well as multi-dimensional subject-matters as we move-on further in our ongoing discussions in this chapter itself, and in addition also in various contexts in the subsequent chapters as well.

But in these larger perspectives, the historic aspects concerning the medieval-era vast continents of the Americas and the subsequent European conquests over them are totally different, from those of the Indian subcontinent as since very briefly touched upon in the above few paragraphs. We may recall that since earlier in the last chapter, we have also briefly discussed some interesting historic aspects of the **age-old Maya civilisation, then flourishing in the northern and central parts of the South American continent**. Ironically though, not much is now known to the world today about this; but some remaining ruins over there gives perhaps somewhat indirect glimpses of that ancient-era glorious and once flourishing South American civilisation. Historically, that's a different matter that when the Europeans landed in those regions and subjugated them especially the Spanish, they found these areas ruined and largely barren and moreover also since those invading Spaniards ruthlessly preoccupied with their principally military-oriented imperial pursuits, unlike the British they never cared to scientifically explore and excavate the bygone era's historic aspects of that ancient human-civilisation for the next generation. These paradoxical sides of some of the medieval-era European Imperial Powers did also substantiate that **there are indeed some multiply consequential and strange human dimensions too, in the multi-dimensional phenomena of the making of man's history in this world**.

In the above contexts, Jawaharlal Nehru has very knowledgably mentioned that, "It appears that there were three centres of civilisation: in Mexico, in Central America and in Peru. It is not clear when they started, but the Mexican calendar began with the date corresponding with 613 BC. We find in the early years of the Christian era, the second century onwards, already many cities growing. There is stonework and pottery, weaving and very fine dyeing. Copper and gold are abundant, but there is no iron. Architecture develops and the cities vie with each other in building. There is a special kind of rather intricate writing. Art, and especially sculpture, is much in evidence and is of considerable beauty. There were several States in each of these areas of civilisation. There were several languages and a considerable literature in them. Well-organised and strong governments existed, and the cities contained a cultured and intellectual society. Both the legislation and the financial system of these States were highly developed. About 960 AC the City of Uxmal was founded, and it is said that this soon developed into a great city comparable to the great cities of Asia in those days. There were also other large cities: Labua, Mayapan, and Chaomultun. The three leading States of Central America formed an alliance, which now called the league of Mayapan.... So a millennium after Christ there was a powerful combination of civilised States in Central America. **But all these States and the Maya civilisation itself were priest-ridden.** (***What a coincidence indeed here; as in the ancient Indian subcontinent, historically otherwise highly developed in the multi-dimensional civilisational domain the numerous glorious Hindu Kingdoms too, were so regressively and self-damagingly priest-ridden: courtesy, the powerfully entrenched and craftily scheming 'Brahmin priest-class'. We may recall that earlier in the fourth chapter, while discussing multiple aspects of the ancient and glorious Indian civilisation, we touched upon these aspects in some detail, both geostrategically and politico-militarily, since very ironical and self-destructive facets of the ancient Indian civilisation and its societies***).... The Aztecs (of *Mexico*) were a military nation. They had military colonies and garrisons, and a network of military roads. It is even said that they were clever enough to make their dependent State to quarrel with each other. It was easier to rule them if they were divided. That has been the old policy of all empires. **Rome called it: *Divide et impera.*—divide and rule. The Aztecs, in spite of their cleverness in other matters, were also priest-ridden**, and, worse still, their religion was full of human sacrifice.... For nearly 200 hundred years the Aztecs ruled their empire with a rod of iron. There was outward security and peace in the empire, but the people were ruthlessly exploited and impoverished. A State so built and so carried on could not endure ...This is one of the most amazing examples of the collapse of an empire. And this was brought about by a Spaniard, Hernan Cortes, and a small troop with him. Cortes was a brave man, and daring enough. He had two things which were of great help to him—firearms and horses. Apparently there were no horses in the Mexican Empire, and there were certainly no firearms. But neither the Cortes's courage nor his guns and horses would have availed him if the Aztec Empire had not been rotten at heart." (Nehru, 1934, pp. 212 & 214)

Well, while critically looking deeper into the last line of the above quoted paragraphs from Nehru's book especially in our historically rather cold, discerningly calculative but at the same time, imperatively pragmatic too: and contextually in our larger geostrategic and politico-military perspectives, it appears perhaps quite naïve to digest its indicated wisdom in totality. It may perhaps be a specific perception of conjecture only and hence, could also be debatable as well. Since Nehru after 1947, remained independent and democratic India's first Prime Minister for nearly the fourteen long years and at the same time, he indeed being a great scholar and visionary, was also phenomenally instrumental in laying the dynamic, democratically enduring and lively foundations for the many kinds of systems and Institutions in India's national governance as well as decision-making processes. But at the same time, it would also be in the fitness of things that in order to substantiate some varying perceptions too in this regard, historically some befitting examples need to be precisely cited too, for putting the relevant points in a perspective. Geopolitically, the pivotal points in these indeed epoch-making, and rather in universal contexts always remain that in any given and particularly in all the more threateningly demanding situation, during the actual clashes of rival arms in the battlefields; superior military strategy and tactics, military wherewithal made available for superior fire-power and military-mobility and then no less important, the aspects of rigorous training-cum-discipline of troops and the professionally superior and dedicated military-leadership aspects at all levels—mostly carry the day.

It is first suggested that indeed along with so many other books and writings, concerning post-1947 era's short and alive history of the independent and democratic India, India's first Deputy Prime Minister-cum-Home Minister, **Sardar Patel's Biography** ('Biography of Sardar Patel', B. Krishna, Rupa & Co., New Delhi, 2005) may also be thoroughly referred to, for comprehending as much historic facts as perhaps substantially possible in this regard. **Hence coming to the precise point, soon after the end of first 1947-48 Kashmir-War between India and Pakistan, and certainly the uncalled-for and naively unilateral declaration of the 'ceasefire' by Nehru certainly not to the liking of India's 'Iron-Man' Sardar Patel for vital and long-term Indian strategic interests, did create a number of highly complicated problems for India with Pakistan, over the Indian State of Jammu and Kashmir. But historically in retrospect both geostrategically as well as politico-militarily it stands now conclusively established that certainly this border mess-up with Pakistan was decisively preventable by India alone in the first place for the permanent peace not only in the region, but even as indeed a phenomenal contribution towards the larger cause of the world-peace as well. We may even emphatically maintain that indeed, it remains a great irony of the recent history of independent India in a context, courtesy: Nehru again, in the present modern-era.**

As a second example we may here come to the episode that when little later, **Patel took a decision to send a column of Indian Army to Hyderabad in 1949, to end the then Nizam's rule over there** then another development began unfolding in the corridors of power in the Indian capital of New Delhi. Lord Mountbatten then on the Indian Government's request, still functioning as the Governor General, quite naively tried to frighten the Sardar, at his own initiatives since understandably then and perhaps compelled by some hidden British geopolitical strategic interests vis-à-vis Pakistan in South Asia, which turned out to be humorously an exercise in futility for the British. As per the above mentioned well-researched book, **Mountbatten** told the Sardar that it could be very dangerous for India then, since the Nizam of Hyderabad (*now the capital city of the Indian State of Andhra Pradesh*) was having a military-pact with Turkey, and also Turkey had some fighter-bomber squadrons which could be launched against India, via Pakistan and the fledgling Indian Air force then, did not have the necessary military wherewithal to counter this menacing threat. **But to the utter dismay of Lord Mountbatten, instead of Patel getting frightened, he just laughed at it while taking this rather dubious advice in his usually confident stride, like a seasoned Statesman.** Then he in turn replied to Mountbatten that in any case in a kind of geopolitical bargain in India's larger national interests if few Indian cities near western coasts like Bombay and Ahmadabad receive few bombs, so be it and such a big country like India in any case, won't perish and moreover; if the Indians want their freedom and geopolitical sovereignty to endure, then they must remain ready to shed some blood as well. In addition, to the great amazement of Mountbatten; the Sardar also added in his reply, that during the days of Second World War, Great Britain being a relatively much smaller islands' nation, was relentlessly bombed by **Hitler's Air Force Luftwaffe;**

did in fact survive and also finally achieved a victory over the Nazi Germany. And hence in any case, what is so serious a danger lurking over India that, the Indians should ever get panicky about it? Hence, firm and strategically prudent leadership at both political and military levels and in addition also the aspects of brilliant political statesmanship always hold the key, in the business of statecraft and the matters-military.

Now as a third example, we may then like to come to the main historic facts of the India-China border war of 1962 when Nehru was democratically 'reining' supreme since Sardar Patel was no more. Ironically this short war on India's northern and north-eastern borders, remains a tragic politico-military saga of the independent India's indeed a preventable humiliation in the hands of Communist China. History of this tragic border skirmish is already since a fairly well-known long-story with a lot of historic literature available, conveying different shades of opinion and hence we would only be relevantly discussing its salient features in our larger perspectives, in the later chapters. **Here precisely the principal point remains that after 1947 at no given point, India's democratic and sovereign Nation-State has never remained 'rotten at heart' but subject to the highest-level leadership aspects, victory or defeat in independent India's short clashes of arms at her borders have been shaped by the nation's geopolitical and strategic vision, advance planning, and multi-dimensional as well as well-coordinated all-round preparations and not the least, the firm and visionary statesmanship at the highest political level.** So far as the above referred aspects of medieval-era Mexican-Empire in north and central regions of the South American continent, being 'rotten at heart' is concerned, these have very scholarly been explained and indeed also with great historic precision by Nehru, in his since well-known book on the vast subject-matters of World-History.

Now in these larger perspectives, we may say that and as also borne by the man's history too **that if any Empire or Nation-State, at any given historic point, gets 'rotten at heart' sooner or little later it would inevitably perish on its own, irrespective of its politico-military invasion, by any foreign power**. That in any case is a different story and hence, already historically remains a long forgone conclusion. But at the same time, in the earlier quoted paragraphs from Nehru's book on *Glimpses of World History*, he indeed as an accomplished scholar and author, has clearly brought out contextually in a sharp focus, the pivotal fulcrum of the processes of firming-up as well as enduringly staying power with geostrategic and politico-military strengths of any particular foreign imperial power's multiple aspects of the concepts and practices of its ambitious imperialism, over a totally subjugated country. And specifically in this regard, **that pivotal-idea-cum-fulcrum happens to be geopolitically crafty dimensions of the imperial aspects of 'divide and rule'**. These exploitative imperial concepts always being so compulsively connected, to all dimensions of the imperialism; have been universally true since the very ancient of times. Hence as we proceed further in our ongoing discussions; we would also very interestingly be coming across the different situation-based manifestations of that pivotal-idea-cum-fulcrum, i.e. since obviously being as 'divide and rule' wherever imperialism pertaining to any advancing Imperial-Power, could take its roots and endure.

Man's living history has been showing that since the beginning, there have never been any substitutes for the already earlier discussed essential prerequisites and at the same time; it is needless to say that these in turn, are always linked to the multiple aspects of **'essential strategic-culture and thoughts'**. Moreover, so far as India's vast living-history since the ancient times is concerned, numerous examples to substantiate these factual aspects could always be cited in the regard. In these far-reaching contexts, the central point remains that the **multiple-aspects of 'essential**

strategic-culture and thoughts' do not pertain to the matters-military alone, but equally concern all areas of the national governance and multi-dimensional developments as well. First and foremost, if these aspects are duly cultivated, evolved, developed and fruitfully imbibed then they in turn make the peoples and their ruling-elites essentially conscious and equip them with motivation, to continuously carry out their nation's situation-based 'SWOT analysis' and subsequently also, lead them to proceed with the **most important exercises for comprehensive 'threat-perceptions'**. And then, these immediately trigger the processes of proactive planning in the all concerned areas of nation-building and governance, and so on. And needless to say that, multiple aspects of the matters-military; since already discussed earlier in various emerging situations, are by themselves very important segments of these areas. We may also recall that the aspects of 'SWOT analysis' too, have been discussed in the earlier chapter.

We are since already aware that man's vast and complex living-history remains perpetually in the making, in integrative as well as multi-dimensional manners. In the contexts of the above discussed subjects-matters we find that **Richard Overy** has written that, "The Indian historian K.M. Panikkar described the period from 1489 to 1947, from Vasco da Gama's discovery of the sea route to India to Indian independence, as the European age in history. Others have pointed out the element of exaggeration in this definition: if, in 1750, the Europeans had abandoned their isolated settlements on the coasts of Asia and Africa, they would have left few traces. Nevertheless, around 1500 the balance, which hitherto had weighed on the side of Asia, began to change, and after 1750 that change was momentous. Before 1500 civilisation had been essentially land-centered, and contacts by sea relatively unimportant. If the year 1500 marks a new period in world history it is because thereafter direct sea contact linked the continents. This resulted not only in the integration in a global system of regions which hitherto had developed in isolation, but in a challenge to the age-old, land-centered balance between the Eurasian civilisations. Nevertheless, the impact of the European expansion should not be exaggerated. The 16th century saw a remarkable resurgence of the Muslim power in the Ottoman Empire, Safavid Persia and Mughal India. China, meanwhile, remained the world's most advanced state. Indeed to many in 18th century Europe, Turkey and China were the exemplars of civilised living. The industrial revolution put Europe ahead; but the fruits of that process—some of the poisonous—were only garnered in the 19th century. Europe between 1500 and 1815, for its thrusting novelty, was still an agricultural society of Lords and peasants, closer to its agrarian past that to its industrial future." (Overy, 2008, pp. 149 & 150)

But in these contexts, we have observed that even few years before the discovery of sea-route from Europe to the Indian Peninsula by Vasco da Gama, no less significantly in the year 1492, Christopher Columbus had already set-foot on the American continents' coasts across the Atlantic and thereby, historically triggering a chain-reaction of many epoch-making developments. First and foremost, the geographical existence of the previously unknown and very vast twin continents of both North and South Americas were discovered and there-by introduced to the medieval-era's leading European Imperial Powers, since already having a dynamic tradition for adventure and sea-faring and along with, also possessing their powerful Armies and Navies. In addition at the same time, all of them were increasingly harbouring ambitions for their numerous kinds of overseas imperialism too and hence as a consequence, they were rather compulsively restless and searching for the new openings as well as 'virgin' territories. At the same time then these leading European Imperial Powers had already fruitfully cultivated and equipped themselves, with multiple-aspects of the earlier discussed, **four essential prerequisites for the imperialism**, in their own differently varying degrees as their assets in this regard.

Hence, as we begin to delve deeper into these historic scenarios then emerging before those leading European Powers, we also very interestingly notice that recipe for the unchecked imperialism, with its all sorts of subsequently possible manifestations in the far-off 'virgin' continents of the Americas was already complete, for unprecedented phenomena of struggles among those European Imperial Powers. And thereby, totally unprecedented and new saga began getting added to the man's living-history, and in the process they also caused almost complete changes to the face of world, for the entire humanity's forthcoming momentous future. History shows that in this historically emerging scenario, the **five lead history-making players in these unprecedentedly expanding games of imperialism, happened to be the sea-faring England, Spain, France, Holland and Portugal**. Interestingly at this precise point, we also like to bear in mind that back in their homelands in Europe they were no friends but bitter rivals, with many historic legacies of rivalry among themselves, and hence it is needless to say that as they separately moved across the Atlantic for relentless pursuits of their unchecked imperialism in the twin continents of Americas, they perforce carried the heavy baggage of their age-old legacy of rivalries too!

As the old saying goes that **two swords can't be accommodated in one scabbard and hence, it would have been rather naïve to expect these imperial European powers to remain friends with each other. Further since, it is also said that politics is a game of possibilities and so in many ways it goes too in the pursuits of no-hold-barred imperialism, in other peoples' territories**. But at the same time in these contexts, we do also notice that when there happens to be more than two rival imperial powers moving onto the scene with their separate agendas, and hence as in this case four ruthless rival powers then being present in the medieval-era twin continents of Americas, and so as the emerging situation demanded at any given historic moment or as per their own geostrategic and politico-military calculations, any two or three could also temporarily club-in against the rival fourth, in the clashes of their different imperialistic interests. Naturally any kind of perceived ethics in possibly larger or perhaps regarding any higher human-level considerations, naturally seldom have any place in these games of the pursuits of ruthless imperialism. And in these contexts, we may also add that these historic pursuits of the games of imperialism, were since always driven by the desire to build and expand the empires of the concerned Imperial Powers.

In these rival games of imperialism, then being played by those four European sea-faring Powers, we also prominently notice the highly complex applications as well as interplay of the already earlier discussed four essential prerequisites, regarding the pursuits of imperialism and thereby, building of the overseas colonial empires, across the Atlantic in the twin American continents. As we proceed to delve upon these aspects even deeper; we also interestingly notice that aspects of the fourth prerequisite, i.e. multi-dimensional conduct of the art of highly complex international game of the rivalry-infested geopolitics and diplomacy unlike the Asian Indian subcontinent, was hardly relevant or applicable during the pursuits and furtherance of imperialism in the medieval-era American continents. We have since already touched upon the historic aspects of Spain's ruthlessly imperialism-oriented military initiatives in the medieval-era Mexico and other central regions of the South American continent.

Historically in the nations' both individual and collective contexts, and also as per the relevant socio-cultural as well as geopolitical considerations the inherent attributes of the Spaniards, were not like those of the British people and then even to a lesser degree, like the French too. The historic complexities and criticalities of these aspects during those momentous years of the medieval-era in this regard, are indeed like cobwebs but regardless of them, these aspects too had their own decisive role in the subsequent promotion of imperial colonialism and the integrative history-making phenomena then, in the American continents. In her medieval-age pursuits of imperialism anywhere Spain largely relied on her powerful military assets, for achieving maximum possible favourable results regardless of the consequences for posterity, concerning the civilisational aspects of the mankind.

In these contexts, concerning the historic aspects of the very complex and ruthlessly competitive games regarding the pursuits of imperialism by rival Imperial Powers, we may even be inclined to consider the above mentioned almost total disregard to multiple kinds of damages/destructions caused to the much older civilisational legacies and heritage, in different far-off regions of the world as a great irony or even rather a tragedy in this regard. Man's history has since conclusively shown that after Columbus's historic voyage across the Atlantic in 1492 when the huge American continents were thrown open to those leading sea-faring and militarily strong rival European Powers; and as earlier since brought out, **particularly in the northern and central regions of the South American continent, fairly advanced and cultured human-civilisations were already well-established and flourishing. But these sea-faring European Powers in course of their rather blindly imperialism-driven and lustful military invasion, particularly Spain and Portugal, never bothered to spare any time and inclination to care for the conquered territories' civilisational past and heritage**.

And tragically in historic contexts, lot of such collateral as well as civilisational damages which were since largely preventable, had perhaps been callously allowed to take place. Nevertheless, these medieval-era South American fairly advanced civilisations, and along with some magnificent facets of civilisational marvels and also in the same context, had few rather unacceptable shortcomings too and which needed to be abolished, such as cult of 'human-sacrifice and so on. But ironically as a great and avoidable tragedy, those invading European Imperial Powers never cared to pause and think about these delicately finer civilisational aspects in course of their imperial military campaign. Well, at this precise point, man's living-history too provides no clear reasons as to perhaps what all could have been their own historically compulsive reasons, for their imperialism-triggered military callousness. Hence needless to say that, **there are many legacies and facts of the humanity's historic past, which since remain buried in the sands of time as well as many kinds of lingering mysteries and ironically thereby, these never got any light of the day so far for posterity**.

Well, with respect to broader civilisational considerations since the ancient ages, it is also true that **historically in course of various imperialistic military invasions, for the building/expansion of empires during different historic times and in different regions of the world, the above discussed ironically rather certain unbecoming dimensions have been largely true in different manners, with almost all imperial powers**. In these larger perspectives, we have since discussed in the previous chapters, many such historic events of the past, where ruthless sacking of various cities and many other kinds of civilisational assets as well as military forts have tragically been resorted to by the irrational destruction by the victorious powers, while establishing and expanding their empires. We may recall that the **Egyptians, Greeks, Romans, Indians, the Chinese, the Europeans and almost all others, resorted to the same practices in their different ways. We have since also noted that the Mongols and later-on the Muslims too, were even terribly more ruthless during their own historic times of imperialistic pursuits as well as building/expansions of their big empires in the ancient and medieval-era continents across the world**. These happen to be rather extremely dark sides of man's history, since the very beginning itself.

In the larger perspectives in this regard historically indeed, another very significantly far-reaching point emerges here. And precisely it happens to be that, during the long passages of time comprising of many ravishing centuries **for any gloriously ancient civilisation to be able to endure the savages of both, the centuries as well as ruthlessly inflicted by various invading armies from time to time and still survive it has to have certain deep-rooted beliefs, very strong socio-cultural value-systems, rich traditions of spiritualism, multi-dimensional intellectual pursuits and heritage, and last but not the least, proactively vibrant as well as dynamic multiple-aspects of the matters-military, so on**. In addition, all these socio-cultural, socio-political, and politico-military attributes in totality should also essentially enable the peoples concerned, to socio-strategically cultivate, groom, develop and imbibe the much needed judicious pride among themselves, and their multiple-heritage as well as also, their age-old many other civilisational aspects too. These aspects perhaps exclusively hold the key to the answers, as to why some ancient civilisations have resiliently been able to survive and advance enduring both the cyclonic savages of time and also the ruthlessness of the invading military forces. Then paradoxically some others with different kinds of historic tragedies, got buried in the perpetually piling sands of time.

In these precise perspectives, we do also note, that the **magnificently ancient and multiply-glorious civilisations of both China and India**; enduringly do surface over the unending 'canvas of time', with lively continuity. These two world's unique and great civilisations since thousands of years have been surviving and spreading their own distinctly different civilisational fragrance across the oceans and continents; but at the same time, they do also command their mutually enriching but, totally different historicity. And in addition we may also recall that, in the present times and precisely **during the recently passed 20th century China was violently overwhelmed by the ravaging waves of communism which in turn, have even acted ironically in a way, as her highly devastating civilisational-disconnect from her heritage and past too, for the modern Chinese generations**. But luckily for the Indian populace consisting of numerously mind-boggling communities, races, caste, creed, cultures, languages, value-systems and so on, India as a unique ancient civilisation and country, have luckily escaped any such kind of politico-military turmoil.

While delving deeper into these highly engaging historic aspects in a perspective, we may also be able to clearly visualise that **on multiply complex socio-cultural plain, while surviving with still a vibrant and lively civilisational antiquity, heritage, traditions and richness, is historically one thing; but at the same time along with also maintaining socio-political sovereignty with the requisite as well as multi-dimensional geostrategic and politico-military strengths is different and hence, historically quite another matter altogether**.

And therefore while objectively discussing, we do also find it as no wonder that due to numerous geopolitical naivety as well as a kind of rather embarrassing lack of the long-term strategic-vision, on the part of prominent political leaders **on the eve of India achieving political freedom in 1947, her extremely violent partition was caused and imperially manipulated by the departing Imperial-Power itself, at in the cost of human-blood, socio-cultural sufferings and along with, affecting many permanently irreparable damages to the age-old Indian civilisation and the then fledgling Nation-State.** Islam was born in Arabia to begin with; and even during those times Arabia since with her quite a long pre-Islamic existence, she too had somewhat fairly developed civilisational assets and traditions; but ironically all these were swept away and swallowed by the **powerful currents of the spread of Islam**. Islam before expanding in some parts of Africa towards its southern regions and then making some inroads into Europe, Asia and elsewhere, first spread and took firm roots in the Middle-East. We may recall that we have since briefly discussed in broader perspectives various subject-matters concerning Islam, in the previous chapters. Historically we have also seen that during her expansion in almost all geographical directions, this then newly born religious faith gradually took the shape of **'Political-Islam'**. This rather not at all so peaceful politicization of Islam in turn triggered many kinds of politico-religious clashes of arms, devastating violence and mindlessly relentless thirst for the human-blood.

And historically since all these historic phenomena were geopolitically driven by the overwhelmingly ambitious aspects of ruthless imperialism as well as the increasing urges to build-cum-expand trans-continental Islamic-empires, on the part of rather newly-converted ruling-elites of certain Middle-Eastern Kingdoms and also, of the Turks across the Mediterranean, the aspects of political-Islam perforce became the main driving-force behind their devastating politico-military campaigns. And hence, almost throughout the medieval as well as also the initial periods of the presently ongoing modern eras; man's living-history contains numerous human-blood soaked saga of the violent Islamic incursions into almost all continents of the world. Egypt was one of the world's oldest and most glorious civilisations but as consequences of such violence and politico-religious upheavals, the old and **once magnificent Egyptian civilisation, stands devastated and reduced to only some monuments and pyramids**. We may also recall that **Mesopotamia too**, suffered the same fate. Many such examples, with their different historicity, can be cited in this regard. We have since also discussed that when newly-converted to Islam, the masquerading Turkish herds also entered the Indian subcontinent, they encountered a much more colourful, rich and ancient civilisation and religion which they were not prepared to accept with any civilisational magnificence, but the Turks were militarily better organised and ruthlessly focused with their swords and horses and these eventually, led to shedding of human-blood and a series of the destructions of temples and other assets. In totality and in course of time **these violence-ridden Islamic aspects since got almost completely divorced from the multiple magnificence and grand spiritualism of the Sufi-Islam, resulting in a cult of religious fanaticism and mindless violence. Sadly, this is a kind of historically prominent negation of Islam as well**.

Now again coming back to the multiple phenomena of the building, expansion and imperial consolidation of their medieval-era empires to begin with, in the North and South American continents by the competing European sea-faring Powers, also in turn, resulted in the beginning of many kinds of geostrategic and Politico-military struggles and clashes among themselves and in fact, not only in the Atlantic and the American continents, but almost all over the world. Interestingly; wherever they came face to face with each other, during their compulsively relentless and rival pursuits of imperialism. Historically in fact, for more that hundred years, beginning from the commencement of 16th century to right into almost the mid-17th century, this checkered historic period may also be known as the ending medieval-era to already an initial one and half to two centuries of the present modern-era; as relatively short and crisp age of the momentous struggles for empires among those four imperially fanatic, rival overseas European Powers. Interestingly in our present contexts, these struggles among them during this period historically did create a multiple saga of the geostrategic, military and naval maneuvers, crafty geopolitical and military alliances, coercive diplomacy of the ruthlessly imperial bargains—leading to at times amusing gives and takes of imperial 'monkey-business' and finally, if any compelling necessity did arise and then, even resorting to a series of the bloody clashes of arms as well.

These imperial struggles and even violent clashes among themselves interestingly had a number of dimensions in Europe and hence in their home waters, in the continents of Asia, Africa and Americas and thereby, in the surrounding seas and oceans as well. Man's living-history of those centuries, since rather amusingly shows that if somewhere England and France fought, then somewhere Spain and Portugal clashed; and then at times interestingly, as per their geopolitical and strategic imperial calculations any two of them craftily and temporarily allied themselves, against any other remaining rival Power in the field. And some European schools of thought went that far, as to even call their naked imperial ambitions and historic initiatives as towards civilising the peoples in various Non-European continents of the world! It appears in a context interestingly that, these medieval-era historic aspects of the European imperialism; perhaps may have motivated **Kipling** to call them as the **'white man's burden'.**Whereas the hard historic facts remain that, the principal motivating and driving forces at the back of those overseas European Imperial Powers (*in a way, mostly led by the British.*) were their pressing economic compulsions for overseas-trade, resettlements of their growing populations and along with, the dominating pressures by the Papacy back home, for opening-up of the new avenues and fresh territories for propagating Christianity. We may therefore, always need to bear in mind, that historic myths and historic facts do remain totally different.

While keeping in mind these factual aspects; we naturally find that when the earlier noted medieval-era's leading European Imperial Powers were multiply making the imperialistic inroads across the Atlantic into the American continents they were simultaneously too, fast consolidating their different kinds of imperialistic holds and making further inroads into the various countries and areas too, in South as well as South-East Asia. As in North America and so here too, Great Britain finally triumphed and thereby then the truly global British-Empire was fast getting fructified and subsequently, triggering the **proverbial and now interestingly historic, saying that 'the Sun never sets in the British Empire'.**

Now in the above historic contexts, while again referring back to **Richard Overy's** book he writes that, "From 1713 to 1815 Great Britain and France fought for global supremacy in India, North America and the Caribbean. Britain with great naval strength was always at an advantage because France was first and foremost a continental power. Control of the seas allowed the British to defeat her enemy in North America and India. European territorial expansion overseas continued throughout the 18th century and led to serious clashes between Portugal and Spain in the Banda Oriental (Uruguay), between Spain and Great Britain in Georgia and between Great Britain and France elsewhere in North America. Trading monopolies proved an even greater source of friction. Spanish attempts to suppress British and Dutch smugglers reduced the Caribbean to a state of undeclared war. Further north British attempts to enforce similar restrictions upon its American colonists provoked resistance and finally, from 1775, open revolt (***i.e. Historic American war of independence and its own causative-factors there-in.***). During 1739-40 the fragile peace brought about by the Treaty of Utrecht in 1713 collapsed as Great Britain and Spain went to war in defence of their trading rights. European states, American settlers, Native American chiefs and Indian princes all fought as subsidized and dependent allies of these two great powers. The Treaty of Aix-la-Chapelle (1748) settled none of the outstanding questions, and fighting began again in North America in 1754. France quickly achieved local military superiority with its strategically sited forts preventing further British expansion. The outbreak of the seven years' war in Europe (1775-83) transformed the situation. France was handicapped by its continental commitments and Britain took control of the Atlantic, decisively defeating the French fleet at Quiberon Bay and Lagos in 1759. With French forces in Canada cutoff from reinforcements Louisbourg fell in 1758, Quebec in 1759 and Montreal in 1760." (Overy, 2008, pp. 199 & 200)

Further in the then eventful North and South American continents, we do find even more such struggles and relentless games of imperialism. Portuguese, since already present in the northern and central regions of South America; with some smart initiatives, took control of Brazil and then interestingly, the Spaniards certainly not prepared to be left behind took possession of Mexico and Peru—in fact and since already discussed little earlier. Spaniards were the first to land on the Mexican shores in these ruthless games of naked imperialism and then with her superior military wherewithal and leadership of her military commanders, she could subjugate and conquer the Mexican empire and even adjacent territories. Consequently, man's living-history also informs us that these imperial colonies of principally the Mexico, Brazil and Peru formed the pivotal nucleus, for subsequent and now the globally acknowledged, formation of the Republics of South America. Interestingly further in similar contexts in addition, we do also notice that the Dutch, who were the first to land on the shores of North America and made some historic inroads even before arrivals of the rival British and the French. But unlike in some areas in South-East Asia the Dutch could not imperially compete against the British and the French in the vast and relatively quiet 'virgin' territories of North America and hence, they were pushed out of this imperial race over there.

Well, if contextually the earlier noted and rather imperially-fixated and also, since then prevailing some late medieval colonial-era European 'schools of thoughts', subscribing to now perhaps the amusing idea regarding the so called "the white man's burden"; may justifiably be termed as historically untenable and hence not even perhaps debatable. But at the same time with the sole exception of Japan, since whole of the Asian and African continents then having missed the unprecedented, highly progressive as well as multi-dimensionally path-breaking developmental aspects then, being triggered by the **historic 'Industrial-Revolution'**, and these also being capable of providing so much impetus to the man's civilisational advancements too and therefore in these precise contexts, **it is also an undeniable historic fact that these numerous Nation-States of Asia, Africa and even of South America won't have been what they are today—but for the somewhat prolonged colonial-rule imposed over them, by those medieval-era overseas rival European Imperial Powers**.

In other words and in our larger perspectives we may also say that, but for the long spell of since now bygone centuries' imperially dominating colonial-rule, these numerous Afro-Asian and Latin American countries, would have been left far-behind in so many of the present modern-era's socio-cultural, socio-political, socio-economic-cum-industrial as well as geopolitical and matters-military related spheres and which are indeed now, the most vital aspects in this regard. At this precise point, we may also recall that we have since earlier discussed historically the path-breaking phenomena of the dawn of democracies in the medieval-era Europe and, also along with their principal causative-factors, such as the sprouting of revolutionary ideas of the Magna Carta in 1215 in England and so on, which in turn as consequent of the birth and gradual development of democracies over there, multiply changed many of the European societies and their Nation-States forever.

Interesting further in these contexts, so far as the then numerous Afro-Asian and Latin American Colonies were concerned since democracies in various forms were fast sprouting, taking firm-roots, and developing into the homelands of their Colonial-Masters in the Western Europe, these historically unprecedented revolutionary ideas and ideologies concerning, since the earlier unknown socio-political philosophies subsequently had to perforce travel far and wide, and hence into their Empires in the overseas continents as well. The timeless universal point in this regard and, as since also shown by the man's living-history remains, that the Empires and Nation-States in the world may naturally have their political or territorial boundaries; but humanity's thoughts and ideas of any kind, cannot be confined within any limiting boundaries anywhere. And it is needless to say that, no Monarch or dictator has ever succeeded in doing so, in any part of the world. As few more centuries further passed and hence with the changing times; **one after another these Colonies were able to gradually discard the yoke of colonial-rules since imposed over them, and achieve their political independence, and consequently during the last 20th century itself, many of them began their history-making 'trysts with the socio-political and socio-economic revolutionary ideas of democracies' since having come in the first place, from the West itself.**

We also find **altogether new kinds of globally geopolitical-oriented games of the 'hegemonic-imperialism'**, now since already being played and wherein; the sole hyper super-power the USA, obviously remains in the lead and also far-ahead of others. And in addition, now the fast emerging China too is rather even overstretching herself to geopolitically show-off that she is a new and menacingly powerful competitor of the USA in this regard. **These new and historically unprecedented games of the multiply complex and geopolitically far-reaching as well as globally influencing hegemonic-imperialism are indeed simultaneously being played in different directions and with different but, otherwise inter-connected as well as inter-supporting rather some invisible game-plans and also since as per the overall geostrategic and politico-military national interests and ambitions of these new-era's hegemony-obsessed Imperial-Powers such as the USA.** We would also be discussing the multiple-aspects of these totally new and the present modern-era's highly complex and fast-moving games of the hegemonic-imperialism in the subsequent chapters.

So far as the historic phenomena of colonization of the American continents are concerned, we find the North and South Americas have very different historic epics with their distinct geostrategic and politico-military saga of story-lines. We have since earlier noted that the northern and central parts of South America had some reasonably advanced and flourishing civilisations such as the Aztecs and the Mayas. But even regarding them, not much is known now, as to how, where and when; these even somewhat ancient American civilisations precisely originated and in addition also, how they historically developed? But it also stands authenticated that in a context; fairly advanced civilisations were already firmly rooted and flourishing in those northern and central regions of South America since a number of centuries, before the Spaniards and the Portuguese from Europe, were able to reach their shores. We have since also touched upon as to what were the geopolitical and military consequences for the peoples of those fairly older American civilisations; when they came face to face with the newly arrived alien overseas Europeans, and who had also already brought their savage and powerful imperialistic military for conquering their lands. These highly consequential developments in turn, phenomenally triggered the historic processes of the birth and thereby, gradual consolidation of a number of modern Latin American Republics, in the vast continent of South America. But in these precise contexts, the history and subsequent developments in North America, were entirely different.

The initial overall historic scenario in the North American continent even before the arrival of the rival European imperialists like the English, Dutch and the French as noted above, were almost entirely different from that of the sister continent in down south, where the Spaniards and the Portuguese were already making substantial inroads. In these perspectives, the North American continent, which now consists of two big and highly developed countries of the USA and Canada; was then relatively a vast 'virgin land'. Though there were humans already dwelling over there since many centuries, but they were relatively few and far in between and also being **relatively**

primitive tribals, now popularly known as the 'Red Indians'; then living as nomads and hence, they could hardly claim to command any lively civilisational legacies. But they were the primitive tribal-warriors too, which were definitely of some concern for the European Colonialists and moreover, with their strongly fixated socio-psychological as well as socio-cultural traditions, were not the right kinds of reservoirs for the much needed man-powers. Historically this particular fact and the then situational reality, existing in the North American continent perforce induced those newly arrived, European Imperial Colonialists; to devise ways and means to import the Africans as the necessary pool of labourers for their vast plantations, constructions, mining and so on. Over the years, these immigrant Africans then, became the forefathers of the present Negro population in the modern USA.

In these historic perspectives, **James Truslow Adams** in his book has mentioned that, "No dates marks the beginning of our tale (*i.e. America's*). With the exception of the Aztecs and the Mayas, no aboriginal American knew a calendar, and for those who lived within the bounds of what is now our country (*i.e. The USA.*), time was dateless. It simply flowed as it had always flowed from time immemorial, marked by seasons, by birth and life and death. How long the Indians (*i.e. the Red-Indians here*) had been here or whence they came, we know no more than they. The continent on which they dwelt, on which beyond the furthest reach of native myth they had forever worked and played, loved and warred, had remained unknown to all the world except themselves.... Their weapons were bow and arrows, tomahawks, and clubs. As we pass southwestward toward the Indians of New Mexico and Arizona we find an increase in skill in such arts as pottery and weaving, although the finest designing on the northwest coast.... The country was sparsely settled, considering its size and the number of inhabitants, but we must recall that it takes a large area to support its people in the hunting stage of culture.... War between tribes, save when treaties had been made for a period, was a normal state of existence, and, almost as much as gathering food, was the chief occupation of the men. For the purposes of both the peaceful migrations and the war expeditions, the Indians had clearly marked trails extending practically over the continent, through the forests, over the plains, and along portages connecting their navigable streams and rivers, the routes chosen being scarcely capable of betterment by a modern engineer. The length of time taken to locate them must be reckoned by centuries when we consider the vastness and difficulties of the land and the fact there were thousands upon thousands of miles of these trails." (Adam, 1931, pp. 1, 5 & 6) (***These lines in few words, indeed gives a vivid picture of the then prevailing scenario in the vast and in a way 'virgin' North American continent and the amazing lifestyles of its aboriginals. And today, the world's most advanced, industrialized, economically the most developed and militarily the strongest Nation, i.e. the USA forms part of the same North American continent.***).

The pivotal-point in this regard, happens to be that the **renaissance means awakening** of the socio-cultural spirits and the dynamic urges to get reconnected with the marvels of one's past, and especially since the multiply creative legacies of some civilisational societies historically having a glorious past; and certainly with laudable intentions to revive the art, culture and other creative activities, again in the bygone era's classical styles. In fact, in these civilisational phenomena the larger ambitions happen to be inspiringly as well as proactively, to get connected with one's past and in the process again creatively make efforts to achieve the past glory in the present times' multiple socio-cultural-cum-intellectual domains, and also possibly at the same time rather reincarnate them, as creatively motivating for the future as well. We may recall that in some contexts earlier, we have since mentioned the **civilisational aspects of socio-cultural and socio-political phenomena of the**

renaissance, reformation and revolution too, and hence contextually in other words all these with their multi-dimensional totality, may be identified in some manners; as also the paths for the ongoing indeed, the multi-directional civilisational advancements. In any case, in the various contexts and also in course of the discussions, these may be taken as even some matters of conjectures.

As we delve further deep, we may notice that during the geopolitical and politico-military turmoil and travail ridden times in the medieval Europe of the 14th and 15th centuries and accordingly, **in the larger civilisational perspectives, the principal causative-factors behind these renaissance-oriented awakening in the then European societies, were to inspiringly revive the multiple historic linkages of the European civilisations with the ancient era's once glorious Greek and Roman civilisations**. In fact, the increased and more discerning knowledge of those two glorious and ancient European civilisations on the part of peoples in Europe; did create a highly motivating awakening among them, especially among their thinkers, scholars and the intelligentsia and thereby in the process; momentously triggering a movement during those times, also known as the renaissance in the medieval-Europe. **Interestingly with the great and glorious ancient-era civilisational past, in the areas, now constituting the modern and sovereign Nation-States of Greece and Italy, the Renaissance in Europe, first flowered in Italy. And subsequently from there on, the historic movement of renaissance spread as since a multiply great civilisational awakening, all over the European continent.**

It has generally been found that the learned historians all over the world have mostly identified this medieval-era historic movement of renaissance, in almost whole of Europe; as mainly with the very enchanting revivals of the old classical styles of art, literature and even sciences, since belonging to the already bygone ancient-era in Europe. But the pivotal point is that indeed, the gloriously historic and inspirational sources for the multiple-facets of this renaissance movement in Europe, have principally been the multiple-legacies of the ancient Greek and Roman civilisations. And hence at the same time in these contexts **we also cannot ignore, certainly the no less important and far-reaching civilisational geostrategic and politico-military cultures concerning the matters-military as well as historically the war-fighting oriented legacies of the ancient Greeks and Romans**.

Hence when this historic movement of the renaissance in Europe began and gathered momentum the highly inspirational multiple-facets of the ancient Greek and roman civilisations, also perforce contained their age-old aspects of the matters-military as well. Moreover at the same time, we may also need to remember that **historically during their times; the Greeks and Romans too, had cultivated and nourished their different kinds of strongly fixated imperialistic ambitions; for relentlessly building/expanding their vast empires**, as then appearing to be feasible. And historically therefore, these geopolitical as well as matters-military-oriented highly inspiring multiple-legacies of the ancient Greeks and Roman too formed indeed powerful segments of the principal sources of inspirations for medieval Europe's, themselves the extremely feudalistic and imperialism-fixated Nation-States.

In addition, another no less relevant historic aspect in this regard, and particularly at this precise point emerges that, during the period when renaissance movement in Europe was gathering momentum, the leading European Imperial Powers had already mastered their substantial sea-faring technologies and capabilities for building their powerful Navies, as well as big and strong armed ships. European explorers, such as Christopher Columbus and Vasco da Gama, were already taking path-breaking initiatives for exploring the oceans. And in the process, had begun charting the long sea-lanes towards other distant continents of the world. During these momentous times in the medieval Europe too, man's living-history was being simultaneously made in many dimensions and thereby also, covering so many civilisational aspects as well. And logically therefore in a perspective, the movement concerning renaissance in Europe, was also rather subtly contributing perhaps some kinds of inspirational inputs; into now the historic aspects of the then uncontrollable imperialistic zeal of the already earlier mentioned five leading and rival European overseas Imperial Powers, since making substantial inroads with their different geopolitical-cum-military strategies, into the overseas distant continents of the Asia, Africa and the Americas.

Man's living-history since interestingly informs us the **famous historic city of Florence in Italy, was the Mecca for the renaissance movement in the medieval Europe**. In these contexts, **Jawaharlal Nehru** has indeed very beautifully mentioned that, "Florence had already produced in the thirteenth and fourteenth centuries, **Dante** and **Petrarch**, the two great poets of the Italian language. During the middle Ages it was for a long time the financial capital of Europe, where the big money-lenders congregated. It was a little republic of rich and not very admirable people, who often ill-treated their own great men. 'Fickle Florence', it has been called. But in spite of the money-lenders and the despots and tyrants, this city produced, in the second half of the fifteenth century, three remarkable men: **Leonardo da Vinci, Michelangelo, and Raphael**.... Most amazing of all were Leonardo's attempts at aviation or flying in the air. He did not succeed, but he went a good way towards success. There was no one to follow up his theories and experiments. Perhaps if there had been a couple of Leonardos to follow him, the modern aero plane might have been invented 200 or 300 years ago. This strange and wonderful man lived from 1452 to 1519. His life, it is said, 'was a dialogue with Nature'.... So the Renaissance flowered in Italy from the fifteenth century gradually travelled to other Western countries.... During this period—the fifteenth to the seventeenth centuries—science also gradually forges ahead and comes into its own. It had a stiff fight with the Church, for the Church did not believe in making people think and experiment. For if earth was the centre of the universe and the sun went round it and the stars were fixed points in the heavens. Anyone who said otherwise was a heretic and might be dealt with by the Inquisition. In spite of this **a Pole, named Copernicus**, did challenge this belief and **proved that the earth went round the sun. Thus he laid the foundations of the modern idea of the universe. He (*i.e. Copernicus*) lived from 1473 to 1543. Somehow he managed to escape the wrath of the Church for his revolutionary and heretical opinions**. Others who came after him were not so fortunate. **Giordano Bruno**, an Italian, was burned in Rome by the Church in 1600 for insisting that the earth went round the sun, and the stars were themselves suns. A contemporary of his, **Galileo**, who made the telescope, was also threatened by the Church, but he was weaker than Bruno, who thought it more expedient to recant.... Among the prominent men of science in the sixteenth century was **Harvey**, who finally proved the circulation of the blood. In the seventeenth century comes one of the greatest names in science—**Isaac Newton**, who was great mathematician. He discovered what is called the law of gravitation—of how things fall—and thus another of nature's secrets. So much or rather, so little could be for science. Literature also forged ahead during this period. The new spirit that was abroad affected the young European languages powerfully.... But the new spirit, and paper and printing, pushed the languages ahead.... The languages of Europe progressed and gained in richness and power, till they became the fine languages they are today.... In England there was the famous **Shakespeare** from 1564 to 1616; and immediately following him in the seventeenth century was **Milton**, the blind poet of *Paradise Lost*. In France there were the philosopher **Descartes** and the dramatist, **Moliere**, both in the seventeenth century. Moliere was the founder of the Comedie Francaise, the great State theatre in Paris. A contemporary of Shakespeare in Spain was **Cervantes**, who wrote *Don Quixote*.... **Machiavelli**, another Florentine, was just an ordinary politician in the fifteenth-sixteenth centuries, but he wrote a book, called *The Prince*, which became famous. The book gives us a glimpse into the minds of the princes and politicians of the day.... Even today the Imperialist Powers behave much like the prince of Machiavelli. Beneath the appearance of virtue, there is greed and cruelty and unscrupulousness; beneath the kid glove of civilisation there is red claw of the beast." (Nehru, 1934, pp. 322-326)

Contextually in our larger perspectives, we find that the multi-dimensional expansion of human-knowledge as well as emotionally surcharged awareness and, also overwhelming desires to inspiringly return to their roots extending up to even the multiple-aspects of once highly glorious ancient Greek and roman civilisations the medieval-era European peoples were indeed driven towards now historically a far-reaching movement, since known as the famous and also **history-making 'Renaissance in Europe'**. Hence in the overall civilisational contexts what all were indeed the multiply-enriching creativities then in medieval Europe and also along with, many other peaceful and many not so peaceful activities with contextually, even far-reaching historic dimensions and destinations in this regard, have been briefly elaborated in the few above preceding pages here. In addition further then interestingly, we find that the medieval-era phenomena of the renaissance in Europe were simultaneously also able to trigger the epoch-making processes of indeed the geopolitical as well as politico-militarily, a gradual consolidation of the States over there. At the same time historically; a kind of rather unchallenged, both the socio-cultural and socio-political supremacy of the Papacy then over the masses, bordering on even with the politico-religious imperialism in many ways were indeed decisively curtailed, in the medieval European societies then once for all.

This certainly was no trivial historic development; in the sense then looking into the nauseating aspects of prevailing Papacy power and culture of cruelty, since meted out to the great and highly creative intellectuals who were always keen to propagate scientific-temper and eradicate all sorts of mental darkness and also lingering superstitions, among the religious masses in the medieval Europe of those times over there. At the same time, while objectively looking deeper into these revolutionary aspects of their great thoughts and scientific discoveries then we may clearly see that, these highly gifted people were never critical of the Supreme Reality of All Mighty as well. All they persevered to advocate, was some sort of common-sense intelligence and the scientific pragmatism. As we further delve with our broader strategic criticality; then contextually the man's living-history also shows in this regard, that the highly inspiring republican traditions of the ancient Roman civilisation too, had a tremendous impact in the sense that a substantial secularization on a fast-track basis, of the medieval European States was also triggered. And as fallouts of great historic significance, in the medieval societies and Nation-States in Europe then, a culture of exploring as well as multiply-enriching debates too commenced and thereby, coupled with the path-breaking scientific advancements and promotion of the new traditions of book-printing and reading; an unprecedented dawn of the culture of cultivating newer frontiers of knowledge, awakening and also enlightenment, began as well. And historically, Europe was never the same again.

And with all the more historicity and simultaneously in a parallel manner the phenomena of gradual, but also progressively firming-up of the multiple-roots of democracy in different shades and, with varied political ideologies as socio-cultural-cum-socio-political systems of governance too, began in the Nation-States of the Western Europe during those times. These revolutionarily momentous historic phenomena which were indeed destined to significantly acquire the path-breaking and also, as the developments of a series of socio-political revolutions for doing away with the age-old and highly feudalistic systems of governance; in the whole world later, and along with and during the dawn of highly dynamic and very fast-advancing, the present modern-era. The renaissance in the then medieval Europe too, did indeed contribute many decisive inputs in these historic developments.

Interestingly for India in fact, finally then leading to her political independence in 1947 the phenomenal movement of the **historic Indian renaissance** too indeed had far-reaching and many everlasting multiple-impacts on the Indian societies and, her Nation-State as well. Moreover in many ways, it also contained many kinds of the well-considered and enchantingly crafted multiple-aspects of the unprecedented **'spiritual-practicality'** as well. Besides that, the multiple-aspects of the Indian-renaissance was also decisively instrumental in triggering as well as boosting-up in many ways, the historically prolonged Indian freedom movement too. Though imperially colonized and geopolitically subjugated by the overwhelmingly strong and powerfully overbearing, an alien Imperial-Power; during the recently bygone **eighteenth and nineteenth centuries in India, a galaxy of indeed great men and women, with very great intellect, highly gifted minds, unparalleled sense of both history and destiny; were born one after the other and, who individually as well as collectively both triggered and at the same time steered too, the above mentioned historic phenomena of the Indian-Renaissance and also in addition, the Indian movement for political independence as well.**

Now coming back to the historic aspects of the large twin-continents of the Americas, by the already earlier identified principal sea-faring and rival medieval-era European Imperial Powers, we also find that interestingly these vast continents across the Atlantic, were extensively rich in both the varieties as well as huge reservoirs of the so many kinds of indeed highly valuable natural assets and resources. Naturally, the multiple kinds of the huge natural resources and wealth were available both above the ground, and well as beneath the earth. In fact, during the medieval times and when these highly adventurous and substantially prepared as well as equipped Europeans, of the already earlier mentioned different nationalities and also having their quite different kinds of rather 'compulsive' imperialistic agendas; landed on the earlier uncharted American coasts, they were indeed unexpectedly mesmerized and amazed by even an initially partial glimpse of those kinds of natural wealth and bonanza, being largely virgin and obviously waiting to be exploited.

Hence now in the historic perspectives, it was naturally inevitable that the phenomena of the extremely ruthless struggles for dominating and securing the necessary geostrategic and politico-military supremacy over as large of the areas in these vast twin-continents as since possible, for taking control of the vast resources and wealth there-in had to commence at the earliest and multi-dimensionally intensify as well. And these phenomenal historic developments in the process did also trigger a prolonged saga of the momentous making of now so well-known, the American living-history. Historically, phenomena of the **ambitious programme and multi-dimensional conducts of the imperialism as well as establishment/expansion of the colonial empires anywhere perforce also have three dimensions, i.e. (1) Securing the decisive geopolitical, geostrategic and politico-military supremacy over potential rivals, (2) propagation of one's religion in the colonially subjugated territories and, (3) exploitation of the natural resources/ assets as well as other kinds of wealth and riches as since available for multiple benefits of the concerned imperial power's home country and thereby, mostly regardless of any kinds of adverse consequences for the subjugated peoples and their colonised lands**. These are by and large in fact, the universal 'core-truths' of the imperialism and though historically, there have been exceptions as well.

But at the same time while delving deeper, specifically into the above discussed rather inevitable aspects of historic drives of the conducts of imperialism and colonization of alien lands we very interestingly do also notice that historically, all are not just 'black and white' in this regard. Though the basic point with respect to the multiply complex aspects of imperialism, remains always well accepted and that is; that there is seldom and that too if at all, any possibility for the any kind of **'human-level sincere benevolence'** towards any colonised population, its aspirations, its landmass and the natural resources and wealth there-in. At the same time in these precise contexts, we may need to bear in mind that these relatively not so apparent but also no less important aspects, concerning the human-level sincere benevolence during a rather thick of the multiple conducts of imperialism anywhere, also largely depends upon the situational realities then prevailing on the ground and along with all the more important—socio-psychological, socio-cultural and socio-political traditions and historically civilisational-heritage on the part of that concerned Imperial Power, since ruling over those colonised territories and their peoples. These precisely are the reasons, as to why specifically the colonial-era living-history of any particular colonised country, during any historic period and in any continent, is colourfully quite different as well as contains very different kinds of complexities as well.

Now so far as the continent of South America is concerned, we have noted earlier that, from among the overseas European Imperial Powers, the Spaniards and the Portuguese were the first to set their foots on its shores on the Atlantic side, for making inroads as much deeper as possible first in the northern region for conquering territories and firmly establishing their imperial colonies. And in the process, it was principally the Spaniards, who seemed turning out to be smarter and more adventurous in the vast continent of South America. Similarly at the same time, and during the different years of medieval-era sixteenth century, in both the North and South American continents, at many places on their Atlantic coasts a number of overseas expeditions of the white men from Europe, were making their landings and also rather inevitably confronting the natives since already residing over there. Historically in fact, all these sea-faring phenomena from the far-off Europe, resulted in multiple kinds of gradual but fast and steady, rather human-level swarming by these white men of the twin American continents, with all kinds of their imperialistic programme and intentions, resulting in the far-reaching and unprecedented consequences for Americas and hence naturally the Americas were not going to be like their older selves again from then on, and for the all times to come.

From the early decades of the sixteenth century itself, the Spaniards had begun their momentous landings on the coast of Mexico in search of the gold and new territories for their empires. Portuguese too, came and landed at different places on the vast coast-line with similar intentions. And little later the French and the English too came on the North American coasts, during the second half of the sixteenth century and naturally in totality, all these resulted in a somewhat complex and rivalry-ridden scenario and interestingly, with rather two-dimensional clashes over there. Hence historically in these contexts, one pattern of clashes inevitably surfaced between the overseas European Imperial Powers, at the places they landed and also, as they moved deeper into largely the unwelcome territories and hence with the Native Red-Indians since settled there and as the further years rolled-on, the other pattern of rivalry-driven clashes began between those European imperial-expeditions themselves. We do also notice in this regard, that since these rival European Imperial Powers were also sea-faring nations and obviously their expeditions too, were always accompanied by their powerful Navies and consequently at a number of times, such clashes among themselves also took place in the sea as well. In addition interestingly, during such historic naval-battles among those rival European Powers in the Atlantic at times the **savage elements of sea-piracies too** surfaced during those naval clashes of arms. Moreover, British Navy was always able to maintain a decisive edge over her rivals in the Atlantic, as in the European waters the English were mostly able to siege victories, in those naval engagements and thereby the Great Britain finally managed to push-out her rivals and colonise the North American continent, and hence consequently the modern large Nation-States of the USA and Canada were born.

But before the final phases of the processes of colonization of the North American continent, the times for the English over there, were extremely difficult and in fact, even always full of risks and danger to their lives from the highly volatile and intolerably hostile native Red-Indian warriors. Many English and other European lives were lost in these clashes and, they had to repeatedly come in more numbers, with more arms and stores over a number of years from their homelands across the Atlantic; before they could begin dominating the North American continent and **thereby, commence the process of establishing their colonies and gradually plant there-in, the British systems of governance. Here, the firearms since already available to the Europeans, were decisive against the Red-Indians**. Consequent to the gradual implanting of the multiple-aspects of these British systems of governance, also historically resulted in the far-reaching and multi-dimensional impacts in the subsequent continuous making of the living-history of both the modern, very big twin Nation-States of the USA and Canada. In this specific perspective, we find that the USA happens to be the world's most developed, industrialized and, technologically the most advanced as well as both, economically and militarily most powerful Nation. No wonder in a way that both interestingly as well as quite ironically too **in every corner of the world today and thereby, in every nation's almost everything of any consequence, whether directly or indirectly in various degrees and as per that nation's any relevant situational scenario, USA's influence and impacts, whether clearly visible or even invisibly apparent, are always felt**. We would be discussing them in more details, in the forthcoming chapters.

Now so far as the historic aspects of Spanish colonisation of the Mexican region were concerned they were in fact, quite different from those of the North American continent. In these contexts, the living-history also shows that when the imperial Spaniards arrived there, they certainly were not expecting to encounter the natives living over there, since possessing a fairly advanced civilisation and though, definitely of a kind of barbaric splendour. But subsequently with imperial colonisation of Mexico the Spaniards systematically didn't allow anymore, the natives' old civilisational aspects to exist and in return interestingly they were bestowed with a **'gift' of a new religion, i.e. Christianity**. This had multiply far-reaching impacts and the Native Mexicans found many new developments in their areas and in their lives too, and thereby; the Spaniards brought many new kinds of animals, such as horses, cattle and poultry and in addition also, introduced some new crops, vegetables and fruits and so on. The age-old barbaric custom of human-sacrifices as a highly superstitious religious-ritual was totally abolished for the good and hence, the socio-cultural as well as socio-political systems of living and political-governance too, of the Mexicans; as a new Nation-state, were almost totally transformed.

Moreover, during the Spanish imperial-era in Mexico; the Mexican natives and their lands were exploited and at the same time, though having subjected to a subordinate status, they were given constitutional rights to always claim to be an integral part of the main social stream. This was no trivial matter in Mexico then since as a result, we also find many new kinds of additional developments, then taking place, on the Mexican soil. By 1580s or so, more than two hundred new towns on Spanish patterns and having sizeable populations were established and as a natural consequence, a gradual but steady urbanization of the Mexican natives too, had begun. Contextually during the momentous sixteenth century more and more and new kinds of far-reaching developments in Mexico had materialised. New schools and colleges, including many new village-schools in the rural areas were established by the Spaniards. On top of it **in 1551, even the University of Mexico was founded in the capital Mexico-city which catered for higher studies in the native languages as well, besides teachings of the Latin and the other subjects of sciences**. It was indeed a path-breaking development. Cultivation of cotton and sugarcane was introduced and, even much before 1590 itself and then as an ongoing process a number of sugar-mills were established over there, for quickly commencing the mass-production of sugar. And hence before the end of sixteenth century itself, they had begun exporting more than 200 to 250 hundred thousand pounds of sugar every year.

But most precious metals of gold and silver were still the principal targets of the imperial Spaniards in Mexico. And hence, they took the situational geostrategic and politico-military initiatives to penetrate further deep into the landmass of South America where the modern Nation-State of Peru is now located and as a result to their extreme surprise they found Peru to be equally rich like Mexico, in both gold and silver. Historically, it turned out to be altogether a new bonanza; since again waiting to be exploited by the imperialistic Spaniards, for the multiple benefits of their home country during those momentous centuries; and hence thereby the sizeable Spanish fleets were able to carry then, more than fifteen million dollars' worth of gold and silver to Spain in Europe. This too is always another very important as well as decisively motivating and, rather since universally applied dimension of the age-old historic aspects of imperialism.

As we further interestingly move-on the highly engaging pages of man's living-history, with our contextually critical objectivity in this regard; we do also notice that during the zenith of medieval-age multiple-aspects of the imperialism, since then being geopolitically conceptualised and vigorously practiced by the already earlier listed European Imperial-Powers, in almost all continents of the world; historically we do also notice that among themselves, the British happened to finally excel over her potential European rivals with her unique arts and crafts of coercive-diplomacy, geostrategic shrewdness as well as politico-militarily at times, brilliant-cum-proactive applications of their matters-military and assets. In a perspective we may also say that, in these the then relentlessly competitive games of the medieval-era international and overseas imperialism, the British were perhaps wittingly or otherwise, able to somewhat fruitfully apply the ancient-era's, since **timelessly universal concepts and theories of the both, China's Sun Tzu and Chanakya of India** in always the focused pursuits of her agendas of the imperialism. In our contexts here, we would be covering these historic aspects of the medieval-era British imperialism; since spilling-over also into the beginning periods of modern-age and at the same time, having some kinds of rather global ramifications, as we move-on further in our continuing discussions.

So far as the historic aspects of the consequentially far-reaching and, indeed epoch-making colonisation of the North American continent by the British are concerned, **James Truslow Adams** has written that, "**The English mind had long been accustomed to the triple combination of King, Lords, and Commons; and in general this type was reproduced in America, with local variations**. In the experimental stage, when the colonies were first being planted, various sorts of government were imposed or developed, but, after the return of the Stuarts from exile in 1660, there was strong tendency toward increasing the imperial control and making the governments more uniform, a tendency which was notable in the **forfeiture of the old Massachusetts charter**

in 1684. In general, though Rhode Island and Connecticut remained almost independent, there was in each colony a Royal Governor, appointed by the Crown, who represented the King; a Council, or Upper House of the Legislature, who tended to be undemocratic owing the methods of selecting its members, and for other reasons; and an Assembly, or Lower House, which was popularly elected, as popular election was then understood.... If these governments, however, in their triple form resembled the King, Lords, and Commons of England, there were new factors involved which greatly affected the comparison. For one thing, the Royal Governor, too often a needy and broken politician from home (though there were notable exceptions), was invested with more of the sanctity and reverence which hedged the monarch himself.... **Several important results flowed from these conditions**. In the first place, the ordinary man represented in the Assemblies, who, as we have seen, had risen considerable in his sense of independence and self-esteem in America, as contrasted with England, felt a good deal freer to fight a parliamentary battle against a governor than his fellow commoner in England did to fight on against the King.... In this same situation was evolved the germ of another and even more serious trait in American character. A sense of law and respect for it is one of the deep-rooted traits of Englishmen. It was amply displayed in the Mayflower Covenant and other acts of the first comers.... Under better economic conditions, crimes against person or property became extremely rare in the colonies—so rare that, in spite of the lonely roads through the woods connecting almost all the settlements." (Adam, 1931, pp. 42-44)

Further in our larger perspectives if we may pause and critically begin to look back into highly checkered, USA's history we may have glimpses of the epoch-making and highly consequential phenomena of the **American War of Independence (1775–1783)** and subsequently within a century itself, the **American Civil War (1861–1865)** too. And historically we may even say that highly consequential developments across the Atlantic, were indeed decisively able the influence as well as globally change subsequently; the multiple-courses of the making of man's living-history in this world. We would be discussing these highly phenomenal historic American episodes, in some detail and in various contexts little later.

And in this precise context, we may recall that the analogy pertains to the aspects of timelessly perpetual, incomprehensively complex, infinitely vast and in addition; the multiply and unendingly continuing phenomena of evolution of the Mother-Nature; since the very beginning of Creation itself and, within the infinite Universe. And that too **perhaps, if any such hypothetical point indicating this 'beginning of the Creation itself'; can ever be conceptualised and then scientifically fixed or identified, with any tangibly conclusive precision**. Well, in any case at the most, these elements of thoughts do remain matters of philosophical conjectures only. In these precise contexts, we may recall that in the initial chapters only we have already quoted certain very illuminating paragraphs, from the earlier referred scholarly book, on 'The Meaning of Evolution:...'; authored by the **US Professor George Gaylord Simpson** and delved upon these subject-matters in various contexts; and hence, we may need to refer to the same book again, precisely with respect to the present point of our continuing discussions. In his book, Professor Simpson has indeed shown us, with his brilliant scholarship, that how each and every phenomenon of the Mother-Nature's unendingly continuous systems of evolution in the infinite Universe never happens in isolation and so, they are all integrative and interactive as well.

While keeping the above discussed perspectives in mind, we also learn that along with the historic phenomena of the American twin-continents and then, in the integrally developing process of birth of the Nation-States of USA and Canada the simultaneously parallel but quite different in patterns and no less historic, phenomena of the colonisation of Asia too, were in progress. And in the very vast continent of Asia as well as the Indian Ocean and other adjacent Asian seas, the same leading seafaring and rival European Imperial Powers, were very much competitively in action with their respective strategic ambitions and imperial agendas. Interestingly in a context at this precise point, we may say that since the world is round, and so the various human-civilisations with their differently historic advents also keep-on advancing multi-dimensionally. And historically further, one of the very far-reaching as well as overwhelming ambitions in this regard, happens to be age-old multiple-aspects of the geostrategic and politico-militarily crafted imperialism anywhere, on the part of concerned Imperial-Powers. And hence, we find historically the different kinds of almost simultaneous manifestations of the same multiple-aspects of imperialism; of the above referred rival and ruthlessly competitive seafaring European Imperial-Powers, in the American and Asian continents and at the same time, as highly complex historic consequences; we do also notice that subsequently altogether, very different but also integrative and mutually influencing colonial-era living-histories of these geographically far-off located, American twin-continents and that of Asia too, were gradually made and progressed for posterity. Further, in those European Imperial-Powers' history-making aspects of their ambitious imperialism, in the continents of both Americas and Asia, we find that their numerous path-breaking as well as epoch-making imperialistic but substantially, the civilisational 'thoughts and actions', in those since already colonised continents, many multi-dimensional and geopolitically almost everlasting consequences too, materialised for the native populations in the erstwhile colonies. Well interestingly in a context, we may also say that in a way this too, is another distinct path of the making of man's living-history in the world. More specifically therefore, in our larger perspectives we interestingly find that though in altogether different patterns but, precisely as per the above spelled out historic 'causes and effects' systems of the making of colonial-era living-history, did take place both in the American and Asian continents.

We may recall that if in 1492, **Christopher Columbus** launched his historic discovery voyage of the twin-continents of Americas and then; just after six years in 1498, the Portuguese adventurer named **Vasco da Gama**, was successful in discovering the sea-routes to the south-western coasts of Indian peninsula, for the medieval-era European overseas Imperial-Powers. Moreover interestingly, these historic discoveries took place even more than a century before the imperial Mughals were able to set their feet on the soil of Indian subcontinent. We further notice that since the **Portuguese were the first to arrive on the southern-Indian shores** with their specifically ambitious imperialistic agenda and, along with their in-built civilisational flexibilities even with some multiply encompassing and befittingly desirable, while some unbecoming ones too—the socio-cultural and socio-religious thoughts/fixations came into actions as well, in their colonised territories. The Portuguese were able to politico-militarily conquer Goa on India's south-western coast even before Babar, founder of the historic Mughal imperial-dynasty in India, was able to reach the north-western borders of the Indian subcontinent. In fact, after these historic events took place; the imperial Dutch, British and the French could also reach India, with their different imperialistic programme and agendas. All these rival overseas European Imperial Powers did arrive in the medieval India, first as traders; since they were already aware of the vast Indian subcontinent's substantial riches as well as age-old civilisational glory. Over time, they encouragingly found numerous avenues, so valuable to them, gradually opening up historically on the Indian soil for them to make further imperial in-roads into the Indian-subcontinent, not only territorially but also, socio-culturally, socio-politically as well as politico-militarily too.

Specifically in the above perspectives and also precisely concerning the Portuguese, the well-known historian and author **William Dalrymple (*White Mughals*,** 2004, p. 11), writes very informatively that, "Until the 1830s, there was every sign India would as dramatic a transforming effects on the Europeans who followed the Mughals. Like all the foreigners before them, it seemed that they too would be effortlessly absorbed. This 'crossing over' was a process that dated from the very beginning of the Europeans presence in India. The Portuguese were the first to make the transition. After the conquest of Goa in 1510—sixteen years before the arrival of the Mughals in north India—the **Portuguese commander Alfonso de Albuquerque** made a point of ordering the men to marry the widows of the Muslim defenders they had massacred during the taking of the city. Albuquerque himself presided at the weddings of these 'fair Mooresses of pleasing appearance' and provided them with dowries. The fair Mooresses were then forcibly converted to Christianity, and after baptism, many were made to receive the rudiments of the Catholic faith. **But this crude attempt at force feeding unadulterated Portuguese culture to India proved short-lived, and as unsuccessful, as previous attempts to impose unadulterated Turkish, Sassanian, Persian or Greek culture had been during the preceding centuries**. Over the course of the next fifty years, the women, the environment and sheer distance of Goa from Europe all worked on the new arrival, so that gradually, generation by generation, the conquistadors began abandoning the ways of Portugal and taking on instead the customs of India."

In the above quoted paragraph William Dalrymple has indeed been able to very captivatingly paint consequentially important and far-reaching glimpses of the medieval-era Portuguese imperial in-roads and their some short-sighted misadventures on the Indian soil, in his lucid literary-style as well as with his remarkably precise historicity too. Historically in the larger perspectives he has conveyed volumes regarding the multiple-advents of European imperialism along with its numerous historic facets. And hence, as we further delve into some salient features of this highly readable book historically as well as very interestingly we find that, during initial phases of the in-roads into the Indian subcontinent, then since being made by the multiple-aspects of European imperialism; with the first-go itself quite fascinatingly, instead of those European imperialists rather successfully submerging Indian natives into their kinds of socio-cultural systemic thoughts and lifestyles; they themselves got substantially overwhelmed by the age-old value-system depths and multi-coloured socio-cultural richness and customs of the Indian peoples.

Hence historically during the initial phases of European imperialism coming into India we find similar things happening with the British and French imperialist too, interestingly much more similar results gradually began coming onto the surface particularly with the British, since finally it was the British who managed to secure the successful imperial sway over the Indian landmass. So concerning the above mentioned medieval-era initial phases of historic in-roads of the European imperialism into the Indian-subcontinent contextually and quite befittingly, we may also say that **instead of colonizing the natives; amazingly those European imperialists in India got first themselves overwhelmingly colonised**. Soon, there were some reversal in such initial trends too and more so, as the overseas British Imperial-Power began consolidating its geostrategic and politico-military imperial-holds over the vast expanse of India. But in spite of such periodical geopolitical imperial-holds by the different foreign imperial-powers, during the bygone historic-ages the multiple-strengths of remarkably enduring, gloriously ancient and enchantingly rich Indian-civilisation, could never be shaken by any kind of imposing thrusts of the different alien Imperial Powers. That is a great civilisational mystery of the very ancient land, since known as India, which has indeed, been able to maintain her remarkably unbroken and multi-dimensional civilisational continuity.

In the above specific contexts and, especially concerning the British in India during that bygone colonial-era again, **William Dalrymple** has mentioned that, "The deeper I went in my research the more I became convinced that **the picture of the British of the East India Company as a small alien minority locked away in their Presidency towns, forts and cantonments needed to be revised. The tone of this early period of British life in India seemed instead to be about intermixing and impurity, a succession of unexpected and unplanned mingling of peoples and cultures and ideas**.... This was something I became increasingly sensitive to when, in the course of research, I discovered that I was myself the product of a similar interracial liaison from this period, and that I thus had Indian blood in my veins. No one in my family seemed to know about this, though it should not have been a surprise: we had all heard the stories of how our beautiful dark-eyed Calcutta-born great-great-grandmother Sophia Pattle, with whom Burne-Jones had fallen in love, used to speak Hindustani with sisters and was painted by Watts with *rakhi*—a Hindu sacred thread—tied around her wrist. But it was only when I poked around in the archives that I discovered she descended from a Hindu Bengali woman from Chandernagore who converted to Catholicism and married a French officer in Pondicherry in the 1780s. It also became increasingly clear to me that the relationship between India and Britain was a symbiotic one. Just as individual Britons in India could learn to appreciate and wish to emulate different aspects of Indian culture, and choose to take on Indian manners and languages, so many Indians at this period began to travel to Britain, intermarrying with the locals there and picking up Western ways. **The Mughal travel writer Mirza Abu Taleb Khan**, who published in Persian an account of his journeys in Asia, Africa and Europe in 1810, described meeting in London several completely Anglicised Indian women who had accompanied their husbands and children to Britain, and one of whom had completed the cultural transformation so perfectly that he 'was some time in her company before I could be convinced that she was a native of India'. He also met the extraordinary Dean Mahomet, a Muslim landowner from Patna who had followed his British patron to Ireland.... Only seventy-five years after the death of James Achilles Kirkpatrick (*Then, the British Resident in Royal Court of the Nizam in Hyderabad*), and indeed within the lifetime of his Anglo-Indian, Torquay Hyderabadi, Islamo-Christian daughter, **it was possible for Kipling to write that 'East is East and West is West and never the twain shall meet'. There is a tendency to laugh at Kipling today; but at a time when respectable academics talk of a Clash of Civilisations (*Well, courtesy indeed: Prof. Samuel Huntington.*) and when the East and West, Islam and Christianity appear to be engaged in another major confrontation, this unlikely group of expatriates provide a timely reminder that it is indeed very possible—and has always been possible—to reconcile the two worlds**." (Dalrymple, 2004, pp. xl, xli, xlii & xliii) (*Indeed perhaps in a most befitting and globally timeless manner this remains the central theme as well as the most enchanting fragrance of all the glorious and enriching human-civilisations, since the very beginning.*).

In our larger perspectives and, also concerning the above spelt-out contexts at this juncture the precise point historically coming to our minds, clearly happens to be that during the 16th century itself how a relatively very small number of British soldiers, traders and missionaries from the far-off Islands of Great Britain, situated in the European North-Sea could so fast manage to secure the vast landmass of Indian subcontinent, belonging to the then already few million Indian peoples; and since also, belonging to one of the very glorious ancient civilisations in the world? It remains one of the most amazing historic facts and also at the same time, indeed a very fascinating historic display of the triumphs of multiple-aspects of the well-crafted British Imperialism into the medieval India. Historically in this regard, one singularly decisive fact clearly stands out and also emphatically mentioned earlier that India since ages and so then too ironically, waiting to be geopolitically colonised as she had always been callous and also, never been able to fruitfully cultivate, evolve, develop and imbibe the timeless universal aspects of the **'essential strategic-culture and thoughts'**. Wherein the British then prominently a strong and overseas European Imperial-Power, had already cultivated the necessary expertise and, indeed become a global-master in the arts and sciences of the concepts and proactive applications of those aspects of the 'essential strategic-culture and thoughts', in all their strategically consequential matters of the statecraft and hence in addition also; in similar manner, in the international geopolitical games of the extremely violent and rivalry-ridden—imperialism as well.

And therefore, if we critically peep deeper into the concerned historic chapters of the highly checkered earlier periods in India we may clearly notice the already earlier fructified seeds of the concerned causative-factors and hence, the 'die was already cast' in this regard for the forthcoming British imperial colonisation of the Indian subcontinent. And thereby, the **paradise of the historic Raj for the Imperial British in the already colonised India** was established then and consolidated, indeed with a substantially enduring success. In these engaging contexts, the man's living-history very clearly shows us that in 1837, when Queen and Empress Victoria came on the British royal throne in London the European population and that too mainly the British, living in India happened to be just around 45-47000 only and out of them; nearly 40000 were the British military men including officers and, nearly another 1000 were the **ICS (Indian Civil Service)** British officers, plus some European Missionaries, academicians and traders as well. Along with that they also had, willingly handy population of nearly 35000 of the mixed parenthood at their imperial disposal, since later-on known as the Anglo-Indians. And historically then, they were since already imperially ruling over the vast and rich landmass of the Indian subcontinent with nearly two millions of the native Indian population and thereby, as a manifestation of the British-Raj with their own kinds of intended as well as commendable success too. And all these in turn as a great combination, were indeed able to imperially transform India as the historic '**Jewel in the British Crown**'.

These developments then in the colonial India were indeed the zenith-period of the phenomena of success, powerful consolidation, glorious expanse and magnificent grandeur of the multiple-manifestations of, now the historic era of British-Imperialism in India. Well, while looking into these historic aspects pertaining to the times of colonial India, with certain critical objectivity in this regard; we do find that this period of the British imperial-grip over colonial India; was indeed a golden-age of nearly 300 years for the British, both in India as well as back-home in Britain too. And hence no wonder that for very many years even after India had already gained political independence and, fast got transformed into a rapidly developing, the sovereign democratic Republic, the surviving English men and women in the modern UK, were still looking back and remembering the bygone era of the British imperialism in India nostalgically.

Interestingly therefore, and with historic objectivity as well as perhaps, for the intellectual-cum-cultural enrichment of the humanity at large another phenomena concerning—the writings of many books, plays and quality-theatre programme, movie-making and so on also materialised in the, already tremendously transformed and politico-economically herself struggling Great Britain, who was then herself occupied indeed again, with commendable zeal and focused national efforts to repair fast, the colossal damages and, healing the wounds, she received while magnificently fighting the Nazis in Europe, Africa, Asia and then in their own continental seas and skies too; and at the same time, along with massive helps and support provided by the giant USA and, to her erstwhile numerous big and small colonies since then, located almost all over the world. Historically, these aspects are by themselves, a big as well as multi-dimensional saga and hence specifically in our larger perspectives quite befittingly and appropriately in more detail, we would be discussing them and much more, when we come to the next chapter and there specially, when we come to the multiple historic aspects of both the World-Wars, during the last 20th century. Now, so far as the multiple historic-aspects of the period of British Imperialism in the colonial India was concerned, they might have been a period of golden-times for the Imperial British but this was an age of highly mixed results and even at times, years of great sufferings, humiliation and prolonged turmoil-ridden political struggles for the native Indians.

In the colonial Indian-era belonging to the British imperial-rule, how India was looted; how her age-old multiply magnificent cottage-industry was systematically crushed to make massive markets for the exploitative promotions and sales of the British manufactures, then being dumped into the occupied India; how her massive forests, vast natural resources, huge deposits of various mineral-ores and riches were exploited and shipped back to the imperial UK for her massive economic-cum-industrial enrichment and so on, are by themselves again a long saga and, would briefly be taken-up later. But at the same time, we may need to bear in mind the other historically linked but also, somewhat different historic aspects too in this regard. As the saying goes that, every coin has two sides and hence here too perhaps, the historic saga in its substantial totality and pertaining to the very colourful as well as both sweet and sour British imperialism in colonial India, also have two interesting and no less important and epoch-making sides for the modern and sovereign Indian-Nation.

While reviewing these highly historic and indeed path-breaking aspects, with the requisite geopolitical and socio-cultural criticalities, and as guided by our larger and specifically predefined perspectives we do also clearly notice **that during their erstwhile imperial-rule in India, the British brought many new ideas, thoughts, system of education as well as initially the basic processes of developments/modernisation as well. And in addition then, they also laid solid foundations for the many valuable, unprecedented and highly enduring Institutions too and, some such wonderful institutions were even built and made to thrive as well, in India. To begin with, we may not forget that the very idea of the parliamentary democracy as a system for the national governance too, came from there itself**. If medieval Europe, including the Britain, were so much feudalism-oriented then India too as an age-old civilisation, was all the more extremely exploitation-ridden and regressively feudalistic, as well as mindlessly caste and creed based society, with hardly any sense of human-equality at all. In what manners India was exploited and looted and at the same time, in what all ways India unprecedentedly gained and got gradually enriched too, we would be discussing these far-reaching historic subject-matters suitably later and in more details, in the subsequent chapters.

We may recall that even before the arrival of rivalry-ridden leading overseas Imperial Powers, with their different kinds of imperialistic programme and agendas, onto the shores of the twin American continents as well as the Indian subcontinent, many kinds of historic developments and upheavals in Europe, preceded these adventurous sea-voyages across the Atlantic and the Indian Ocean too as their obvious causative-factors. And therefore, relevantly in these contexts, we may prefer to critically again look into certain aspects of the Europe's living-history of the 15th and 16th centuries. During those times, Europe had already begun breaking away from its traditional and turmoil-ridden medieval past and thereby, many kinds of highly unsettling restlessness and, dissatisfactions were cropping up among the larger European masses, covering many nations over there. These by themselves, triggered a kind of medieval sub-era of the revolutions and socio-politically highly surcharged conflicts and hence in different contexts, we have since discussed some of the prominent ones earlier. The main platform for them to fructify, was indeed provided principally, by the all sorts of **peasants revolutions against the ruthless exploitation by the feudal lords and also, against the tyrannical diktats of the Churches in Europe. In fact, the main conflict took the forms of clashes between the old and traditionally exploitative European feudal-classes and the then newly emerging, wide-awake as well as highly conscious of their legitimate rights and privileges, i.e. the fairly educated middle-classes, having tremendous leadership potentials** at all the socio-cultural and socio-political levels as well.

New times were since dawning over almost the whole of Europe and hence, the main sheet-anchor for the older systems of feudal-power and wealth, i.e. the ownership of vast expanses of the lands; had gradually begun to give substantial spaces to the newly emerged wealth generated by the manufacture of goods and the innovative creation of services and the trading activities, both for the domestic and overseas markets. And further to the utter shock of then the powerfully entrenched old-order, these new and fast advancing wealth-generating activities again, came largely under the control of peoples belonging to the newly emerged and also, independent-minded middle classes and consequently, Europe fast began to rather irreversibly change and rapidly transform in many ways. These by themselves are no less path-breaking and magnificently historic saga of the continuous as well as multi-dimensional advancements of the human-civilisations. Such historic and revolutionary smashing of the strong and ruthlessly exploitative shackles of the old-order in turn triggered many new developments, like the new discoveries, advances in the fields of science and technologies, building constructions and architecture, new explorations and adventures and so on.

Jawaharlal Nehru has very informatively recorded that, "**Spain and Portugal took the lead in the voyages of exploration**. Spain was just then driving out the last of Moors or Saracens from Granada. The marriage of Ferdinand of Aragon and Isabella of Castile had united Christian Spain, and in 1492, nearly fifty years after the Turks took Constantinople on the other side of Europe, Granada, of the Arabs fell. **Spain immediately became a great Christian Power in Europe. The Portuguese tried to go east, the Spaniards west**. The first great advance was the discovery by the Portuguese in 1446 of Cape Verde. The cape is the westernmost point of Africa....as one sails down from Europe towards this cape, one has to go south-west. At Cape Verde one goes round the corner and begins going south-east. The discovery of this cape was a very hopeful sign, for it made people believe that they would be able to go round Africa towards India. (***Enchanting attractions for the India's riches and unique wealth; had continued to lure the Europeans since the ancient-age and hence, they were being relentlessly driven towards India.***). It took another forty years, however, before Africa was rounded. In 1486, Bartholomew Diaz, a Portuguese, went round the **southern tip of Africa—that is, what is called the Cape of Good Hope. Vasco da Gama reached Calicut on the Malabar Coast in 1498. So the Portuguese won the race to reach India**. But meanwhile great things were happening on the other side of the world and Spain was to profit by them. Christopher Columbus had reached the American world in 1492.... Columbus started with three little ships and eighty-eight men. It was a brave and adventurous voyage into the unknown, for no one knew what lay ahead. But Columbus had faith, and his faith was justified. **After sixty-nine days of sailing they had reached land. Columbus thought this was India. It was, as a matter of fact, one of the West Indies. Columbus never reached the American continent, and to the end of his days he believed that he had reached Asia. This strange mistake of his has persisted to this day. These islands are still called the West Indies, and the original inhabitants of America are called Indians or Red Indians even now**.... It was soon after this that **Vasco da Gama** hastened his eastern voyage and reached Calicut. As the news of fresh discoveries came from east and west, the excitement in

Europe grew. The two rivals for dominion over these new lands were Portugal and Spain. The Pope then appeared on the scene.... In 1493 he issued a Bull—the Papal announcements or edicts are for some reason called Bulls—call the Bull of Demarcation. He drew an imaginary line from north to south 100 leagues west of the Azores, and declared that Portugal was to have all the non-Christian land to the east of this line and Spain the lands to the west of the line.... The Azores are islands in the Atlantic Ocean, and line drawn 100 league—that is, about 300 miles—to the west of them would leave the whole of North America and most of South America to the West. **Thus, practically, the Pope made a present of the Americas to Spain, and of India, China and Japan and other Eastern countries, as well as the whole of Africa, to Portugal. The Portuguese set about taking possession of this vast dominion. This was not so easy. But they made some progress and continued to go east. They reached Goa in 1510; Malacca in the Malay Peninsula in 1511; Java soon after; and China in 1576. This does not mean that they took possession of these places**. They just got some footings in a few places.... In 1519 Magellan (*a Portuguese*) stated on his western voyage, which was going to be the greatest voyage of them all. He had five ships and 270 men. He crossed the Atlantic to South America and continued going south till he reached the land of the continent. He had lost one ship by shipwreck and another had deserted; three ships remained. **With these he crossed the narrow strait between the South American continent and an island, and came out into the open sea on the other side. This was the Pacific Ocean, so called by Magellan because it was very peaceful compared to the Atlantic**. It had taken him just fourteen months to reach the Pacific. The strait he passed through is still known after him—the Strait of Magellan.... This was the most terrible part of the voyage. No one knew that it would take so long. For nearly four months, 108 days to be exact, they were in mid ocean with little to eat or drink. **At last, after great privation, they reached the Philippine Islands**. The people they met there were friendly to them and gave them food and exchanged gifts. But the Spaniards (*Magellan's companions.*) were offensive and overbearing. Magellan took part in some petty war between the chieftains and was killed. Many other Spaniards were killed by the people of the island because of their overbearing attitude.... The Spaniards and Portuguese of those days were proud and overbearing and cruel people; but they were wonderfully brave and full of the spirit of adventure.... Where the great explorers and discoverers had gone, hordes of adventurers followed, eager for loot and plunder. Spanish America especially suffered from this crowd, and even Columbus was treated very badly by them. At the same time **gold and silver flowed unceasingly to Spain from Peru and Mexico.... The success of Portugal and Spain naturally fired the imagination of the people of other countries, especially of France and England and Holland and the north German towns.... New discoveries came one after another, oceans and new continents, and wealth beyond measure, just waiting for the magic call—'open sesame'**. The very air must have breathed of the magic of these adventures." (Nehru, 1934, pp. 278, 280 & 281)

Historically the above quoted informative lines from Nehru's well-known book do give us very brief but, indeed panoramic views of the epoch-making phenomena of discoveries of the twin-continents of the Americas and their adjacent islands across the Atlantic and then, also unfolding of the sea-routes for reaching the southern coasts of India by encircling the African continent as well as subsequently, the sea-routes to Philippines and in addition; a cluster of many other island States of South-East Asia. These indeed historically turned out to be very far-reaching and revolutionary initiatives on the part of the Portuguese and Spaniards to begin with, during the momentous 15th and 16th centuries and; almost immediately triggered a series of further adventurously explorative activities on the part of other European sea-faring Powers and hence in a perspective, these may also be called a ruthlessly competitive initiates by the already earlier discussed late medieval-era rush for gold and similar riches by those then, towards India as well as comparatively more virgin territories in the South-Eastern Asian region and then; subsequently to China and the Far-East as well, via the newly discovered sea-routes.

In fact, as we further proceed to critically review these activities, of the then rival European overseas-Powers in our larger perspectives; we do also notice many simultaneously parallel as well as inter-related and momentously history-making phenomena, getting initiated too. We do also find the same aspects getting manifested, during the historic phenomena of highly adventurous and exploratory activities across the seas and oceans since highlighted above, on the part of medieval-era leading but also, irreconcilably rivalry-ridden, overseas European Imperial-Powers. Further historically another interesting dimension in this regard; since coming to the notice remains that, all these unprecedentedly epoch-making developments took place before indeed **the far-reaching advent of the Industrial-Revolution**. And hence since the historic Industrial-Revolution was yet to dawn, on the otherwise certainly expanding horizons of the continuously advancing as well as multiply enriching human-civilisations, there was no question of man during those times so-far, getting aware of the scientific aspects of the conversion of any form of energy, into any advantageously applicable motive-power then, in any manner.

So during those highly historic times ships were still sailing across the seas and oceans, by means of the age-old and wind-power oriented sails and hence, aided as well as strenuously controlled and managed by the manual muscle-power. Naturally therefore, it is not that simple and easy now, to even imagine as to how much the man's those historic launchings of the unprecedented and adventurous activities for exploration and discoveries; towards the totally unknown and which in the process then, being inherently dangerous and full of extreme hardships and also at times, causing deaths as well. It is also undeniably true that, since the very beginnings of civilisational times there have always been various kinds of causative-factors, since acting as compelling forces towards those historically far-reaching discoveries and exploratory initiatives on the part of man but at the same time; it is also equally true that while getting unstoppably engaged with them, the supremely powerful mother-Nature then as per her own perpetually ongoing evolutionary schemes of multi-dimensional manifestations, has always been extracting very heavy prices from the man in this regard. Hence, while keeping the above discussed larger evolutionary historic facts in mind, in the present precise contexts, we do notice that the zeal for new adventures and the irrepressible imperial urges to get hold of more and more new overseas territories as well as, to acquire more gold, valuable gems and other riches from there, on the part of those European Imperial Powers happened to be relentlessly so powerful that they happened to be highly restless to launch more voyages for the new explorations and discoveries in the far-off continents of the world.

It is also true that, historically to begin with during the medieval-era itself, among the various leading and rival imperial European-Powers; commendably Portugal and the Spain, and since both also already then, being the highly accomplished sea-faring, as well as militarily quite strong and experienced Nation-States definitely took the unprecedented bold initiatives and thereby, established initial lead in launching their highly adventurous sea-voyages towards Asia and also the newly discovered Americas in this regard. Those were indeed very momentous times then in Europe and hence **all these revolutionary developments, as being then pioneered by the Portuguese and the Spaniards in turn had to shake-up other no less ruthlessly rival and also interestingly all the more ambitious, sea-faring, as well as militarily strong and accomplished**

European imperial-Powers; such as the Great Britain, France and Holland too. These three European Powers were in no mood to be left behind and hence they too then, prepared and organised themselves with a feverish pitch for participating in these almost free-for-all competitive adventures but at the same time also, since these being ruthlessly competitive, and hence consequently therefore at times; even leading to the clashes of arms among those rival European Imperial-Powers, both in the sea and on the far-off lands as well.

We may also recall that little earlier in this chapter itself, we have since briefly discussed with some necessary criticalities too that the historic as well as imperially formative-stages of these unprecedented and almost global phenomena of the building of overseas Empires, by those European seafaring Imperial-Powers in the far-off continents of the Americas, Africa and Asia resulted in their own kinds of epoch-making living-history as well. So indeed therefore, **these globally ruthless great-games of highly ambitious imperialism with possibly all kinds of no-holds-barred fallouts** and obviously, on the part of those rival and also highly ambitious European Powers since as mentioned above, had already since commenced with maximum possible fervour. Consequently then-on, as further results the world and simultaneously the subsequent paths of making of the man's living-history, were never going to be the same again. Prominently as some of their multiply epoch-making and historically far-reaching resultant fallouts even perhaps the multi-dimensional pace of the man's civilisational advancements and enrichments, since the very beginning of the living-history itself too, had indeed increased as accompanied by their enhancing complexities as well.

At this precise point, as we further begin to delve into the phenomena of the simultaneous building of Empires by those rival and ruthlessly competitive overseas European-Powers in the newly discovered American and then, soon after in the Asian and African continents were indeed also seen to be fast gathering intensely competitive momentum as well. In these contexts, we do also notice that shortly later the **historically revolutionary advent of the so-far unprecedented and even perhaps unexpected, dawn of the 'Industrial-Revolution', and along with its multiple-fallouts too, indeed phenomenally contributed in many ways, into these rather unchecked imperial games then being played with the sole ambitions for building of the overseas Empires, garnering all possible riches; and in addition also thereby in their processes, consequently encompassing almost the whole of the world**. Historically, these unprecedented phenomena of the extremely rivalry-ridden games of imperialism and, the simultaneously on-going building of overseas Empires and then, devising and imposing various systems of imperial-rules over them; and along with their own kinds of indeed, quite overwhelmingly powerful global-reach as well as the multi-dimensionally enduring influences onto their historic colonies and also, over their erstwhile colonised native populations too, were perforce imposed in the various far-off continents. Well consequentially in this regard we may also observe that historically, these happen to be few of the important segments of indeed, the so many history-making fallouts of the perforce imposed aspects of imperialism; as since overwhelmingly guided and also perhaps compulsively driven by the numerous kinds of imperialistic interests, of since the conquering overseas Imperial-Powers; during their imperial-rules, in their geostrategically and politico-militarily subjugated alien colonies.

These path-breaking historic phenomena interestingly in turn had their own kinds of long-term multiple-impacts, over those colonised peoples and their societies, and hence consequently in the processes from then-on all these did gradually trigger many irreversible changes in the subsequent courses of the making of their history. In these historic perspectives we have since also discussed that the above mentioned multiple-aspects of the erstwhile ruthlessly competitive games of imperialism, as well as the simultaneously gradual building of Empires in the far-off overseas continents were indeed not any kind of sudden developments, as may perhaps generally be perceived in certain shades of prevailing opinions. But as per the authentic historic realities; there indeed were certain kinds of preludes having already gradually taken place and then, we also notice that as these in turn were then since piloted by those erstwhile European Imperial-Powers themselves in their respective areas of influences in the overseas continents then, also created various kinds of rather quickly surfacing opportunities for the building, expansion, as well as fast consolidation of their Empires.

Well in these perspectives in addition the man's living history also shows that, since the beginning of the phenomena of perforce imposing the multiple-aspects of various kinds of imperialism, as well as perforce thrusting of the imperial-rules over the colonised territories and along with on to their native populations, by any conquering Imperial-Power all these are indeed the inherently logical and also, rather obvious consequences of the imperialism and, its multiple-manifestations in this regard. Then at the same time, it is also equally true historically that, the above discussed influences as well as enduringly long-term impacts of the imposition of imperialism, and the forced imperial-rules onto any overpowered colonies in the alien territories, such multiple influences and impacts are naturally able to gather numerous dimensions like—socio-cultural, socio-psychological, socio-political and then in addition geostrategically, politico-military as well. Hence, we would be discussing these aspects too suitably in some more details little later, as we proceed further in our continuing discussions.

In fact, as we further go on to, critically delve even deeper into these historic aspects, in our larger geostrategic and politico-military perspectives we interestingly notice that to begin with, even those sea-faring and highly ambitious overseas European Imperial-Powers, during the commencement phases of their mutually rival voyages across the Atlantic, the Indian Ocean did have their own compulsions. In spite of the inherent aspects of their overpowering imperialism thus, the initial plans and intentions leading to these adventurous sea-voyages to begin with, were to get hold of relatively the new and open territories for making fresh settlements for their increasing populations, who were since turning out to be quite a drag in their home lands back in Europe. And simultaneously at the same time; they were in need to explore new markets for their trades, as well as if suitably any new opportunities coming in their ways then, not to be left behind in getting hold of any precious metals/gems, other valuable natural resources and riches as well. Hence in their larger perspectives, man's history also shows us that the multiple-aspects of the age-old imperialism and their numerous kinds of manifestations, during different historic-ages as well as, in different geographical locations in the world; have also critically been seen to be displaying their different kinds of situations-based complexities too.

Hence historically in this regard, we do also find that the multiply numerous aspects of the highly complex and universal manifestations of the age-old imperialism within the imperially expansive domains the ancient Greeks, Romans, Indians in South-East Asia, Mongols with their vast continental reach earlier and subsequently then, during the much later times those of the seafaring European-Powers, with their unprecedented global dimensions and also at the same time, with substantially far more multi-dimensional as well as almost permanently enduring impacts, onto their far-off located numerous colonies across the world have indeed been creating many long and interestingly captivating historic saga and also, enriching literature along with, highly informative theatrical-shows and films in differently relevant perspectives in so many different languages even now across the world. Hence, it is naturally obvious that all these by themselves are vast as well as indeed the multi-disciplinary subject-matters even now, for the extensive studies and research, for the scholars and expert academicians the world over.

We may recall that, at a number of stages we have emphasised that the causative-seeds of the ultimate demise of empires and, as well as the multiple-aspects of imperialism universally have always remained integrally hidden in their own imperial rise and consolidative manifestations since the very beginning of the man's various civilisations and along with, their vast and continuing living-history in the world. And therefore in our these precise perspectives, the same historic aspects are obviously found to be equally true, in case of the overseas colonies as well as their integral imperialism since belonging to the earlier mentioned medieval-era leading European Imperial-Powers, in the then newly discovered twin American continents across the Atlantic as well. Hence, in these specific contexts and also at the present precise point if may critically refer back to the already earlier quoted few highly informative paragraphs and also care to read in-between the lines, from **James Truslow Adams's book, *The Epic of America*,** then we may interestingly find them to be equally factual too, as in case of the various well-known empires, since the very ancient times the world over. In this precise context and also we may contextually prefer to critically review now, those multiply unprecedented aspects of the British imperial-rule, especially in precisely very large territories in the North-American continent, which now very significantly happens to be the United States of America. And here too, and also since analytically brought out in the James Adam's above referred book, **the causative seeds for the highly historic American War of Independence (1775-1783)** are clearly found to be integrally hidden, in the imposing British imperialism as well as, now the bygone aspects of the British colonial-rule over there.

Concerning historically the unique and also, multiply geostrategic and politico-military aspects of the epoch-making 18th century American War of Independence the book **Lynn Montross** has since written with great historic precision that, **"Boys as young as 14 were fighting in Washington's army during the American Revolution. The little army had poor clothes and most of the time received no pay. It was often beaten army in battles against the larger armies of the Kings redcoats. But it would not stay beaten. And in the end it gained American independence in an eight year war against the most powerful nation in the world of that day (*i.e. the then British Nation across the Atlantic in Europe.*)**.... If, said George Washington in 1783, writers of history were to tell the true story of the American Revolution, it would not be believed. No one would believe that the large armies that the Great Britain had in America for eight years could be beaten by a much smaller army. And this army was made up of men who often had not enough to eat or wear, and who experienced every kind of suffering which human nature knows.... Washington had an idea. There is nothing in war more powerful. And Washington's idea in the end led to victory and independence. Washington saw that the British must hold the cities on the seashore. They received most of the supplies in ships from England. Now that the Americans had the help of French sea power, he dreamed of a land and sea attack to close the British lines of supply. Then, without food and the materials of war, they would have to surrender. That would probably mean the end of the war. Another defeat like Saratoga would force the present government out of office. When the long overdue ships of Count d'Estaing (*i.e. the French Navy*) arrived, the American commander hoped to use his idea to beat the British in New York. The French ships could keep the British from escaping by sea. The redcoats might try to fight their way out. Monmouth had proved that trained Americans were equals of the best British regiments. **Washington had an army he could trust. The enemy could not have been beaten without the help of French money and ships. But if the cause of victory could have been any one thing, it was the character of the American commander. George Washington stood between the army and defeat during the dark years. George Washington alone held the army together** until the time came to strike at Yorktown. The last Tories and redcoats left American soil on December 4, 1783. Thousands of Tories sailed to England, Canada and the West Indies, leaving their American homes forever.... The British ships were getting ready to sail when Washington took leave of his officers. He planned to ride quickly to his home and take up life where he had left off as a Virginia planter. **He did not realise that he was world famous, highly thought of in Europe and even in England. He was the father of his country, and never again would be able to have a quiet and peace he desired more than anything else on earth. In a few more years he would be called upon to help form a plan of government to hold the new nation together. He would become its first President. Famous people from all over the world would visit his home. His mail each day would more than any man could answer.**" (Montross, 1962, pp. 1, 2, 85, 115 & 116)

So consequently and indeed with great path-breaking historicity the American-Revolution and the War of Independence (1775-1783), under the brilliantly bold and charismatic leadership of George Washington; indeed led to birth of the then territorially large and fledgling, but soon going to be subsequently; the modern-era's most powerful as well as, industrially most advanced Nation in the world, i.e. the United States of America. But immediately after the violence-ridden birth of this vast and new Nation in the North-American continent; things were not at all going to be any easy as well as, both socio-politically and politico-militarily far from being peaceful and smooth, for this newly born nation. In fact, as we proceed to critically review these historically highly checkered developments occurring in the far-off North American continent then, in our larger perspectives; we do also significantly notice that lot of water had to flow down the river of Mississippi and then subsequently, that nation perhaps inevitably in a context; had to perforce get engulfed by an intense four year civil-war, before the USA came of some age and, began to proceed with her own process of settling-down as a viably enduring and truly democratic American Nation-State.

The last few decades of the 18th century when America had momentously succeeded to gain political independence, under the brilliantly visionary leadership of George Washington, we do also notice that simultaneously many far-reaching historic developments were indeed perhaps independently, taking place in the far-off continents of both Asia and Africa. In fact these long-term, enduring and, so much epoch-making developments in these far-off continents, across both the Pacific and the Indian Oceans were indeed taking place as per the different kinds and, locally the situation-based manifestations of the multiple-aspects of imperialism, of those European overseas Imperial-Powers themselves as brought and imposed by them over there. These historic developments were principally taking place, in course of the geostrategic and politico-militarily well-crafted initiatives of the European Imperial-Powers, that were then being taken in the imperially attractive vast territories of the South-African region, the unique Indian Subcontinent, the Island territories of South-East Asia and, subsequently in the Far-East and China as well.

During these highly momentous times, a new but still fledgling and vast Nation-State of the USA was already born, after the eight years' long American War of Independence against the die-hard imperial British but irrespectively at the same time, those leading overseas European Imperial-Powers had already begun making imperially deep inroads in the vast landmass of Africa, both

South and South-East Asia and also along with subsequently at a later stage into the Far-East as well. In addition, in fact of much more historic significance the unprecedented and also perhaps, the most far-reaching and subsequently the unparalleled history-making birth of **'Industrial-Revolution'**, had already taken place in Europe. Hence naturally, its multi-dimensional progress too in Europe, consequently as a chain-reaction had already revolutionarily commenced. Many kinds of previously unthinkable and indeed totally irreversible changes in the entire life-patterns as well as the thought-processes of the mankind, were soon going to get manifested altogether in a different mould and thereby the whole mosaic of the age-old systems of the continuing advancements of various human-civilisations in the world too stood entirely altered forever.

We may recall that the pioneers among the leading, seafaring and bitterly rival European Imperial-Powers in the first historic instance, in undertaking the unprecedented and dangerously adventurous exploratory sea-voyages towards the unknown twin-continents of the Americas across the Atlantic, were indeed the Portuguese and the Spaniards. And again they were also first, in discovering the sea-routes through the Indian Ocean to the coasts of South Indian Peninsula and subsequently along with; charting of the sea-routes all across the Pacific, first to Philippines' Islands in the South-East Asia and then to some other Islands in the Pacific, on the other side of Globe. But we are also aware that, the man's living-history since continuously in the making, always has in the process its own kinds of strange and also perhaps, the historically inevitable ironies as well. **Strange are the ways of the perpetually on-going making of the man's history, and also further thereby stranger are its ironies too in this regard**! Hence precisely at this present point of our continuing discussions, concerning the proactive geostrategic and politico-military thoughts-processes as well as, the adventurous overseas initiatives, on the part of those leading European Imperial-Powers, in the other far-off continents, though Portugal and Spain took their own significant kinds of imperialism-based geopolitical leads to begin with, they were soon not at all in a position to maintain the same as well as any other kind of monopoly in this regard.

As quite sensationally the news of their historic breakthrough spread in the whole of Europe substantially the British and, soon followed by France and Holland too, joined with great vigour in these ruthlessly competitive games of their different kinds of agendas and imperialism-based sea-voyages and also along with their different kinds of rivalry-oriented geopolitical, politico-economic and strategic-cum-military agendas. So historically in those far-off continents the great imperialism-driven games of those rival European-Powers, had already commenced with maximum ruthlessness and vigour. Hence quite ironically in these highly historic contexts history also shows that, irrespective of who initially took the lead to begin with, in the earlier discussed dangerously adventurous and explorative sea-voyages across the various oceans it was Great Britain who, with her perhaps unrivaled geostrategic and politico-militarily visionary shrewdness, could largely corner the substantial sizes of the huge cake of their imperialism-driven multiple fruition of ultimately the vast overseas Empires in those continents, since being so far-off from the Europe. In these historically very significant perspectives, we find that the imperial British was able to finally colonise, with their historically unrivaled and perhaps smarter, geostrategic and politico-military shrewdness almost the whole of North-American continent and thereby, significantly making it the first flag-bearer of the then still expanding British-Empire into the zenith of its glory, it generally became proverbial in a context, that the 'Sun was then never setting over the British-Empire'. In the larger contexts of these historic aspects there is perhaps no point for anyone to get emotional in any manner in this regard, since now in retrospect these always need to be dispassionately taken, with the desired historicity as well as the essential academic objectivity too.

Hence during those historic times in almost southern half of the North-American continent, which now constitutes the present United States of America also then being under the British imperial-rule and where in fact, the first consistent struggle to secede from that unprecedentedly global Empire had begun with a geopolitical aim to get transformed into a sovereign political-entity and thus, to be the first independent Nation-State of the new world, i.e. in the twin American continents. This relentless struggle for the political-freedom ultimately resulted in the historic American-Revolution and which in turn, finally culminated in the eight-year American War for the political independence (1775-1783), from the British imperial-rule. In our larger perspectives, we have since also briefly discussed the various significant aspects of this historic war little earlier, in this chapter itself. But in these precise contexts, as we continue to delve even deeper into these unique and truly epoch-making aspects of the man's living-history, we do also quite significantly notice that historically the successful conclusion of this American War of Independence, during the latter half of the 18th century and then subsequently, during the beginning of the 19th century itself resulted in many kinds of far-reaching repercussions, as well as numerous developments in the South American continent too. In fact, as we begin to critically review the historic processes of the colonisation of South-American continent, in our larger perspectives we do find them to be quite different to those of the North-American continent, which now quite interestingly constitute the twin large Nation-States of the USA and Canada.

We may also be able to observe that, naturally in a larger historic perspective, the distinctly different and numerous imperialism-based aspects of the geopolitical culture and traditions; on the part of different conquering Imperial-Powers, anywhere and in any given situation during any particular historic-age might perhaps have had **some decisively significant inputs in the making of socio-cultural, socio-psychological, socio-political, politico-military and many other aspects of their colonial-empires and the imposition of their different kinds of imperial-rules over them; with respect to the making of subsequent colonial as well as, even the post-colonial-era living-history of those erstwhile colonies in this regard**. We do find that the historic manifestations of the above mentioned numerous dimensions of the many important aspects of the making and consolidation of the various colonial-empires and then, imposition of the imperial-rules over them, since specifically by those different European overseas Imperial-Powers; in the continents of North and South Americas too, are naturally found to be fairly different in many significant history-making ways. In these specific contexts, we do find that the North America was principally colonised by the imperial British finally and then; the South America mainly by the Spaniards and, also then by the Portuguese as well and hence, in these specific contexts; we do find the historic manifestations of the above mentioned precisely relevant and indeed very important aspects, taking altogether their different paths in this regard.

Historically the minds of Imperial-British have both traditionally as well as functionally, been accustomed to their unique kind of the triple combination of King, Lords and the Commons. But at the same time, it is also true that historically, as this unique kind of triple combination have been continuing in UK even now and by the time, the North American continent was colonised by the British, also a pioneering system of the Parliamentary democracy had already dawned in their Islands-Nation in Europe and, also along with its own kind of enduringly vibrant consolidation of democracy onto both the socio-cultural and socio-political canvas of the typically traditional British societies as well. In fact during those bygone years; these aspects concerning the concepts and practices of the ideas of democracies were yet to take any significant roots, in Spain and Portugal as then and also in many other nations in the continental Europe.

In these historically interesting contexts and though these may perhaps sound as matters of conjectures but as befitting examples in this regard, we may even mention that the gradual evolution of the principal cultural aspects as well as, the courses of subsequent multiple-processes of the ongoing making of the histories, of the two distinct groups of Nation-States, since as belonging to the two separate North and South American continents, have also been clearly found to be totally different. And in these continuing contexts and, at this precise point, the earlier discussed and even highlighted **historic aspects of certain significantly decisive-inputs in this regard since remaining interestingly hidden but if delved deeper then, may even be seen as proactively participating in the continuing phenomena of the making of multi-dimensional living-histories of different Nation-States, since belonging to the two distinct American Continents**.

In this precise context, we would later be able to perceive through the above mentioned historically momentous aspects, concerning any particular Imperial-Power vis-à-vis its colonised territory/nation, its societies and the peoples as well; when we begin critically delving deeper and discussing the **multiple historic-aspects of the recently bygone few centuries of the British colonial-rule, specifically over the vast Indian subcontinent wherein, since the very ancient of times numerous 'nations' and societies in fact, have been commendably existing together, with their traditionally perceived and highly vibrant civilisational-assets and also along with, their continuing social-culture of the peaceful coexistence**. And subsequently then, we would also be delving upon the very recent historic-aspects of the making of specifically the two independent modern Nation-States, within the Indian subcontinent itself and at the same time in addition even preserving the essentials of the highly rich heritage of their gloriously unique ancient civilisational-aspects, even now in the present modern world. Hence broadly in these precise perspectives, as we may wish to further critically review the **similarly concerned aspects of the histories of colonial as well as the post-colonial times of both the USA and Canada, we may also notice that in the multiple processes of indeed the magnificent flowering of the truly strong and vibrant democracies in these two big Nation-States in the North-American continent historically, as per the different kinds of emerging situation-based and with both direct and indirect consequentially decisive inputs rendered by the imperial British then have truly been phenomenal**, irrespective of the other kinds of differently perceived dimensions in this regard. Hence in these historic contexts the prudent objectivity also demands that, indeed the unassailable as well as timeless truths of the man's vast and multi-dimensional history needs to be accepted as they in any case would always continue to remain unaltered, and the same irrespective of the any kinds of factually perceived or otherwise motivated and nurtured opinions, anywhere and any in school of thoughts.

In precise contexts of the above discussed **aspects of the imposition of highly greed-based imperialism and simultaneously as well the making, ruthless expansion and, the ambitious consolidation of numerous colonial-empires in the far-off continents of the Americas, Asia and Africa, during those momentous periods and at the same time, we do also find that finally it was the imperial British, who decisively excelled over the other no less ambitious and, geopolitically as well as militarily capable, sea-faring European Imperial-Powers since then relentlessly engaged in these historic great-games of their multi-dimensional imperialism**. In addition and also all the more, in our larger geostrategic and politico-military perspectives the man's living history also shows that behind any historically momentous phenomena, there are always very decisive but, generally hidden specific causative-factors, and in turn they continue to proactively trigger those phenomena as well. And hence, in these historic aspects of the great games of multi-dimensional imperialism too, since then as being played by those rival and ruthlessly competing European Imperial-Powers, in the far-off overseas American, Asian and African continents, there were powerfully decisive causative-factors, which were indeed triggering the multiple British successes, finally in this regard. Hence, it may here appear appropriate now to focus on those causative-factors since decisively influencing and shaping then a series of historic successes on the side of Imperial-British, in the then great-games of the ruthlessly competitive-imperialism; in the above already identified overseas continents, during the medieval-era in the world.

We may recall that in our larger perspectives in this regard we have since already identified and discussed the important aspects of certain specific prerequisites and, indeed the needs for their proper and proactive placement, for precisely the decisive successes in the great-games of imperialism in any geographical territories, to be both geostrategically and politico-militarily subjugated and colonised, as different segments of the expanding empires of the concerned Imperial-Power. As since already spelled-out earlier in the broader contexts, principally these prerequisites are four in numbers and hence at this precise point of our continuing discussions as we begin to again recollect them, we would interestingly be finding them to be rather self-elaborating and explanatory too, in this regard. In fact, precisely they happen to be: (i) Proactive presence of the universally timeless aspects of the **'essential strategic-culture and thoughts'**, (ii) Strategically the imaginative development, well-rounded and dedicated nursing of the all aspects the matters-military, (iii) Possession of the proactively geopolitical and strategic skills for the statecraft as well as the essential aspects of governance at home and finally, (iv) The essential strategic acumen for the multi-dimensional conduct of the art of highly complex international games of rivalry-infested geopolitics and even at times, resorting to the need-based coercive diplomacy as well.

Well, in broader contexts of the above elaborated four prerequisites in this regard, we do specifically find that especially the first one, i.e. the aspects of 'essential strategic-culture and thoughts' by themselves, are so comprehensively all-encompassing in these precise perspectives that indeed in the womb of particularly this first prerequisite as per the above in fact, do contain all the essential aspects of the remaining three prerequisites which in turn, have to understandably form the inseparably integrative segments of essentially the first one since for meaningfully as well as proactively, transforming all its comprehendible aspects; i.e. obviously, total aspects of the 'essential strategic-culture and thoughts'. **In fact, all the analytically conceivable aspects of the 'essential strategic-culture and thoughts' by themselves; are so multiply-vast and complex that they are befittingly multi-dimensional as well as indeed capable of including almost all kinds of perceivable dimensions and aspects (*of almost anything and everything concerning any Nation/ State*)** of the remaining other three prerequisites in this regard. But, in order to be more elaborative and also for defining them perhaps more in contextual details we have since tried to compartmentalize and list them in the above mentioned specific contexts here, into relevantly four essential prerequisites.

Now coming back to historically the momentous aspects of colonisation of the twin-American continents and consequently at much later times, the births of a number of modern-era Republics over there we again do interestingly find that, naturally as per its indeed inherent ways and also in these specific contexts as well the continuing phenomena of since the making of man's living-history in the twin-American continents, during the now bygone colonial-era too; remained multi-dimensional in numerous ways. In fact interestingly, be it North or South America, Africa or even Asia simultaneously in multiple contexts of the imperialism and also building of the colonial-empires, by those overseas European Imperial-Powers in these far-off continents, many different kinds of historically momentous developments were fast taking place, indeed in far-reaching and epoch-making manners.

Hence, in larger contexts of the briefly above discussed aspects of the imperialism and building of the colonial-empires, we do also historically notice that during those momentous times then in South America, since as the Portuguese were able to occupy Brazil and almost simultaneously then, the Spaniards were able to conquer and colonise Mexico and Peru and hence with the continuing passages of few centuries, subsequently these historic colonies were indeed able to develop into such history-making pivotal-centres, which in turn were also able to trigger the gradual phenomena of the historic fructification of a number of modern-era Republics, in the vast South American continent. During those momentously historic times interestingly, we do also notice that in fact it were the Dutch who could first put their feet on the coasts of North America but at the same time, they were principally interested in the East, and moreover in this imperialism-driven ruthless race they couldn't continue since they were decisively overwhelmed by the more overpowering British and the French. But, before leaving the Dutch in their own specifically distinct imperialistic manner too, had also established their new American territorial settlements; such as New Amsterdam and further in fact **a British named Hudson being then in service as an employee of the Dutch, was first able to explore and discover an important river way back in 1609; since now famously known as the Hudson-River in the modern USA**. But soon the vast and relatively more virgin continent of North America was in fact wide-open for the same highly greed-based and ruthlessly competitive imperialism; among the historically well-known old European rivals, i.e. obviously, the British and the French seafaring Imperial-Powers. Historically, they were irreconcilably old rivals in Europe too since centuries and hence naturally they were just not expected genuinely to be even least cooperative in, as well as across the Atlantic then also, in the vast continent of North America as well.

In the multiply imperialism-driven competitive race then on, in the North American continent the quicker, geopolitically more strategic and shrewd English, were advantageously able to occupy the already settled territories since perforce vacated by the Dutch, and hence the settlements of **New Amsterdam** was occupied by the **Duke of York**, the younger brother of the **British Monarch Charles II** and named them interestingly as, the **New York** and **New Jersey**. **These favourable developments for the British, further facilitated establishment of the numerous English colonies in many places in the relatively 'virgin' North American continent historically founded by the earliest British peoples reaching there and since interestingly now, known as the 'Pilgrim Fathers' who were indeed pioneering Englishmen in this regard and who sailed in the erstwhile famous ship named 'Mayflower' in the year 1620 itself, i.e. almost six years even before the founder of the historic Mughal-Empire named Babar, could reach the frontiers of the coveted Indian subcontinent.**

These English people then arriving in North America, could occupy substantial stretches of territories for their expansive settlements which in turn grew into a powerful group subsequently, since now known as the New England in today's USA. But we may also recall and as discussed earlier that such occupation of increasing territories were seldom easy and peaceful over there, since the British contingents then arriving there; had to fight hard against the native 'Red-Indians', before achieving decisive supremacy over them for their settlements. During those times in Europe; both British and the French were pitted against each other, in their almost all out decisive seven-year War for supremacy and finally, the British came out victorious and thus ending this war in a favourable 'Treaty of Paris', signed by them in the year 1760. This historic Treaty could also seal the fate of French in many ways from then on; since competitively aspiring for their imperial-supremacy against the British, in the already above named various overseas and far-off continents. In fact interestingly in retrospect, we may also contextually say that whosoever had finally come out victorious and, so during those bygone historic centuries of the zenith of imperialism and building of the numerous overseas empires, would have been decisively calling the shots, in the historic global phenomena concerning the multiple processes of subsequently, indeed the path-breaking making of the then man's largely colonial-era's history in the world.

But the man's vast, highly complex and the multi-dimensional living-history is also able to always act in many ways interestingly, as a vast and lively mirror, for the man and his every kind of deed concerning any conceivable area of human-endeavour as well. In fact since as discussed in the different contexts earlier and hence, as the time keeps-on rolling-by, and in the different far-off continents; as also the various human-civilisations keep-on advancing in their multi-dimensional ways, and also simultaneously at the same time along with them; in the processes of since always the living-history's on-going making too, the man's living-history spares no one at any given moment of time and also in addition, irrespective of the consequences anywhere and in any manner in this regard. Hence in the larger perspectives, during any on-going exploratory discussions concerning the man, his continuously advancing various human-civilisations since the beginning of his civilisational-history itself and along with that, covering any specific areas of the man's relentless endeavour again, the continuously emerging objective records of his living-history, while acting as an all-inclusive vast mirror since simultaneously also always having been able to perform as a valuable medium, for bringing-out the creatively analytical display of the apparently hidden—the various causative-factors, concerning subsequently the consequential occurrences of the various kinds of thereby, the inevitable historic events. These also very valuably happen to be one of various principal roles of the man's living-history as well.

As since also discussed in various contexts earlier we have always seen that, the continuous multi-dimensional making of the man's living-history do always simultaneously keep taking-place in fact in numerous far-off areas in the different continents of the world and interestingly, at the same time; they mostly do remain inter-related as well as, mutually influencing too. Hence, during those momentously historic times, belonging to the 16th, 17th, 18th and 19th centuries since witnessing the multiple-advances of clearly the unprecedentedly new kinds of imperialism and simultaneously along with, founding of the many overseas and expanding colonial-empires by the leading and ruthlessly rival and imperially obsessed European seafaring Imperial-Powers, we do also find interestingly as per the above perceptions in this regard, that quite the captivating phenomena of the highly eventful as well as, both mutually influencing and interrelated as well as simultaneous and almost global, the imperial-era making of the man's history, in the different far-off continents of the world. So the highly consequential, historic as well as substantially, the epoch-making fallouts of the above mentioned 'Treaty of Paris' in the North American continent, besides the immediate manifestation of the British imperial supremacy were that territories of both the present USA and Canada, became part of the Britain's fast expanding and almost the global colonial-empire.

But, **all these historically didn't mean that the French were completely wiped out from the North America; since they had entrenched themselves in and around Quebec in Canada and, continued to flourish there as well. Hence no wonder that even today interestingly, a substantial**

part of the modern Canada around Quebec; happens to be inhabited by the French-speaking Canadians. And also further, since as we contextually keep critically delving even deeper in this regard we notice that during later part of the 18th century then, the highly historic American War of Independence (1775-1783) too, took place in the vast British imperial-colony in North America. Though in this historically far-reaching War the principal geostrategic and politico-military aim of General George Washington, since leading the valiant American freedom-fighters against the British was to achieve unconditional and total political independence for his country, which was interestingly then at the stage of being born, but historically this war's path-breaking geopolitical tremor was indeed soon going to be felt far and wide, including even in the European continent itself. In fact, man's living-history since covering these highly consequential historic-era; further shows that this 18th century American War of Independence ended with the highly eventful Treaties of Paris and Versailles, signed in Europe by the British and French but at the same time its successful conclusion historically also ensured the total political independence of the settlers and, birth of the USA as a sovereign nation as well.

In these highly historic contexts interestingly further we may also need to bear in mind that **these 18th century Americans, then since fighting for their independence from the British imperial-rule in fact, were no original inhabitants of the North American continent, but they were settlers having later migrated from Europe** only when the rival European overseas Imperial-Powers had already launched their discovery voyages across the Atlantic and other oceans/seas, for multi-dimensional furtherance of their own different kinds of overwhelming imperialism. At this precise point of our continuing discussions and also, within our larger perspectives we may say that, if the perpetually ongoing making of the highly complex as well as multiply vast man's history, have its own strange ways of manifestations then at the same time man's typical and inherently and also, amazingly complex psychological make-up and also then along with, his emerging situation-cum-circumstances-based responses too; are by themselves no less complex and unpredictable as well. Hence in these interesting historic contexts even if those migrant European settlers in the North American continent, were initially of the European origin and though, majority of them were originally from the Great Britain as well therefore consequently, **they gradually got both historically and socio-psychologically disconnected then, from their British-cum-European origin and also at the same time; got both geopolitically as well as politico-militarily, almost totally converted into the 'die-hard Americans'**. That's the way the USA got first manifested. And so as per the above perceptions, these are the ways for the above phenomena to manifest and hence the man's living-history does consequently act as a mirror too, precisely reflecting the results of his those multiple responses as well.

Now, with respect to the monumentally global consequences of the historic eight-years American War of Independence (1775-1783), **Professor S.R. Sharma** of India has indeed given a brilliantly graphical account of these historically path-breaking events for easily perceptive understanding, "This eventual victory of the settlers had important and varied consequences: (1) it created the independent U.S.A. (2) it precipitated the Revolution in France; (3) it brought to an end the 'old colonial policy' in England no less than the last bid for personal rule made by the English Monarchs. **Turgot's dictum that 'colonies are like fruit which drop off from the stem when they ripen'** was proved true at least in this important case. More than anything else, the **American Revolution convinced England of what Chatham had meant when he warned his countrymen saying: 'We may bind their trade, confine their manufactures, and exercise every power whatsoever; except that of taking their money out of their pockets without their consent.'**

The great hero of the American triumph was **George Washington about whom the English historian John Richard Green has written: 'No nobler figure ever stood in the forefront of a nation's life...first in war, in peace, and first in the hearts of his fellow countrymen.**" (Sharma, 1948, p. 383)

In addition to the above and has since also brought out earlier too, that the historic American Revolution had many kinds of fallouts in the South American Continent as well as in other far-off places as well, like in some ways in the substantial parts of both Asia and Africa. That's also the way too that man's multi-dimensional history is made simultaneously at many places, as a highly complex and also continuing phenomena. In fact quite appropriately in many ways, it is generally said that America is a new world and initially due to certain kinds of typical circumstances beyond anybody's control, different levels of violence then were only incidental. Well, whether it was just incidental or not or perhaps, perforce turned out to be regressively lingering and socio-psychologically, a rather ironical segment of inevitably the negative sides of the multiply fast-evolving American societies may also justifiably be taken in the larger perspectives, but remains debatable too. In any case, it is also very healthily true that the multiply powerful as well as, indeed the far-reaching cultural aspects of America particularly those of the modern USA and Canada, though largely on the British-lines but mostly happens to be indigenous to America as well, Hence, if we may care to delve deeper into these highly complex but multiply unique socio-political and cultural aspects of particularly the today's USA, we may perhaps be able to have indeed some enchanting glimpses of the apparently somewhat hidden as well as inherent causative-factors, triggering the multiple manifestations of their strengths and phenomenal progresses in almost all areas of the present-era's human endeavour. We would appropriately be discussing these indeed quite engaging modern American aspects, in the later chapters.

Further in the larger perspectives, the historic points since briefly delved upon in the above few paragraphs concerning the **historically famous and unique 'Monroe Doctrine'**, again Professor S.R. Sharma has very informatively written that, "The victory of the U.S.A. had also important repercussions in South America. Napoleon Bonaparte had overthrown the Bourbon ruler of Spain and seated his own brother Joseph on that throne, but the Spanish colonists in South America refused acknowledge the usurper. Under the leadership of Simon Bolivar, Venezuela, Columbia, Ecuador, Bolivia, Peru etc asserted their independence even when the old dynasty was restored in the mother country (under Ferdinand VII). Mexico became independent in 1821, but fell a prey to continued disorder, The Portuguese colonists of Brazil likewise set up an independent state in 1822 with Don Pedro as their King.... But two fact may be noted: (1) all the colonies set up republican governments before the close of the century; (2) **the U.S.A. proclaimed the famous 'Monroe doctrine' when the European Powers tried to meddle in their affairs**. It lay down:

> 'In the wars of the European Powers, in matters relating to themselves, we have never taken any part, nor does it comport with our policy so to do. It is only when our rights are invaded or seriously menaced that we resent injuries or make preparations for our defence. With the existing colonies or dependencies of any European Power we have not interfered and shall not interfere. But with the Governments who have declared their independence and maintained it, and whose independence we have on great consideration and on just principles acknowledged, we could not view any interposition for the purpose of oppressing them or controlling in any other manner their destiny by any European Power in any other light than as the manifestation of an unfriendly disposition towards

> the United States. The occasion has been judged as proper for asserting as a principle in which the rights and interests of the United States are involved, that the American continents, by the free and independent condition which they have assumed and maintain, are henceforth not to be taken considered as subjects for future colonization by any European Powers." (Sharma, 1948, p. 383 & 384)

The above reflected historically path-breaking and universally famous doctrine was indeed able to very effectively as well as permanently, put an end to the any possible manifestation of the any kind of possibly overwhelming and ambition-driven desire to launch henceforth, any generally quite menacing imperialistic misadventure for fulfilling any hidden or apparently very greedy geopolitical aim towards colonisation by any overseas European Imperial-Power, in any part of any of the twin American continents. This was indeed a great turning-point from then on, in the multiple-phenomena of the history of America. But in addition to this, the **'Monroe-doctrine'** also then perhaps, might have had even if indirectly; the substantial impacts in numerous ways on the subsequent thoughts-processes in this regard, also in the other continents. **James Monroe was the fifth President from 1817 to 1825 of the United States of America; as then this newly born nation called USA was still evolving, both geostrategically and politico-militarily consolidating and at the same time, socio-culturally firming-up, as well as had begun to settle-down as an independent sovereign democratic Nation-State in the North American continent**. So far as geopolitically the multiple international aspects concerning the multiply complex subject-matters pertaining to indeed the multi-dimensional domain of the games of international power-politics, matters-military and also, the aspects of multi-faceted threat-perceptions were then concerned and also in addition then, since all these as a combination, rather menacingly confronting the President of the USA as nation's Chief Executive both the historically extraordinary strategic vision and brilliant statesmanship on the part of Monroe as the US President then, lies in his great sense of history and statecraft in penetratingly seeing through the thick fogs of the various kinds of imponderables and unforeseen possibilities concerning the emerging threats, as per the above mentioned paramount factors in this regard.

At this precise point of ours, since the multiply continuing discussions we may perhaps appropriately maintain that at times, these naturally otherwise rather hidden, but also of great importance history-making facts pertaining to the recently bygone era, can be very revealingly visible contextually in our larger perspectives, if we may care to multiply dig and delve deeper and deeper into the specifically concerned aspects, mostly remaining apparently buried into the vast and highly complex wombs of again, the man's living-history itself. While in continuity of our ongoing discussions at present, as we may visualise the then prevailing overall geopolitical scenario, as confronting the US President and his core-level team of advisers and policy-makers, we find that their brilliant strategic-acumen lies in formulating the historic 'Monroe-Doctrine', with such long-term geostrategic and politico-military vision and precision. Thus this historic document, during those momentous times then indeed performed the path-finder's role as a very valuable beacon-light, in the highly complex and rather a kind of very 'rough-seas', concerning the formulation as well as conduct of the then American foreign-policies. Why we may historically maintain that, if the epoch-making and historic 'Monroe-Doctrine' had not so precisely been formulated and proclaimed in time the way it was indeed done in the Americas, under the leadership of the fifth US President James Monroe; because then with a reasonably fair amount of historic certainty, we may also visualise that consequently, the subsequent paths of the making of the histories of both the American continents perhaps, would have been totally different.

Hence, very appropriately in the larger perspectives concerning the great historic importance and the far-reaching contributions of the 'Monroe-Doctrine, during those turbulent times especially in the twin American continents then; Late Jawaharlal Nehru (page 660) has written that, "The effect of the widespread net of railways and telegraph was to knit together this huge country (*i.e. The United States of America*). This would have been impossible in the old days, when it took weeks and months to travel from one coast to another. In the past we have seen that there were often great empires in Asia and Europe. But these could not be closely knit together because of the difficulties of communications and transport. Different parts of the empire would be practically independent, leading their own separate lives, except that they acknowledged the supremacy of the emperor and paid tributes to him. They were loose associations of different countries under one head. There was no common outlook about them. The United States, however, because of railways and other methods of communication, as well as uniform education, developed this common outlook amongst its different races. The races were gradually assimilated into a common stock. The process is by no means complete; it is still going on. There is no other instance in history of assimilation on such a large scale. The United States tried to keep away from European entanglements and from the intrigues of European Powers, and they wanted to keep away from America, both North and South....the 'Monroe Doctrine', the rule which President Monroe of the United States laid down when some European Powers—the 'Holy Alliance' wanted to interfere in South America to preserve Spain's empire. **Monroe declared that the United States could not tolerate armed intervention in the whole of America by any European Power. This declaration saved the young South American Republics from Europe.** It almost led to war with England once, but America has successfully stuck to this policy for more than 100 years now. South America was very different from the north, and hundred years have not lessoned the differences. Canada in the north is becoming more and more like the United States, but not so the southern republics."

It is true that in course of our continuing discussions and, in different perspectives at a number of times earlier, we have since been emphasising that the man's history is always made continuously in the different areas differently, and in different contexts simultaneously and at the same time these processes do also remain mutually influencing, as well as interrelated too. But in addition to these, the highly complex and vast phenomena of the ongoing making of history in certain cases indeed, especially concerning the very far-reaching, as well as historically epoch-making emergence of the certain creatively visionary ideas/doctrines during the different historic periods do phenomenally remain enlightening and connected, for receiving very valuable as well as enriching inputs, in the areas concerning the subsequent thought-processes in this regard. In these wider historic contexts, now precisely coming to the conceptual-framework concerning the multiple making of the above discussed **aspects of the historic 'Monroe-Doctrine'**, we may now recall that we have also discussed, the path-breaking and historically very far-reaching **important multiple-aspects of the 'Magna-Carta (1215 AD)' as well**.

Contextually at this precise point very interestingly in this regard what we are aiming at is to delve deeper and then, be able to visualise historically the multiple interconnectivity between these two very far-reaching and indeed revolutionarily and conceptually great documents. **Moreover, the further outcome of the greatly creative and age-old and also since the very beginning, the unique genius of the mankind does remain functioning as always within naturally, the overall highly complex and multi-dimensional evolutionary phenomena of the all-encompassing, timeless and universally infinite mother-Nature.** We may recall that in these perspectives, the man's living-history shows us that during the extremely tyrannical reign of the

British Monarch John (1199–1216) indeed consequently did trigger a historically path-breaking and both socio-culturally as well as socio-politically unthinkable new trend in the entire thought-processes of both the then British feudal-elites as well as the larger English intelligentsia during the medieval age.

Though historically to begin with the powerfully entrenched British feudal-class, otherwise ironically themselves multiply functioning quite exploitatively under the royal supremacy of the British Monarch, in fact had become so fed-up with the all sorts of mindless extreme tyranny of King John that, they quite unconsciously initiated many kinds of reforms in the then prevailing, rather unquestionable autocratic systems of the royal governance of medieval-era British Kingdom. In addition to these, they almost quite effectively forced the King to accept certain unprecedented reforms and in the process the powers of the British Monarch in fact, were substantially curtailed. **Historically, the Magna-Carta did contain the first ever initial seeds regarding the concepts of indeed the revolutionary aspects of democratic polity and, the peoples-oriented participative systems of governance, and hence the original ideas concerning the laying of foundations for the British Parliament then quite unexpectedly fructified, which subsequently and indeed so fascinatingly soon, was going to be known as the mother of all Parliaments in the world.** In our larger perspectives therefore, as we may begin to perceptively reflect deeper into these, momentously far-reaching as well as path-breaking historic developments, we may also even very fascinatingly visualise that from then-on historically, **the seeds for the very soon forthcoming dawn of democracies in the West**, had indeed got sown at this precise point of the man's living-history in Europe. But at the same time, we may also need to always bear in our minds that historically soon in any case; the subsequent dawn of democracies in the West, was hardly going to be any smooth development.

Briefly the above discussed aspects of the origins of concepts of indeed then, initially the first ever conceived ideas concerning some sort of democratic polity and, the peoples' participative systems of governance as well as also, concerning the beginning of the conceptually fledgling ideas of the Parliament were then totally unknown to the whole world, and therefore, these also happened to be unprecedentedly revolutionary as well. Though historically at the same time it is also equally true that the then vast majority of the very powerful British feudal-elites in fact, were in no position to ever be capable of being visionary at all, and hence obviously in the larger perspectives they could not ever be aware of the much bigger and subsequently, the quite inevitable multiple-fallouts of these indeed the totally path-breaking reforms. And therefore very interestingly, they were also soon going to be decisively instrumental in revolutionarily getting trimmed, their own age-old and powerfully entrenched socio-economic and political power-base in the British social hierarchies, and along with the gradual but also fast-emerging British democratic polity began to evolve, for the first time in Europe and in the whole world. But as at many a times since discussed earlier in the different contexts, the man's living-history at times has very fascinatingly and quite suddenly, and indeed also strange as well as circumstantially guided and sometimes even perhaps quite unknowable in advance, the ongoing and highly complex systems of its own making as well.

Hence, quite fascinatingly in precisely these larger contexts both socio-culturally as well as socio-politically, the greatly thoughtful vision contained in the historically path-breaking ideas of the Magna-Carta, though originally materialised during the early medieval-age in the England only, but indeed remains equally relevant and conceptually applicable to the entire mankind, even in the present modern-era too. If we may care to delve deeper into the above discussed two conceptual ideas, concerning first historically the earlier **'Magna-Carta'** and then, that of the historic **'Monroe-Doctrine'**, objectively together and precisely in our contexts, at the present point of our continuing discussions, we may clearly see indeed the greatly valuable and enriching inputs from the multiple aspects of the preceding 'Magna-Carta' in a very influencing manner; into again the similarly conceptualising and path-breaking framing of the 'Monroe-Doctrine' as well. And therefore, in our perspectives it is precisely here that, the multi-dimensional history-making connectivity between them does clearly emerge in a very fascinating manner.

Well, at the same time in our larger perspectives we may also need to always bear in mind that in fact, this is not the only example in this regard or in any other numerously allied contexts there are so many such mutually enriching examples in the fields of sciences and technologies as well as in the vast areas of matters-military, the social-sciences and aspects of the spiritualism as well, since ancient of times. In fact historically, we may also befittingly say that these aspects since the very advent of various human-civilisations of the unaccountable, ongoing and, multiply enriching as well as influencing phenomena have also been some of the many causative living-factors, multiply involved in triggering the continuing multi-dimensional advancements of those human-civilisations around the world. Hence, while always maintaining the desired objectivity on the multiple-paths of our continuing discussions, we may also say that the man's very vast and complex living-history since the beginning, also does contain the interesting as well as perceptively reflecting saga of the above mentioned phenomena too in this regard, in fact as important segments of also again the much larger phenomena of the continuously enriching of indeed the enchanting fragrance of the perpetually advancing various human-civilisations as well.

Well in addition and in the much broader contexts, we may also say that the multiple-aspects of the imperialism since the very ancient ages too have been in fact, inseparable segments of the same perpetually advancing human-civilisations in the world with the rather mixed legacies, of historically positive as well as negative baggages. It is a different matter that many may or may not perhaps agree specifically in the above contexts, with any kind of such perceptive opinions. Precisely in these larger perspectives while delving more, into the very vast and expanding canvas of the man's living-history since its beginning itself and hence in other words, we may also say that **the numerous kinds of the multiple-aspects of indeed the age-old imperialism have always been manifesting themselves; simultaneously along with the continuing multi-dimensional advances of the various human-civilisations, during different historic-periods and thereby in different geographical territories of the world. In fact, one historic reality is indeed clear that contextually the multiple manifestations of the very much lively as well as lingering aspects of the imperialism, since ages have by themselves also been highly complex too.**

Also, we do also historically notice that imperialism by itself has numerous kinds/varieties as well. And therefore, in the broader geopolitical contexts and during different historic periods, we do find that any particular Nations-State having developed and acquired the requisite geostrategic and politico-military wherewithal, for very ambitiously getting transformed into a credible Imperial-Power, subsequently do launch some sort of the imperialistic venture towards her targeted alien territories or, onto the any other geopolitically as well as perhaps imperialistically vulnerable Nation-State for subjugating as her colonies and thereby, imposing her imperial-rule over it according to her own pre-planned but also having imperialistically, shrewd and flexible long-term agendas. Hence historically, the multiple manifestations of precisely these specific kinds of the imperialism are prominently driven towards the intended targets, by the dominating wherewithal regarding the geostrategic and politico-military capabilities on the part of concerned

Imperial-Power. But interestingly, in these precise contexts as we proceed further we do also find that within concerning both the Nation-States, i.e. the Imperial-Power as well as the conquered and imperially subjugated Nation or even some kind of territorial entity too, there are different kinds of indeed the internal and at the same time in addition, also mutually-influencing manifestations of the **imperialism. In this precise context, rather one most important point to be kept in mind is that any such internally manifested aspects of the imperialism, are principally triggered by the internally entrenched and lingering aspects of the different kinds of feudalism; since always socio-culturally dominating the peoples' lives in any concerned Nation-States**. And interestingly further, while discerningly delving deeper into these precise phenomena we do also observe that such specific kinds of the imperialistic manifestations as since emphasised here, do contain their own kinds of the history-making capabilities too, as indeed the additional contributing segments towards much larger aspects of continuous making of the man's vast and multi-dimensionally complex living-history.

Now coming back to the twin-continents of the Americas we may also recall that we have since discussed and noticed that, with the successful conclusion of the historic American War of Independence, era of the imperial colonialism in both the American continents was historically made things of indeed the bygone past. But the above identified internal and feudalism-driven imperialism was also then getting manifested over there, and all the more specifically in the newly independent sovereign Nation-States of the South American continent. Moreover, since these specifically different kinds of fast-sprouting and forcefully driven by the lingering aspects of feudalism and thereby the different kinds of the internal aspects of imperialism; were naturally then getting manifested, onto the socio-cultural as well as socio-political canvas of those South American societies. Well contextually, that is a different matter altogether, but relevantly the more important point here is that during the 18th century, eclipse of certain kinds of the geostrategic and politico-militarily triggered imperialism since historically imposed by the European overseas, powerful and rival seafaring Imperial-Powers onto both the North and South American continents from as early as the 14th century onward by itself just did not mean that these kinds of in fact, continuing phenomena of the multiple manifestations of imperialism, had ended in the world. Historically it is a fact that those rival European Imperial-Powers, after their largely forced withdrawal from the Americas, interestingly got all the more imperialistically focused, onto the other already actively targeted continents of the world. Hence, if we may pause a little at this precise point of our continuing discussions, to rather discerningly reflect on these historic issues and then the man's vast and multi-dimensional living-history, may perhaps indirectly but apparently quite convincingly too, show us that it is never all that easy ever to wish away the overwhelmingly powerful and historically lingering existence of various kinds of the multiple-aspects of imperialism.

Further, delving into the concerned historic aspects of specifically, the geostrategic and politico-militarily driven imperialism we do interestingly notice that in fact, consequent to the final outcome of the historic eight-year American War of Independence (1775-1783) though the leading and imperially highly obsessed seafaring and rival European Imperial-Powers were conclusively forced to withdraw from the American continents, but that did not at all mean that they were finally retreating to their home countries in the overseas European continent. In these perspectives, we may recall that somewhat earlier in this chapter itself we have since briefly discussed that historically, when those rival and powerful seafaring Imperial-Powers were already imperially consolidating their numerous colonies in both the American continents almost simultaneously they had also begun making, imperially substantial inroads, as a parallel imperial phenomena into the far-off continents of the Asia and Africa since located across the Pacific, on the other side of the globe. It is a fact that, initially to begin with historically the rival European Imperial-Powers had first proceeded towards the then newly discovered vast territories of the twin American continents. The highly checkered living-history of the why and how their adventurous and highly motivated explorers were organised and equipped to first proceed across the Atlantic and towards the far-off discovery of the so-called new world, and i.e. now known as the North and South American continents have since already been discussed in the few pages earlier, in this chapter. But then since one after the other, the highly adventurous ventures, to chart-out the sea-routes towards reaching the southern Indian coasts in Asia were also launched by those rival European sea-faring Imperial-Powers for exploring more and more new territories for their imperial colonisation. Hence interestingly in other words, making of the parallel colonial-era living-histories, pertaining to the differently targeted various continents of the world had already since simultaneously, begun, and as then being triggered by the multiple-aspects of their own overwhelmingly fixated imperialism.

Now again coming back specifically to the initial historic phases of the newly born and still fledgling Nation-State, the United States of America (i.e. USA); as we critically begin to delve deeper, we do notice historically that in the beginning; the many years during her subsequent multi-dimensional processes of socio-cultural, socio-psychological, socio-economic, as well as the socio-politically; multiple integration-processes, were far from being peaceful and smooth at all as a vast democratic Nation-State, consisting of so many provincial States within her international boundaries. Further, during those initial years of the highly complex phenomena of the making of USA we do notice and as also discussed earlier, that the first highly obsessed and fixatedly ambitious seafaring and rival imperialists, belonging to almost all the leading European Imperial-Powers, had relentlessly begun to set-feet on the Atlantic-shores of the both American continents and thereby, also competitively commenced to get hold of maximum possible territories for their advantageously immediate colonisation. In the process, these rival European seafaring Imperial-Powers also repeatedly fought among themselves both in the sea as well as, in both the South and North American continents. In addition to these, the many kinds of geostrategic and politico-military rivalries among them, back-home in Europe too had multiple-effects—as the all kinds of new opportunities then, being thrown-up for some and also simultaneously as some rather crippling compulsions on some others; in their 'no holds barred' competitive-struggles in the twin American continents, for promoting their different kinds of the multiple-aspects of imperialism and also, for establishing their colonial-empires over there. These far-reaching and substantially history-making historic phenomena indeed do give us some new kinds of glimpses of manifestations of the same age-old multiple-aspects of imperialism. Thereby, the ongoing making of the living-histories of both North and South American continents, got totally altered forever.

And as these historically highly complex and also violence-ridden phenomena then fast getting unfolded in the Americas and contextually, as since already briefly discussed earlier; so far as the North American continent was concerned, the final as well as hence, the highly consequential outcome, of the rivalries among those competitive European Imperial-Powers indeed, conclusively brought the Great Britain as the supreme imperial-victor over there. Hence, gradually the entire North American continent now since constituting the twin-giant Nation-States of the USA and Canada to begin with, was brought under the then unchallenged and indeed sweeping sway of the Union-Jack. Historically, therefore the subsequent and highly checkered gradual phenomena of the births of vast nations of the USA and Canada, had since commenced. As we have since already noticed that in the very beginning many imperialist-settlers of various European nationalities and in numerous shiploads; and with their all kinds of belongings as well as even the livestock, then so far not seen in the American landmass; had begun arriving for settling over there. In USA, finally the British settlers were able to commandeer their almost total monopoly for the settlements, and then even in Canada too, they established their dominance but, the French were indeed able to make some inroads there for their settlements. We have since already touched upon these aspects, in course of our discussions earlier; during our critical survey of the history and then, rather a fast-paced colonisation of the North America.

So as this phenomena of the colonisation in North America commenced, the gradual shaping-up of the demography of the regions now constituting the USA resulted in the substantial population of the European imperialist-settlers and that too mainly the British, and subsequently the large numbers of their black African-Negro slaves and then also added to them in some pockets here and there, were some surviving and indeed the original inhabitants; but now totally marginalized in their own ancient lands, i.e. the 'Red-Indians' as well. So, the recipe was historically getting complete and in the process, **the seeds for the then future but forthcoming American Civil-War (1861-65) too, were getting sown for their fast sprouting in the near future**. So far as the cruelly violent and highly-checkered aspects of the entire historic gamut of systems of slavery, in those North American territories, now constituting the USA are concerned, this is indeed a very pathetically painful saga, since recorded in the annals of man's living-history, and we would critically be delving upon them in some detail more appropriately, as we keep proceeding further, little later in this chapter itself. At the moment, we may first more befittingly prefer to conclude our brief and critically analytical discussions, on the multiple historic-aspects of the American Civil-War; in the larger geostrategic and politico-military perspectives in this regard. And also preferably, we would do so, before coming to the multiple-aspects of the multi-dimensionally historic geopolitics of again ruthlessly competitive making of the multiple-inroads by the same leading rival European Imperial-Powers, into the vast African continent across the Pacific for establishing additional and all the more far-reaching colonial-empires. At the moment, we may just prefer to bear in our minds that the historic saga of entire phenomena of the ruthless practices of the multiple-aspects of imperialism as well as building of the numerous colonial-empires in the vast and otherwise highly rich in natural resources and minerals, and so in the quite 'virgin' African continent, is no less momentous and historically interesting as well.

Now coming back to the multiple-aspects of the historically highly consequential **four years American Civil War (1861-65)**, appropriately in the fitness of things first and foremost, we may have to critically look into the central causative-factors that were principally instrumental in triggering this indeed momentous Civil-War. In these larger perspectives, it is generally understood that right from the beginning of now the remotely bygone ancient-ages so far as the so many wars, historic-battles and the skirmishes, already having taken place in different regions of the world are concerned since even before the advent of human-civilisations as well as, the very beginning of man's living-history besides the various kinds of geostrategic and politico-military reasons, **there have mostly been powerfully compulsive economic-cum-survival reasons, compelling the human-beings to fight among themselves; and which were also decisively able to trigger any wars, in almost all the historic ages**. And hence, as we proceed to critically delve little deeper into the momentous history of this American Civil-War; we broadly find very interestingly, the similar no less historically overwhelming reasons in this regard too.

Therefore, which in turn somewhat amazingly, compelled the various newly constituted provincial-states thus forming the then fledgling and also in addition, rather very reassuringly already being democratic the sovereign Nation-State as the USA, decisively encompassing the entire vast southern region of the North American continent, to get violently involved in some sort of the all-out civil-war among themselves and thereby, so as to some states wanting to secede from the Union, while others were than bent upon maintaining the unchallenged unity of the same Union at all costs. The authentically recorded and also, since already well-researched history of this war therefore, further very informatively but also rather amazingly shows us that accordingly as per the above then, the battle-lines were firmly drawn within the sovereign territorial limits of the USA herself. Ironically, a irreconcilable situation had developed there in spite of the USA already being quite a healthily functioning democracy, with more than half a century of standing; since the successful conclusion of the American War of Independence under the charismatic leadership of General George Washington, who immediately after the independence from the then British colonial-rule; became the first and in fact, Founder-President of the USA. It has perhaps been rightly said that, the man's living-history sometimes does have such strange ways of its making that at times, even certain historic-facts within their given historic-situation, amazingly do look stranger than any fiction itself.

In the above perspectives in this regard; the Second World-War era senior British Army officer, **Major-General J.F.C. Fuller** has very perceptively written that, "**Accustomed as they are to accepting the verdict of the ballot-box, the people of a democracy do not find it easy to comprehend the causes of the American Civil War. There is a tendency to oversimplify the matter by saying that the war was fought to abolish slavery in the United States**.... It was not the abolition of slavery on the one hand, nor its retention on the other, that dictated the grand strategy of the belligerents; it was the question of the right of one or more States to secede from the Union.... Democracies do not fight like that for abstract aims. Their peoples need to be moved by powerful emotional impulses, to be convinced that the cause for which they fight is so transcendently important that it is more precious than the life itself. In the calm, detached view of history it is difficult to recapture that essential element of civil war in a democracy, to recreate the deep, passionate conviction that the cause was not only just, but that it was the only conceivable justice. It is not necessary to agree with either view, but it is essential to see the situation as they saw it, to feel as they felt. Otherwise we shall never understand why or how they fought. **The framers of the American Constitution deliberately left two problems for the solution of time—States Rights and Slavery. To have forced either of these to an issue would have been fatal to the adoption of the Constitution**. Slavery was too strong an economic factor to be solved with the financial resources then available to the Union, while most of the States were unwilling to surrender more that the barest minimum of their sovereign rights to the Federal Power. It was hoped that in the course of time slavery would gradually die out, and that as the advantages of the Union became more fully apparent the States would become willing to accept and extend the authority of the Federal Government.... From the beginning the concept of State sovereignty was more firmly entrenched in the South than in the North. As time went on, divergent material interests other than slavery, and the different structures of the two societies, tended to extend the concept in the South, and to restrict in the North. The character, institutions and interests of North and South were as different as there of two distinct nations. Southern society was an aristocracy based upon a rigid observance of class distinction. The North was a plutocracy where the wealth or merit constituted the only claims to advancement. The South was an agricultural community and, enjoying as it did a virtual monopoly of the cotton market, demanded free trade. The North was largely devoted to manufacturing industries, and in its own interests enforced a protective tariff, which told heavily against Southern prosperity. Nevertheless, slavery was the only non-adjustable issue which could not be resolved by compromise." (Fuller, pp. 11-14)

In the above quoted lines from his book, the Second World War era Learned British soldier, thinker and also gifted author on the matters-military and military-history too, General Fuller has brilliantly put-across the basic genesis of the seventeenth century's American Civil-War, which was in an entirely path-breaking manner; indeed been able to change the multiple-paths for the subsequently ongoing making of living-history of the United States of America. Historically, it was definitely a big turning point in man's living-history since always perpetually in the making; which in turn, was indeed going to multiply bring-in many kinds of impacts interestingly perhaps, both positive for some and rather perceptively negative for some others in the international arena then-on for centuries to come. And specifically in these broader contexts, both interestingly as well as, in many ways quite engagingly perhaps; we find multiply no-less both directly and indirectly and continuously too, the numerous kinds of US impacts on almost all Nation-States of the world, even in the fast-paced twenty-first century. The fundamentally central point in this regard, remains that at the roots of these ongoing and both deliberately or even factually otherwise, the many kinds of US-influenced developments in almost all corners of the today's world to start with in many ways the bygone seventeenth century's American Civil-War too, have been a decisively history-making contributor.

In fact, quite understandably at the apparent surfaces of all these above hinted globally multiple-kinds of the US-influenced developments today, indeed we may also objectively perceive their basic causative-factors as to why the USA has truly been able to perhaps, an unrivaled hyper-super Power in multiple ways such as, the economic, industrial, scientific and technological, geopolitical and, above all principally due to all these in the matters-military as well. And further on top of all these, since **her multiply-safe and far-off strategic location across the Atlantic and, also coupled with their continental-size territory situated in the North America have indeed been able to always provide her geostrategic and politico-militarily unique as well as, substantially advantageous position**. Well, in contextually larger perspectives at the present precise point of our on-going discussions, the above highlighted points concerning the present modern-era's, a particular Nation-State the USA internationally are largely factual, well-taken and indeed understandable as well. But in these specifically precise contexts as we may begin to further critically delve deeper, we may also be able to historically see as to, why and how the USA was indeed able to multiply become the truly unrivaled hyper-super Power as since brought-out in our above discussions? The precisely focal point is that all these could happen because, in order to become the international hyper-super Power; first and foremost the USA had to perforce remain politically as well as constitutionally united completely and also, quite desirably remain very large an American federal Nation-State, for democratically ensuring the healthy survival of this federal socio-political and constitutional unity of the USA and certainly not get divided into two Northern and southern sovereign nations. To begin with, this was indeed inescapably mandatory for the USA, as fundamentally the first prerequisite and hence thereby, it remains historically undeniable that it was the American Civil-War since occurred in the seventeenth century itself, and though extremely violent too which in fact ultimately once for all, ensured this US unity of paramount importance and also as subsequently turned-out to be both historically as well as futuristically and as also in addition, borne-out by the man's living-history, to be far-reaching for the larger world as well.

At this precise point, it may appear fairly relevant, to very briefly touch upon the basic core values, which are indeed constitutionally the central theme, towards making of the USA and in addition also, since they do historically in fact, provide permanently the main compass and anchorage; towards the socio-psychological, socio-cultural, socio-political, geostrategic and politico-military stability and national well-being to her, as a sovereign Nation-State in the world. In these specific perspectives, it may appear to be essential that we may very appropriately prefer to look into the 'Preamble' of USA's Constitution and hence, fruitfully be able to receive some sort of desired enlightenment in this regard. In these precise contexts; as we are aware and also, since the man's living-history clearly shows; that the nations are never enduringly made or born, with the 'drop of a hat'. In fact interestingly, we may also say that even any kind of a 'Banana-Republic' in any corner of the world too, has its own geopolitical, at times perhaps quite a humorous and superficial system concerning its making or any kind of birth.

And therefore in these specific perspectives the initial paragraphs of the internationally indeed famous **'Declaration of Independence'**, as since adopted by the founding fathers of the United States of America, under the charismatic leadership of **General George Washington**, immediately after conclusively winning the eight year long, historic American War of Independence (1775-83), very emphatically and precisely declares that, "When in the course of human events, it becomes necessary for one people to dissolve the political bonds which have connected them with another, to assume among the powers of the earth, the separate and equal station to which the Laws of Nature and Nature's God entitle them, a decent respect to the opinions of mankind requires that they should declare the causes which impel them to the separation.

We hold these truths to be self-evident, that all men are created equal, that they are endowed by their Creator with certain unalienable Rights that among these are Life, Liberty and the pursuit of Happiness. That to secure these rights, the Governments are instituted among Men, deriving their just powers from the consent of the governed, That whenever any Form of Government becomes destructive of these ends, it is the Right of the People to alter or to abolish it, and to institute new Government, laying its foundation on such principles and organizing its powers in such form, as to them shall seem most likely to affect their Safety and Happiness. Prudence, indeed, will dictate that Governments long established should not be changed for light and transient causes; and accordingly all experience hath shown, that mankind to more disposed to suffer, while evils are sufferable, than to right themselves by abolishing the forms to which they are accustomed. But when a long train of abuses and usurpations, pursuing invariably the same Object evinces a design to reduce them under absolute Despotism, it is their right, it is their duty to throw off such Government, and to provide new Guards for their future security—Such has been the patient sufferance of these Colonies; and such is now the necessity which constraint them to alter their former Systems of Government." (*Source: 'New Webster's Dictionary and Thesaurus and Medical Dictionary', Book Essentials Inc. New York, 10016, 1991 Edition*).

The above quoted a couple of initial paragraphs of the 'US Declaration of Independence'—— after winning of the American War of Independence and, along with framing of the independent and sovereign USA's Constitution; did give the necessary facilitators towards the historic dawn of the democracy in this newly born nation and also thereby provided the mandatory anchorage to the American societies, in their best all-round national interests and well-being. The 'Constitution' always and at every moment, plays multiple roles in the entire gamut of governance, maintenance of the unity, integrity, security and diversity, as well as the very survival in the life of any concerned sovereign Nation-State, and inescapably all the more in case of any democratic nation. But at the same time, man's living-history also emphatically shows that, in case of any newly born nation to begin with most of the time, just putting-up a comprehensively well-framed Constitution may not be sufficient as, in the overall life of any such nation in the comity of nations in the world, may not be adequate at all since in any given internal as well as, overall geopolitically external situation the life is seldom simple and linear.

Hence in these larger perspectives; as we critically look into the living-history of the beginning years of the USA, as a newly born, sovereign and democratic Nation-State we notice that within few decades of her independence and, for her continued some sort of externally the threats-free survival, and as already discussed earlier the fifth US President James Monroe, during his Presidency from 1817 to 1825 and, indeed with his own duty-bound, pragmatic as well as quite legitimate intention in this regard; he had to issue and declare enforcement of the famous 'Monroe-Doctrine' in fact, covering both the American continents. The man's living-history also shows that in almost all historic-ages so far as the Kingdoms/Empires or the Nation-States are concerned, the multiple-influences of the various kinds of specifically situations-based overriding scenario since as prevailing from time to time around them; have been dictating the overall geopolitical and then the internal socio-political life of any concerned nation. In the broader framework since ages in this regard, the ground realities as always continuing to prevail, have always been such that in fact, irrespective of the man's perpetually ongoing civilisational-level advancements each nation is largely compelled to adequately safeguard her desired external as well as internal security, peace and enduring tranquility in the comity of nations. So in a way, it also appears that a nation is still perhaps forced to largely live in a kind of the state of Nature. That's the way the world has been since the beginning of various human-civilisations and also, that's the way world continues to remain so far. **Hence during the somewhat beginning of the 19th century itself, if the historic enforcement of the famous 'Monroe-Doctrine' was able to effectively ward-off any kinds of possible external-threats against the USA; and then even after less than forty years or so, in order to effectively ensure the no less supreme and vital internal security and unity of her internally numerous provincial States-oriented National-Union; she was to perforce get engaged in the four year fierce Civil-War (1861-65).** Therefore, in comparatively a very short living-history of the relatively young nation—the USA; all these far-reaching as well as momentously history-making events need to be seen and critically perceived, in their desired perspectives.

In our larger perspectives and also at this precise point of our multiple-discussions concerning the USA; it may quite relevantly appear the necessity to contextually bring-in indeed, a very far-reaching as well as, momentously history-making point here. And that historic point in this regard, has precisely been in fact; instrumental in so famously as well as enduringly, in the making of USA herself, as a multiply-unique Nation-State that she is today. Hence, that historic point perhaps obviously has been the **simultaneous and though highly checkered and indeed fast-track; cultivation, evolution, development and imbibing in her multi-dimensional body-politic as well as societies too, up to a considerable extent the multiple-aspects of 'essential strategic-culture and thoughts'**. We may again recall here that we have already discussed earlier, quite a few number of times and since in various contexts, the historically far-reaching and timelessly of paramount importance, those multiple-aspects of 'essential strategic-culture and thoughts'. Hence, while keeping in our minds the above mentioned discussions as we may commence to further delve deeper, into crisp the history of this relatively young and very vast Nation-State the USA, which is now the sole hyper Super-Power in the present modern-era; we contextually find that besides she inherently having certain unique and location-based strategic advantages, she has also been critically pragmatic in fruitfully displaying the requisite geopolitical prudence, while proactively observing the multiple-tenets of these timeless 'essential strategic-culture and thoughts'. USA has been able to achieve the multiple-successes in both geostrategic and politico-militarily conducting her foreign-policies. But at the same time, whenever she has been found somewhat wanting in this regard, the end results quite embarrassingly for her, have not been very palatable. We would be discussing critically in more details; these multiple-aspects of indeed substantial historic reach, in the subsequent chapters.

Now while again **coming back to the multiple-aspects of American Civil-War,** as briefly discussed above in its some socio-political, socio-cultural and socio-economic perspectives, now it may appear to be quite appropriate and relevant to also briefly touch upon the key areas of no less important and indeed decisively dominating aspects of its matters-military as well as, of the military-history. And there-in we find some very astonishing historic-facts; which were really instrumental in piloting the course of this war and thereby, determining its final outcome. Further therefore, the final outcome and subsequent fallouts of this civil-war were certainly no small matters since these were singularly the most important decisive-factors permanently, in shaping-up of the multiple-paths for subsequent ongoing making of the living-history of USA. In the specific contexts of matters-military in this regard, at the outset we may clearly say that, though the Union forces under the supreme political-level leadership of **President Abraham Lincoln** and along with finally the top-level military leadership and command, of **General U.S. Grant**; were indeed conclusively able to win this historically path-breaking civil-war, but certainly it was no easy matter of them. World's history of the warfare since ages shows that there are always numerous kinds of imponderables in war and hence any contesting side can perforce seldom be in a position to confidently predict, or be sure of final victory in advance. Though it is a different matter that most of the time, while historically taking any kind of perceptively detached views of any warlike event; much after its occurrence and also after the heat and dusts have settled down, that a kind of professionally and also, intellectually well-analysed academic-views generally begin to get currency; that the final results of any war mostly get decided interestingly rather, well in advance of the actual fighting having taken place. Well, that may generally be true academically in retrospect but contextually, any kind of advance judgment, is seldom possible objectively.

The interesting point in the beginning itself appears to be that, almost all the leading military commanders at all levels and, on both sides were in fact; trained and groomed as officers in the same US military training institution. And hence; they were since graduates of the same **United States Military Academy**, located in the city of West Point in the State of New York and thereby **famous throughout the world interestingly, as the WEST POINT**. Historically therefore, the qualitative aspects of the training-inputs and also in addition, the professionally allied grooming-systems for the officers on both sides, were indeed absolutely identical. But, though the military professional training-inputs concerning the two opposing sides being identical, the matters regarding aspects of similarity in fact, just ended there. We may in fact, need to bear in our minds that the professional ethos as well as, the matters concerning the loyalty and human-level thought-processes within any particular Army or the entire Armed Forces, are crucially guided and shaped by the prevailing social, cultural, economic and political environment and currents, as prevailing in the much larger any concerned society, within which that particular military-force exists, remains a part thereof and also, operates accordingly.

And therefore in the overall geopolitical as well as matters-military perspectives the larger hard realities remain, that any kinds of major social, economic and political turmoil as well as, larger socio-psychological fissures in turn are bound to perforce bring-in greatly unsettling and politico-militarily far-reaching multiple-impacts within the concerned military-force. In fact in these specific contexts, man's inherent socio-psychological make-up and thereby, his vast living-history since the very beginning has consistently been showing that, in certain kinds of far-reaching, highly influential and also fast emerging circumstances it seldom remains possible to keep and professionally maintain the concerned military-force, desirably isolated or perhaps insulated from the multiply-larger and prevailing social currents of the time, in any given bigger and dangerously fluid situation. In fact historically we may also find that at times, even a very attractive and greatly socio-cultural message-oriented sentimental literary-piece of writing could also come in a situation-based, somewhat history-making syndrome. Hence, for any kind of war in any given far-reaching and emerging situation, including a civil-war within any nation or a socio-political entity; historically there are several different kinds of simultaneously materializing multiple-influences, events and the inputs; which at times, may in fact, act as apparently or as being hidden or even indirectly; different kinds of proactive causative-factors, towards doing the necessary ground-work for and then, also triggering that war. Man's substantially vast and highly complex living military-history in fact, does always maintain a saga of many such events and episodes, down the ages in this regard.

As per the above discussed broad-based historic perceptions and also, specifically concerning the multiple-aspects of American Civil-War (1861-65) in this regard, General Fuller in his already earlier referred book has since written with great precision that, "Then **Harriet Beecher Stowe wrote 'Uncle Tom's Cabin'. In the interest it aroused and the influence it exerted, this novel must be classed as one of the greatest ever written**. It was quite fair. It was also very clever. The softer and lighter aspects of slavery described in the early part of the book threw into sharp relief the horror and infinite sadness of succeeding chapters. The story and the manner of its presentation gave the book a vogue, not only in the Northern States but in many overseas countries as well. Great Britain bought a million and a half copies. Two London theatres produced the play. Three daily newspapers in Paris produced it as a serial, and the Parisians filled two theatres nightly to weep at the fate of Uncle Tom. Fifty years later the book was still being read by nearly every boy and girl in the English-Speaking world. As late as 1929, audiences surreptitiously dried their eyes as they filed sadly out of the theatre. The effect in the North was tremendous, and gave to the abolitionist movement the fervour of a great religious crusade. The South thundered back that the conditions of the slaves were no worse than those of the juvenile workers in northern factories, an assertion that wounded deeply because it contained an element of truth. Reasons flamed into passion. Hate flowed down the Mississippi and met a tide of hate coming up.... Fresh fuel was thrown on the flames by **John brown's raid on Harper's Ferry, Virginia. John Brown was the leader of the Free State gunmen in the Kansas fighting**. He was a man of extraordinary fanaticism and limitless determination, and he had the moral and material support of several rapid anti-slavery societies. In 1859 arrived with 21 followers in the neighbourhood of Harper's Ferry with the object of establishing a collecting post for fugitive slaves escaping en masse. In effect he hoped to raise a general insurrection of the slaves. To set the rising in motion he seized the arsenal at Harper's ferry. This was of course an open attack on the authority and property of the United States, and Federal Government promptly replied by sending a small force of marines to restore order. After a brisk skirmish John Brown was captured, tried and duly hanged for treason against the United States." (Fuller, pp. 14-16)

It is generally quite appropriately said that the leadership at all levels and also in the all areas of human endeavour, is largely situational but at the same time; the multiple-aspects of leadership do also contain many kinds of numerous dimensions as well as, complexities too; precisely in their any particular kind of the situation-specific manifestations at all levels in any affected-cum-involved areas. Interestingly, so far as the aspects of top strategic-cum-tactical level military-leaderships, as since being proactive on both the opposing sides are concerned; most of the time the multiple-aspects of their leadership potential as well as, the professional military-acumen, in any given scenario; though being extremely rivalry-oriented, are largely found to be more or less matching; but obviously in this regard, there also happen to be some historic exceptions as well. But as we further continue to delve deeper, we may be able to notice that in fact most of the time; the no less **crucially important aspects of the top political-level leadership, and then these being sometimes even overriding too, are seen to be largely decisive** in so far as the aspects of shaping the course of war and thereby, its final outcome are concerned. In fact, it is natural that in case of a democracy, the multi-dimensional canvas at the apex political-level leadership perforce get more and more complex and enlarged too. Hence at this precise point, we may prefer to add and as since the man's vast and continuing living-history also substantiates that historically the same has been largely true, even during the Monarchy's times as well and moreover it remains equally true in case of any military-dictatorship regime too, where the multiple-aspects of both the political as well as military level leaderships, get perforce vested in one person, or say in one singular power-centre itself. And so in fact, in any kind of war-like scenario in spite of the essential aspects of professional military-level leadership being quite competent but, if the political-level visionary statesmanship is found missing or perhaps, geostrategic and politico-militarily seen to be not up to the mark then the end results could even unexpectedly land-up in a jeopardy. As we all since know that, **it is generally said that war is too serious a business to be left to the Generals but at the same time it remains equally and irrefutably valid that the war is too serious a business to be left solely in the hands of naïve and clueless politicians**. We would since be discussing these historically interesting and debatable, but also somewhat controversial aspects with more befitting and actual examples, in the next forthcoming chapters.

Well in these larger perspectives, so far as the specifics of the 19th century's momentously history-making American Civil-War are concerned, USA's living-history emphatically informs us that, as we rather somewhat closely and discerningly go into the personality as well as, the then critically emerging situation-based political-level leadership attributes of the **Union President Abraham Lincoln** and then, those of the **Confederate President Jefferson Davis,** we do indeed find a number very important and decisive contrasts between these two historic personalities. In fact, the game of politics and the multiple-aspects of the political-leadership too are quite strange and the matters of high complexities as well. These political-level complexities in fact, do perforce acquire some new and additional dimensions; during any fast emerging war-like situations and hence in turn, do also manifest some unexpectedly new, burdensome and even all the more threatening challenges; before the concerned top-level political leadership. And in these highly weighty and interesting contexts; man's living-history in general, also shows that at times; irrespective of the educational, political as well as even, other kinds of backgrounds, specifically concerning the two opposing and directly facing politicians in power in their respective rival countries/territorial groupings, in any particular wartime scenario the end results could turn out to be quite surprisingly unexpected as well.

Hence, in any particular live and somewhat fast-paced wartime situation at times, the man's vast living-history shows that **any unique and some singularly uncommon and, inherent streaks of genius** in case of any one particular top-level political leader, could even turn-out to be decisive and make far-reaching differences towards final outcome of the concerned war. Well, at this precise point, we may also quite befittingly say that, it remains equally important and appropriate to emphasise that the above mentioned aspects of the some kind of inherent streaks of genius, are also always found to be all the more crucial and decisive, in case of the professionally proactive and any particular situation-oriented multiple-manifestations of the top-level military leadership aspects and hence always being much more so, during the wartime situations. In these specific perspectives we would contextually be discussing these important aspects in more historic details, in the next chapters. At this precise juncture of our on-going discussions it appears to be rather critically inescapable to bring-in, in a context; a crucially important and factual, but at the same time generally not so palatable point for the top political-level leadership that at every stage before any particular war or even during the peacetimes, it is the concerned political leaders, and all the more so in a democracy who at times are more often found to be wanting due to their many kinds of naivety towards highly important and complex matters-military as well as ironically, somewhat also suffering from unexpectedly strange limitations in many ways, when it comes to the conduct of statecraft along with visionary statesmanship as well. Man's vast aspects of the living military-history too, since ancient of times are ironically in fact littered with many kinds of embarrassing and even at times, historically tragic examples in this regard.

But at the same time, it does always remain equally true that **whenever and wherever the concerned top-level political leadership has been able to fruitfully display, indeed the far-sighted and national-cum-international level political sagacity, as well as the highest kinds of positive and valuably broad-based understandings of the matters-military and also along with, the requisite statesmanlike political awareness of the overall geostrategic and politico-military imperatives and thus in the process, substantially be cultivating and enjoying the confidence of his/her military-chiefs and also top field-commanders and thereby ensuring decisive successes and so in addition, the end results could even turn-out to be some kind of a wind-fall**. Will it be perhaps considered as somewhat overstretching the point if we may say that the above mentioned multiple-aspects, concerning the top-level political leadership if allowed to fructify and proactively put in place, then **these too in turn, may even multiply function as force-multipliers as well as also ensure in fact in numerous ways, the creation of multi-dimensional national-level synergy as well**? Ironically even during the present so much fast-paced and high-tech modern-era, man's on-going living-history has been clearly showing that often **in the independent, modern and sovereign India too** instead of paying rather the inescapably essential and learning-oriented attention to the above emphasised highest-level and indeed far-reaching and crucial leadership attributes, meant for the concerned political personalities since placed at the helm of affairs from time to time they very embarrassingly tend to forward all sorts of lame and misguiding excuses, mainly for the consumption of innocent masses and if required quite unhesitatingly pass the buck to the concerned military-brass and further then, even try to very craftily find some escape-goats among them. In these broader contexts, while being truthful to the perpetually on-going as well as in the making—the man's multiply vast living-history, always with critical objectivity and thereby we find that as a result of the above identified pit-falls or slip-ups, at the top political-level and at times even at the top military-level leadership functions, in the end tragically the soldiers are driven towards naïve initiatives of futility and lose their lives; some persons in uniform are sacked and then, nation is made to suffer; but seldom top-level political personalities are made to account for their all sorts of omissions and commissions. Well, that precisely indeed, happens to be another irony of man's living-history and hence; the same history has many such examples since recorded in its pages, for posterity.

Now coming back to the war-time comparative studies of the multiple leadership-aspects of the **Union President Abraham Lincoln** as well as, the **Confederate President Jefferson Davis**, in the then turmoil-ridden and also still evolving USA, we interestingly notice that Jefferson Davis was formally more educated and well trained, for the highest-level public office. Besides getting educated in civil institutions, he had received military education and also graduated from the **U.S. Military Academy, West Point** in the year 1828, i.e. many years before the actual occurrence of American Civil-War. Then he had also served in action, while in the Mississippi Regiment, during Mexican-War. On the political side too, after resigning from the Army; he had gained considerable experience, as being an US Senator and in addition then before switching sides and assuming charge as the Confederate President, he was working as the Chairman of Military committee of the US Senate. These advantages on the part of Jefferson Davis, did give him a sort of considerable head start in favour of the South and hence in the beginning, on a number of fronts the South's forces were able to achieve initially, some stunning victories. The ground-level realities were that as the Confederate President, Jefferson owing to his sound military background; was indeed able to organise and put in place the proper hierarchy-based military-command structure and thereby, he could successfully bring-in and maintain much better discipline in his Army, since launched against the Union forces. But ironically, these were not going to endure since he at times, couldn't resist his unnecessary temptation to interfere, with the operational-plans of his field commanders. Moreover, his principal war-aim being naively defensive in nature, just to maintain the existing statusque, he couldn't even dream-of achieving any decisively conclusive military-victory over his rivals. Man's multiply vast living military-history since the beginning shows that for any kind of the decisively conclusive victory in war, it is seldom possible without well-conceived military-offensives if one wishes to firmly deny his adversary, any opportunity for his military offensives against himself in any particular war.

Whereas the overall scenario in the North in this regard, happened to be totally different and substantially encouraging in spite of the many kinds of rather quite understandable limitations since existing in the beginning only, on the part of the Union President Abraham Lincoln. Little earlier, we have since talked about quite a weighty point regarding the aspects of unique, some singularly uncommon and inherent streaks of genius, on the part of any particular top-level political and/or military leader, and which indeed go a long way in decisively maintaining a visionary edge over the concerned adversary and then, securing a final victory in war. And hence, while critically looking deeper into the life-sketch as well as, saga of remarkable achievements on the part of the then US President Abraham Lincoln, we may clearly see some similar kinds of the inherent streaks of genius in his, indeed perhaps the inborn leadership-oriented mental-cum-intellectual framework and the personality. And historically, as already so well-known that subsequently made all the difference decisively in favour of the Unionists, under his top-level political leadership in the American Civil-War and hence; USA did indeed remain united firmly forever.

In these larger perspectives and at this precise point, in order to contextually substantiate the above mentioned outstanding leadership attributes concerning **Abraham Lincoln**, it may perhaps be sufficient to look into the following quoted paragraph of **General Fuller's** book, "In 1858, a comparatively unknown man, Abraham Lincoln (1809-65) appeared upon the scene. In his contest with Stephen A. Douglas for the senatorial seat of Illinois he poured no little common sense on the burning question and won ears of the whole country when he proclaimed: **'A house divided against itself cannot stand. I believe this government cannot endure half slave and half free. I do not expect the house to fall, but I expect it will cease to be divided. It will become all one**

thing or the other.' The election of 1860 (*i.e. subsequently, the then Presidential Election*) took place in an atmosphere of extreme tension. Each side had conducted its campaign with utmost vigour, and its propaganda had been couched in the most violent language. No political device, however doubtful its legality or morality had been left untried. The results went up on 6 November 1860. The Republicans—the extreme abolitionists—had won. Their candidate, Abraham Lincoln, had been elected President of the United States. In the South (*i.e. principally the agricultural States within the US*) passion suddenly subsided into firm resolve. So far as the ruling class could see, secession was now the only means by which they could avoid personal ruin, the only means by which they could preserve the society to which they were so deeply and genuinely attached. South Carolina took the lead. On 20 December her State Convention passed an Ordinance declaring: 'The Union between South Carolina and the other States under the United States of America is dissolved. 'Mississippi, Florida, Alabama, Georgia, Louisiana, and Texas followed the lead in rapid succession." (*And so the die was proactively and firmly cast finally, for the extremely violent American Civil-War.*). (Fuller, p. 17)

We do also notice that, more outstanding the leadership-oriented personification of that particular historic personality, threateningly more serious are the challenges surfacing and also magnifying, from all sorts of directions to confront the leader since on his way to remarkably make a memorable history in his own right as well as, indeed in his own historic ways. And hence, the same historic aspects are indeed so outstandingly true in case of Abraham Lincoln as well. Lincoln since came on the scene with his very humble background and therefore, he had to struggle a lot to keep himself afloat and perforce had to engage himself in all sorts of very ordinary jobs to survive and then, also keep-on grooming himself, for a career-oriented progress too. Unlike his adversary Jefferson Davis, he had no military education and also no experience in the art and crafts of soldiering, especially as a commissioned officer in the Army. And yet, within a broader historic framework, as a top-level political authority, he had a focused mind with clear and well-gelled highest national-aim, unique streaks of genius, and his unwavering concepts with inspiring confidence, concerning the war and its highest level national-aim and then in addition also including his visionary grip over the no less multiply complex areas of the statecraft as well as, the finer points concerning the governance of his country.

When the die was firmly cast and battle-lines were finally drawn, as we begin some critically discerning geostrategic and politico-military assessment between the two ready to confront groupings—as the Northern Union and the Southern Confederate; and thereby at the first such glance, North appeared to be more formidable and hence, displaying considerable edge over the South. Firstly the population-wise, the North was indeed having much more human resources with nearly 22 million people whereas the South was having just around 5.50 million people and in addition, nearly 3.50 million black Negros as slaves; since earlier very craftily transported from Africa, in all sorts of manners. Interestingly as a considerable military bonanza; in fact with some very negligible exceptions, the then entire US Navy quite dedicatedly lined-up on the side of the North and hence, the Union authorities were very effectively in a position to enforce a naval-blockade against the South. Hence, in terms of the military-cum-strategic as well as socio-psychological considerations, it was indeed a sizeable blow to South during the war. In addition, the North was much more industrialized and consequently had a far stronger manufacturing base for the numerous warlike equipment and consumables than the South and hence, the Union forces were able to equip and militarily organise as well as, to provide for themselves in much better manners.

In these precise contexts, as we continue to discerningly proceed with our critical assessments;

we do also notice that the South too, had some sort of unique advantages as well. Though, the South was primarily engaged in various kinds of extensive agricultural pursuits, as their socio-economic occupation, its population, in spite of being numerously much lesser; the peoples were more rugged, earthly, better horse-riders since their childhood and hence in many ways, more accustomed to hardy and rustic lifestyles. Further, surprisingly and rather quite shockingly to the North **when the fighting began with negligible exceptions, most of the slave-population in the South, instead of resorting to any kind of mutiny or rebellion; remained steadfastly loyal to the Southern white population and hence, when the men folk left for fighting the war, their black slaves loyally managed to look after their agricultural fields, women folk and their children as well**. In the beginning, all these did remain somewhat disturbing for the Northern Army's military commanders and leaders. At the time of confrontation, substantial majority of the States, i.e. twenty-two of them got lined-up in favour of the North as against almost half the number on the side of the South and hence here too, South was facing another handicap. In addition, North was also having comparatively, better infrastructural facilities; as considerably more wartime assets in fact at times, acting as even valuable force-multipliers.

Further, though the US Military Academy, West Point, was located in the Northern side during the commencement of Civil-War in fact, great predicament with lots of embarrassments, began to confront some of the West Point Graduates more that 1200 of them, since available for service in the Army. Amongst them there indeed was a great tussle of conscience and hence ironically approximately one-fourth of them with heavy hearts, were driven to resign their commission; and thus making themselves available to fight, on the sides of their respective States on the Southern-side. Understandably it was indeed very heart-splitting for them since they were professionally as well as apolitically, trained at the West Point, in their sacred culture and military traditions of the US Military Academy and also nationally, in their proud traditions of the US Army. Moreover at their personal level, they were also putting their future in jeopardy in fact, by losing an assured and prestigious military career in the US Army since quite understandably they were unable to bear the burden of fighting against their own peoples. In the beginning of this US Civil-War itself and, specifically in these precise perspectives what could be more prominently historic and painful an example, than that in the personal capacity of **General Robert E. Lee** himself?

General Robert Lee was such a senior and well-respected General-rank officer, that in fact during those momentous times he was quite befittingly offered command of the most important US Army, by the US President Abraham Lincoln himself. But with a torn-heart, he had forced himself to resign his commission in the Army, in order to make available for service in his home State in the South and thus naturally later on, he was appointed Chief of the Confederate Army for taking-on the powerful Union forces in the war. Therefore, it has been very appropriately said that, the man's multiply-vast as well as highly complex and, continuously on-going living-history has indeed at times, very bizarre and even strangely painful and then precisely some very exceptional kinds situations-specific ways of its making too, both during the periods of peace and war in the world, where a person irrespective of his/her station in life and also, having been trapped in some overwhelmingly powerful situation; is perforce torn at heart with no escape and thereby, gets painfully forced to act as well. So at the time of Civil-War paradoxically, **General Lee** too found himself trapped in some kind of an identical situation and hence, forced to take a decision this way or the other. Well, he was a well reputed and highly professional soldier and hence, much later after the war began he was appointed Chief to lead the Confederate Army, with still an unhampered professional soldier's frame of mind. His wartime operational planning and decisions and his conduct in war, are indeed historically praiseworthy and hence thereby; his life and professional conducts and actions during the war, still remain a very valuable storehouse for studies and lesions to learn, for the military professionals as well as, the students of military-history. And as subsequently the destiny would have it when **General US Grant**, since being quite junior to Lee, was appointed Chief of the Union Army the tide of war had already decisively started turning, in favour of the Union forces.

Union President **Abraham Lincoln** had indeed found an exceptionally brilliant and right kind of top-level military commander **in the persona of General Grant**. And as his good luck would have it, later he was able to surround General Lee and his forces from all sides; in the siege of Petersburg and hence, he succeeded in shutting all doors for Lee towards taking any initiative against him and historically, this turned out to be the final moment of reckoning. Consequently, **General Grant accepted the surrender of General Lee at the Appomattox Court House on 9th April 1865,** thereby; bringing an end to this ruthless Civil-War in the USA, and hence, heralding a great turning point in the history of this new nation. **The American Civil-War did ultimately succeed in achieving the two unprecedentedly far-reaching results in this regard: (i) first and foremost, the USA finally excelled in remaining united as one vast, sovereign and democratic Nation for all times to come, and (ii) consistently the gradual eradication of since regressively entrenched and highly inhuman systems of slavery was initiated and ensured**. These two indeed positively unique and history-making fallouts in turn, began to substantially contribute in the making of the most developed and super-power Nation, i.e. what the USA happens to be in the world today.

In the above discussed broad-based perspectives of the historic four year American Civil-War (1861-65), **Field Marshal Montgomery** has written that, "Lincoln considered that slavery was an issue which time and common sense would solve. But once the Union was split it would never be restored: North America would become like Europe, a continent torn with jealousy, economic rivalry and war. He fought to preserve the Union. Lincoln had no practical military experience, and his appointments were too often made on political grounds; when popular outcry demanded the recall of a defeated general, he generally yielded. Gradually the best men came to the top, but several good men disappeared. The General, whom Lincoln eventually found, **Ulysses Grant**, was a soldier's general; thoughtful of his subordinates and fitting his orders to their experience and skill. Of the generals on either side of the war, he alone demonstrated the capacity to command small forces as well as large ones in battle under a great variety of conditions—and finally to command and direct the operations of several armies.... Davis in the South was well qualified to handle the material at his disposal; he was a graduate of West Point, had served several years in the regular army, and had been Secretary for War in Washington. Not only did he select the right men, but he supported them in adversity and did not at once remove a general because he met with a reverse. Furthermore, he had in **General Lee** possibly the best American military thinker and organiser of those days, and during the first year of the war he retained Lee as his Chief of Staff in the Capital, a decision which in my view had much to do with the initial success of the South. But as a commander in the field Lee was not sufficiently firm with his corps commanders, and he was not a good picker of men. His own judgment was generally sound, but he didn't like to give firm orders to unwilling subordinates; this was brought out very clearly at Gettysburg, in which battle he was at his worst." (*We clearly notice one historic fact again coming to light that* ***at the any level of military command as well as the staff-work, within the well-defined military hierarchy in a professional and apolitical Army, it is not necessary that any exceptionally good and efficient***

staff-officer in a headquarter set-up, can also be a very good and effective commander too, in the field. Though, and as also the man's vast and highly complex living military-history shows that, luckily there could be exceptions as well. This crucially important and substantially far-reaching decisive aspect, concerning the conducting and directing a war always needs to be borne in mind, by both the apex level political-leaders at the helm of affairs, as well as by the concerned military-chiefs too). (Montgomery, 1983, pp. 257 & 258)

Now while proceeding further and, also before coming to the Asian continent; we first need to similarly look into another vast continent of **Africa, historically was also commonly known as the 'Dark Continent' of the world**. Interestingly, as in case of the twin American continents and subsequently to begin with, in the vast African continent too, a number of the adventurous as well as highly motivated European explorers had already commenced making the substantially explorative inroads for gathering more and highly useful information and data, for the ready uses by their own respective countries in their imperial pursuits into this additional newly found continent.

Specifically in these imperialism-related historic aspects, concerning the African continents, **Professor S.R. Sharma** has written that, "**David Livingstone (1849-73), a Scotch missionary who crossed the entire Dark Continent from sea to sea, is one of the best known of Africa's explorers. Mission work went hand in hand in Africa with geographical discovery**. While Islam made its home in North Africa from Morocco to Egypt, in Abyssinia, Siberia, and South Africa Christianity succeeded in establishing itself: the rest of Africa remained heathen. **Almost all the European nations participated in the exploitation of Africa. Particularly in the last quarter of the nineteenth century, there was a regular scramble for its tempting profits. In the past, Africa had provided the richest quarry for slaves; in more recent times it has given valued rubber, ivory, diamonds, gold and other rich natural products of a tropical continent**." (Sharma, 1948, p. 384) (*In these perspectives, whatever may be the different perceptions but civilisations had then, begun multiply-entering deep into the African continent and for this kind of ruthlessly uncaring imperialistic civilisational invasions ironically, Africa was just not prepared. And consequently the Africans were exploitatively forced to pay all kinds of prices to the crafty European imperialists, as then earlier already entrenched in the Americas.*).

Similarly as in the twin American continents earlier there was indeed a kind of the competitive mad-rush for acquiring as much riches as possible and simultaneously, for acquiring as much territories for the intended colonisation on the part of those rival European Imperial-Powers, also in the vast and all the more 'virgin' African continent too. History shows us that, as the then on-going games concerning the relentless pursuits of the multiple-aspects of imperialism, had begun progressing in the vast landmass of the African continent, the Spaniards were the first to take possession of the northern coasts of Morocco, while the Portuguese were able to imperially control Angola and sizeable territories in the eastern-region of Africa.

During those rather fast moving historic-times, when dawn of the historically unprecedented **'Industrial-Revolution'** had already taken place in Europe to begin with, and in addition its fast-paced advancements too had commenced; perhaps it did interestingly appear to be a sort of the games of 'free for all', among those ruthlessly rival and leading European Imperial-Powers in the vast African continent; for relentlessly maximizing their own kinds of the highly competitive imperialistic gains. So as the relentless pursuit of imperialism progressed in Africa, even Belgium was able to colonially take possession of Congo. France as a diehard competitor of the Great Britain, was indeed able to get hold of the maximum territories of the vast African continents. She therefore, as historically expected in turn fast proceeded to establish her imperialistic ownership of Algeria, Tunisia, almost all of Morocco, the mountainous valleys of the Senegal as well as Upper Niger and then further-on even before her other no less fast moving European rivals. France also proceeded to establish her imperial supremacy over the Somaliland and Madagascar too. Germany and Italy were quite late in setting their feet in Africa.

Contextually therefore, at this precise point of our on-going discussions, as well as since, according to Professor S.R. Sharma, the German living-history informs us that **Fredrick the Great** had declared that, "All distant possessions are a burden to the State. A village on the frontier is worth a principality two hundred and fifty miles away." Even indeed **another great history-making leader of Germany, Bismarck** too considered himself, a 'no colony man'. But in spite of the then historically such geopolitical perceptions on the part of famous German leaders when unification of Germany did materialise, she took her own imperialistic initiatives in the vast African continent, and took colonial control of the coastal South-West Africa, north of the Orange River and then, also the Cameroons and the East Africa as well. Italy was able to enter rather very late into the vast and indeed, quite challenging playing-fields of the highly competitive games, then already in full momentum; for the multi-dimensional pursuits of the different kinds of imperialism, on the part the leading European Imperialistic-Players. But even then quite interestingly, Italy was able to imperially get hold of Eritrea, Italian Somaliland, Libya and Abyssinia. But both the Germans and Italians were later forced to abandon their African colonies, consequent to the Great War of 1914-18, and little later due to the war's inevitable fallouts too.

Man's very vast, highly complex and multi-dimensional living-history, as since always being steered by its own inherently powerful attributes, never favours or opposes anyone and in the colonial-era history of Africa we quite revealingly find that though France was able to take imperial possessions of the maximum possible stretches of the African territories, ultimately over there it was the Great Britain, who did finally succeed in somewhat enduringly establishing her unchallenged imperial supremacy, contextually both in colonial reach-cum-power as well as, in the imperial importance over the vast African landmass. But as we delve deeper into these aspects of the Africa's colonial-era living-history, contextually in the larger geopolitical perspectives, we find these historic but also futuristically indeed far-reaching developments then, comparatively going much in favour of the British, as a no surprise at all. We may recall that in specific contexts of the multiple-aspects of the pursuits of imperialism as well as the substantial building-cum-expansion of colonial-empires and, more so in the far-off overseas continents, we have since spelled out in some detail the four essential prerequisites, on the part any concerned ambitious Imperial-Power. And at the same time moreover these four essential prerequisites must always be able to decisively establish a competitive-edge over the other rival Imperial-Powers already since no less actively involved, in any give situation. And hence precisely in these specific perspectives we do find that the British had then since commenced achieving historically far-reaching as well as, indeed the long-term colonial-era history-making successes, in pursuits of the multiple-aspects of their typical British-imperialism over her other European imperial-rivals, since they too all the more then, were already proactively challenging her, both in Africa and Asia as well.

It may interestingly, be relevant at this precise point to recall again very briefly the specific details of the earlier mentioned four essential prerequisites, which do indeed have great historic significance in this regard. And hence, **these mandatory four prerequisites are: (i) Essential strategic culture and thoughts, (ii) Strategically, the imaginative development, well-rounded and dedicated nursing of all aspects of the matters-military, (iii) Possession of the politico-**

strategic skills for the statecraft and all aspects of the governance at home and, (iv) Possibly most successfully, the multi-dimensional conduct of the art of highly complex international games of always the rivalry-infested geopolitics and diplomacy. And hence contextually, while keeping the decisively principal aspects of above mentioned essential four prerequisites in our minds, if we may critically delve deeper into the overall historic processes of the colonisation of Africa and also even earlier, the same historically imperialistic phenomena in even the bigger Asian continent, we may then be able to visualise the principal reasons, as to why finally the British triumphed over her other no less ambitious European rivals in these 'no holds barred', multiple imperialism-driven games for the establishment of colonial-empires with indeed the great and far-reaching positive and negative consequences subsequently, for the colonised peoples, their societies and also their Nation-States in the modern-era world.

In the African continent therefore then it was no wonder that during the bygone 19th and the early 20th centuries the British were able to conclusively establish their colonial possessions of the **highly strategic territories on the coasts of Guinea and in addition, continuously unbroken stretches of lands from the Cape of Good Hope to the no less strategically significant, North African coast-lines of the Mediterranean**. But during those indeed highly consequential history-making times in Africa, the great imperialistic games for the continued expansion as well as a series of new territorial additions to the British-empire, did not interestingly end here. The British acquired the greatly attractive **gold mines of Transvaal** and then, quickly took the unchallenged possession of **diamond-mines situated in Kimberley** and thereby their these imperial holdings transformed their African colonies along with their supremacy over the **Union of South Africa**, indeed as the invaluably precious segments of the already fast expanding and consolidating British-Empire.

Then further on, **Rhodesia** was added to the British-Empire as the noted **British Imperialist named Cecil Rhodes** first acquired it and hence, this African territorial part was named as Rhodesia. In addition then, during the Great War of 1914-18, the famous British **top-level military commander named Kitchner**, victoriously conquered the German colonies in South-West and East Africa and added them to the vast British-Empire. The English imperialists further on, proceeded to also take colonial possessions of and, control over the famous **Suez Canal**. Interestingly, the **Suez Canal was designed and constructed with great efforts and engineering skills by the French engineer named Ferdinand de Lesseps and completed in the year 1869, but finally came under the decisive control of the British authorities** since located over there. **The historic nearly 7000 miles long railway-lines from the Cape-Town in the southern tip of Africa all the way to the city of Cairo since constructed with great pioneering efforts by Cecil Rhodes** and on completion provided strategically great infra-structural assets in the African segments of indeed the global British-Empire and even later-on, during the Two World Wars in the last 20th century. Historically as very luckily it turned out to be that, during these two momentously unprecedented World-Wars, indeed truly on the global scale and so devastating, the Union of South Africa too, became a valuable reservoir for the supplies of resources and manpower, for workforce as well as troops to fight under the British military-officers.

The historically far-reaching British colonial conquests in Africa too perhaps conclusively establishes **that the British had excelled in these since ruthlessly competitive and, indeed geostrategic and politico-military games concerning the multiple-pursuits of imperialism almost everywhere, mainly due to their decisive edge in the proactive skills, in the areas of crucial applications of the earlier mentioned four essential prerequisites** in this regard. In fact contextually, the crux of the matter historically remains **that, in both the absolute terms of strategic resources as well as, the politico-military wherewithal the Great Britain was never in a position to command any kinds of vast superiority, over her other rival European Imperial-Powers** in the highly competitive great-games of their ruthless pursuits of the multiple aspects of imperialism, in various continents of the world. But still the great historic fact also remains that, the British decisively excelled over her rivals; in the multiple-pursuits of imperialism as well as building of the vast colonial-empires, the world over. And therefore, it is needless to say that contextually; the crucial key-factors were indeed the Great Britain's unsurpassed geopolitical vision as well as the multi-dimensional expertise with respect to the already earlier discussed, the decisive aspects of the essential four prerequisites in this regard. In these contexts in addition to these, if we may so desire interestingly perhaps, we may even like to relevantly add, the dimensional kind of 'luck-factor' as well.

But in the larger contexts and hence, at this precise point of our ongoing discussions, we may quite appropriately look into perhaps the darkest as well as certainly very inhuman and ruthlessly exploitative side, of those rival European seafaring Imperial-Powers in the pursuits of their 'no holds barred' very arrogant and multi-dimensional games of the imperialism as well as, building of their colonial-empires in the vast African continent. And historically **that darkest side in this regard, had precisely been the humanly nauseating trades in the human beings as then being treated as a tradable commodity or perhaps even as livestock, for the sale and purchase as slaves**. And all these were happening not only during the distantly bygone ancient-era or even during already the past medieval-age, but shockingly rather, during the beginning periods of the fast-paced present modern-era in the 17th, 18th and 19th centuries interestingly, as a kind of 'slave-trafficking' between the continents of Africa and the Americas and at the same time, duly patronized by the same rival European Imperial-Powers, especially when the **Industrial-Revolution** had begun fast advancing and the Europeans claimed themselves to be rather somewhat extraordinarily civilised. This slave-trade as then being practiced by them, where the selected portion of the African black negro-population considered fit for manual-work, became a 'commodity' or a kind of the 'livestock' for trading as slaves in this rather cruel business but nevertheless; this slave-trade then, had become a very profitable and a very monopolized money-earning business for those European imperialists. In a way historically, we may also say that these aspects of the slave-trade as well as, exploitatively employing the slaves for all sorts of labour-work with extreme inconsideration and cruelty too, must have substantially aided and abated the **white race's highly regressive fixations towards their colour-cum-racial prejudices and apartheid, towards all kinds of non-white native populations all over the world**.

In the above precise contexts here, concerning the slave-trade and the slavery, **Jawaharlal Nehru** has written that, "**The story of how these Negroes were brought from Africa is a very sad one. The slave trade began early in the seventeenth century and a regular supply of was kept up till 1863. At first, cargo-boats passing the West African coast—a part still called the 'Slave Coast'—picked up the Africans, whenever they could do so easily, and carried them to America. Among the Africans themselves there was very little slavery: only prisoners of war or debtors were so treated. It was found that this carrying of Africans to America and selling them as slaves was a very profitable business. The slave trade grew, and was subsidized as a business chiefly by the English, the Spanish and the Portuguese. Special ships—slave-traders were built with galleries between decks. In these galleries the unhappy Negroes were made to lie down, all chained up, and each couple fettered together. The voyage across the Atlantic lasted many weeks, sometimes months. During all these weeks and months these Negroes lay in these narrow**

galleries, shackled together, and all the space that was allowed to each of them was five and half feet long by sixteen inches wide. Liverpool became a great city on the foundation of the slave trade. As early as 1713, in the Peace of Utrecht, England extorted from Spain the privilege of carrying slaves between Africa and Spanish America. Even before this, England had supplied slaves to the English territories in America. An attempt was thus made in the eighteenth century to make the Africa-America slave trade an English monopoly. In 1730 Liverpool had fifteen ships engaged in this trade. The number went on growing till 1792 there were 132 ships employed by Liverpool in the slave trade. The early days of the Industrial Revolution led to a great advance in cotton spinning in Lancashire in England, and this led to a demand for more slaves in the United States. **For the cotton used by the Lancashire mills came from the great cotton plantations of the southern States. These cotton plantations were rapidly extended, from slaves were brought over from Africa, and every effort was made to breed Negroes. In 1790 there were 697,000 slaves in the United States. In 1861 the number rose to 4,000,000.** Early in the nineteenth century the British Parliament passed stringent laws against slavery. Other countries in Europe and America followed. But when the slave trade was thus outlawed, Negroes were still carried from Africa to America, with this difference, that the conditions of their journey were far worse. They could not be carried openly, so they hidden away from sight on loose shelves, one on top of the other. Sometimes, an American writer tells us, 'one crowed on to the lap of another, with legs on legs, like riders on a crowded toboggan." (Nehru, 1934, pp. 652 & 653)

Specifically in these contexts, the above quoted lines paint a very painful and cruel, but indeed a factual picture, concerning plight and extreme sufferings inflicted by the white European imperialists upon the captured African Negros for their inhumanly selfish, highly motivated and flourishing slave trade then, since being so much profitable in the Americas. At this precise point of our on-going discussions; interestingly and while considering these aspects of the highly inhuman slave-trade perhaps philosophically, we may even say that the **multiple-aspects of the continuously advancing human-civilisations since the very beginning have always been having indeed two inter-linked sides: one so white and dazzling, while the other one so repulsive as well as equally black too**. We may further even quite relevantly add that, these indeed so mutually contrasting aspects; since always present in human-beings, are perhaps inherently linked with their ongoing evolutionary aspects under the supreme command of the mother-Nature. **Infinite mother-Nature's timelessly ongoing, all-encompassing and universal; even without any beginning or end, and also incomprehensively vast and complex aspects of the endless as well as multiply countless processes of the continuing evolutions are indeed infinite too**.

History since its beginning has been a saga of the multi-dimensionally advancing human-civilisations since their advents with marvellous human achievements as well as the parallel narrations of the human sweat, blood and tears—all put together. So it remains obvious as to where in these contextual templates of the man's complex history, does the highly painful story of the slave-trade and ruthless profiteering by the seafaring European Imperialists fit in, by forcibly involving the African Negros as a sort of trading commodity, during their colonisation processes in the twin American continents? But factually, this historic-story does not end here. We may precisely recall that little earlier in this chapter itself we have since discussed the four-year American Civil-War (1861-65). And during this historic war too these black, no less brave and by then, a sort of naturalized American Negroes, were ruthlessly exploited and unjustifiably differentiated upon, by the white Americans in spite of their superb soldierly attributes and bravery in that war, which subsequently helped in the unification and firm consolidation of the then fledgling USA, in the North American continent.

In the above context, the internationally well-known, the first ever black American and now Retired Chairman of the US Joint Chiefs of Staff as well as, a former US Secretary of State and a former diplomat **General Colin L. Powell** has written that, "A certain ambivalence has always existed among African-Americans about military service. Why should we fight for a country that, for so long, did not fight for us, that in fact denied us our fundamental rights? How could we serve a country where we could not even be served in a restaurant and enjoy the ordinary amenities available to white Americans? Still, whether valued or scorned, welcomed or tolerated, hundreds of thousands of African Americans have served this country from its beginning in Massachusetts, where I was now serving, blacks, free and slave, were inducted into the militia as far back as 1652. During the Revolution (*i.e. eight-year American War of Independence: 1775-1783*), over 5,000 blacks served under General Washington, helping the country gain an independence that they themselves did not enjoy. Nearly 220,000 blacks served in the Union ranks during the Civil War: 37,500 of them died. Blacks were emancipated, but they returned home to suffer bigotry, the rise of Ku Klux Klan, and lynchings. After the Civil War, the Congress authorized four coloured regiments, the 24th and 25th Infantry and the 5th and 10th Cavalry. They became known as the **'Buffalo Soldiers'**, so called by the Indians (*i.e. Red Indians*), according to legend, because of their dark skin, kinky hair, buffalo-pelt coats and courage in battle. The creation of these regiments, however, was no act of racial enlightenment. Washington merely wanted white settlers protected from the Indians as the West was settled. The Buffalo Soldiers were to help white folks acquire and defend land that blacks, for the most part, were not allowed to own.... **Why have blacks, nevertheless, always answered the nation's call? They have done so to exercise their rights as citizens in the one area where it was permitted. They did it because they believed that if they demonstrated equal courage and equal sacrifice in fighting and dying for their country, then equality of opportunity surely must follow**." (*Well, in the above quoted lines from General Colin Powell's autobiography we do find that the General has clearly put across his perceptions in their relevant contexts such as—historically, presently, as well as even futuristically as being undeniably factual, indeed praiseworthy and truly magnificent too.*) (Powell, 1996, pp. 59-60)

Now at this precise point, we should give a brief look into the short and swift historic **Boer-War (1899-1902)**, since fought in the South African region. In these specific contexts, we may recall that in the then 'Dark-Continent' of Africa too, it were again the Spaniards who were first able to touch the uncharted African coastlines. The Spaniards first managed to take possession of the northern coasts of Morocco and shortly afterwards even the Portuguese had succeeded in reaching East Africa and thereby, took control of Angola and in addition, quite sizeable territories of the Eastern Africa as well. By that time, additional groups of the Dutch adventurers too then, took to the sea and managed to reach, all the way the South African coasts for making their first settlements. In the contextual historic perspectives interestingly, all these quite unprecedented developments had since taken place in the African continent in fact; even before the arrival of relatively cool, calculative and shrewd and then, imperially better skilled British into the African continental scene; to play their own kinds of role in this highly competitive great-game of the ruthless imperialism. We may also recall that, we have since already touched upon the interestingly, historic aspects of the multiple imperial-shrewdness of the British little earlier; while briefly discussing as to **how and why the British were able to achieve substantially better success globally, in this bygone colonial-era's multi-dimensional great-games of the highly competitive and rather forced**

imposition of their imperialism in the subjugated and colonised vast territories in different continents of the world.

The historic Boer-War was fought in the territories of South Africa for nearly three years in fact much later, during the years 1899 to 1902 when the imperial Britannia had already commenced almost globally ruling the waves and historically, the unprecedented British-Empire had reached the pinnacle of its imperial grandeur and glory. And hence the mighty imperial British were in no mood to reconcile with the fact that some others had already independently established settlements in South Africa and, were also refusing to accept British imperial supremacy and pay tributes to them and thereby come under the sway of Union Jack. Historically, in fact, the **Boers were settlers of the Dutch origin** and understandably from their points of view, they happened to be fiercely opposed to accepting the British supremacy and live under the British imperial-rule over their territories in South Africa. In course of their since relentless colonial-expansion with their all sorts of highly intoxicating doses of British imperialism when the British reached South Africa they in fact, mentally and even politico-militarily, were not expecting to encounter with any groups of the other European settlers, like the Boers over there. And in any case, since possessed by their highly intoxicating 'imperialism', and also the already multiply successful imperial-British too then, were just not willing to tolerate any peoples like Boers to keep breathing down their stiff-noses. So the die was already cast once and for all, for the final and decisive encounter concerning the imperial supremacy in South Africa. Since centuries, the Boers had already settled in South Africa, and since thev considered that they were in fact, God's own chosen people to be there. This generation-old mindset was in fact, so powerfully embedded in their psyche that they were since ready to quite fearlessly face the British military-might and hence, they fiercely fought two bloody wars with the British forces, before coming to terms.

In the above specific contexts, Ann Kramer has mentioned that, "**The first Europeans to settle in South Africa arrived at the Cape of Good Hope in 1632. They were Dutch merchants and they founded a vital supply post for their ships on the Cape. But as more whites arrived inland, seizing land from the Africans. They called themselves 'Boers' (farmers) and later 'Afrikaners' and they developed their own language, Afrikaans.** In 1795, the British captured the Cape from the Dutch and formed a colony there, the Cape Colony. Unhappy with British rule, some Boers trekked inland, taking land from the Zulu, Xhosa and Ndebele peoples. They created two Boer republics, the Orange Free State and the Transvaal. In the late 1800s there were major discoveries of diamonds and gold—and both British and Boers wanted to profit from them. This led to terrible conflict. **The British were the victors and took over the Boer republics, uniting the country as the Union of South Africa in 1910. The British government in London allowed the South African whites to rule themselves. Blacks formed the majority of the population, but they had few rights.** Black people could not own land outside the crowded reserves allotted to them, and black men needed a pass (written permission) to move about." (Cramer, 2005, p. 24)

If we may care to read in between the lines of the above quoted highly informative paragraph we are indeed be able to notice some very relevant historic facts in this regard, such as about the basic genesis of the Boers and; as to how **a new language, since known as the 'Afrikaans'** got its birth. In addition at the same time, we may also clearly notice as to how the extremely ruthless kinds of apartheid was born and then so forcefully enforced upon the hapless black population, who constituted the vast majority and also already settled since thousands of years and then, imperially demarcated by the victorious British, as the geographical territories of the Union of South Africa. The types of totally unjustifiable and quite nakedly 'colour-cum-racial prejudices and apartheid' based and highly cruel white-minority rule since imperial-British had so shrewdly and craftily imposed, during the years of their unchallenged global imperial sway and glory. And hence; in the first decade of the last 20th century itself, in the Union of South Africa, we certainly find perhaps a typical classic example of their skillful imperial-machinations as well as, their geostrategic and politico-military vision-cum-geopolitical shrewdness too in this regard. How strongly all these cruel aspects of apartheid were already entrenched in the South Africa, since powerfully converted into a kind of solid fortress that even during the present high-tech and fast-moving modern-era, it took years of relentless struggles and sufferings for the South-African black majority population **under the charismatic leadership Nelson Mandela** to achieve their political freedom, social-equality, social-reconciliation and their now much cherished democracy.

In our larger geostrategic and politico-military perspectives; in case of the historic Boer-War I (1899-1902) too, when the war ended and the British came out decisively victorious and then; we notice as to how certain no less momentous aspects then, was since seen to be emerging in this regard—**the first one being specifically concerning the matters-military and hence, obviously the second one being principally about the political as well as matters concerning the subsequent political-governance** of the Union of South Africa. Historically, it may seem to be important that both aspects in fact, have since also thrown-up important lessons for the mankind in today's modern-era, in their own specific domains and hence these too, need to be contextually looked and suitably brought into, at this precise point to further our ongoing discussions. We may again say that the events, their contexts and the different kinds of developing or emerging causative-factors and then their some fallouts/consequences and thereby in-turn, as the continuing phenomena again based-on those fallouts/consequences, the same cycles getting repeated but again in perhaps totally different manners, contexts and results, are all variously influencing and interrelated as well. Hence, **at no point of the man's history we can probably isolate any particularly specific event or a set of events, their different kinds of interrelated causative-factors, their consequential fallouts or any other aspects and then, put them in any kind of perhaps conceptual 'watertight compartment'; for our analytically interpretive discussions in any given specific perspectives**.

Pertaining to the historic Boer-War (1899-1902) **Field Marshal Montgomery** writes that, "**The Boer-War of 1899-1902 was a much larger scale, but it contained many of the features of the previous irregular imperialist wars**. The Boers, hardy farmers and excellent marksmen with Mauser rifles, were unhampered by traditional military methods and they resorted to guerilla warfare on a large scale. They fought as a people-in-arms, determined to preserve their lands and way of life. 85,000 names were inscribed on the Boer commando lists; they were first-class horsemen, with a natural sense of minor tactics; although nearly all were mounted, they fought on foot; they were unexpectedly strong in artillery. Their weakest feature was indiscipline: they disliked being organised and their officers could never count on all the men on the muster-roll being present to go into action. Fighting over their own country they had no problem about supply. The British forces in South Africa in 1899 did not exceed 10,000 men; in small armed they were well-equipped, but the Boers had the better artillery. The British soldiery was inadequately trained to deal with fighters so well armed and as capable as the Boers. As one staff officer wrote: 'We must make the soldier in many cases a fool because we start with the assumption that he is a fool, and gradually teach him that he is thus to regard himself.... **Britain's chief lesson from the war was to be the need for keeping a larger and better trained reserve**, and by 1914 she was in a reasonable position to undertake a major war. Besides the reinforcement a new commander-in-chief arrived: Field Marshal Roberts, a veteran commander of north-west frontier of India. Roberts brought with him as chief of staff a first-class organiser, General Kitchener, and veteran railway engineer, Colonel Girouard. The transport was reorganised; Roberts manoeuvred skillfully; the besieged towns were relieved and the Boers were defeated at Diamond Hill and Belfast. Kruger, the Boer President, fled to Europe; the war seemed virtually over and Kitchener was left to finish it off. The war was in fact far from over. Under an outstanding guerilla leader, Christian de Wet, the Boers carried on an irregular war for nearly two more years. Moving secretly and rapidly they executed highly skilful raids, sabotaging the railways, cutting off British detachments, and evading pursuit. **Kitchener** had repeatedly to call for reinforcements. To strike at the enemy lifeline **Kitchener** instituted the hated system of destroying the farms and interning the civilian population in camps, where inevitably the living conditions were terrible. Even that did not end the Boer resistance. Eventually Kitchener found the solution: he systematically partitioned the country into great areas enclosed by barbed wire. Slowly but thoroughly, British columns drove through each enclosure, and flushed out the enemy. **At the peace of Vereeniging in 1902 generous terms were given—and kept**." (Montgomery, 1983, pp. 266 & 267)

Historically it was indeed a high degree of the British geopolitical as well as military-oriented cool strategic shrewdness and maturity that, they never naively wanted to get bogged-down in some sort of costly and perhaps unacceptably bleeding and also some unconventional pattern of military-stalemate, in spite of the first decisive military victory over the Boers in their own areas and hence resulting in some kind of non-starters on the paths for further pursuits of the multiple-aspects of their highly ambitious imperialism in the Union of South Africa. And thereby consequently, very far-reaching, history-making and indeed highly momentous ideas of the 'peace of Vereeniging in 1902, with very generous terms and then sincerely keeping those terms too', were resorted to by the British in the Union of South Africa.

Now so far as multiple-aspects of the political-matters as well as the then gamut of the political-governance of South Africa are concerned **Jawaharlal Nehru** (page 650) has very concisely written that "Meanwhile at the end of the century (*i.e. 19th century*) England had a little war of her own in South Africa. The discovery of gold in the Boer republic of the Transvaal led to this war in 1899. The Boers fought with amazing courage and perseverance for three years against the leading Power of Europe. They were crushed and had to acknowledge defeat. But soon after the **British (the Liberal Party was then in office) performed a wise and generous deed by offering a full self-government to their recent enemies. A little later the whole of South Africa became a free Dominion of the British Empire**." (*So from then on, the subsequently emerging white-minority populations in the South Africa; were indeed a hotchpotch mixture of the Boers, British and also possibly any other later day European white migrants, having entered into from time to time, and permitted to settle-down over there*).

In course of the present brief discussions pertaining to the various aspects of historic Boer-War (1899-1902), having been fought between the militarily strong and dominatingly seafaring British Imperial-Power, and the Boers initially of the Dutch origin in South Africa, we have highly relevant lines and thoughts, from some of the well-known books as since precisely commenting on this subject too and hence, included them as few of the above given paragraphs; and which in turn, have indeed given us certain valuable inputs for putting the various dimensions of this historic war in a desired perspective, to further facilitate our discussions. Now in our larger perspectives at this precise point, if we may have a critical look into the entire subject-matters specifically pertaining to this war, we may clearly notice that the British then, while having been able to solve their certain and quite pressing empire-related problems and some hick-ups vis-à-vis the Boers in South Africa, with their fairly developed geostrategic and politico-military oriented long-term vision coupled with their typical strategic shrewdness, they were also able to very callously ensure sowing of the highly fertile, but **humanly so unbecoming and demonizing seeds of the 'colour-cum-racial prejudices and the apartheid'**, as indeed universally a great civilisational-negation on the virgin and still evolving socio-cultural-cum-political soils of the then fledgling South African societies.

If historically in a perspective, it remains acceptable that the British did successfully put in action—the multiple-aspects of their indeed well-developed 'essential strategic-culture and thoughts' for achieving theirs, a kind of imperial peace and understanding with the Boers, in the larger long-term interests of the then firmly consolidating rather, already a kind of the unprecedentedly global British-Empire as well as, for furtherance of the multiple-aspects of their highly ambitious imperialism; then at the same time in course of all these, it is also all the more true that, the British as being totally uncaring for certain inevitable and extremely damaging fallouts/consequences, and thereby ceaselessly driven by their highly arrogant imperialism, they just did not hesitate to herald the so much suffocating and humanly negating demonic aspects of their 'colour-cum-racial prejudices and apartheid', over the South African societies. We have since already brought the historic reality into our conceptual framework that, the very old multiple-aspects of imperialism are indeed a kind of by-products of all the more ages-old and in certain ways still lingering, the multiple-aspects of feudalism in almost all the societies in the world, and at the same time wherever they happen to exist and thrive together, they are always mutually supporting and nursing too. Another aspect that comes to light is the 'colour-cum-racially prejudices and the apartheid'. These demonic colour-cum-racial fixated prejudices and the apartheid have many dimensions and also, numerous ways of their multiple manifestations—on the overall socially structured mosaic of any particular multi-racial as well as the multi-coloured society. It may very strongly and also, all pervasively be seen to be multiply manifesting as well as very arrogantly operating; at that particular society's socio-cultural, socio-political and also socio-psychological levels. These are various kinds of indeed very unbecoming, dehumanizing and suffocating systems of responses and interactions: between different groups and constituents existing, and rather perforce made to live together within a larger social framework, duly carved-out as a distinct political entity as since the man's living-history little earlier, has shown us onto the social canvas of the Union of South Africa.

Though for the our present discussions sake, we have clubbed-in together, otherwise the two somewhat distinct segments of the multiple-aspects of both, the colour-cum-racial prejudices and then, those of the apartheid; but in fact we may have to bear in mind that these indeed happen to be two different and hence, very age-old sociological phenomena since historically situations-based and rather always, overwhelmingly manifesting and operating in the life of man in his any particular society. In fact in these contexts presently in hand, the colour-cum-racial prejudices are commonly identified as appropriately indeed, very criminally illegal and totally unacceptable aspects of racialism; whereas the similarly unbecoming and no less demonic, aspects of the apartheid are more complex, socially much more deep-rooted and having numerously more dimensions of its all the more overwhelmingly much stronger manifestations. Hence, in order to let the multiple-aspects of specifically the 'colour-cum-racial prejudices' to crop-up and endure, any particular society has to be multi-racial and in addition; it has to have peoples with different colours of their skins, such as since generally seen in the world—black, white, somewhat yellow and then even of racially mixed kinds of races.

Whereas the no less humanly unbecoming multiple-aspects of the **apartheid can even take**

deep and strong roots in a society, composed of the peoples of the same race as well as, having identical background and obviously therefore, the aspects of apartheid within a single racial society are generally seen to be much more complex, strongly entrenched, highly regressive and also, quite overwhelmingly capable of maintaining quite a vice-like tenacious grip, with its multiple tentacles over almost the whole of the concerned society in this regard. In other words, the different kinds of manifestations of any situations-based multiple-aspects of the apartheid; as generally found to be entrenched and thriving since ages, in any particular society that they in turn, are able to affect and substantially distort the thought-processes, mindsets, almost the entire gamut of the socio-cultural psychologies and then in turn, even indeed the multiple value-systems of the peoples concerned. And consequently therefore, we do also find that regarding such aspects of any specific case-based apartheid in any concerned society in this regard; no amount of even powerfully sustained multiple social revolutions and reforms in course of time, are rarely able to uproot those kinds of apartheid and thereby, desirably change the concerned society for the better. And obviously therefore as indeed a ready example, comes in our minds, **the very ancient Indian societies in this regard, wherein the highly regressive as well as totally dehumanizing multiple aspects of the tenaciously lingering and since age-old systems of caste, class, creed and also social-hierarchy based, multiple manifestations of the above mentioned aspects of apartheid are still existing as rather, embarrassingly a very big and powerful drag, on the very ancient as well as multi-dimensional Indian social-mosaic, even in the present 21st century too**.

Now coming back to the multiple-aspects of the 'colour-cum-racial prejudices and the apartheid', since as specifically pertaining to South Africa we do notice that though it is true that, after a kind of successful conclusion of the Boer-War, as the Union of South Africa was then geopolitically included as a free dominion within the larger framework of the then British-Empire; indeed a system of multi-party democracy too was heralded over there, as its multiply viable and enduring systems for the political governance. But historically, it is needless to say that possibly any kind of the 'rosy-picture' in this regard, just ended there itself. The obvious reason being that in a way, the above defined and since already discussed, the highly dehumanizing, demonically regressive as well as socially, extremely draconian multiple-aspects of the 'colour-cum-racial prejudices and the apartheid' had since already taken firm roots into the multi-racial and multi-coloured South African societies and also deliberately and very callously, made to thrive through even the official and legal tenets over there. It was indeed ironically, a great negation of the human civilisational values in the present modern-era. During those years of apartheid in action, in the last 20th century in South Africa; among the Political Parties there was quite a fanatically powerful one, then known as the **Afrikaner National Party**, which had since already ruthlessly built itself on the all kinds of policies towards the skin-colour based discriminations as well as, the extreme kinds of racial prejudices. And hence in any case their coming to power meant all sorts of the much more sufferings for the black and other non-white populations, as since they were already suffering and inhumanly subjugated, in spite of being substantially the majority segments in the South African societies.

Apartheid means 'apartness' and *The Compact Oxford Dictionary* very precisely spells out its meaning as 'separateness' and hence, defines apartheid as 'the official system of segregation or discrimination on racial grounds'; and then goes on to emphasise as being formerly in force in South Africa, and also identifies its origin as that of 'Afrikaans'. And therefore, it gets irrefutably established that at the roots of all these historically perhaps unprecedented cruelties against the humanity by the fellow human-beings, irrespective of the inherent colour-pigments of their human-skins, and thereby criminally causing and enforcing a nauseating kind of the great civilisational-negation, onto the multi-racial societies in South Africa; were indeed the political leadership since belonging to the Afrikaner National Party. As we go deeper into the historicity of draconian enforcement of the multiply demonic aspects of apartheid in South Africa, by the so-called ruling white political authorities sitting in the corridors of power over there; ironically we may clearly notice that these were in fact, the pivotal-fulcrum of principal agenda for the political-governance then; for the seats of power in the capital-city of South Africa.

Historically, in the above highly momentous contexts precisely; Ann Kramer (page 38) has indeed very valuably written that, "In 1948, when the Afrikaner National Party came to power under Dr Daniel Malan, Mandela was stunned and dismayed. He knew that under the Nationalists conditions for black would deteriorate. Malan's government introduced ('apartness'). Laws were passed that formalized the separation of black from white. Black Africans were forbidden to go into the white-only areas, they could not marry white people, and blacks and whites could not go into the same schools. In 1958, Hedrick Vervwoerd became Prime Minister. He made apartheid even stronger. His parliament passed a new law; the Promotion of Bantu Self-Government. Blacks were moved and resettled in so-called tribal homelands. **Apartheid affected every part of the lives of black people. Africans, Indians (who had originally come from India for work) and some whites resisted, but protests were put-down brutally. Apartheid continued until 1990.**

Apartheid Laws

1949: Prohibition of Mixed Marriages: People of different racial groups were not allowed to marry.

1949: Immorality Act: It was illegal for people of different racial groups to have sexual relations.

1950: Population and Registration Act: South Africans were divided into four racial groups: 'Coloured' (People of mixed race), 'Africans', 'Indians' and 'Whites'.

1950: Group Areas Act: Different Areas were set up in towns for different racial groups.

1953: Reservation of Separate Amenities Act: Blacks and whites were segregated, or kept apart, in all public places."

(Consequent to sum-total of the above; what an overall scenario indeed it was during the times of no-holds-barred apartheid, since then in action in South Africa; thank God, it's all over as a great relief to the entire humanity.*)*.

Well, in the above precise contexts while dealing with the multiply nauseating and draconian negation of almost all the magnificent tenets as well as marvels of human-civilisations since their inceptions, as triggered by the multiple-aspects of 'colour-cum-racial prejudices and the apartheid' since indeed so demonically created by the man himself and that too, during the present modern-era itself. One legendary Indian who comes in mind in this context is Late M.K. Gandhi, since universally now known as **Mahatma Gandhi**. We would be discussing about him and his truly unprecedented and historically far-reaching activities in South Africa along with then his all other massively history-making initiatives and hard work, more appropriately in the subsequent chapters.

The evolutionarily inherent but negative seeds of rather the demonic and strongly lingering aspects of the **'colour-cum-racial prejudices and the apartheid'** too, got fructified fast. It is true that ironically the factual historicity of the multiply-complex manifestations of these prejudices and apartheid cannot be wished away but at the same time, their powerfully overwhelming manifestations are indeed the matters of deep and great civilisational-negation for the man too, and hence cannot ever be denied in any manner. In these contexts, we may also recall educative lines from Professor George Gaylord Simpson's book that one of the greatest achievements of early Christianity and some other religions too, was the clear recognition in principle, the universal brotherhood of man. Hence, the **demonic aspects of colour-cum-racial prejudices and apartheid could even trigger the almost totally unbecoming and also, substantially ruinous negation of the man's otherwise magnificent civilisational manifestations** and then in addition also, of the fundamentally timeless and universal tenets of Christianity as well as of the other prominent religions; since already pronounced as the universal brotherhood of man, as per the universally believed Divine Will of God for the entire humanity.

If that be so and at the same time, since as the mankind without any hesitation always claims to be highly civilised and multiply advanced as well and then, the obvious questions arises as to in man's socio-cultural and socio-political life; where can the **all sorts of lingering aspects of the feudalism, imperialism and the colour-cum-racial prejudices and the apartheid**, can ever find a place and survive; in the otherwise indeed remarkable manifestations of the magnificence of always multiply advancing human-civilisations and, the man's societies all around the world? These indeed being very weighty questions and, along with also the many other kinds of extremely disturbing questions, since as sorts of rather lingering pin-prick ones are always found to be embarrassingly staring into the man's eyes and, paradoxically at the same time most of them are his own creations as well as unfortunately, the inevitable products of his own various kinds of deliberate or otherwise situations-based, omissions and commissions from time to time as his great civilisational betrayals and at the same time, as being part of his own highly complex and multi-dimensional living-history in its continuing course of the making, onto the infinite 'canvas of time' in the world. We may recall that in course of our continuing discussions earlier on a number of occasions we have rather conclusively come to the point that, man's living-history since continuously in the making and unfailingly too, has always found to be never sparing anyone person, society, Nation-State, or a combination of the some of these or even all of them; as per the man's any individual or collective actions and also the emerging situations. Hence, as per the any situations-based man's thoughts and actions, if the living-history rewards him and then at the same time, depending upon the certain factual facets of his some thoughts and deeds otherwise; the same living-history forces him to face the appropriately inevitable consequences as well. We would be contextually elaborating and also substantiating these interestingly perhaps even appearing to be philosophical matters of conjecture; but otherwise realistically practical too, at the different stages of our on-going discussions in the larger perspectives, as we continue to proceed critically further in this regard.

It is a historic fact that during the unprecedentedly momentous and prolonged-era, in somewhat the later period of the medieval-age, and then also quite substantially spilling into earlier period of the presently ongoing and all the more complex and earlier unforeseen high-tech driven modern-era; of the unique, truly enchanting, multi-dimensional, paradoxically colourful as well as deeply soaked in pleasures, human tears, sweat and blood; the historic-saga of the building of numerously vast and varied overseas colonial-empires in almost all the geographically far-off continents of the world by the earlier already identified and briefly elaborated upon, the ruthlessly rival, militarily strong and aggressive and, no-less seafaring European Imperial-Powers; have been always found to be multiply engaging, for the historians and scholars of the different academic-cum-professional disciplines the world over. The multiple-reasons for such always persisting interests on the part of international historians, scholars on different subjects, thinkers, the military-professionals and writers; are indeed understandable, and logical too in this regard. Hence at the same time, it is also a historic fact that in these bygone era's multiple-contexts concerning all kinds of struggles between the then rival European seafaring Imperial-Powers, for establishing the all-round imperial-supremacy, building of secured colonial empires in the different far-off continents and there-by, furtherance of the multiple-aspects of their different kinds of overwhelming imperialism and, with desires for cornering as much riches as feasible from their colonies **it was ultimately the British who came out victorious and, began commanding sweeping supremacy over her other rivals**. In this chapter itself we would be coming to since these multiple historic aspects again and again in various related contexts, in course of our further on-going discussions.

Specifically in these larger perspectives; **Stewart Bins**, a well-known British TV Series producer and writer has written that, "The history of the civilised world has, in large part, been a history of empires. Greek, Roman, Inca, Han, Islamic; each has left an indelible mark on the modern world. However, it is arguable **that no empire has had a greater presence in history than the British Empire. It touched every continent and every major religion, held domain over hundreds of millions of people and played a significant part in shaping the destinies of at least three score of the world's nation states. Perhaps most significantly of all, it gave to the world the English language—the global lingua franca of commerce, entertainment, science and politics**.... Many motives drive the British in the search for and conquest of their Empire. There were idealistic and altruistic intentions; there were materialistic and megalomaniac imperatives; and there were patriots and zealots as well as healers and missionaries. Similarly, imperial rule had many faces; sometimes it was benign, often it was cruel, occasionally it was malicious usually it was patronizing. For the Empire's subjects there was little or no choice but to accept the governance of London. Many embraced it wholeheartedly, at least for a while; most put up with it as best they could. But eventually, all but a few rejected it and rebelled against it.... The British themselves were and are a multifaceted people, while their empire had as many varieties and complexions as it had islands, territories, dominions, colonies, protectorates and mandates. From the mundane—like its street names and postboxes—to the magnificent—like its art and architecture—the legacy of the Empire will last for many generations. **Britain changed forever the lands it ruled, but Britain was also changed irrevocably, as it moved from being the Anglo-Celtic imperial homeland of the world's most powerful empire, to a multicultural, decentralized middle-ranking nation, firmly anchored to its European neighbours. Of course, Britain is also a paradox, for there are many in Northern Ireland, Scotland and Wales who would see themselves as inappropriately and unacceptably ruled by the English. Whether this paradox will be resolved by the establishment of national parliaments in Belfast, Edinburgh and Cardiff remains to be seen**.... Britain's growing empire, like all empires, was in part an empire by conquest and in part an empire by economic control. Famously, the arteries of the Roman and Inca empires were their extensive road systems. For the British Empire, its lifeblood by carried by sea. Although it extended over vast lands, **it was essentially a maritime empire, forged and maintained by military and merchant fleets, the scale of which the world had not seen before. The British Navy was the envy of the world**. For 350 years, from the defeat of the Spanish Armada in 1588, the Royal Navy was envy of the world. It allowed Britain to impose its will on its empire, as well as to thwart any foreign challenge to its possessions. The British merchant fleet held similar sway over international trade, facilitating an exchange of goods that granted the Empire its economic imperative." (Binns, 2002, pp. 9, 10 & 11)

The above quoted lines indeed historically, do give us highly informative and critically analytical panoramic multiple-views of the very genesis as well as unprecedentedly global fruition of the vast and intercontinental British Empire. The motives behind all these truly far-reaching and indeed history-making multi-dimensional phenomena, as since clearly mentioned above too, had many structural facets. As we may proceed to critically delve deeper into the multiple-making of now the historic and truly global British Empire, where the Sun never used to set and Britannia once happened to be ruling the oceanic-waves in almost all corners of the world; we may also perforce begin to realise that during now, the historically bygone era of since the powerfully overwhelming multiple-aspects of the rivalry-cum-jealousy driven imperialism and thereby, building of the overseas colonial-empires; the then British, both geostrategically and politico-militarily perhaps, might have had certain unique visionary-dimensions into the multiple-aspects of their fairly well-groomed and appropriately-stationed—**'essential strategic-culture and thoughts'**; at all the necessary levels of their since situation-cum-opportunity based, multiple processes of the decision-making and initiative-taking, in the pursuits of their multiply imperialistic-agendas in this regard; which did enable them to ultimately command decisive edge over their other European rivals, in the multiple-fields of these then the great imperialistic games. At the same time in these contexts we may need to bear in our minds that at no stage, the British ever had a kind of any 'cake-walk'; towards the above graphically described historic successes in pursuits of the multiple-aspects of their distinct types of imperialistic-agendas as well as, building of their once so vast and truly intercontinental British-Empire; with all its colourful imperial grandeur, and then also along with its indelible stains of human-exploitation, sufferings, sweat, tears and blood; in its so many imperial-colonies.

In fact, before the British could come to the stage of a kind of the historically somewhat magnificent flowering of their British Empire, they had to first deal with since the threats posed by other no-less ambitiously rival European Imperial-Powers. Among those historic European rivals in this regard, **France then happened to be rather most powerful and also multiply threatening rival to the British. In fact France at every stage did not only challenge the British Naval Fleets in the different seas and oceans across the world but she was also menacingly threatening the British home-country islands too, across the English-Channel**. The French, besides their indeed a 'Grande Armee' and powerful Naval assets and military wherewithal in the seas, they had a historically famous and outstandingly charismatic leader too, in the persona of **Napoleon Bonaparte**, who was known to achieve unimagined surprises as well as, did actually command a highly motivating sway over the entire spectrum of the French military forces and, also the peoples at large in France. Luckily for the British, **Admiral Nelson** could decisively beat the French Fleets in the **naval battle of Trafalgar in 1805** and shortly after this the **Duke of Wellington** could also finally overwhelm the French military, in the historic battles of first the **Spanish Peninsular-War (1808-14)** and then that of **Waterloo in 1815**, since giving ultimate relief to the British Nation. In our larger perspectives specifically in this regard we may need to bear in mind that historically these were since no random kinds of routine wars or clashes of arms, between rather ordinarily rival-cum-localized military powers or comparatively small and feudalistic chieftains in any corner of the world, but instead these were indeed very far-reaching and momentously history-making wars, between the then two great, multiply-powerful and ruthlessly competing leading European seafaring Imperial-Powers across the world and with multiply global consequences, for the mankind and its many both ancient-cum-medieval human-civilisations. And hence as a result too at this precise point here, we are since discussing their multiple historic fallouts as well as, consequentially the numerous kinds of indeed momentously history-making turning-points in the subsequent paths for the on-going advancements of various human-civilisations and also the numerous Nation-States, situated in almost all continents of the world.

We may recall that in the appropriate contexts, we have already since discussed the above mentioned historic clashes of the French and British arms then having taken place in Europe. We have also briefly sketched earlier in this chapter itself as to how these battles between France and the Great Britain then, could be so far-reaching to even change the multiple-paths of indeed the subsequent making of the world's momentous living-history in this regard. In fact, consequent to the British victory over Napoleonic France in the above mentioned historic battles and afterwards; all these did go a long way in paving further-on, the paths for subsequent promotion of the British imperialistic-agendas all over the world, as well as multiple-making of the global British Empire. **We may recall that there were in fact, a number of extensive military clashes having already taken place, even before the wars of Nelson and the Duke of Wellington with Napoleon in Europe; between the French and British Forces in different parts of the world, in the multiple pursuits of their ruthlessly rival and irreconcilably competing and distinctly different aspects of their imperialism**. Historically, the highly checkered period between the 1755 to the 1763 was in fact, the years of extensive clashes the world over, between then imperially competing—the French and British Armed Forces; for ensuring their supremacies in the far-off different continents across the seas. These were historically in fact, a kind of the ruthless battles of naval attrition between the French and British Navies. Man's living-history does inform us that on 9th March 1757; there was a fierce clash between these two very powerful and ruthlessly rival European Imperial Navies in the sea across the Cape of Good Hope, for securing penetration into the vast African continent, via its strategically very important southern entry-point. **A series of such historic British naval victories over the no less powerful French Navy in fact; compelled France to abandon her claims onto Canada, in favour of the British through the Treaty of Paris in 1763 and thereby, finally paving the way for the subsequent decisive global supremacy of the British Empire**.

At this precise point of our ongoing discussions, it may appear quite relevant to recall again that, for any substantially ambitious Imperial-Power in order to be capable of establishing a very vast and global colonial-empire, consisting of numerous colonies located in different continents across the world; she perhaps first must have **certain exclusive kinds of essential prerequisites** properly evolved, cultivated, developed and imbibed and then; effectively stationed in their total national-being and, the gamut of their governance as well as, proactively in the entire decision-making-cum-implementation processes too. We have earlier listed these essential prerequisites for the highly ambitious kinds of imperial-pursuits, concerning any particular powerful imperial-power; as four in numbers and they have since been as follows: (i) Multiple-aspects of the 'essential strategic-culture and thoughts, (ii) Strategically, the imaginative development and, well-rounded and dedicated nursing of the all aspects of matters-military, (iii) Possession of the essential politico-strategic skills for the statecraft at home and finally, (iv) The paramount capabilities to successfully conduct the multi-dimensional art and crafts; concerning the highly complex and rather always challenging 'games' since perforce to be played, in the international arena of the extremely rivalry-infested geopolitics as well as diplomacy.

In these specific perspectives as we may critically go deeper into the so many multiple-facets of since the already identified and briefly discussed, and again here contextually listed the above four prerequisites, they in turn do enable us to indeed have a momentous feel of their enormous complexities and also thereby, their situation-based changing dimensions and challenges as well.

For further facilitation as a suitable example in this regard, we may pause a little and precisely look into the interesting historic-aspects of the entire colonisation process of the New Zealand, since located in around somewhat south-east of Australia in the Pacific by the then imperial British, during the mid-nineteenth century. In fact historically, after their pioneering arrivals on the coasts of New Zealand and then along with their many multiply-proactive initiatives over there which were certainly no easy matters in any case—so much principally aimed at a number of somewhat quite volatile groups of tribals, living for many centuries in the New Zealand's Pacific islands, the British had to be strategically imaginative and also 'think out of the box', to imperially rein-in the various Tribal-Chiefs, while coming face to face with them, if they ever wished to strongly firm-up their footholds and thereby, bring-in those Pacific islands within the folds of then, the unprecedented and globally expanding British Empire. And all these subsequently resulted in a greatly historic turning-point over there, when **in 1840 the British Captain William Hobson was indeed able to imperially very successfully conclude the Treaty of Waitangi with a large number of indigenous Maori Tribal chiefs. Well, with this mid-nineteenth century historic Treaty; which was no small matter, and by this since mutually agreed upon formal instrument of agreement; the British were indeed able to finally ensure the formal assumption of their imperial sovereignty over the far-off but strategically important Pacific Islands of the New Zealand**.

Even then the times were certainly not at all easy for the British since it all resulted in quite a number of years of turmoil and forceful resistance to the British settlers, and hence in turn, the British had to put their own quite reasonably well-groomed skills concerning the already earlier above listed four essential prerequisites, to the extremely challenging tests with respect to the relentlessly multiple-pursuits of their highly ambitious and since sharply-focused aspects of the British imperialism, for finally achieving decisive successes in the multi-dimensional historic phenomena of in a way, the superbly enduring colonisation of the New Zealand. And hence as since also clearly seen today; historically the New Zealand was in fact, going to totally change forever. In a context very befittingly we may also say that, **the entire colonisation process of the New Zealand in the Pacific may also look like perhaps a kind of test-case example with respect to historically, the multiple aspects of the unique and multi-dimensional British imperialism-based, their zeal for indeed the global pursuits of colonialism during now the bygone overseas empire-making era**. But at the same time in each continent and so perhaps in each different Nation-States also located in those far-off continents too, the multiple-pursuits of their numerous imperialistic-agendas by the then imperial British; in fact did assume different facets and dimensions in this regard, and hence would since be seen, as we proceed further specifically on these subject-matters, regarding our continuing discussions.

At this precise point, so far as the multiple aspects of the medieval-ages, building of overseas empires and the matters relating to the dawn of democracies in the West are concerned, perhaps a kind of somewhat suitable necessity is now felt, **so as to have a very quick and fleetingly reviewing glance over the range of our these specific discussions, since so far as already having taken place** to rearrange and bring the concerned principal issues in focus and hence to further facilitate our presently on-going discussions.

We have first discussed the many momentously historic various aspects of the **Mongols' ruthlessly cruel imperialism** and then, unprecedentedly very vast territorial conquests in Asia, Europe and even in some parts of North Africa for the relentless building and almost in continuity, the all possible expansion of their humanly very harsh and **many autonomous Mongolian-Empires**.

And then contextually in our larger perspectives; we have proceeded to touch upon and also briefly discussed the **many historic facets of the imperialism in the medieval-era China, Korea and then in Japan too**, which since as per their times and multiply historic situations, were no less momentously checkered than the rival and seafaring European Imperial-Powers; and their overwhelming zeal for establishing as well as, the expansion of their colonial-empires in different far-off continents of the world. Historically, there was one important development during the bygone Mongolian-era; was the coming of **Marco Polo the great traveller** on the scene and, we have since also discussed multiply many interesting aspects about him and his intercontinental travelogue; later becoming a valuable source for information and guidance.

In these historically specific contexts; we have also noticed that the multiply exploitative and highly ruthless aspects of the **Japanese-imperialism** in fact, as more strategically focused and imperially calculative too, and at the same time; being totally callous towards even some horrible consequences to their colonised nations and other Asian territories. Here and as already noted earlier that, a very special case regarding Korea comes in mind; where then the Korean peoples terribly suffered in many ways, under the prolonged Japanese occupation and hence in this unfortunate process; they lost some of their ancient heritage and, the unique cultural values and ethos as well. In fact, and we have since earlier discussed too; that during those highly checkered medieval-era times in the Far-East; among all the concerned as well as affected nations and their various Monarchies, it was principally Japan alone, which had been able to display and multiply put in-action; their reasonably well-developed and well-stationed, the multiple-aspects of their **'essential strategic-culture and thoughts'**. Historically during those times, and especially much before the advent of certainly the most momentous, far-reaching, history-making and also, the unprecedented **'Industrial-Revolution'** in the world; this was indeed unique and no mean achievement on the part of Japan and the Japanese peoples, during those largely isolated and the medieval-era times in the Far-East.

As we have then begun to delve deeper in the multiple-aspects of the Japanese living-history a decisively important point coming in our minds, is since the deep-rooted and also **powerfully flowering aspects of the Japan's unchecked militarism**. Hence historically, Japan's the then multiple-aspects of their overwhelmingly powerful as well as, also the geostrategically and politico-militarily manifesting; the curious mixture of their **age-old feudalism, imperialism and along with their highly ambitious and also since overseas spilling culture of the ruthless militarism** turned out to be indeed a very **'lethal-combination'**. It is also a fact of the mother-Nature, that no Nation-State can ever be in a position to change her geography and the geographical location. Hence, since not very far-off situated Korea and China from the location of Japan; had to perforce fall victims to the then extremely menacing, multiply manifesting and also ruthlessly advancing multiple-aspects of the above mentioned 'Japanese lethal-combination'. We have also noticed that both geostrategically and politico-militarily, the territorially very large size and geographical vastness of any big Nation-State do remain always in many ways, a big asset as well as a great source of strength. First and foremost, **a very large territorial size having different kinds of geographical features and a combination of the many topographical terrains do always provide very valuable strategic-depth**, from both geopolitical as well as military considerations. Hence, much more than big and vast China, historically Korea unfortunately fell as a total victim of the multiple-aggressions of medieval Japan and the ruthless subjugation caused by her imperialism and, all the rest is now part of the Korean living-history, since already briefly discussed earlier.

Well, in the course of our discussions in the earlier part of this chapter itself, concerning the medieval-era multiple historic upheavals, various kinds of the clashes of arm and also the imperialistic-aggressions in the Far-East largely then among the indigenous Nation-States situated over there; as per the dictates of again the multiply on-going making of since the man's highly complex living-history, simultaneously in almost all corners of the world; we have also discussed the historically far-reaching and highly momentous salient features of all the more historic and tectonically powerful upheavals and competitive-clashes, as then seen to be already taking place in the far-off European continent. Here and as also we have since been critically observing that the depths, dimensions, complexities and the multiple-range of the history-making manifestations of these medieval-era developments and a series of important events then taking place in Europe, were indeed much more far-reaching and momentous too, in comparison to those in the other far-off continents. All these then had many historic dimensions, and hence during those turmoil-ridden times in Europe new kinds of the highly complex and far-reaching twists and turns were seen to be surfacing and multiply influencing in fact, much beyond the Europe itself on the multiple-paths of the ongoing making of the medieval-era European living-history.

Hence as we now proceed further, if on the one side the initial seeds of the then path-breaking democracy in the European continent, principally including the British islands too, were seen to be struggling to sprout through the far-reaching advent of the **Magna-Carta** in the medieval England, then on the other side, the already earlier identified and discussed, the ruthlessly rival, seafaring and even globally highly ambitious European Imperial-Powers too, were momentously fast manifesting and powerfully coming onto the scene. And in these highly historic processes, many new kinds of paths were seen to be paved for the multiple-making of indeed unprecedentedly the new kinds of man's living-history in almost all continents as well as, all corners of the world. The multiple capabilities and overwhelming aspects of the imperialism-cum-greed driven, momentous ambitions on the part of those since already earlier identified and discussed the leading, competitively rival and seafaring European Imperial-Powers, subsequently then multiply began to proceed towards their historically the no-holds-barred power-struggles, first in their homelands and in the seas around, in the continental Europe itself; and then these relentless tussles for power and the unchallenged imperial-supremacy also consequently thereby, seen to be spilling-over and momentously expanding fast, into both Atlantic and the Pacific and naturally then; into the distant continents of Americas, Africa and also the vast Asia as well.

We have critically observed that as **those rival seafaring European Imperial-Powers, since overwhelmingly possessed by their multiply powerful and historically a Circe-kind of the overbearing aspects of their ambitions, military-powers and insatiable types of their extreme greed-driven imperialism**; they first took historic initiatives to launch a number of the highly adventurous, daring and indeed the unprecedented and exploratory sea-voyages across the Atlantic as well as the Pacific, in search of new lands to meet their multiple-needs. History of again their extremely jealous and the rivalry based these certainly path-breaking explorations and discoveries of the new continents and charting of their sea-routes, are indeed highly checkered as well as turmoil and violent clash-ridden too; and we have already discussed them in some brief graphical details earlier, in this chapter itself. In a context historically, the **famous travelogue of Parco Polo** too, was quite instrumental in giving them valuable information as well as impetuous in this regard. All these and much more subsequently; led to historically the extensive and almost all encompassing colonisation of the twin-continents of Americas and then, the other far-off continents of Africa and Asia. **These multi-dimensional historic phenomena gradually turned out to be so**

much far-reaching and path-breaking that, both the numerous countries situated in these far-off continents, as well as those leading seafaring Imperial-Powers themselves subsequently, have indeed been found to be substantially never the same again. And in all these highly competitive games of the multi-faceted imperialism as well as, building of the vast and overseas intercontinental colonial-empires; we have since interestingly also seen that it was principally the British, who ultimately came out to be decisively the victorious and hence, far ahead of others.

We have since seen that it being then principally the Spain and Portugal, who were able to substantially achieve their multiply much deeper imperial-penetration into the vast interiors of South America and hence, could also largely succeed in its ultimate colonisation. The British mainly concentrated onto the North America where they found largely the French, as their principal contenders in these historically far-reaching and the multi-dimensional great-games of the imperialism and colonialism. But, we have also seen that the British, to their good luck and perhaps also due to their more fine-tuned geostrategic and politico-military acumen and the imperialistic shrewdness; that they could outclass the French and hence, finally succeeded in decisively establishing unchallenged sway of their multiply imperialistic supremacy over almost the whole of North American continent; which now principally constitutes of the big and sovereign Nation-States of the United States of America and, the Canada. In fact as we proceed further in our historically critical and contextually, an analytical review in this regard; we have since also quite interestingly observed that the same kinds of perhaps all the more intense rivalry, multiply came into play among those leading and seafaring European Imperial-Powers; in their similar great-games again in the geographically far-off continents of both Africa and the Asia as well.

And here too, due to the various kinds of historic twists and turns and, also since cropping-up of the all kinds of reasons and complexities it was mainly the Great Britain, who again succeeded in decisively achieving their imperialistic edge and, also their colonial-supremacy over her other European imperial-rivals. We may recall that we have since already discussed in some details, the various historic aspects of the highly checkered and multi-dimensional saga of the imperial colonisation of the vast African continent. And in addition, we have also touched upon the similar, but having some historically very important as well as highly consequential dimensions therein concerning the much more heritage-based complexities and then, along with having substantially even quite remotely the ancient-era historic pasts in the multiply vast Asian continent being all the more colourful, complex and highly checkered saga of the extreme rivalry and also intensely struggles-infested colonisation phenomena; as then being triggered by those European Imperial-Powers in Asia. Interestingly, as per the datelines-based structural-framework of the man's living-history we have also at a number of times noticed that, these highly historic phenomena since being mostly multi-dimensional, are in many ways very commonly simultaneous in their manifestations too; in different continents and also in the different locations therein, and they all happened to be mostly interrelated and mutually influencing as well. And hence from here on, as we begin to proceed further historically in these precise contexts we would be seeing more of these multiply imperial power-games and the expansion as well as consolidation of their colonial-empires, in course of our on-going discussions in this regard.

As since the British largely succeeded in finally achieving the decisive imperial-supremacy in these historic great-games, in almost all corners of the world, the unprecedented and truly far-reaching and also the globally intercontinental British-Empire with all kinds of its imperial grandeur began to fast fructify and flourish as well. **On the economic front, multiple international Trades were in fact, the pivotal lifelines of this vast empire**. And hence suitably and also with their typical long-term strategic shrewdness and vision; the **British took historically far-reaching and indeed the well-planned steps in setting-up the East India Company in the year 1600 in India and the Hudson's Bay Company in the year 1670 in Canada, under the Royal Charters, which did substantially succeed in providing great openings and the large fields to profitably operate the various trading and manufacturing activities, to the proven entrepreneurial skills of British businessmen and the industrialists** in this regard. The earlier already talked about since the making and launching of the great and also superbly powerful British Naval Fleets and in addition, also the unsurpassed and so large British Merchant Shipping-Fleets with global reach and endurance did very efficiently ensure indeed the highly profitable and secured global movements of the goods, various kinds of raw-materials and highly influential as well as also the much sought-after British expertise and services too. So the massive and historically unprecedented British-Empire soon began to reach indeed its imperially global dimensions and its American colony began agitating to secede from the empire, leading to the bloody American War of Independence during the years 1775-83 itself resulting in the loss of large American colony to the empire and, which also ensured the subsequent historic birth to the sovereign Nation-State of the United States of America. Loss of this huge American colony, did result in multiply a severe blow to the British-Empire; but the strategically shrewd and visionary **British soon succeeded in establishing highly productive and multi-dimensional trade-links with this big nation the USA; which compensated for this in fact, quite serious British imperial-loss and also greatly facilitated to ensure their continued strategic dominance over the Atlantic, from Canada to the West Indies**.

In fact the multiple aspects of Britain's multi-dimensional, long-term and, also far-reaching overseas imperialistic initiatives and building of the numerous imperial-colonies in the far-off different continents of the world again had multiply strategic as well as, well-planned purposes. We may recall that we have already discussed earlier, these multiple-aspects and dimensions concerning the highly ambitious pursuits of imperialism, by the then European seafaring Imperial-Powers since generally **as known the 'great-games of imperialism'**. We have also noted that during the later periods of the historic medieval-era itself, in these ruthlessly competitive great-games, then having been played by the leading European seafaring Imperial Powers, the British ultimately came out decisively in the lead and thereby, then the comparatively more victorious British imperialists, ultimately succeeded in establishing the historically unprecedented overseas colonial-empires across the world; being much bigger than those of the Britain's other European imperial-rivals. We may now note that **one of the very important purposes of acquiring overseas vast territories for their imperial colonies across the world was to purposefully shift their fast expanding population from Europe and also, to make substantially beneficial to the all concerned, the quite productive and economically paying penal-settlements**.

In the above multiply historic contexts, **Professor SR Sharma** has indeed very interestingly mentioned that, "The surplus population, including a large number of criminals, after being swept into the army and navy (for which there was great demand on account of the various wars) was still available for colonisation abroad. The **epoch-making discovery of Captain Cook (1769-79) made Australia readily available for the purpose. Before the United States became independent America had been used as the 'Andamans' of Great Britain. Australia soon received such a large population of criminals that crime offered no means of livelihood for the immigrants there. Hence the deportation of undesirables from England proved a double blessing: it blessed them that they went, and them that sent**. The well-known words of **St. Bernard of Clairvaux** with reference to the recruits for the Second Crusade may very well be applied to the founders of

Australian colony: 'In the countless multitude you will find few except the utterly wicked and impious, the sacrilegious, homicides, and perjurers, whose departure is a double gain. Europe rejoices to lose them and Palestine to gain them; they are useful in both ways, in their absence from here and their presence there." (Sharma, 1948, p. 388)

All these collectively in turn during the later part of the eighteenth century had in fact been substantially great years for the British imperialists and their massive empire. Hence, as we critically proceed further to delve deeper in these far-reaching and highly historic and checkered saga of now the bygone overseas empire making-era, and indeed the multi-dimensional great-games of the no-holds-barred and perforce impositions of the so many different kinds of imperialism and colonialism in different corners of the world by those rival and seafaring European Imperial-Powers, even during the Post-American War of Independence the British with their typical geopolitical shrewdness as well as strategic-vision, were again able to very daringly turn the numerous threats and challenges into great imperial-opportunities for the British adventurers, colonizers and the settlers. In fact both geostrategically and politico-militarily, they were particularly quick in setting their eyes onto the vast territories into the interiors of Africa, the Far-East and then almost simultaneously towards the huge virgin landmass of Australia, since all these were not earlier occupied by the Europeans.

Contextually during our these ongoing discussions in their specifically precise perspectives, we may need to pause a little and reflect over since the fundamental genesis behind all these highly historic global phenomena and thereby, we find a dramatic increase of population in Europe and also, a kind of the multiply overwhelming imperialistic urge to secure domination of possibly the maximum areas in almost all continents of this world, for satisfying their expanding greed, desires for more imperial-power and then nonetheless, their increasing needs for settling their excessive populations in the newly acquired overseas imperial-colonies. In fact somewhat precisely substantiating these perceptions; **Richard Overy** has written that, "**Between 1815 and 1914 Europe thrust out into the world, impelled by the force of its own industrialization**. (*In fact; the leading, ruthlessly rival, seafaring and highly ambitious European Imperial-Powers had already relentlessly begun making their multiply imperialism-driven geostrategic and politico-military inroads since the sixteenth and seventeenth centuries only, into the interiors of the twin-American continents and side-by-side, very soon into the vast continents of Africa and Asia as well*). Millions of Europeans poured overseas and into Asiatic Russia, seeking and finding new opportunities in the wider world. Between 1880 and 1900 Africa, a continent four times the size of Europe, was parceled out among the European powers. And when in 1898 the United States of America, following Europe's lead, annexed Puerto Rico, the Philippines and other islands of the Pacific, and asserted a controlling voice in Latin American affairs, **it seemed as though Western expansion had secured the domination of the white race over the non-white majority. But expansion carried with it the seeds of its own destruction**. Even before European rivalries plunged the continent into the war of 1914-18 (*i.e. The First World-War*), the beginnings of anti-European reaction were visible in Asia and Africa, and no sooner had the USA occupied the Philippines that they were met by a nationalist uprising... **Today on retrospect we can see that the age of expansive imperialism was a transient phase of history. Nevertheless, it left a lasting European imprint**. The world in 1914 was utterly different from the world of 1815, the tempo of change during the preceding century greater than in the entire millennium before it. **Though industry in 1914 was only beginning to spread beyond Europe and North America, and life in Asia and Africa was still regulated by age-old traditions, the 19th century inaugurated the process of transformation** which dethroned agricultural society as it had existed for thousands of years, replacing it with the urban, industrialized, technocratic society, which continues to spread into even the most remote parts of the globe—for good or ill—more than a century later." (Overy, 2008, pp. 215 & 216)

We do find that contextually in many ways, the above quoted paragraphs are historically quite informative as well as, indeed go a long way in putting certain importantly relevant events and fact in their proper historic perspectives. But, if we may emphatically maintain that by the year 1914 or so, very meagre industrialization had since taken place outside the continents of Europe and the North America it may perhaps mean almost totally ignoring the many kinds of somewhat far-reaching industrial-developments, then already underway fast in the Far-East Pacific Islands of the Japanese Nation-State. **The larger fact remains that the Japan was already far ahead of other Asian or African nations. It is also true that, when the unprecedentedly historic 'Industrial-Revolution' was momentously manifesting fast in the Western Europe as in the beginning, led by the Britain, Japan since located in the Far-East Asia, was not at all part of this loop, but at the same time in this precise context; Japan was not at all prone to be a sort of habitually sleeping or rather continue to be socio-culturally as well as socio-psychologically, living in the past; like the other Asian countries such as then, even like giants China or India.** We may recall at this precise point that we in a rather appropriately relevant context, have already since discussed these historically important aspects of Japan and more so, in the logical comparison with other big nations like China and India in this regard, in the preceding earlier pages of this chapter itself. Japan since the very beginning of the nineteenth century itself, was superbly imaginative and farsighted enough to send suitable young men to the western countries for higher education and valuable training in the then fast expanding subjects of the sciences and technologies; with as far as possible, a particular emphasis most valuably, on the extensive practical and on-the-job training wherever feasible, in the Europe's factories.

Hence, even much before perhaps the inevitable commencement of First World War Japan largely on its own, was indeed able to build very powerful and modern Army and the Naval Fleets, since always ready to go into action if necessary. Why and how Japan got so obsessively engaged in the 20th century's both World Wars one after another, and since fought alongside the Axis-Powers; we would more appropriately be discussing these aspects and, all their related dimensions, in the next seventh chapter. During the bygone times of 18th and 19th centuries, when the leading, rival and seafaring European Imperial-Powers were competitively and fiercely engaged making their vast overseas colonial-empires in the different far-off continents of the world; the Asian Imperial-Japan, in any case as being both socio-culturally as well as geopolitically of the entirely different moorings and also, not being part of the white-race nations; couldn't have been, in that imperially historic and globally being played, the great-games' league. **But in the Far-East and within her domain of influence, historically Japan was found to be no less in her own specific kinds of imperialism-cum-militarism-obsessed and also, since quite ruthless in establishing her colonial-empire.** However, while critically delving deeper into these multi-dimensional aspects of historically, the highly proactive imperialism and colonialism; the multiple-aspects of the great-games since pertaining to those leading, ruthlessly rival and seafaring European Imperial-Powers in this regard; had many more different facets as well as, global dimensions, and naturally encompassing the all far-off continents of the world; which in any case, were far beyond the geostrategic and politico-military capabilities of the then Imperial-Japan. And moreover, when these ruthlessly ambitious European Imperial-Powers and also later joined by now the highly resourceful, superpower USA in course of their multiple great-games, as subsequently reaching the eastern Chinese coastlines and the various islands in the Far-east; Japan was herself found to be cornered and, forcibly put at the receiving-end. We may recall that, we have contextually since earlier, already discussed these aspects in some details, in this chapter itself.

And consequently to these ruthlessly competitive and, since the imperially triggered great-games historically a farsighted and indeed very daring initiatives on the part of British imperialists, military men, colonizers, planters, businessmen, settlers and the missionaries; the British gradually began to flutter their flag in more and more far-off territories of the world and thereby, Britannia's rule over the oceanic-waves globally got further consolidated. Another historic dimension to it, which then **being instrumental to greatly facilitate these indeed far-reaching, history-making and superbly the Britain's imperialism-based and multiply imaginative endeavours was in fact, the historic 'Jewel in the Imperial-Crown of the British-Empire', i.e. INDIA—since acting as an unique and highly valuable strategic-base, sources for supplies and, greatly advantageous centres for both the logistical management as well as military command and control too**. It has indeed been well acknowledged that historically; the vast Indian landmass with its great multi-dimensional and highly complex history, cultural-moorings, the mind-boggling socio-cultural heritage and complexities and also its population acting as the great treasure-houses for military-recruitments, since the remotely ancient-times; have always found a very special place in the British-psyche and hence, also in the multiple living-history of since now the already bygone-era of British-Empire. So we would briefly be taking-up later, the analytically critical-survey of India, and her since making of the historic 'Jewel in the Crown' during the British imperial-era in this regard within the larger frame-work of our geostrategic and politico-military perspectives.

Now in the above multiple-contexts of the global expansion of the then British-Empire, **Stewart Binns** has mentioned that, "The Empire took many forms. There was direct rule and occupation, but by contrast, many areas were governed by treaties with local potentates and princes who would be reminded of their responsibilities by a visit by the Royal Navy in the policy that became known as **'gunboat diplomacy'**. There was also an 'informal' empire. British military influence and economic power meant that many parts of the world that were not formally a part of the Empire fell within the British sphere of influence. This was particularly true in South America and for what called Britain's 'practical protectorate' over the Turkish Empire'. But like the Roman Peace, ***Pax Britannica*** was not consensual. Rule, either paternal or repressive, was enforced. Broadly speaking, British imperialists fell into two philosophical positions: the authoritarians, often those like servicemen and district administrators, who had to deal with the workings of empire at the sharp end, and who felt that their colonial subjects were inferior and should be treated accordingly, and the idealists, often missionaries and educationalists, who felt that the Empire could provide a model of culture and enlightenment for native peoples to aspire to and better themselves by. The two schools of thought fought out their differences over the issue of slavery. To its credit, Britain abolished the slave trade in 1807, and 1808. Sierra Leone in West Africa became a Crown colony and a model of paternalism as it strove to introduce civilisation among the natives. Squadrons of British warships were dispatched to arrest slavers not only off the West African coast, but also in the Indian and Pacific Oceans. However, it proved to be more difficult to abolish the institution of slave labour in the empire. Resistance from the Caribbean plantations in particular meant that it wasn't abolished until 1833. Ironically abolition led to a shortage of plantation workers, resulting in the reinstitution of the practice of indenturing servants—many of them poor Chinese and Indians—who were dispatched to all corners of the Empire as cheap labour. (*Interestingly, we may also recall that in similar contexts, we have earlier briefly discussed the historically many rather humanly* *quite unbecoming, the multiple-aspects of the Black Africans since brought to the North American colonies; as slaves to work in the colonists' farms and the plantations as well.)....* **With the notable of exception India, where British rulers occupied lands previously governed by Hindu princes and Mughal emperors, the greater part of the Empire had to be carved out from lands regarded as 'virgin' territory**. Men like **Fredrick Lugard** in East Africa, **Cecil Rhodes** in southern Africa and **Stamford Raffles** in the Far East, used their tenacity and harnessed their personal ambitions and dreams to carve out new territory for the Empire. **The 'scramble for Africa' was typical**. It became the **'white man's burden', as Kipling called it**, out of its supposedly savage ways along the path towards civilisation.... **In 1897 Queen Victoria celebrated her Diamond Jubilee—an event which with hindsight stands out as the zenith of the British Empire**." (Binns, 2002, pp. 22 & 23)

Hence, the above quoted paragraphs in this regard are quite graphically able to give us a globally macro-level panoramic-view of the very fascinating and also the multiply-complex facets of this truly intercontinental and, indeed the unprecedentedly vast and historic British-Empire. We may recall that we have since earlier contextually noted that in these multi-dimensional, ruthlessly competitive and also numerously clashes-ridden great-games of promoting the distinctly varying multiple-aspects of their extreme and also jealousy-driven imperialism and simultaneously securing as much of the territories as possible in the different continents of the world, for establishing their very ambitious and also highly profitable imperial-colonies, the Great Britain was seldom in a position to command any unchallenged supremacy over her other rivals. But in these contexts both geostrategic and politico-militarily; one verv far-reaching and also somewhat decisive fact has emerged that the British Navy was indeed having multiple-edge in overall qualities and sea-worthiness of their warships, multiple-aspects of the professional-skills and soldierly acumen of its sailors and officers during those times, over its other rival European Navies.

Moreover, the multiply key-fulcrum in these entire rivalry and violent clashes-driven great-games among those leading seafaring European Imperial-Powers, **for securing conclusive and finally the unchallenged, multi-dimensional imperial supremacy in the far-off continents; were principally indeed based of the military dominance of the crucially important—the numerous shipping-lanes across the world's seas and oceans**. And hence needless to say that militarily, the multiple wherewithal of the rival Navies; were indeed holding the keys in this regard. And so in this precisely crucial domain and also, from the so many other relevant considerations the British Navy was then indeed commanding many kinds of encouraging advantages over her rivals. Hence, the decisive naval-superiority, along with other contextually valuable attributes in this regard and since already elaborated earlier, were finally responsible for substantially the ultimate global British successes in these great-games of the multiple aspects of imperialism, and building of their vast and indeed intercontinental British-Empire, and the rest is now all part of the man's living-history, since always perpetually in the making.

Historic 'ills' of India and the Indians

Hence contextually, specifically coming back to the **multiple-aspects of establishing the overseas colonial-empires in the largest continent of Asia**, by the above already identified and, since ruthlessly rival and seafaring European Imperial-Powers, during those historic times; we also notice that the different societies in the vast continent of Asia and the **multi-cultural societies, with the all kinds of their mindboggling complexities in the Indian subcontinent** have indeed been highly feudalistic, since the very ancient of times. And hence, the **socially entrenched,**

extremely complex and fixated aspects of the age-old feudalism in the Indian societies, were then adding some new dimensions and complexities to the entire subsequent phenomena of their colonisation. All these were also especially **for the sake of the riches of India and for commanding imperially as well as geopolitically, the military supremacy first in the Indian subcontinent and then using India as vast and grand strategic bases as well as sources for their multiple-logistics and thereby, to ensure their strategic superiority over the entire Indian Ocean region, the middle-East, the South-East and also simultaneously, along with the Far-East regions** as well all indeed for the multiply vast, varied diversified—the Asian continent.

The vast riches and indeed the magnificently multi-dimensional unique and great civilisational-cum-material resourcefulness of India have always been known to the world, since the very ancient of times. And so it has always been a huge 'mouth-watering cake' for the militarily-oriented Monarchs, rulers as well as invaders since thousands of years, from any possible corner of the world. And then the subsequently numerous developments and events then taking place in India and also often triggered from across her north-western borders had repeatedly been showing that the various Indian rulers, as well as their so much regressively-fixated feudalistic Indian soldiers and their military-commanders, in spite of being very brave and great fighters were almost perpetually lacking in the **crucially important and hence, decisively far-reaching multiple-aspects of the 'essential strategic-culture and thoughts'**. And even today during the presently ongoing as well as superbly advancing high-tech modern-era perhaps most paradoxically for India and the Indians, belonging to the all kinds of cultural-shades and also the various Indian social-cum-professional classes and, along with their rulers/decision-makers have contextually been mostly foolhardy and thereby, regressively lacking in fruitfully cultivating, evolving, developing and imbibing the multiple-aspects of 'essential strategic-culture and thoughts'. So historically the vast and otherwise very rich Indian-subcontinent has always ironically been a very embarrassing-kind of rather openly inviting territories down the ages, for the all kinds of foreign invaders. We may recall that we have since repeatedly taken-up these crucially important aspects in the variously different contexts, from time to time in the earlier chapters too, during our ongoing discussions.

So when historically the rapid decay and decline of the once very powerful and glorious Mughal-Empire, commenced during the second half of the seventeenth century very ironically a kind of a similar pictorially emerging situations also again perhaps began to surface, on the then always so much regressively feudalism-fixated Indian geopolitical horizons, as indeed greatly attractive invitations to this time since already earlier discussed: the leading, ruthlessly rival and also powerfully seafaring: the European Imperial-Powers. Hence, historically now during a very different kind of the later medieval-era times and, in a very different overall scenario **the vast, ancient and so much varied Indian landmass, where the masses have always been very tragically nothing but highly dispensable pawns on a grand chessboard; was again perforce turning into a huge arena for now, the totally different kinds of the multiply imperialistic and no-holds-barred great-games for imperial-subjugation, loot and plunder to begin with and then, also for the earliest building of vast colonial-empires; and then unprecedentedly this time, sponsored from the far-off European continent**. Man's very vast, multi-dimensional and highly complex living-history always had many different paths of its simultaneously continuous making, in different corners of the world and at the same time they mostly remain inter-connected as well as, mutually-influencing too. So during the 15th and 16th centuries, when the European imperialists since belonging to their different European nationalities, were setting their so much history-making feet on the shores of India, via the highly adventurous sea-routes, simultaneously many different kinds of the momentous historic developments were also taking place, back in their overseas homelands located in the far-off European continent.

In the above multiple contexts we also notice that **Jawaharlal Nehru** with his scholarly insight (pp. 270 and 271) has written that, "**Shah Jehan** was contemporary of **Louis XIV of France**, *le Grand Monarque,* and the Thirty Years War was then ravaging central Europe. **As Versailles took shape, the Taj Mahal and the Pearl Mosque grew up in Agra, and Jama Masjid of Delhi and the Dewan-i-Aam and the Diwan-i-Khas in the imperial palace. These lovely buildings with a fairy-like beauty represent the height of Mughal splendour. The Delhi court, with its Peacock Throne, was more magnificent and luxurious than Versailles, but, like Versailles, it rested on a poverty-stricken and exploited people**. There was terrible famine in Gujarat and the Dekhan... Meanwhile the naval power of England was rising and spreading. The only Europeans that **Akbar** knew were the Portuguese. **During his son Jehangir's time the British navy defeated the Portuguese in Indian seas and Sir Thomas Roe, an ambassador of James I of England, presented himself in Jehangir's court in 1615. He succeeded in getting permission to start factories. The Surat factory was started and Madras was founded in 1639. For over 100 years no one in India attached any importance to the British**.... When the Mughal Empire was visibly weakening during **Aurungzeb**'s reign, the British made an organised attempt to increase their possessions in India by war. This was in 1685. Aurungzeb, weak as he was growing and beset by enemies, succeeded in defeating the British. Even before this the French had established footholds in India. **The overflowing energies of Europe were spreading out in India and the east just when the political and economic condition was rapidly declining**... In France **Louis XIV** was still continuing his long reign, laying the seeds of future revolution. In England, the rising middle classes cut of the head of their king, **Cromwell**'s brief-lived republic had flourished, **Charles II** had come and gone, and **James II** had run away. Parliament, representing to a large extent a new mercantile class, had curbed the king and established its supremacy... It was during this period that Aurungzeb succeeded to the throne of the Mughals after a civil war, having imprisoned his own father, Shah Jehan. Only an Akbar might have understood the situation and controlled the new forces that were rising. Perhaps even he could have only postponed the dissolution of his empire unless his curiosity and thirst for knowledge led him to understand the significance of the new techniques that were arising, and the shift in economic conditions that was taking place." (Nehru: *Discovery of India*, 2002)

It is a well-known historic fact and would also be seen while critically glancing over again the highly turmoil-ridden European living-history that indeed traditionally, the French were no less imperially obsessive and overwhelmingly ambitious too than the British in the earlier talked about in some detail and also in various contexts, the ruthlessly rival great-games of the competitively varying, the multiple-aspects of imperialism and building of the overseas colonial-empires. **No wonder that the French as so much in the European continent itself and so in all the other continents of the World, had always been great rival of the British and hence interestingly, during now the bygone overseas empires-building era, we find a somewhat similar kinds of highly momentous and rivalry-ridden scenario then, fast emerging in the Indian subcontinent as well**. But it is also a historic fact that to begin with in India of the 16th and 17th centuries, all these since very obsessively outward-looking Europeans had landed on the Indian shores at different places and during different times, first as traders. As we begin to critically delve into these highly historic times in our contextually larger perspectives; we also very interestingly notice that neither the Indians nor those Europeans, knew much in any detail and depth about each other in advance and obviously therefore, they both were heading gradually for lot many surprises and, also coming face-to-face with the all sorts of rather totally unexpected situations, and hence this perhaps principally happens to be one of the important reasons as to, why and how the subsequent making of the living-history of the British-period Indian subcontinent, turned to be so multi-dimensionally complex, highly-checkered as well as colourful too. So far as the French were concerned they were comparatively late arrivals in India. Even the British landed in India, much after the arrivals of Portuguese and the Dutch but the French reached India even after the British. And needless to say that these imperially obsessed Europeans, since as they were both ruthlessly as well as irreconcilably, the arch rivals and traditionally diehard enemies to each other in the Europe itself, it was never expected of them to being even slightest friendly to each other, in India as well.

It is historically, a different matter that in these rivalry-ridden great-games of their multiply competitive aspects of imperialism; it was ultimately the Great Britain, who came out victorious in India too, due to various rather geopolitically overriding reasons; and these have since already been discussed in various contexts, earlier in this chapter itself. And these Europeans as duly attracted since by the age-old and well-known names and fame concerning the riches, marvellous heritage and quite unique and even valuable strategic resources of the Indian subcontinent and thereby, after first arriving in succession as highly ambitious traders to begin with, they essentially needed India's shore-based port-facilities and, lands for setting-up their warehouses and the factories. **To indeed their super diplomatic as well as the various kinds of human-level interactive skills; it was to their historically far-reaching credit that, they successfully managed to fast secure these strategic facilities; for their both commercially as well as imperially, highly ambitious future activities and also subsequently then were able to gradually make multiply substantial inroads, into the vast Indian heartlands**. For their all these multiply commercial and imperialistic activities; they had already very imaginatively planned and established now the interestingly historic, **their respective 'East India Companies' in the Indian subcontinent**. Hence, the **French *Compagnie des Indes*** though late, turned out to be quite comparable to the Dutch and British East India Companies. And historically in these precise contexts; that these rival and also imperialistically obsessed Europeans again came face to face with each other in India too, with no love lost amongst themselves!

Overall collectively at the macro-level, it was all fast developing into a highly momentous as well as, extremely turmoil-ridden scenario in the Indian subcontinent. As we may critically view this scenario, then prevailing in the expansive Indian subcontinent in its historic perspectives, we may also rather be able to very clearly see the multiply-historic enormity of the all kinds of its emerging as well as, fast-moving and multi-dimensional complexities as well. Broadly in these historically precise perspectives, concerning these unprecedented, truly intercontinental and multiply-competitive imperialistic great-games, since then being played by those leading, extremely rival and seafaring European Imperial-Powers; one crucially important point that may need to be borne in mind remains that, overall the multi-dimensionally situational frame-work then prevailing in the Indian subcontinent in this regard, was in fact totally different and much more challenging, than those in the American and African continents. Now, so far as the historic aspects of these multiply imperialism-driven great-games, then being played in India, by those rival seafaring European Imperial-Powers the Dutch and the Portuguese sooner or later, became out of the scene, except the Portuguese in fact, were able to retain their footholds on the port-town of Goa and its adjacent areas, situated in the western coast of India. But, so much unlike the contextually earlier discussed continents of Americas and the Africa, India has been one of the oldest and very much vibrantly continuing ancient civilisations of the world.

In these historically precise contexts since thousands of years, multiply different aspects of the much more feudalistic imperialism, whether indigenous or foreign wars and periodical invasions down the centuries and then, building of different empires, establishment of different Monarchies and Royal-Dynasties from time-to-time, were in fact nothing new for India and the Indians. So for the remaining two and since the ruthlessly competing European Imperial-Powers, the France and Britain, the challenges facing them in the Indian subcontinent, were in fact much more complex and multi-dimensional as well. Interestingly in a way, finally there appeared to be three main contenders, left in the field in this regard. Since the once mighty Mughal-Empire was fast getting weak and disintegrating the British, the French and then, a couple of imperially feudalistic Indian Chieftains, were left in the vast Indian arena with a rich and highly checkered past to both geostrategically and the politico-militarily, decide among themselves, for finally be able to establish one's imperial-supremacy. **Among the Indian imperial chieftains there were the Nizam of Hyderabad, the Maratha Confederacy, Sikh-Empire headed by Maharajah Ranjit Singh in the north and finally, Hyder Ali in the South and also each having different kinds of their imperialistic ambitions**. And moreover interestingly these Indian imperial chieftains were no friends to each other and hence, they all had their own kinds of highly rivalry-ridden historic past in India. At the same time as we proceed further to critically delve deeper, we do also interestingly notice that, there was another historic dimension to it in the sense that these imperial and highly feudalistic chieftains were in fact, having very limited and short-term vision, with respect to their respective imperialistic ambitions unlike the British and the French, who were much more strategically visionary as well as, were always clearly focused with respect to the long-term multiple-aspects of their since overseas, highly ambitious and intercontinental imperialistic-agendas.

In addition historically in these precise contexts in the Indian subcontinent then, one more new and highly lucrative dimension since as being great opportunities, along with the fast disintegration of the Mughal-Empire was seen to be emerging for the imperialistically obsessed French and the British, and they were very craftily quick to grab them, to their great commercial as well as, strategic advantages too. It was then soon turning out to be broadly multi-dimensional as well as highly checkered scenario then, as fast manifesting into the much later medieval-era Indian subcontinent. In these contextually precise perspectives of during those highly momentous times in India then; even regarding the traditional rivalry between British and the French, a new kind of specifically situation-based development too, was seen to be taking place in the sense that in the eastern India, if the **British succeeded in making the then Calcutta as a hub** of their highly lucrative multiple-trading as well as, for the governance of their acquired territories in India, along with their other imperially ambitious purposes then interestingly; the **French too established and developed Chandernagore in Bengal, with similar motives**. Further at the earliest opportunity, as the **Imperial-British could acquire Madras as their multiply strategic-base on the eastern-coast** in the southern side of the Indian peninsula; and then interestingly in an identical manner, the **French too succeeded in getting hold of the nearby Pondicherry** (*now since called Puducherry*), for their in fact multiply rival strategic-base in the South India. And in these precise historic contexts, even if the British finally succeeded in establishing their unchallenged imperial-supremacy over almost whole of the vast Indian subcontinent; **the French were able to retain their imperial hold over both Chandernagore in Bengal, as well as the Pondicherry onto the southern Indian coast in the east, and effectively kept them outside the then imperial domain of the British India**.

In the larger geostrategic and politico-military perspectives; we do also notice that with the then arrivals of these rival and seafaring European Imperial-Powers into the Indian subcontinent and along with their equally rival, obsessively ambitious as well as long-term and multiply imperialistic-cum-commercial agendas; India's ruling princes and their military-chiefs also began to have some mesmerizing glimpses, of so far totally unknown to them and hence at the same time; **a new kind of better disciplined, more regimented, militarily better drilled—aspects of the European soldiering, their armaments and weapon-platforms**. Since the early ancient-ages itself, India has had a long historic-records and traditions of the many kinds of regressively lingering aspects of feudalism, highly obsessive imperialism, warfare, chivalry and also, the traditionally conservative aspects of a military-culture. Though along with all these, it is also undeniably true that since ages as India and the Indians being conservatively and perhaps even mindlessly being resistant to much otherwise essential changes as per the powerful signals of the man's dynamic living-history, since perpetually in the making; India has not so far sadly been able to meaningfully evolve, cultivate, develop and imbibe the universally timeless **multiple-aspects of the 'essential strategic-culture and thoughts'**. We may recall that we have since discussed these aspects in their various contexts, in this as well as earlier chapters too.

Now, those rival and always warring Indian ruling-princes and other feudalistic lords, obviously began eyeing for the European aspects of soldiering, better formulated and more precise systems of military-training as well as, armaments and other military-gear for getting hold of them with the sole aim to further their own rivalry-based imperialistic purposes and then, to settle issues with their other indigenous rivals. **Whereas interestingly again, these rival European seafaring Imperial-Powers, i.e. principally the British and French; were viewing these fast developing situations in the feudal-India of those times to begin with as in fact, God-Sent opportunities as then a highly lucrative All-India military-bazaar, with many kinds of very advantageous opportunities for themselves.** Hence in retrospect now, we may even say that during those times in the post Mughal-era medieval-age, historically altogether new and totally unprecedented kinds of the highly complex situations, were fast developing into since this always greatly prized Indian subcontinent for already the earlier discussed but all the more challenging kinds of the extremely complex and also multiply imperialism-driven great-games, to secure the imperial-supremacy over possibly the entire length and breadth of this subcontinent. In this complex, fast emerging and multiply feudalism-cum-imperialism soaked scenario, during the ending periods of the medieval-era India; since a new dimension was frequently being seen by both the French and British and, that being the clashes among the eligible contenders for successions to the throne, in almost all the leading and big Indian princely States. And here too, they wanted to enter, both very craftily as well as advantageously in support of one or the other contender and thereby, to be able to multiply firm-up further, their advancing foot-holds in this highly prized and historic subcontinent. Meanwhile, as some years went by and **Robert Clive** managed to become British East India Company's Governor-General with his seat in Calcutta and interestingly, **Joseph Dupleix** too had similarly become the highly ambitious French Governor in Pondicherry. And at the same time, their East India Companies too were already flourishing in India, in their own respective ways. **If the French and British had traditionally fought directly in the European continent for years, due to the various kinds of their rivalry-cum-jealousy-ridden historic reasons, till the peace was restored in Europe in the Indian subcontinent they many a times, indirectly fought their something like proxy-wars, through the rival Indian princely-rulers.**

In course of these and with respect to the multiple-aspects of the matters-military through these rather proxy-wars, they were also able to try and fine-tune their military thoughts, doctrine and skills of soldiering and the professional tenets of their respective military-leadership in war.

The last historically noted Mughal-Emperor Aurungzeb had died in the year 1707 and then, during almost for the one hundred years; it was in fact a century of extreme turmoil and, uncontrolled feudalistic-cum-short-sighted-oriented clashes between many princely States for some or other kinds of supremacy over each other or even over some disputed territories. And as per the above discussions, obviously we do also notice that those European seafaring Imperial-Powers and their lieutenants in India; had already since begun very profitably fishing in highly troubled waters, of the nineteenth century as well as mostly as usual, the highly regressive and divided India and along with, her strategically naïve and foolishly feudalistic ruling-princes. It is true at the same time that, seeds of the rather quite fast decay of once the mighty and historically so glorious Mughal-Empire in India had already been sown during Aurungzeb's time itself, and also mainly due to his own faulty styles and some short-sighted systems of governance and thereby those seeds had begun fast sprouting when, the aging and disenchanted Aurungzeb was still on the Mughal imperial-throne. In this case, **the dynamic, highly mobile and successful military adventures of the great-Maratha military-strategist and ruler Shivaji, against the Mughal-rule and also for establishing his imperial-supremacy over the whole of Deccan and much more and all these in-turn, were since decisively instrumental in establishing a mighty Maratha-Empire, in quite major portions of the South and Western India**. Hence the back of Mughal-Empire was almost broken and the consequent demise of Aurungzeb too, was accelerated. Meanwhile, military-hollowness of the Nizam of Hyderabad in many ways stood very pathetically exposed, both to the French and British and in addition, also to the Hyderabad's other Indian feudalistic rivals. In sum-total both the geostrategic and politico-militarily, it was all a very queer scenario principally for the imperially ambitious Europeans since already powerfully entrenched in India and thereby then, opening many multiple-kinds of the new and highly inviting opportunities to them.

In the above multiple-contexts, while elaborating these highly historic scenario then prevailing in India, **Jawaharlal Nehru** has written that, "The Mughal-Empire rapidly fell to pieces and imperial viceroys and governors began to function as semi-independent rulers, though so great was the prestige of the descendent of the Mughals in Delhi that a formal allegiance was paid to him even when he was powerless and a prisoner of others. These satrapies had no real power of importance, except in so far as they helped or hindered the main protagonists for power. The **Nizam of Hyderabad**, by virtue of the strategic position of his state in the south, appeared to have a certain importance in the beginning. But it soon transpired that this importance was entirely fictitious and the state was 'straw stuffed and held upright' by external forces. It showed a peculiar capacity for duplicity and for profiting by the misfortune of others while avoiding all risk and dangers. **Sir John Shore** described it as 'incorrigibly depraved, devoid of energy....consequently liable to sink into vassalage.' The Marathas looked upon the Nizam as one of their subordinate chieftains paying tribute to them. Any attempt by him to avoid this and to show independence met with swift retribution and the Marathas put to flight his feeble and none-too-brave army. He took refuge under the protective wings of the growing powers of the British East India Company and survived as a State because of the vassalage. Indeed the Hyderabad state enlarged its area considerably, without any remarkable effort on its part by the British victory over **Tipu Sultan** of Mysore.... The real protagonists for power in India during the eighteenth century were four: two of these were Indian and two foreign. **The Indians were the Marathas and Haidar Ali and his son Tipu Sultan in the south; the foreigners were the British and the French.... In Bengal, Clive, by promoting treason and forgery and with very little fighting, had won the battle of Plassey in 1757, a date which is sometimes said to mark the beginning of the British Empire in India**.... In south India, the struggle between the English and the French, a part of the world struggle between the two, ended in the triumph of the English, and the French were almost eliminated from India... With the elimination of the French power from India, three contestants for supremacy remained—the Maratha Confederacy, Haidar Ali in the south, and the British. In spite of their victory at **Plassey** and their spreading out over Bengal and Bihar, few, if any, people in India then looked upon the British as a dominant power, destined to rule over the whole of India." (Nehru, 2002, pp. 273-275)

Historically in a specific perspective, the above quoted paragraphs do indeed give us a broader glimpse of the state of affairs, extent of the extremely unsettled situation as well as, the geopolitical-turmoil then so menacingly prevailing over the post Mughal-period Indian subcontinent and along with that, also the depth and range of the power-vacuum then prevailing in India. As also hinted earlier and since quite unlike the other continents of the Americas and the Africa in precisely the multiple-contexts of already earlier discussed aspects of the **overwhelmingly imperialism-driven great-games**, as then almost being globally played by those leading seafaring European Imperial-Powers; the multiply feudalism-fixated overall geopolitical situation in the ancient and always highly complex Indian subcontinent, happened to be totally different. Here **a number of the no less imperialistic as well as, politico-militarily quite powerful and the age-old indigenous princely-states too, happened to be traditionally warlike and at times, even irrationally ambitious as well**. They were obviously in no mood to give any easy time to the British or the French and hence; they too were having their own very different kinds of imperialistic ambitions and, wanted to even craftily enter in a rather their narrow, localized and, shortsighted manner into this multiply imperialism-driven great-games, as then since the Indian subcontinent itself was made an arena and thereby, imperially profit from the power-vacuum since created with the demise of the Mughals. But these Indian princely-states had their own age-old and regressively lingering fault-lines in the entire geostrategic and politico-military framework of their mostly distorted thought-processes, then always fixatedly prevailing amongst them, which were based upon the all kinds of mistrusts and suspicions; since always being triggered so damagingly by their all kinds of jealousy, selfishness and the typical short-sightedness, as well as their traditional lack of the any fruitful understanding of the **paramount and always crucial multiple-aspects of the 'essential strategic-culture and thoughts'**.

At this precise point and in our larger perspectives; it may appear rather relevant to note that **historically, Afghanistan since ancient of times has always been part of the Indian subcontinent**. But again since the ages, precisely this regressively lingering lack of the multiple-aspects of 'essential strategic-culture and thoughts' have been always principally instrumental in causing many kinds of the crippling damages in many spheres, to the Indians and their Nation-State India and thereby this singularly the highly damaging lacuna has also been responsible towards the failure of the Indians, to be able to successfully cultivate obviously the **multiply-valuable 'unity of purpose and thoughts'** amongst themselves. If we may recall the multiple-aspects of our strategically critical discussions concerning the ancient Indian-civilisation in the fourth chapter and then subsequently, on the Indian subcontinent-related multiple-aspects of the matters-military in the fifth chapter, and then in these specific contexts here, we may visualise as **to why and how, India has been both continuously as well as pathetically under some kind of the foreign-power sponsored subjugation and its rather suppressive imperial-rule for almost last one thousand years**. In some kind of the similarly quite damaging chain of historic consequences, the **extremely savage military-invasion of an Afghan warlord Nadir Shah in the year 1739**, gave a death blow to the already crumbling Mughal seat of imperial-power in Delhi and at the same time, it also ensured annexation of Afghanistan from India. And the British then, though perhaps apparently as only traders in India, were watching and studying these highly unsettling developments, with their own very valuable futuristic vision, since always fine-tuned by their already reasonably well-groomed multiple-aspects of the 'essential strategic-culture and thoughts'; which would in any case, was going to be highly beneficial for their multiply imperialistic agendas in future, in the Indian subcontinent. As we may critically delve deeper into these historic aspects, we may also notice them as some different segments in the already earlier talked about, **relentless and the overwhelmingly imperialism-driven 'great-games'; principally intentioned towards acquiring the unchallenged imperial-supremacy as well as, building of the vast colonial-empires in the so lucrative Indian subcontinent.**

John Keay has written that, "The British would often think of their conquests in India as fortuitous. It gratified a cherished conceit about the Englishman's amateurish innocence and it obviated the need to confront awkward questions—like how such aggression could be justified. **Clive** himself, normally neither temperate nor cautious, would agonise long and hard over whether to assume what he called 'the sovereignty over Bengal'. Twenty years later the polymath, **Sir William Jones would marvel at how Bengal had, like the over-ripe mango, 'fallen into England's lap while she was sleeping'**. Even Warren Hastings, Jone's patron and the first British Governor-General of India, would shy from the idea of the all-India dominion with whose foundation he is rightly credited; it was something 'which I may not mention without adding that it is what I never wish to see'... Picking up such breezy disclaimers, historians of the British Raj have generally explained its triumph less in terms of the push of conquest and more in terms of the pull of chaos. The Company was 'sucked into' the 'power-vacuum' left by the declining Mughal Empire. No native regime rose to fill this 'black hole', authority fragmented, and economic decline threatened. The lawlessness of Afghans, Marathas and other 'warlords' brought cries for protection which necessitated an elaborate British-sponsored system of 'ring-fencing' based on subsidiary alliances which would embrace most of the subcontinent. Territorial acquisitions and local forms of resistance, as also administrative and fiscal anomalies, invited: political, judicial and revenue 'settlements'. Suggestive of a pillowed repose, these 'settlements' were supposedly designed to restore a traditional order which, while advantageous to the British regime, would bring security and prosperity to all.... Such views have since been revised. **It is argued that the 'chaos' and the 'vacuum' were in part the Company's own making, in part an invention of its apologists, and in part the result of the misreading of India's history....Regional regimes were not inconsistent with India's political and cultural traditions**. And powerful successor states, like those of Bengal, Hyderabad (the Nizam), Pune (the Peshwa), Mysore (Haider Ali) and Lahore (the Sikhs), were waxing impressively beneath the vault of Mughal authority until extinguished or suborned by the expansionist ambitions of the Company.... **British rapacity attracted much less attention than it did in England. Under the later Mughals as under their 'Great' predecessors, power and prestige depended on conquests and access to revenue. Conversely conquests and access to revenue depended on power and prestige. Greed was as much the essence of government as it was of commerce**. Merchants who became rulers happily adjusted their sights without experiencing any great conflict of interests. And although **it would be hard to prove that either the Company or its servants espoused loftier ideals, it would also be hard to prove that any of their Indian rivals were motivated otherwise.... As Clive would notoriously aver to a parliamentary committee which would eventually investigate his conduct in India, the opportunities which awaited him after Plassey had been almost unimaginable.** ***Clive has said on record*:**

'A great prince was dependent on my pleasure; an opulent city lay at my mercy; its richest bankers bid against one another for my smiles; I walked through vaults which were thrown open to me alone, piled on either side with gold and jewels. Mr. Chairman, at this moment I stand astonished at my own moderation." (John Keay, 2000, pp.383, 384, & 386)

The above quoted lines do give us a very revealing and indeed historically, the multiply complex picture of a rather surprising phenomenon of the gradual and a kind of the perplexing success of British imperial-subjugation of the Indian subcontinent. No wonder that there are indeed numerous kinds of varying view-points and perceptions both in India and the abroad, regarding the very structure of the historic authenticities of this phenomenon; which ultimately did **phenomenally transform India, into the then a 'Jewel in the British Crown'**. In this specific context, and more so at this precise point of our continuing discussions, Robert Clive's above recorded statement, then submitted before a parliamentary committee of the British Parliament, is a particularly revealing statement with its own historicity in this regard. But at the same time, it is also true that during now the bygone historic-era of the then, great-games of the so much ambitious and also the relentlessly imperialism-driven building of the overseas colonial-empires by the seafaring and rival European Imperial-Powers, since already discussed in some details earlier, the **British were astute and farsighted too, in these multiply challenging overseas great-games of imperialism and in addition to that they were also superbly skilled in practicing the situation-based and, sort of hidden meanings of the old proverbs—of 'divide and rule' as well as interestingly that, 'everything is fair in love and war'**.

And hence, as we may critically delve deeper, concerning the historically multiple-aspects of a kind of the fast but, gradually triumphant establishment of imperially the British colonial-rule in India we may also observe that the British, with their typical long-term strategic-vision, did indeed find the vast Indian subcontinent, with its multiply such prolonged panorama of the historic past and the highly complex and multi-dimensional diversities as lucrative for very craftily putting into a consistent as well as judiciously well-calculative practice; their geostrategic and politico-military resources-cum-acumen and, also the highly flexible imperial-skills in this regard. In these specific contexts, we have since observed that once the powerful and marvelously expansive historic Maratha-Empire; was able to command indeed the zenith of its imperial-glory, during the reign of **illustrious Maratha ruler and brilliant military strategist Shivaji**, then being firmly on his imperial-throne. But after his demise, many kinds of dissensions, jealousies and internal-rivalries soon began to crop-up, as dangerous fault-lines within the Maratha-Empire itself. One may even say that in course of few decades only, it took the form of a great betrayal of the historically glorious legacy of the great Shivaji by the so strategically naïve, short-sighted badly divided Maratha feudal Chieftains. And the imperial British in India then, were waiting to grab this opportunity and began to militarily defeat these internally divided Maratha chieftains one by one and, forced them to accept their alien imperial-supremacy.

After the strategically visionary **Hyder Ali's** demisehis **competent and great warrior son Tipu Sultan**, was also initially proving to be militarily quite powerful to the imperial-army of the British East India Company and its British commanders. But he too was treacherously betrayed by few of his own mean associates, since very craftily conspired with the British against him and, this resulted in his killing by those very traitors. Like the two above talked about Indian imperial-seats of power, something on the similar lines happened to the unfortunate fate of once the mighty Sikh-Empire in the north, with Lahore then as its seat of imperial-power. Historically this once vast Sikh-Empire in Punjab as well as the north-western India, was in fact founded and expanded by the **legendary Maharajah Ranjit Singh**, but after his death in 1839, unfortunately the similar fate was awaiting for the Sikh-Empire as well. There was an immature and incompetent heir here too besides others and, **Maharajah Ranjit Singh's youngest son, Dalip Singh** came to the throne but he also proved to be quite naïve and incapable of becoming a king, and consequently in this case, too, the shrewd imperial British finally had a field-day; in successfully establishing their decisive imperial-supremacy, over the whole of the erstwhile once powerful and glorious **Sikh-Empire.**

Richard Holmes has written that, "**Like it or not, the British conquest or dominion of India is one of history's great epics**. A vast, populous geographically varied continent half a world away from these islands (*i.e. the U.K.*) was dominated for over 300 years by relatively a small number of British. It is no exaggeration to say that **India was indeed the 'jewel in the crown' of Empire, with a unique place in public and official estimation in general and in the history of the British army in particular**. 'The British conquered India by military force,' proclaimed the distinguished civil servant Sir Penderel Moon, 'and the campaigns and battles by which this was achieved...were historical events without which Britain's Indian empire would never have come into existence... **It is remarkable to see how much was accomplished by so few**. When **Queen Victoria** came to the throne in 1837, the European population living in India numbered around 41.000, about 37,000 of whom were soldiers; another 1000 worked for Indian Civil Service and about another 30,000 of the population was of racially mixed parentage. There was a native population of at least 15 million, less than 200,000 of whom served in **the army of the Honourable East India Company, the commercial corporation which, in what is perhaps the most striking example of the flag following trade**, actually ran the subcontinent until the British government assumed its responsibilities in 1858.... *In Empire: The Rise and Demise of the British World Order*, **Niall Ferguson** declared that: 'The question is not whether the British imperialism was without blemish. It was not. The question is whether there could have been a less bloody path to modernity.' His answer is generally negative, arguing that the Empire brought 'free markets, rule of law...and relatively incorrupt government'. **Linda Colley** is rather less admiring. In *Captives: Britain, Empire and the World*, she identifies a strong undercurrent of imperial racism, not least amongst the 'poor whites' that provided the shock troops of the Empire. There is little unanimity amongst Indian historians, for while some look upon the whole imperial episode as an obnoxious aberration; others see value and an enduring benevolent legacy in the British contribution. The Indian historian **Kusoom Vadgama** writes that '**independent India is richer for the railways, the telegraphic systems, education, legal and parliamentary procedures based on the British models' and suggests that it was in fact the Quit India movement of 1942 that pushed Britain toward premature withdrawal and partition**. But Moon observed, while the Empire 'has been both lauded and execrated with undiscriminating fervour—its establishment was achievement that ought to excite wonder.... **Lawrence James**'s *Raj: The Making and Unmaking of British India* stands pre-eminent as a readable popular survey....to illuminate the **central paradox of the Raj: how did a relatively small number of British soldiers gain and retain control of the subcontinent**? What encouraged them to go to India; what did they make of an environment so very different to their own; how did they live there and, no less to the point, how did they die?" (Holmes, 2005, pp.xxi & xxii)

In retrospect and objectively in their broader perspectives we may also say that it remains historically true that, the imperial-British in their own geostrategic and politico-military manners; did gradually come to the Indian subcontinent, conquered and succeeded in establishing their imperially almost all pervasive colonial-rule with its some notable endurance and then, finally went away to their home country in Europe.

But their exit from India, in comparison to other rival European Imperial-Powers from their respective imperial-colonies since located in Asia as well as other far-off continents, were indeed much more cordial and imperially graceful in the sense that, they ceremonially lowered their Union-Jack, unfurled India's no less colourful and enchanting Tricolour and saluted it with proper military grandeur-cum-honour, shook hands and departed and thereby; sincerely succeeded in creating some historic goodwill and nostalgia as well. (*In this precise context, the last similar historic-initiative was enacted by the British in 1997 in Hong Kong, in the presence of Prince Charles representing the British-Crown when Britain's this great colonial-outpost was ceremonially handed-over to China, as per precisely the then overriding dictates of the times of man's living-history and thereby, after the colourful but understandably touchy too on the occasion, with the befitting military-ceremony; the British shook hands with the Chinese authorities present there, and sailed-off with dignity in the Royal Yacht Britannia. And thus in the process, the curtain of history was indeed lowered on an historic-era of the phenomenal British imperial-rule in those famous and well-groomed oriental-islands in the Far-East*). And the man's living-history in the process, keeps-on marching regardless, over the unending canvas of time. If during the imperial phenomena of the British colonial-rule in India, lasting for nearly three momentously historic centuries, India found herself substantially changed multiply forever and then at the same time, the British and their half the world away Nation-State, as being mother of all modern liberal democracies in today's world too; found themselves indeed greatly influenced and multiply-impacted, by the ancient, rich as well as powerfully enchanting, the great and unique Indian civilisational aspects.

But in the broader perspectives, we have since earlier discussed that the multiple aspects of the so much ambitious British imperialism was then not only India-centric but truly flexible and intercontinental, with respect to the setting-up of their numerous and vast imperial-colonies and, along with their specifically need-based different systems of the imperial governance in the name of the British Crown across the world too, were both geostrategically and politico-militarily visionary-cum-flexible, in the highest long-term interests as well as endurance of indeed the unprecedented global British-Empire with again all its imperialism-driven tenacity and, highly influential-cum-far-reaching geopolitical display of the British military-might to native populations inhabiting the various far-off intercontinental British imperial-colonies in the world, to prudently facilitate sustaining the vast-empires. Hence these multiply complex historic-aspects, concerning once the astonishing making and substantially prolonged sustenance and governance of the various kinds of imperial-colonies of the British-Empire across the world, have perhaps been the multi-dimensionally vast subject-matters, for the interdisciplinary studies and research, for the authors and scholars in many countries, since their historic legacies fascinatingly continue to live-on in many ways. We may recall that while quoting earlier the few paragraphs **from the book by Stewart Binns** in this regard, we have since noted that Britain changed forever the lands it ruled, but Britain was changed irrevocably as it moved from the Anglo-Celtic imperial homeland of the world's most powerful empire to a multi-cultural, decentralized middle-ranking nation, firmly anchored to its European neighbours.

Now while coming back to the Indian subcontinent, we may contextually prefer to delve little deeper first into multiple aspects of the systems of imperial governance in vast and also the multiply diversified India. First and foremost in this regard, **we may notice two strong pillars which did quite effectively sustain and remarkably carried-on to successfully implement the imperial-**

writ for the colonial-governance, across the entire length and breadth of this subcontinent. The first one was the British East India Company's indeed fast coming to shape—the Army and, the second being its multi-purpose Civil Services. But initially to move-on, as since the British gradually began to firm-up and expand their footholds as well as, began making deeper in-roads into the heartlands of India interestingly, the Company's army was the kind of quite a hotchpotch military combination and the civil services were in fact, yet to born. In fact even the army began to get into some shape to start with, happened to be from a kind of militia initially meant for providing security to the Company's warehouses and the allied trading-assets as well as, Company's other commercial interests. In these specific contexts during those momentous times; two somewhat parallel historic phenomena were almost proceeding together in the Indian subcontinent in the sense that, as the once mighty Mughal Empire began to decay and disintegrate fast perhaps the apparently hidden but also getting stronger, the commercial-cum-imperial tentacles of the British East India Company too, began to get enlarged and in the process always being flexible and prudent enough; both geostrategic and politico-militarily, not to miss any emerging opportunities in this regard. The subcontinent had already geopolitically landed in a very fluid situation, and also with a fast emerging power-vacuum and hence there-by, socio-politically the overall conditions too were quite unsettling. With their fairly groomed and well-stationed multiple-aspects of the 'essential strategic-culture and thoughts', the British were always in a position to keep their hands on the pulses of since the multiply conservative Indian societies, with their shrewd strategic-vision and therefore, they happened to put in practice their acumen towards a couple of very old-sayings in this regard, i.e. 'divide and rule' and also 'everything is fair in both love and war'.

Needless to say that for the successfully multiple-pursuits of their since many kinds of the Company's various commercial-cum-imperial interests, the British did need first and foremost, their military-arms to begin with, in the highly competitive paths of their imperialistic pursuits in the extremely complex and vast Indian subcontinent. And for meeting this foremost emerging need, the British were able to very imaginatively begin expanding and training their militia to transform the same into some kind of a military-force and, also to be officered entirely by the British. In these interesting contexts, **Richard Holmes** (p. xxv) of the Introduction chapter has mentioned that, "Three categories of British Soldiers served in India. Firstly there were, what we might now call the mercenaries, serving Indian rulers in eighteenth and earlier nineteenth centuries, although the term mercenary then had few of its modern pejorative connotations, and soldiers of fortune would be kinder description. Next, there were officers and NCOs of the East India Company's native regiments, and officers, NCOs and men of the Company's Europeans, regiments such as the Bengal Fusiliers and the Bengal Horse Artillery. All of these folk had made a conscious decision to serve in India, though many of them lived to regret it: the diarist and sybarite William Hickey no sooner arrived there as a junior officer in the Company's service than he set off back to England only to reappear in India as a civilian, less trammeled by rules and regulations... Finally there were regiments of the British army, horse and foot, who served in India for tenure ranging from a few years to more than twenty-five. To distinguish them from local units they bore the prefix HM's (*i.e. His or Her Majesty's*) in front of their regimental number."

The essential aspects of always the totally uncompromising integrity and, objectivity of one's discerningly prudent thought-processes indeed perforce demand in this regard, that both since as the history-making as well as, being far-reaching turning-points along the multiple paths of the ongoing making of man's living-history; on any momentous historic-issues, concerning any Nation-State/societies at any given point of its history in any context, may not ever be subjective or probably blurred, so that the focused objectivity of the desired analytically critical-studies itself, never stands perhaps even slightly compromised. In these respects therefore, due caution and care always need to be borne in mind that, any possible rhetoric or jingoism is never allowed to creep into one's thought-processes at any stage, along the paths of analytically critical reviews in this regard. Hence historically again, **it was definitely a sort of no cake-walk contextually for any overseas Imperial-Power to be able to enduringly build any such big, largely apolitical and highly professional army, as a well-trained and well-oiled fighting-machine always available for service anywhere both during the peacetimes as well as in war, where the entire rank and file of the soldiers below the officer-ranks, were composed of the native troops**. As the man's living-history too clearly talks, indeed we do notice that, as the British East India Company's fledgling army itself was in the process of fast evolving as well as expanding, it was both geostrategic and politico-militarily fulfilling the multiple-roles as the principal instrument, for effectively establishing, expanding and consolidating the historic British Empire in the Indian subcontinent.

And **precisely here that, historically the momentous foundations of the present modern-era's now the fast-developing, sovereign and democratic India's totally indigenized, apolitical, and highly professional Indian Armed Forces, were initially laid** and again the rest is history so far as the multiple-aspects of the genesis of now the independent India's matters-military are concerned. But it is also a timeless and universal fact that, the roots of anything or any specific phenomena in any context always in the fitness of things, remain buried in the historic-dusts of the past and therefore, while discerningly viewing all these in the overall larger perspectives; we do also notice that the continuing mutual inter-connections as well as, mutually influencing inter-relationships too, do keep getting simultaneously manifested as a parallel process of the on-going making of the man's living-history. After 1947; the sovereign India's official establishments have mostly been caught off-guard, at great costs and even at times, with some national humiliations. What an irony all these have been for the Indian Nation-State, in today's super and fast moving high-tech world. We may relevantly emphasise at this precise point that, these matters do always remain linked to the multiple-aspects of 'essential strategic-culture and thoughts'. And hence again, here 'lies the rub' for the Indians and their nation.

As per above in the larger perspectives as well as in the specific but multiple-contexts of the armies **Richard Holmes** has written that, "Most successful Armies work because they appeal to two side of human nature. A theologian might argue that human beings are drawn between *logos*, the world of the practical and provable and *mythos*, the less easily defined realm of myth and legend. Some people live at extremes: on the one hand as a worldly individual for whom materialism is all, or on the other, perhaps, as a monk, nun, or religious fundamentalist for whom this world means nothing. But most of us, and soldiers are no exception, have a foot in both camps, drawn on by the hope of material reward as well as by a complex mix of motives which tap deep into the human spirit. In 1806, when the future **Duke of Wellington, a major general at last,** was asked how he could bear command a brigade at peaceful Hastings after his triumph at Assaye, he replied:

> ***I am a* nimmukwallah, *as we say in the East, that is. I have eaten of the King's salt*** *and, therefore, I conceive it to be my duty to serve with unhesitating zeal and cheerfulness, when and where the King or his government may think it proper to employ me.*

Underlying the whole notion of military service in India was a sense of mutual obligation, an ill-defined contract bearing on both parties. British officers and men, and the sepoys, who served alongside them, were influenced by both moral and material considerations. If the symbolic gesture of eating one's employer's salt was important, so too the conviction that pay and allowances should be both adequate and prompt. It was such practical considerations induced soldiers to volunteer to spend the best days of their lives away from home in an unhealthy climate." (Holmes, 2005, pp. 264 & 265)

So in these contexts presently in hand, we may say that during the British colonial-era, with respect to multiple-aspects of the imperial-governance of Indian subcontinent, the British Indian Army provided the essential sword-arm, and the civil-services fulfilled the role of executive-arm of carrying out the tenets of British-rule. So far as the subsequent desirable and totally apolitical germination as well as, flowering of the multiply important aspects of the professional military-ethos; now so prominently-cum-proactively visible in the modern democratic and sovereign India's Armed-Forces are concerned, its seeds seem to be very fascinatingly visible in the above quoted paragraphs. It is a different matter that subsequently after 1947, in spite of the so much relentless heat and dust of Indian democracy, since also being vibrant and in action the India's apolitical military-officers since always being loyal to the Indian constitution and the flag, have healthily maintained with dedication, the unquestionably apolitical character and professional ethos of the Indian Armed-Forces. It is no mean achievement for the Indians and their indeed the magnificent military in the prevailing overall geopolitical environment around India. We would be contextually discussing more of these in the later chapters.

In the larger perspectives in this regard we may for further facilitating our discerningly multiple-discussions, divide the imperial-era of British-rule in India into two parts: i.e. (i) the pre-1857 period and the (ii) post-1857 period. Historically in many other different contexts too, these distinctly two phases of the British imperialism in India, with all kinds of its complexities, paradoxes and even contrasts do contextually mean many different things in different perspectives of studies and analyses, anywhere in the corridors of intellectual-pursuits for understanding and the knowledge. **In many ways the 1857, may even appear to be like a water-shed on the historically highly chequered timeline-canvas of nearly three centuries, concerning the multiple-aspects of British imperialism in the Indian subcontinent.** During the pre-1857 era of the British imperialism in India when the imperial presence of the British with all its commercial-cum-politico-military dimensions was under the multiple command and control of the then British East India Company, which was in fact both geostrategic and politico-militarily, as well as also commercially too was calling the shots, on behalf of the Britannia with all its imperialistic ambitions in the orient. At the same time, it is also true that, the **principal intent of any highly ambitious as well as, systematically piloted and controlled, commercial Company operating anywhere; remains making money and also transferring the same to its masters in their home country, and naturally therefore with its all kinds of historicity, the British East India Company was no exception.** In any case, they had not landed in India for charity or for that matter, any other laudable humanitarian causes.

When numerically few ambitiously commercial interests-driven British landed on the shores of India, equipped with the Royal-Charter from Queen Elizabeth I in 1600, they had not even dreamt of even a fraction of the subsequently destined remarkable sway of the British colonial-imperialism, in almost the whole of Indian subcontinent. Moreover, once the mighty Mughal-Empire was at its imperial glory and, even indeed the great Mughal Akbar too was yet to come on Agra's throne, then being seat of the Mughal imperialism. So anything and everything in the man's living-history since continuously in the making, is not even incubated clearly well in advance.

And we have since earlier observed that along with Aurangzeb coming to the Mughal-throne and also coupled with its certain internal contradictions initially somewhat slow but consistent decay, and subsequently with Aurangzeb's demise a kind of gradual but fast disintegration of the Mughal-Empire too commenced, causing a power-vacuum in the Indian subcontinent. In this highly complex, multi-dimensionally fast-moving historic process during those geopolitical and militarily turmoil-ridden scenario then prevailing in the subcontinent, the trading but ambitious British East India Company, found itself surrounded and, almost wading through the rising murky waters of the fast manifesting power-vacuum around itself. But at the same time, they too had expanded their many tentacles and also, somewhat equipped and trained their militia for the security of their commercial interests-cum-assets. Thereby they had quite ambitiously as well as craftily, begun taking sides among the extremely jealous and warring Indian feudal-lords both for suppressing their other seafaring European rivals and also, for expanding their tentacles and footholds in the heartlands of India. **As they began succeeding in these great imperialistic-games, initially from their militia, interestingly, a kind of the hotchpotch type of the British East India Company's British-Indian Army commenced to be gradually formed and manifest itself.**

The Company needed not only some credible military-force in this regard, but in order to generate revenue and also, for some kind of civil administrative control of the subcontinent's areas falling under the Company's supremacy, it also needed a kind of their effective civil services as well. Interestingly here, as in case of the gradual birth of the British East India Company's army; the initial incarnation of its civil services too, largely remained a hotchpotch combination; but it did work then to further the Company's both commercial as well as imperialistic interests in the Indian subcontinent. In the beginning it was mostly the Company's military personnel, who manned the civil duties, in all kinds of the perhaps quite unexpected environment. And some British civilians also came from Europe for the purpose but several times, some of those civilians too were found to be doing the jobs of soldiers, and even took active-part in battles with glory. In fact at individual-level for both civilian and the soldier, like today's car-driving almost all of them happened to be accomplished horse-riders and that in any case, this happened to be effectively quite handy and indeed a great asset for doing one's duties for the Company. So during those historic times, along with both civil-employees and the soldiers on the Company's payroll, the British East India Company had fruitfully since both the **military-civilians** as well as the **civilian-soldiers** too, in its service. Historically, what a queer situation all these must have been during those times for the British in India, but with their strategic prudence and being well-versed in the already earlier mentioned games of the 'divide and rule' as well as, 'everything is fair in love and war'; **they carried on and subsequently, succeeded in converting India, in a 'Jewel in the British-Crown'**.

What we now know of the system of once so well-renowned **'Indian Civil Service' i.e. the I.C.S.**, is indeed historically a phenomenal development, having in fact, professionally materialized during the beginning decades of the post-1857 era of the British imperial-rule in the colonial-India. We would be discussing about the many interesting aspects of the ICS and also, taking note of others' observations in this regard little later, but we may perhaps more relevantly first critically delve into the various now since bygone dimensions of the historic 1857, to put the concerned issues in their perspectives. We may recall that in different contexts little earlier, we have since already discussed the various indeed, the momentously history-making multiple-aspects of the pre-1857 era of the British imperialism in India. Now, we may take our present discussions into **post-1857 British imperial-era of the colonial-rule in India; since this dateline is indeed historically considered a kind of the 'watershed' in this regard**.

First why historically it is called so because, during the historically highly consequential year of 1857 itself the **'Sepoy-Mutiny'** took place in India against the then inconsiderate imperial-British of the East India Company, which is also known as the **first war of India's independence**. At a kind of the outset itself, as we may pause a little and also contextually may care to critically look into the historic-aspects of those times objectively then we may also very clearly notice that this 1857-rebellion was not a sudden development at all. In fact its principal causes, were incubating into the many kinds of deeds/misdeeds since being so arrogantly as well as even perhaps naively resorted to, by the British themselves. It was not just a very simplistic kind of military-command failure, but some of the blind-cum-imperialism-driven policy-decisions of the greedy and arrogant British and, a very complex and somewhat hidden failure at the levels of the inter-racial human-relationships and thereby; a kind of the highly abrasive collapse of the inter-racial socio-cultural dynamics too. And at the root of all these happens to be the women-folk and hence specifically in this case it was largely the British women-folk, who happened to be really skin-colour conscious, and racially prejudiced that it became almost irreconcilable. **It is generally said that women can move empires and then, they can also unsettle an empire and even perhaps destroy it too**. Well, ***Shakespeare has said that jealously is thy name woman***. And during those highly turmoil-ridden historic times in India this iconic British writer's a famous dictum, was indeed proactively in full bloom and thereby, ultimately leading to the bloody upheaval of 1857, with all its tragic consequences for both the peoples of India, and the British officers and other employees themselves then stationed in India, in the service of their East India Company. Hence, it is also true that in the multiple phenomena of the rise and manifestations of the imperial-glory of the geostrategic and politico-militarily hegemonic and the imperialism-based, British-empire and along with, its naked kind of the exploitative colonialism the seeds of their impending demise too, remained in itself hidden there.

During first the pre-1857 and then further-on, during a couple of the momentous years of 1857-58 are concerned, the accomplished British author mostly staying in India for his writings, **William Dalrymple**, has indeed colourfully painted words in his two well-known books—***The White Mughals*, and *The Last Mughal***. In the book the *White Mughals*, the author has painted indeed a very convincing indictment of the imperialism through a moving love-story between a British army officer and a Hyderabad princess and also along with, the intrigue of the royal ladies and the palace-politics, during the times of a Nizam of Hyderabad. In the '*The Last Mughal*, the author has very touchingly described the times of the clashes between the British and their rebellious Indian native soldiers, and in the process when tide turned and the British militarily began to decisively gain the upper hand over the native sepoys, the cruel retaliation and atrocities of the British officers and also along with, the ruthless sacking of the historic Red Fort in Delhi. One thing is certain here that, one can only have really a kind of the pulsating feel of all these historic episodes of poetic romances, tender emotions and indeed many moving as well as the extremely devastating tragedies of the 1857 after one is able to give the requisite time and really in fact, patiently go through these books.

Interestingly, it is also a historic fact that when the British began to land in India, they came without their womenfolk and understandably as they began to settle down for their business and, gradually got accepted by the upper echelon of the highly feudal indigenous societies in India and in addition in the process; they also in fact, themselves got socially 'conquered' and absorbed by the colourful lifestyles of Indian feudal-barons and aristocrats. **In other words, we may also say that instead of colonizing, the British themselves got first socio-culturally colonised by the then prevailing luxurious and colourful lifestyles of Indian *rajas, nawabs and zamindars***. Hence sociologically, it is quite understandable that they too gradually began to have nearness and liaison with the eligible Indian young ladies and naturally in course of time marriages began to take place and, families were also raised. And as years rolled by, the Company's higher-ups in London gradually got worried on many counts and alarmed too and with counter-measures and cooperation of the British societies in the U.K., they commenced sending young and unmarried English ladies in ships to India, for marrying the young British men since stationed in India. And therefore on landing in India, the comparatively privileged among those English womenfolk, naturally preferred to marry the few of the bachelor British ICS officers, the military officers; and the remaining ones settled down with the British Army NCOs and the railway-men. Some of the lucky ones in some ways, landed in Indian feudal barons' *harems* too.

But before the white-skinned, arrogant and, highly skin-colour as well as racially-prejudiced—the young English ladies began to even land in India in relatively large numbers, the interesting but at the same time, socio-biological and socio-culture based historic storylines in regard, largely happened to be quite different. In these factual and interestingly historic contexts **Richard Holmes** has very colourfully narrated that, "**If drink was one major preoccupation for officers and men in British India women were certainly another**. Here the picture is rich and complex, for attitudes to sexual relationships were not simply the result of changing views in Britain, but had a profound effect on the cultural interchange in India itself.... Relationships between European men and Indian women of sort or another were inevitable, especially given the small numbers of European women in India, though an acid-tongued described those 'memsahibs in India as underbred and overdressed' and that the men were 'in general what a Hindu would call a higher caste than the women'.... In the last twenty years of the eighteenth century it had been regarded as entirely normal for one of the Company's officials or military officers to maintain his own *zenana*, and in his *East India Vade Mecum*, published in 1810, **Thomas Williamson** helpfully told youngsters just what they might expect to such an establishment to cost them. In 1810 it took about 40 rupees per month to keep a *bibi*.... One major ended up with a *zenana* of sixteen Indian women, for each *bibi* required two or three attendants, with an allowance for betel nut, tobacco, clothes and shoes.... While some local women were regarded little more than what a later age would call 'comfort girls', others were emphatically not. It was easiest for the military adventurers to blur the cultural boundaries. **Claude Martin** was French by birth, and probably joined the British army after the fall of Pondicherry in 1761.... Martin kept four Eurasian girls in his *zenana*, and had the usual staff of eunuchs and slaves. **Most of his staggering fortunes—he died in 1800 worth 33 *lakhs* of rupees—was used to create the two La Martiniere establishments for orphans, one in Lucknow and the other in Lyons.... William Linnaeus Gardner**, born in 1770, was a nephew of **Alan**, 1st baron Gardner, an admiral in the Royal Navy, and after attaining the rank of captain in the Bengal army, he took service with the **Maratha chief Holkar**. Gardner tells us how he met his future wife: 'When a young man I was entrusted to negotiate a treaty with one of the native princes of Cambay. Durbars and consultations were continually held. During one of the former, at which I was present, a curtain near me was gently pulled aside, and I saw, as I thought, the most beautiful black eyes in the world. It was impossible to think of the treaty; those bright and piercing glances, those beautiful dark eyes continually bewildered me. I felt flattered that a creature so lovely as she of those deep black, loving eyes should venture to gaze upon me'.... **Whether born in Britain or in India, many officers were comfortably Indianised in their private lives, some wearing 'banyan' coats and Moormen's trousers at home, and sometimes outside, and in 1730 a council meeting in Calcutta was held in loose coats, with *hookahburdars* in attendance**. In the second half of eighteenth century many gentlemen dressed in entirely in Mughal style, with loose shirts and trousers, embroidered waistcoats and turbans when at home, and it was only in the early nineteenth century that the Company generally proscribed the practice.... **Sir Charles Metcalfe**, a distinguished and talented civil servant, who stood in as Governor-General after the departure of **Lord William Bentinck** in 1835, had an Indian consort, probably a Sikh lady he had met on a diplomatic mission to Ranjit Singh's court in 1809. However, she was unable to assume her proper role as Metcalfe's hostess, and Lady Ryan, wife of the Chief Justice of Bengal, was acknowledged as the head of Calcutta society. But Metcalfe was wholly unrepentant, leading **Isabella Fane** to tell her aunt that: 'Sir C has the reputation of not caring for [i.e. not caring about] colour in his little amours'.... There had been frequent grand dinners and 'reciprocal arrangements' which brought British officers and officials together with what were often called 'native gentlemen' on more or less equal terms. (*And so the historically factual stories go on and on.*).... **The memsahibs displaced the *bibis*, and a recent historian has claimed that it was indeed the memsahibs who had the effect of: 'spoiling that cohesive relationship which had been so enjoyed by the sahib and his sepoys in the past. With her petty insularity her homegrown prejudices and petulant dependence on her countrymen, she had succeeded, by the time the Mutiny broke out, in divorcing the officer from his man, the collector from his clerk; Britain, in fact, from India'... For a time some gentlemen like Sir John Hearsey maintained two separate domestic establishments. But by 1830s the easy ways of the past were getting condemned.** (***So who says that 'jealousy is not thy name woman'? And also, who says that woman cannot move empires and; if she remains perhaps 'foolhardy' and also, somewhat callous as well as, regressively persistent can she not even destroy empires?***)." (Holmes, 2005, pp. 436, 438, 439, 440, 442 & 446)

All these in turn, naturally gave some new twists and even some kinds of quite unsettling tremors in the then overall combined social-canvas in the Indian subcontinent. These interesting historic episodes and their numerous fallouts in fact, were bound to create some tensions and frictions during the social contacts as well as almost daily interactions; between Indian and the British population then living in the subcontinent. These frequent contacts and relationships couldn't be avoided but, the dynamics of social acceptance and warmth were the inevitable casualties and hence replaced by the bitterness, mutual lack of trust and intolerance. In the British army officers' households and kitchen, the earlier kinds of smooth, trusted and affectionate free-entrances of the native soldiers serving their officers were resented-upon by the uncaring, shortsighted and snobbish British ladies then living there, and similarly in the official residences of the British ICS officers, their official native staff and servants too were often facing the same insults and rude behaviour. These ladies' prejudiced snobbery and conceit had replaced the old mutual charm and trust and thereby, causing immense harm and corrosion of the socio-cultural foundation and human-level bonds of the British-Empire; in the subcontinent since with unique and rich civilisational past and also having, the multiply diversified Indian societies. The British ladies were intolerant of the Indian bibis/begums, not liking relatively their dark skin-colours and, even the body odour of the native soldiers and other workers. **Sometimes, ladies' nose too work strangely perhaps, without even caring for the circumstantial consequences!** These historic episodes of, the early nineteenth century's British India have contextually been very vividly described in the books of William Dalrymple and Richard Holmes as well. And hence in course of time so far as the native sepoys in the British East India Company's army were concerned, a kind of the unimaginative introduction of the controversial so-called animal fats-based greasy cartridges subsequently became the last straw on the proverbial camel's back, towards triggering of the highly consequential and indeed historically far-reaching, the 1857 mutiny.

But besides these, there was another causative dimension to the 1857 violent rebellion and that was irreconcilably upsetting the hereditary-based or otherwise an accepted form of successions through conscious adoption of a heir well in advance, for the feudal apple-carts of the Indian princely states, by the arrogant imposition of certain irrational and faulty British policies; which happened to begin surfacing just few years before the uprising and, which too also acted causes and symptoms in this regard. In this precise perspective; Richard Holmes (page 70), has contextually written that, "It is ironic that within less than 10 years of the triumph at Gujarat, the Company would be fighting for its very life. The causes of the great Mutiny of 1857 are complex, a significant role was played by the **'doctrine of lapse', the policy devised by Dalhousie which said that any princely state or territory dominated by the British and without natural heir to the throne would automatically be annexed**. In 1848 the Raja of Satara, in western India, died without an heir. **Dalhousie** refused to recognise his recently adopted heir, declaring that his state had 'lapsed' to the paramount power. Two experienced political officers, **Colonel William Sleeman** and Colonel John Low, protested that the policy caused great concern. Low warned Dalhousie that Indians asked him: 'What crime did the Rajah commit that his country should be seized by the Company?' When the state of Nagpore was also annexed, he admitted: 'After a very careful perusal of the Governor-General's minute on this important subject—it is with feelings of sincere regret that I found it is quite out of my power to come to the same conclusions as his lordship.'... What happened to the large state of Oudh was even more serious. There was no question of the conduct of its ruler, **Nasir-ud-din Hyder**, the seventh nawab of his line, left a good deal to be desired."

It has been very appropriately said that the causes as to what, why and how—for the numerously ruthless and highly inhuman events, since having taken place during 1857 in the colonial-India; are indeed historically complex and multiply-disturbing and, even dauntingly somewhat frightening too. It is a different matter that in the present times both in India as well as in the U.K., many perhaps naively hurried individuals tend to adopt quite simplistic answers to these factually multi-dimensional and highly complex causes, while discussing 1857 with its indeed numerous dimensions. Anyway, that perhaps happens to be a different matter altogether. While continuing to delve deeper in the multiple-aspects of 1857 contextually now, we do also locate one more no less overwhelming as well as unsettling cause to the historic 1857 in this regard. And that historically important and specific causative-factor happens to be the **ruthlessly back-breaking socio-economic exploitation of the poor, half-starved and semi-naked peasantries** of the heartlands in colonial-India by the arrogant, racially prejudiced and uncaring British imperialists since in the service of their East India Company. It was not that the native feudal-princes and the local zamindars, since ages and within fact few exceptions apart, happened to be any kinder and less exploitative, but at least they remained aware of the local conditions and weather limitations but when the regressively feudal zamindaris at many places, due to certain fairly known the then historic developments, got transferred to the rising and expanding British East India Company, the uncaring and callous revenue collection scenario in the Indian heartlands, got further exploitative and humanly cruel. Poor and hapless peasants suffering since centuries got further marginalized and as an inevitable consequence then, found themselves, landing from 'frying-pan to the fire'!

So a kind of perhaps the **'final battle-lines'** were already drawn for historically indeed the path-breaking and ruthlessly violent struggles of the 1857. In contextually larger perspectives and in retrospect, it appears to be historically perhaps not factually precise that it was a kind of the truly popular uprising. In other words, we may say that it was not all that mass-based, where the masses across the vast lengths and breadths of the Indian subcontinent got overwhelmingly charged and proactively stirred, as in case of during the later period's Mahatma Gandhi's unprecedentedly popular and totally non-violent socio-political movements; with the focused aims to gain India's independence from the powerful tentacles of British imperialism and the quite exploitative colonial-rule. Man's living-history, since continuously in the making always expects from the man, the unquestionable objectivity, during its critically analytical perusal, in his any given academic context or otherwise, for the any differently aimed studies and research or for that matter, for the history-writing itself. **On the one side of this historically momentous battle-lines of 1857, there were obviously the British, their some 'loyal' feudal-lords, few Indian ruling-princes, some merchants and also, even some Indian sepoys (who were naturally considered as unscrupulous traitors by their comrade in arms, then being on the opposing side.). On the other side, there were rebellious sepoys with the native NCOs as their leaders in action, then the feudal ruling-princes and their supporters, whose princely states were either threatened or already snatched-away by the imperially callous-cum-arrogant British and merged into the territories falling under the colonial-rule of the exploitatively active the British East India Company**, in the subcontinent. On this side of the princely India, once their sovereign, but during later-period just for the name-sake only, the Mughal Emperor in Delhi and, since perhaps understandably having serious grievances against the unscrupulously imperialistic British, was also sucked in this all out 'war'.

During the 1857, now fairly well-known and sadly quite ill-fated and by then the largely powerless **Mughal Emperor Bahadur Shah Zafar-II** was on the Mughal throne in the famous Red Fort of Delhi and quite understandably in their lines of thoughts, this then old and inactive imperial-Mughal was made the ceremonial head of indeed the highly motivated and numerically the massive side, since menacingly lined-up against the British, then present in the subcontinent. But besides these, the native Indian warriors and their feudal leaders, were not even remotely aware of the multiple-aspects, of **'essential strategic-culture and thoughts'**; unlike the British who were definitely well-versed and fairly experienced, both in the perceptive concepts and practices in this regard. Again from the contextual considerations of always the historically paramount multiple-aspects of the matters-military, hardly even one of a number of the universally applicable **'principles of war'**, quite unlike the British, was ever understood and applied by the patriotic Indians, in their operational planning and action. In fact, on the Indian-sides, there was never any dearth of the urge to valiantly fight with patriotic sense of sacrifice, and with militarily motivating valour; and even there were never any shortage of numbers in their fighting-strengths. But, almost the total lack of the above mentioned crucial aspects, the military discipline and along with, absence of the important as well as properly-cum-proactively well-placed military 'chain of command and control', finally ensured perhaps the inevitable and sad tragedy for the Indians. In fact in a way it was only, largely a type of the ruthlessly violent armed-struggle between indeed the broadly justifiable mutinous sepoys of the Company's army, along with the feudal-segments of India; and the then arrogant and quite regressively-fixated imperialist British in India.

While critically delving deeper into these historic aspects, and in retrospect relevantly a very far-reaching question arises as to—**was India during those historically now the bygone times of the turmoil-ridden 1857, meaningfully prepared to remain geopolitically united and composed enough, to be able to really gain perhaps the desired political freedom and vibrantly preserve them** as since now globally conceptualised and also, as understood by the modern Indians today. In fact, this also always remains true that **history is not factually made as per somebody's any contextual thoughts, or even as per any 'flights of imagination'**. While continuing to critically delve deeper in these historic matters in their precise perspectives, we do also perceive that the multi-dimensionally complex and since continuously in the making, **the man's vast living-history**

has its own some kinds of the 'sense of timings' and thereby perhaps, a predestined time for everything since incubating and then later manifesting before the world. Well, these may even be termed just some matters of conjectures, but that is again perhaps a different matter altogether.

In context with respect to the above **Dr. Amit Pathak** *1857... Living-History* and therein, while discussing many imperialistically arrogant and regressively certain blind spots then prevailing in the British psyche (page 282) has mentioned that, "Observations of **Thomas B. Macaulay**, 1833, in his minutes of education—

> 'To trade with civilised men is infinitely more profitable that to govern savages *(Indians)*. *(***Well about this 'peculiar and arrogantly erratic Macaulay, less said the better**.*)*.

Franz Boas commenting on the Aboriginal people of the world—

> 'I often ask myself what advantages our 'good society' possesses over those of the 'savages' and find, more I see of their customs, that we have no right to look down upon them. We have no right to blame them for their forms and superstitions which may seem ridiculous to us. We 'highly educated people' are much worse, relatively speaking. A person's wealth should be judged by the warmth or his heart.' *(Well, that's it, indeed very reassuring and typically British too.).*

The colonial age for me is not a distant unrecognisable past. People of our generation have felt its existence in our lives, though we were born after the demise of political colonisation in South Asia. We felt its presence within system of schooling that we were enrolled in during the 1970s. We felt it in the stories we heard as children from old servants about the rule of the *'Sahib'*, their Hindustani being purposefully mixed with a few words of English."

In a context, the above quoted lines do show a young and patriotic Indian's thoughts, with respect to indeed some historically disturbing elements of since the bygone era's overriding British imperialism as then powerfully thrust-upon onto the Indian subcontinent, and that is understandable as well. But we may again recall that in differently relevant contexts, we have since also discussed that the multiple-aspects of imperialism too, has many kinds of different situations-based facets and their overwhelming manifestations. In addition, we may also recall that with respect to the medieval-era's building of a number of vast and overseas colonial-empires in different continents of the world by a number of ruthlessly rival and powerful seafaring European Imperial-Powers, have since historically also displayed the white-races' imperial subjugation of the non-white majority in the continents of Americas, Africa and Asia. But much earlier to perhaps the inevitable colonisation of the Indian subcontinent, there is a classic case of the imperial subjugation of another white race by the ambitious British imperialists in Europe itself. **And specifically indeed, this saga in this regard, pertains to the prolonged clashes between imperial England and the Ireland**. Hardly any other nation has been able to so consistently as well as proudly, fight for freedom and honour, as Ireland against the highly hegemonic British imperialism in the man's living-history. Well, in a kind of rather entirely different historic perspectives we may also cite the incredible Israel and her highly tenacious struggles for even more than two thousands of years, for restoring the distinctly proud Jewish identity, honour as well as her nationhood and then finally reclaim successfully her sovereign democratic Nation-State in the Middle-East, against all possible odds and hostilities. We have since already briefly touched-upon these historic aspects and their indeed unique saga, in different perspectives in the earlier chapters.

Ireland happens to be the south-western neighbour of England in Europe and, it is a beautiful country in the north Atlantic. It has a very rich and old civilisational past in Europe and, the Irish are understandably very proud of their past and civilisational legacies and they do have good reasons to be proud of. In fact, Ireland was already civilised and culturally quite developed, much before most of Europe and then even England. Within nearly six centuries after Christ, Christianity was quite developed in Ireland and in addition, she was a very important centre of learning in Europe. In Europe, the English people mostly consist of the Anglo-Normans race and, while Irish belongs to the Celts. And rather ironically here, belongs historically the genesis of racial prejudices of the English; as they earlier tended to look down upon the Irish as savages though quite interestingly, since both also have white-skins. We may even call it humanly, a kind of the civilisational paradox in Europe. In fact, after the fall of once the mighty Roman-empire; the Vandals and Huns invaded and ransacked almost many parts of Europe and even, destroyed the Roman civilisation and during those times, Ireland was a magnificent beacon of civilisation in the north-western Atlantic fringe of Europe. No wonder that, nearly for three to four centuries after the sixth century, is also known as a kind of the golden-era in the living-history of Ireland. In addition also, the Irish people have been quite brave, fearless soldiers and always great lovers of their cherished freedom. They never wanted to compromise on freedom and, have always remained ready to defend theirs and their motherland's freedom at all costs. Right from the ancient of times to the whole of medieval-ages and even further Europe's living-history, has always been mostly the history of wars and also all sorts of clashes of arms.

And in a context, Europe's medieval-era living-history shows us that during the eleventh century the Normans under their leader and commander, **William the Conqueror**, invaded and militarily took possession of even Britain and thus, the British is also known as belonging to the Anglo-Normans race. After more than a century, these Anglo-Normans invaded Ireland and after fierce clashes, could take-over a part of the Northern Ireland and from then on indeed, a prolonged period spanning over some seven hundred years of the Irish struggles and conflicts against the imperial-Britain began, to regain their honour and much cherished freedom. In these specific contexts, Jawaharlal Nehru (pp. 665 and 666) has written that, "...these Anglo-Normans invaded Ireland, and the part they conquered was called 'pale', from which probably has come the common expression 'beyond the pale, meaning outside a privileged circle or a social group. (*Well, doesn't one smell even some sort of manifestation of the British imperial-arrogance here?*). This Anglo-Norman invasion in 1169 hit the old Gaelic civilisation hard, and it was the beginning of almost continuous war with the Irish tribes. These wars, which lasted for hundreds of years, were barbarous and cruel in the extreme."

Hence historically in different contexts, we have since also noticed that the multiple-manifestations of imperialism too, have had numerous facets. And further a series of violent rebellions against the medieval British imperialism followed there, and the British imperialists devised a novel method to imperially hold-on to the conquered territories of Ireland. Since Ireland has traditionally been a nation of the hard-working agriculturalists and farmers the British transported herds of English and Scottish adventurers with a lure to take possessions of the chunks of agricultural lands for farming and, got them settled in the Irish areas under their control as new kind of the imported or intruder landlords. It was indeed a queer situation wherein, the Irish people since living in the northern part of their territories under the British control, found themselves slogging under the new kinds of imperially exploitative and alien landlords. Naturally the proud Irishmen were not prepared to accept it and hence, renewed their resistance and even more vigorous Irish struggles for freedom from the British imperial-tentacles ensued. Historically these have indeed been multiple kinds of clashes in Ireland since centuries—socio-cultural, racial, religious, and even ruthlessly the armed ones as well. Finally, the British with their typical geopolitical shrewdness; could divide Ireland in two parts, the Northern and Southern Ireland, by instigating the religious prejudices between the Protestants and the Catholics. Northern part was largely Protestant and hence remained under the British control and then, the southern Ireland remained free and principally agricultural too. But now since the last few decades of the twentieth century itself, things have changed for the better since the U.K. constitutes of England, Scotland and Northern Ireland and then, the Southern Ireland remains independent, mainly Catholics and as the sovereign Republic of Ireland. Few volumes of the highly chequered Irish living-history too, spanning over almost more than one and half thousands of years, are also available for one's reference in this regard.

We may recall that in different contexts, while discussing the multiple-aspects of British imperialism and building of their overseas global colonial-empires, we have since noted that **if the 'colonies' in different continents changed in many ways, than in the process Great Britain too permanently changed in many ways**.

Now coming back to the 1857 in India, we find that historically the recipe for this extremely violent tremor then getting materialised fast and as a final burst day by day, the forthcoming 1857 was indeed becoming inevitably a fait-accompli. Historically in fact from time to time, and even much before the 1857 upheavals on somewhat smaller scale, were already taking place as perhaps prior warnings, but all these were of no avail. Much earlier, a number of peasants' and other kinds of revolts had since already taken place in Bengal. Then there was Vellore mutiny in 1806 and then, another mutiny also took place at Barrackpur, in the year 1824 but ironically all these had fallen on deaf years. So **Bertrand Russell had since very appropriately said that, 'Power corrupts and absolute power corrupts absolutely'**. At the same time after 1857, when the British were imperially ruling the subcontinent; India had almost been totally changed for the British and thereby, the fundamental composition of their stay and the systems of their governance too had gone major alterations. **In 1858 under a Royal Charter since decreed by HM the Queen Victoria, the Indian subcontinent came directly under imperial rule of the British Crown and the earlier so-called hotchpotch Government of the imperialistic traders under the British East India Company, was finally abolished.** ***It was indeed a classic case of the flag following the trade.***

But had really anything substantially changed for the poor and ruthlessly exploited Indian masses. It has rightly been said that always in the callously overbearing multiple-games of the obsessive imperialism, the poor and tirelessly struggling masses in the imperially subjugated country, are since like the highly expendable but also like an unending flow of pawns, on a grand chessboard of the imperialism, in its all kinds of manifestations. **And in these multiply great-games in the colonial-era Indian subcontinent indeed, the British then did need a powerful army and for them, very efficient as well as highly obedient covenanted civil-service, who would dedicatedly serve the British Crown and, successfully carryout the imperial-writ of the British Governor-General since stationed in India as the Crown's representative, throughout the vast and highly diversified sub-continental territories.** During those now the bygone historic colonial-era the vast Indian subcontinent was divided perhaps with the imperially-obsessed great strategic prudence, into three territorial Presidencies: Bengal, Bombay and the Madras, for indeed their multiply diversified socio-economic, geostrategic and also politico-military purposes and long-term ambitions even during the times of British East India Company itself. Each Presidency had its own army and distinct seats of governance, with separate Governors and the commanders-in-chief. Then there was at the apex, a Governor-General along with his imperial-India's overall and senior most Commander-in-Chief. Well interestingly with distinctly special and obvious purposes after 1857, when India came directly under the rule of the British Crown, this Governor-General became also known as the Viceroy of British India and thereby, acting as a representative of the British-Crown. Contextually, some scholars have very interestingly called since these the British-sponsored developments as well as subsequently much more; as almost a bloodless 'White Mutiny' of 1858-61 in the colonial-India. Historically, it had its own numerous consequences, for both India and the Britain, differently in their respective ways too, had got changed forever. Interestingly, many of these are even now enduringly surviving and influencing in many ways, in both the societies.

Lord Cornwallis, who was the governor-general in India from 1780 to 1793, during the highly chequered-era of East India Company was in fact visionary enough, to conceive a proactive plan for the well-paid and covenanted civil-service for the imperial governance of the British India. (*Interestingly, we may recall that he happened to be the same Lord Cornwallis; who was earlier chased away from the North-America, by* ***General George Washington*** *in 1783 at the end of the eight-year American War of Independence; thus laying the foundation of a sovereign and democratic USA. Well, as we have since noted earlier, this war in fact, happens to be a momentous episode of the American living-history.*). Cornwallis's aim in India then, was to curb the short-sighted snobbery and arrogance of the Company's officials and clerical staff, since functioning at different levels in India. Later on to further facilitate this policy, a Company's College with a well-planned syllabus for teaching the principal oriental languages, was established at Haileybury in Hertfordshire (U.K.) in 1809, for training and grooming interestingly, the intended imperialism-oriented but impartial bureaucracy who would also remain debarred from entering into the any kind of commercial activities. For the imperially-obsessed British and their many kinds of the imperialistic and even greedy ambitions these moves, as initiated by Lord Cornwallis, were indeed a long-term and strategically a great move forward. **And at the same time in the process, initial and since fast-sprouting seeds of the later period's famed Indian Civil Service i.e. the ICS, was also sown to fruitfully serve the then British imperial-rule in the subcontinent.** The aim was hardly even remotely to address the real problems and the age-old malingering causes for the ruthless exploitation and sufferings of the toiling masses of India, since always struggling to keep their bodies and souls together. But was it all 'black and white', when viewed from differing and perhaps opposing view-points? We would be coming to these aspects little later.

In specifically the above perspectives, again **Richard Holmes** has very informatively mentioned that, "**The Indian Civil Service, 'one vast club' according to one of its members, and 'the heaven-born' by general consent, was tiny, with around 1000 members**. Not only was it well paid, with starting salaries for assistant commissioners of 300 Pounds a year, 2,700 for judges and collectors, and 8,000 for lieutenant governors, but its members qualified for a pension of 1,000 Pounds a year after twenty-five years' service. From 1854 onwards entrance was by competitive examination, and there had been a shift away from what Henry Lawrence had called 'men who will mix freely with the people, and will do justice in their shirt-sleeves', towards clever young men who could deal with examination with heavy classical bias. **The First Indian passed the exam in 1863, but there were no more passes until 1869, when three Indians were successful. Two who were then rejected for being over-age, duly sued the Secretary of State for India and were reinstated. One, Surendranath Banerjee, was dismissed during his first year of service for signing an inaccurate return. It was an indication of the system's reluctance to accept the inevitable**. There were many who saw the writing on the wall. **Sir Henry Cotton** pointed out that:

'Men who speak English better than most Englishmen, who read ***Mill and Comte, Max Muller and Maine****, who occupy with distinction seats on the judicial bench, who administer the affairs of native States with many millions of inhabitants, who manage cotton mills and conduct the boldest operations of commerce, who edit newspapers in English and correspond on equal terms with the scholars of Europe* (Historically, this is quite interesting and, indeed revealing too.)*...could not be expected to salaam whenever they met an Englishman in the street. It was hard for even those who had been liberal and paternalistic to adjust, not because of social and religious difficulties in meeting Indian women or enjoying a meal together... The best that can be said is that the viceroys or the majority of their officials cared passionately about their task, even if they were not sure quite how to let go of the reins.* **Lord Curzon**, *viceroy from 1898 to 1905, and architect of the Delhi Durbar of 1902, publicly and controversially censured a regiment which had failed to punish two of its soldiers who beat a cook so badly that he died. The Earl of Wellington, viceroy in 1931-36, founded the Wellington Club in Bombay as a protest against the racist attitude of his own club.* ***Philip Woodruff, writing after independence, argued that 'if today the Indian peasant looks to the new district officer of his own race with the expectation of receiving justice and sympathy that is our memorial.*** *" Holmes, 2005, p. 84)*

But today when the all sorts of heat, dusts and, socio-cultural as well as certain kinds of the bygone era's peculiar socio-psychological dislocations, etc., have all become things of the past and so in retrospect, can we with our contextually issues-based perceptive coolness and also with the historic objectivity in a precise perspective, and with an unassailable confidence, say that **Philip Woodruff**'s wishful and indeed quite fair dreams, have even partially come true? Well, the pointed and honest answer in this regard, perhaps unassailably remains an emphatic 'NO' and broadly nothing else. In any case now historically in these contexts as per above, Philip Woodruff's some kinds of the well-intentioned and now largely unfulfilled dreams are indeed understandable. But now much more important and, far-reaching moot point precisely in this regard, and with respect to the modern-era's progressive Indians and their vibrant as well as fast developing, sovereign and liberally democratic Indian Nation-State is **that even contextually India's brilliant-cum-visionary, ironman Sardar Patel's dreams too, sadly remains almost totally shattered.** Besides now historically indeed quite overwhelmingly enchanting numerous achievements on his part, **the towering and brilliant leader-cum-statesman: Sardar Patel** was the principal architect of the historic reincarnation of ICS into the independent India's IAS with many truly patriotic wishes and dreams and contextually therefore, this is also a great betrayal of this great man and contextually his vision too who remained always brilliantly capable to command a great sense of history as well as, visionary grip upon the **multiple-aspects of 'essential strategic-culture and thoughts'**. Hence he was by himself not only a great leader but, an institution as well, to whom only the good of India was dear to his heart.

We may recall that in various contexts earlier, we have relevantly emphasised that the man's living-history since continuously in the making, favours none whatsoever and also, never spares anyone, under any circumstances. In these issues' some larger perspectives, can one even without any slight hesitation not say that these do unquestionably remain, **largely a great betrayal of the independent India's indeed great tryst with destiny** in the multiple-areas of public administration and constitutional tenets of the democratic governance concerning the combined and patriotically mandatory functioning, on the part of India's political-class and the bureaucracy. It is indeed a great betrayal and more so, on part of the top-echelon of India's permanent civil-services based bureaucracy, which has indeed very sadly degenerated **from once the so-called perhaps 'steel-frame', into now some kind of the largely self-centered, somewhat unreliable and, unstable**

kind of a 'rubber-frame'. We would more relevantly in some greater detail, be discussing these multiple-issues in the last sub-chapter of the forthcoming 8th chapter. With respect to the overall qualities and, some factual as well as perhaps overriding flaws too, in the somewhat imperially-fixated ICS these varying but perceptively considered opinions, indeed form a mixed baggage.

In these multiple-contexts and also, conceptually concerning largely the multiple-aspects of historically the multi-dimensional promotions of highly ambitious British imperialism and the building of vast overseas British-Empires, a well-known British academician and author **Niall Ferguson** has mentioned that, "The British went abroad in multiple roles: not only as soldiers and administrators but also as businessmen, engineers, missionaries and doctors. Like America's informal empire today, Britain's empire had its non-governmental character; there were Victorian multinational corporations and Victorian nongovernmental organizations. But key point is that whichever role the British played, they generally stayed—until retirement or, as countless colonial cemeteries testify, death.... Of crucial importance in this regard was the role of the **Indian Civil Service, which became magnet of the very best products of the university system**. The proportion of Oxford and Cambridge graduates in the Indian Civil Service was remarkably high, rising steadily after 1880s to over 70 percent. Two-thirds of the ICS men who served in the 1930s had been educated in England's exclusive public schools; three-quarters had attended either Oxford or Cambridge. All but one of the eight provincial governors in India in 1938, were Oxonians. **John Maynard Keynes**, who by the 1920s had become quite disparaging about the empire, experienced one of the few reverses of his dazzling Cambridge career when he came in second rather than first in the ICS examination. Oxbridge products also staffed the less exalted Colonial Service, which administered the British colonies in Africa and other parts of Asia. Of the 927 recruits to the Colonial Service, between 1927 and 1929, nearly half had been to Oxford or Cambridge. There were also significant numbers of the Oxbridge graduates in the other governmental and private sector agencies that operated in the colonies." (Ferguson, 2005, pp. 206 & 207)

All the above cited indeed quite encouraging aspects of the highly privileged ICS-tribe in the then colonial-era British India is understandable and even laudable, but certainly it appears to be more in tune with a very regressively-fixated and of course also, in the humanly not at all so laudable dimensions of the arrogant-cum-exploitative systems of British imperialism. As it is generally said that, 'there are always two sides of a coin'—well, the point is well-taken in a context; but the substantially enduring praise lies in the quantum of consistent efforts since been/being put-up by the people holding the reins or sitting in the driving-seats, in the areas concerning the all possible elimination or even minimization of the negative or the unbecoming side of the 'coin' concerned. Well, the 'taste of pudding is in eating and, certainly not in its frills'. 'Shoes might not have been pinching to the British feet, but all these did prolong and even remained rather in a way, too much for the poor and toiling masses to endure so long; since principally caused by the callousness of British imperialism and, all its colonial manifestations. The heavily burdensome pillars of the pomp and glamour of the multiple-grandeur of the vast British colonial-empire rested, on the empty stomach and indeed also on bent-back of the poor peasant-class in the Indian subcontinent.

In the above contexts, **Jawaharlal** has written that, "Under British rule Government service gave prestige, security, and status, and later, when Indians were allowed to enter the Indian Civil Service, this service, called the 'heaven-born service'—heaven being some pale shadow of Whitehall—became the Elysium of the English-educated classes, specially lawyers, some of whom earned large incomes in the new law courts also had prestige and high status and also attracted young men. Inevitably these lawyers took the lead in political and social reforms movements. The Bengalis were the first to take to the law, and some of them flourished exceedingly and cast glamour over their profession. They were also the political leaders." (Nehru, 2002, p. 333) (*And historically in an appropriate perspective, we may now clearly see here, the genesis of the initial sprouting and then, fast developing, the multiple-aspects of the Indian freedom movements against the imposing British colonial-rule; and along with that also in many ways, the far-reaching phenomena of the* ***Indian renaissance*** *as well*).

With respect to specifically the ICS, Nehru has very reflectively mentioned that. "We hear a great deal about this ICS. They have been a curious set of persons. They were efficient in some ways. They organised the government, strengthened British rule, and incidentally profited greatly by it themselves. All the departments of government which helped in consolidating British rule and collecting taxes were efficiently organised. Other departments were neglected. Not being appointed by, or responsible to, the people, the ICS paid little attention to these other departments which concerned the people most. As was natural under the circumstances, they became arrogant and overbearing and contemptuous of public opinion. Narrow and limited in outlook, they began to look upon themselves as the wisest people on earth. The good of India meant for them primarily the good of their own service. **They formed a kind of mutual admiration society were continually praising each other. Unchecked power and authority inevitably lead to this, and the Indian Civil Service were practically masters of India**. The British Parliament was too far away to interfere and, in any event, it had no occasion to interfere, as it served its interests and the interests of the British industry. As for the interests of the people of India, there was no way of influencing them to any marked extent. Even feeble criticism of their actions was resented by them, so intolerant were they.... And yet the Indian Civil Service has had many good and honest, and capable people in it. But they could not change the drift of policy or divert the current which was dragging the India along. **The ICS were after all, the agents of the industrial and financial interests in England, who were chiefly interested in exploiting India**. (*We notice that Nehru's these perceptive observations were penned by him in the year 1932, when the British colonial-rule in India continued to be in its full glory and hence; in their historic contexts in a way, these are understandable too).... In the same continuity, on the multiple-aspects of the processes of governance, Nehru very thoughtfully goes on to say that*)... **The bureaucratic government of India, grew efficient wherever its own interests and the interests of the British industry were concerned. But education and sanitation and hospitals and many other activities which go to make a healthy and progressive nation were neglected. For many years there was no thought of these. The old village schools died away. Then slowly and grudgingly a little start was made. This start in education was also brought about by their needs. The British people filled all the high offices, but obviously they could not fill the smaller offices and clerkships. Clerks were wanted, and it was to produce clerks that schools and colleges were first started by the British. Ever since then this has been the main purpose of education in India; and most of its products are only capable of being clerks. But soon the supply of clerks was greater than the demand in government and other offices. Many were left over, and these formed a new class of educated unemployed.... Another fact worth mentioning is that even when the government made a start with education, girls were completely ignored. This is not surprising. The education given was meant to produce clerks, and men-clerks were wanted, and only they were available then, owing to backward social customs. So girls were wholly neglected, and it was long afterwards when some little beginning was made for them.**" (Nehru, 1934, pp. 503 and 504) (*In fact, Nehru has a great point here* *but historically, it was all largely according to the overall highly regressive as well as, callously shortsighted-cum-imperialistic lines of thinking of Macaulay in this regard. Any otherwise larger good and beneficial, which subsequently flowed from all these highly damaging policies in India and for her peoples across the subcontinent and, which did indeed flow multiply to a substantial extent little later were just incidental and in a way, quite surprising fallouts and, certainly not at all intentioned by the British imperialists then ruling over India. Whatever but at the same time historically, it is also true that for some long-term good to India,* ***Macaulay in his own manner and with his some distinctive intellectual prowess, he too made some positive contributions, as would be seen little later.***).

It is historically true that when Macaulay mooted the idea of introducing in fact, **a kind of the colonial-system of 'English-education' in the Indian subcontinent** he only arrogantly thought of serving the long-term British imperial-interests, and never for a moment cared for the India's larger socio-cultural interests and, even regarding any desirable process for any gradual enriching and expansion of the necessary mental-horizon-cum-knowledge-base of the Indian educated-class, to dynamically remain in tune with the changing times. The British imperialists' principal aim in this regard, was to produce clerks to be able to serve in the various offices of the British-empire. But regardless of these historic facts, it is also multiply far-reaching and true that the English-education since last few centuries has been bringing many good to our modern, fast-developing, sovereign and democratic nation in the comity of nations. No wonder that Jawaharlal Nehru, 1934 (p. 505) has himself said that, "*English education, intended to produce clerks, also put Indians in touch with current Western thought.*" It is true that the British in the colonial-India, were in fact always clear and focused regarding their overriding imperial-aims and the principal strategic-interests; not only in the Indian subcontinent but also in almost whole of the South as well as South-East Asia, and also coupled with the Far-East; since India happened to be their grand and indeed very befitting strategic-base and, centres for the all kinds of support for the multiple-logistics in this regard. We would contextually be seeing indeed the multiply proactive fruition of their similar strategically visionary concepts and doctrines, during many emerging situations in the next eighth chapter. In fact, much later even the USA and other Powers in the world learnt from and picked-up many of the earlier British multi-dimensional strategic-concepts, thoughts and numerous geopolitical-ideas. Since the ancient-era, India has been a cluster of multi-lingual and multi-cultural societies all existing together in vast continental landmass since known as the Indian subcontinent. And as centuries passed, gradually different religions too came or fructified here making India consisting of multi-religious societies as well. And Historically, India was yet to evolve into a single and united political-entity. Precisely in this respect and much later during the British-era in the subcontinent; 'English-education' began to overwhelmingly reincarnate India, since multiply-acting as an unprecedented 'cementing-force', subsequently with numerous far-reaching impacts and long-term fallouts.

Lord Meghnad Desai has very thoughtfully written that, "A further problem arose with the very notion of alien rulers. If Aryans were seen to be practically natives and not alien, Muslims were handy scapegoats as alien rulers who had ruined the nation by their conquests, oppression and so on, but had then been supplanted by another layer of alien rule. Integrating Muslims into the national narrative, therefore, became a major challenge to the ideologues of nationalism.... Thus constructing a nationalist narrative in a multi-lingual, multi-religious country was a challenge to which the elite had to respond. They were the new Brahmins who, despite their small numbers, would provide a unifying framework to the nation. They had to because they were at the head of the queue wanting to grab representation in the legislative councils and other official positions.

They wanted to be able to compete for the Indian Civil Service and have respectable positions in the army and the judiciary and the executive (*all because of English education.*)... While the representation which was granted remained small and select, with no whiff of election, the elite strategy of mobilizing the like-minded people together from across India was an effective response. The Indian National Congress, founded at Bombay in December 1885, was a creation of concerned Englishmen such as **Alan Octavian Hume** and **William Wedderburn** who brought together what can only be called **Macaulay's children**. The seventy-two delegates came from all parts of India, and this is where the truly revolutionary nature of the Congress lay. No previous such gathering, secular rather than religious, peaceful rather than military, had existed in India. The very words 'Indian' and 'national' were challenges thrown at the rulers who were busy denying that such a thing as an Indian nation existed (*interesting indeed.*)." (Desai, 2009, p. 116)

Now coming back to the different aspects of the ICS, as we have been discussing earlier, we may need to note that the British were clear about these and their principal-aim to begin with, was to train and make the empire's ICS officers to professionally as well as, in their hearts and minds, to get dedicatedly connected exclusively with the all sorts of colonial-interests of the British-empire in India. In fact during the British imperial-era in India, it was largely too far-fetched to expect anything otherwise from these officers, whether British or Indian since employed by the British, then in the subcontinent. Well so far as the army-officers then serving in India were concerned, it was again largely a kind of broadly an identical situation. But so far as their largely different services-based official perspectives were concerned for both the civil and military officers, it was all a binding matters of oath, since already having been administered always, at the time of their joining the respective services. Since these are factually the historic truths and hence, need to be objectively taken in their appropriate perspectives and, in their proper historic pragmatism. At this precise point, it appears perhaps quite befitting in a context, to go in for the very brief and analytically a comparative discussion in this regard; for putting the relevant issues in their proper perspectives.

So far as the Indian army-officers then serving in the British Indian Army were concerned, and in fact the commissioned-ranks were made open to the Indians much later, i.e. much after even World-War I and that too in a very restricted manner. And hence concerning these Indian Army and little later, even the Navy and Air Force officers; principally due to some kinds of the exclusiveness of their military profession of arms and the connected oath, they remained somewhat lucky and also, with their totally apolitical mindset they found comparatively much easier after 1947, officially in their military chain of command, to reorient themselves from the erstwhile British imperialism-driven demands to now the independent India's totally nationalistic and democratic parameters; as the then newly emerging national traditions and also, within its constitutional-framework, to the changed call to their duties to the nation. In these important contexts, one important point needs to be borne in mind is that even during British imperial-rule in India the army remained and professionally functioned under the civil imperial-supremacy at the political-level and therefore after 1947 when a vibrant, secular, liberal and truly constitutional parliamentary democracy dawned on the political-horizon of independent India, the same Indian Armed Forces officers found no difficulty, in automatically and very smoothly switching over their unquestioned loyalty to the Constitution of India and the national tri-colour, in their letters and spirit. **Hence very soon, these tremendously reassuring developments within the Armed Forces, so healthily and smoothly manifesting, in fact immediately after the 1947 were able to make them indeed very dear and close to the multi-cultural, multi-lingual and also multi-religious vast population-**

base of the newly emerged democratic and sovereign Republic of India. In retrospect, we may even visualise that this was no small matter and in addition at the same time we must not forget that most of the British military-officers who remained in India to serve for some time, even few years before and after 1947 starting from Admiral Lord Luis Mountbatten, then still Governor-General and, Field Marshal Sir Claude Auchinleck, earlier Commander-in-Chief in India and downwards, greatly and also quietly contributed in this military-oriented smooth transition, befitting in a democratic-system.

Historically in many ways and also in many areas, this quiet and smooth process of change had indeed begun even few years before the 1947. In these perspectives, speaking contextually with desired historic objectivity, in a way this too points towards typical of the British character. Specifically in these historic contexts, **Lt Gen S.K. Sinha, PVSM (Retd)** has mentioned that, "The new generation of Indian soldiers was not willing to readily accept the superiority of the White Man... **O'Brien** concluded that the Indian Army could not now be relied upon to uphold British rule over India.... I am convinced that the realisation of the British that they could not rely on the Indian Army to uphold their imperial rule over India, was a major factor which influenced their decision to quit India... Soon after I joined GHQ(I). I attended a lecture at Regal Cinema in Connaught Place. There was no auditorium in New Delhi at that time and GHQ (I) used to hire a cinema hall for organising occasional lectures in the mornings. On this occasion **Mr. KPS Menon** had been invited to give a talk on Korea. He was a brilliant speaker and made a great impression on us. He had been on one of the UN Committees on Korea. Much of what he said in 1946 turned out to be true, some four years later. After the talk when we all came out to foyer, we found a traffic jam outside. **Sardar Baldev Singh**, the newly appointed Defence Member of the Viceroy's Executive Council, and **Field Marshal Claude Auchinleck**, the Commander-in-Chief, were standing in the foyer. The Chief's car with five stars and flying the Union Jack had come inside the driveway, Baldev Singh's car without any flag was far behind. The Auk insisted on the Defence Member's car being brought in first. The other cars had to be taken out. This created a traffic jam. He waited to see off Baldev Singh and after he had got into the car smartly saluted him. Only a few weeks earlier, in protocol Auk was second only to the Viceroy in Delhi and Baldev Singh was virtually an unknown figure. After the interim government had been formed, the position had changed. Auk scrupulously adhered to the new protocol, irrespective of delay and confusion this caused during dispersal after the lecture. **This showed the character of the man**." (*Indeed always, there is nothing bigger and higher than the character of a person in this world.*). (Sinha, 1992, pp. 80 & 81)

In case of the Indian ICS officers after 1947, in a way it was not that easy due to their specific job-profiles and thereby, their psychological bent of minds to mend as per the independent and democratic India's changing needs. But in spite of this in fact, they too found hardly any difficulty in switching over their official commitment and loyalty as per the new constitutional requirements; but problems remained in the multiple-areas of altering the psychological and also official connectivity, now proactively with the peoples and their needs, while carrying-out their official duties. In these contexts, we have since also discussed that, during the British imperial-rule in the Indian subcontinent; the ICS and many other civil as well as, even the police and railway officials, in almost all possible ways, were needed to be proactively connected while discharging their all official-cum-social functions, with the British colonial-interests and historically it is indeed true that, the British were substantially successful in achieving their aims in this regard. But the entire scenario and their perspectives had in fact, totally got altered after the 15th August 1947. Hence among all the various kinds of civil-services; most exclusively the ICS, in the larger national interests, needed to wholeheartedly change their connectivity, their mental perspectives and also their old psychological-fixations according to the fast emerging and totally different in fact, people-oriented democratic scenario in the new India of mid-twentieth century. In fact in few precise words, the ICS needed to totally overall their mindset socio-psychologically and thereby, reorient themselves from habitually being 'masters' to the 'servants' of peoples in the independent democratic India. And here lies the 'rub'; and at the same time, it is also true that some ICS officers were indeed able to successfully do the same but in any case after 1947, ICS were a passing-phase and soon going to be rather honourably 'wrapped-up' contextually, into the pages of India's living-history; and thereby for some historic interests from time to time, for the sake of reviews and studies by the interested persons in future.

But in fact in the India's totally changed perspectives, onus had now to befittingly shift to the new IAS officers, who had since been replacing the old ICS. And here ironically, much has still been left to be desired and broadly the overall picture has not been very reassuring. In most of the situations and unless rather overwhelmingly being forced by certain fast emerging circumstances; people remain much resistant to the any kinds of change, and psychologically by the force of habit, prefer to remain in the old and even the dying, some kinds of the systemic grooves. On these important administrative issues concerning the civil-services, much talks and seminars etc, frequently keep taking place but of hardly any avail. In fact, in the important and largely very far-reaching areas of services to the peoples in any nation and all the more so in a democracy, bureaucracy from top to bottom, always has crucial roles to play. The many kinds of developmental plans being formulated at the top-levels and, the decisions being taken at the political-levels of governance had to be carried forward and then fruitfully delivered at the peoples' doors, by the bureaucracy and hence, if these in fact results-bearing aspects are mostly found wanting, the entire bureaucratic-system indeed regressively become a big and highly damaging drag on the system. **It is historically indeed true that even after the fall and final demise of once the glorious but extremely medieval and autocratic Mughal-Empire in the Indian subcontinent which in spite of being a very ancient and magnificent civilisational entity, India was just not yet geopolitically evolved then in the present modern-era sense; to manifest and maintain itself, as an united and enduringly independent Nation-State of the world**. The age-old regressively lingering and the highly fixated multiple-aspects of feudalism, in the otherwise also extremely caste-cum-class-oriented Indian societies, along with their all sorts of historic legacies and even numerously different kinds of myths too, it was indeed then a tall-order for India, to enduringly remain one vast and united political entity. In retrospect, we may also say that historically perhaps, the requisite time-frame then was yet to come for India; to remain politically united and manifest itself, into an independent, sovereign and democratic Nation-State. And hence as per the then multiply developing historic scenario India had to perforce be colonised, by any one of the rival and seafaring European Imperial-Powers, then being so hyperactive in the subcontinent. These rival European seafaring Imperial-Powers then among themselves were seeing this vast and unique sub-continental landmass; as being multiply resourceful and so rich, like an invaluable goose being capable of laying golden-eggs. In retrospect and historically in the larger perspectives; contextual objectivity demands to visualise that, it was indeed lucky for India, to come under the British imperial-rule rather than that of any other then in the race, but not so multiply far-sighted European Imperial-Powers and also being imperially active in the subcontinent. Hence, anywhere and concerning any historic issues in this regard; contextually the geostrategic and politico-military prudence demands that the man's living-history needs to be viewed and critically analysed, always with impeccable historic objectivity.

In the above specific respect, the eminent Indian Jurist and constitutional expert, **Nani A. Palkhiwala** has mentioned that (*while delivering the Third GEC Lecture at Imperial College of Science, Technology and Medicine, London, November 3, 1992*), "There was much that was ugly in those years (*i.e. under the British colonial-rule*). But, on balance, the good far outweighed the evil. The British took enormous wealth out of India, but they left behind legacies which no money could buy. There were six invaluable gifts which you bequeathed on us... It was the accident of British rule which is responsible for the *unity and integrity* of India today. Whereas before 1858 India was never a united political entity, the proclamation of **Queen Victoria** in that year welded us into one country, one nation ; and when independence came, we had been an unified nationality for almost a century under one head of state.... The largesse of the *English language* enriched Indian culture. It was the cement which bound, and continues to bind, our country together; and it enabled us to deal with the rest of the world from a vantage point. There are sixteen major Indian languages listed in the Constitution, and each of them is used by tens of millions. Those who are following the developments in Canada which has only two languages, English and French, can imagine the multiplication of the problems where sixteen state languages are spoken by different groups who are not unfamiliar with linguistic fanaticism.... It is interesting to note that four of the first thirteen awards of the Booker prize were either to Indians or for books about India. ***The Times (London) editorially*** **noted on February 11, 1992—...Indian English is in some ways more correct than British English, because it is taught from old-fashioned text-books by teachers for whom it is not their first language. There may well be fewer unconscious solecism and barbarism in Indian English than in British. Indians are less casual with the shared language**.

The *civil and criminal laws*, apart from personal laws applicable to particular communities, are based upon British jurisprudence. **The main instrument of criminal justice—the Indian Penal Code—was drafted by Lord Macaulay when he functioned as the Law Minister of India from 1834 to 1838**. It is tribute to his superb draftsmanship that the Indian Penal Code is one of the least amended statutes of India... Broadly speaking the *administrative structure* in India is still as the British left it. The Indian Civil Service (ICS) started by the British was, beyond question, one of the finest civil services the world has ever known. **Sardar Vallabhbhai Patel had the good sense to recognize this, and he introduced the Indian Administrative Service (IAS) it was based on the model of the ICS. At the same time it must be confessed that while top ICS officers were outstanding, the bureaucratic structure which the British created left much to be desired... Not the least of the gifts that the British left behind was the institution of the *armed forces* imbibed with the tradition of being totally apolitical**. Today when we see corruption and degradation all round in India, the one institution of which we can be justly proud is that of the armed forces. In point of discipline and dedication, efficiency and competence, it is the equal to its counterpart anywhere in the world... Finally, paradoxical as it may seem, If India is a *free republic* today that is also the consequence of British rule. Indians fought, and fought valiantly, to get rid of foreign domination. But it is probable that up to now India would not have shaken off the domination of Indian rulers but for the notions of freedom imbibed from the days of British rule. **Macaulay foresaw this development. He said, 'by good government we may educate our subjects [so] that they in some future age demand European institutions. Whenever such a day comes, it will be the proudest day in English history**... Fairness begets fairness, and understanding begets understanding. Englishmen have been fair-minded enough to recognize India's right to freedom; and likewise Indians have been fair-minded enough to acknowledge the benefits derived from British rule.... It was a British constituency which elected **Dadabhai Naoroji**, exactly a hundred years ago, as the first Indian member of the House of Commons. Such was Dadabhai's faith in British innate sense of justice and fairness that when he published a book in which he bluntly criticized British rule in India, the title he chose was *Poverty and Un-British Rule in India*. Dadabhai said—

> ...the world may rejoice in a glorious chapter, added to its history, of the regeneration of an old but long unfortunate race, and **India may for ever remember gratefully the benefactors who restored her to more than her ancient splendour and civilisation** *(Historically so factual and true indeed.)*.

The one lesson of history is that sometimes the impossible happens and the inevitable does not happen....'Crowns and thrones may perish, kingdoms rise and wane'. **It was unthinkable to Lord Curzon that the Indian empire could perish or wane. But in less than fifty years, the unthinkable came to pass. In 1947, within two years of the second World War coming to an end, Britain overruled Winston Churchill, the greatest Englishman of the age, and with characteristic graciousness conceded the absolute right of self-determination to India**." (Palkhiwala, 1994, 15, 16, 17 and 18)

Historically in retrospect, for India then, it was momentously far-reaching as well as indeed initiating from then-on altogether unprecedentedly new paths of the making of its living-history in the world. India achieved her political independence from the British-rule, and thereby consequent to the above narrated historic fallouts; the multiply geopolitical as well as geostrategic visionary awareness-levels of the then learned political leaders and statesmen had already reached such heights that among themselves, they could set-up the modern India's historic 'Constituent Assembly', under the chairmanship **Dr. Rajendra Prasad**. In the historic deliberations and indeed so much visionary working of this Constituent Assembly and the scholarly brilliance as well as acumen of **Dr. B.R. Ambedkar** as a political-scientist, reached their zenith, in drafting the India's Constitution, within a span of few years. **Members of this historic Constituent Assembly, did consist of the India's who's who and, they all were not only great learned men but, well versed as well as well-read in the English language too and thereby, fruitfully commanding their brilliant grips over the multiple-aspects of Western thoughts and literature. And no wonder that, India did then receive a most valuable gift as the Indian Constitution**. Hence, as the sovereign peoples of India indeed adopted this Constitution as their most valued doctrine for the India's democratic governance, the nation was declared a Democratic Republic as per the tenets of this very Constitution, and the rest is now history. **The historic saga of the freedom struggles of India is also quite a long one and thereby naturally, volumes have been written on them by many scholars for posterity**. And at the same time, the broad details of India's historic struggles for freedom too, are since fairly known to most of the peoples in India and these continue to remain as highly cherished and also inspiring ones, in their memories.

In the above historic perspectives and also contextually we may recall the **historic address since delivered by Dr. Rajendra Prasad, on his swearing-in as the first President of India, on 13 May 1952** and wherein he is magnificently on record to have said that, "I have just taken the oath of office as President and affirmed my determination to dedicate myself to the service of this great country. As President I stand before you as the sign and symbol of the Republic of India. **In our ancient history we read of republics having been established in different parts of our country and at different times, but their sway was limited to small parts of it and we do not fully know the method of their administration. This is the first occasion when the administration of the**

entire country has come under the jurisdiction of one great democratic republic. The structure of this republic of ours, as laid down in our Constitution, is broad-based on all the adult men and women of this country. Over 170 million of them have elected their representatives to carry on the administration and shape the destiny of India. These representatives have elected me as President and in so doing, they have given now the full effect to the Constitution which we made with such labour.... The establishment of this democratic republic could only take place after the attainment of Independence. It is therefore first and most important duty of everyone to preserve and protect this freedom which we have secured after generations of struggle and sacrifice. It is our earnest aim to improve and raise the masses of our people, but all our plans for improvement and uplift of the country depend upon the preservation of our freedom. Our entire life, national and individual, revolves round that basic freedom. It is my duty, as it is yours, to preserve and protect this freedom at all cost... It will be my first and foremost endeavour in performing this duty to treat all our people belonging to the different part of the country, to various classes, creeds and schools of thought, with equality and impartiality. Another duty, which I share with you, will be to seek the friendship of all other countries and to find ways of cooperation with them. My request to all the people of this country is to treat me as one of them and to give me the opportunity and encouragement to serve them to the best of my capacity. I pray that God may give me the strength and wisdom to dedicate myself in the true spirit of service to the fulfillment of my duties and responsibilities." (*Source: Official publication of the 'Speeches of President Rajendra Prasad, delivered from 1952 to 1956', as published in January 1958 by the Publications Division, Government of India, New Delhi.*).

So the magnificently historic contents of India's first President, Dr. Rajendra Prasad, on 13 May 1952, to the free peoples of the newly born independent Democratic Republic of India shows that India had come of age and from then on, a kind of the continuous and multiple-makings of altogether new beginnings, had indeed commenced in India and new dawn had broken, with fully active and democratic participation of vast multitudes of the Indian peoples. But the man's vast and highly complex living-history has the continuous phenomena of its multi-dimensional makings in the world and thereby each dimension therein, is mutually interrelated as well as interconnected in some manner or the other and at the same time, with their complex roots again phenomenally rooted in the past. Hence what all happened so momentously in India in 1947 and onward up to 1952, since as reflected in the above Indian Presidential address too have their roots sprouting earlier in the past, during the British imperial-rule in India. In retrospect, the critically objective reviews of the concerned historic developments and their multiple-aspects; do show that the British with their own interestingly fixated imperialistic dimensions too, had already begun contributing since many decades earlier; towards these subsequently momentous historic developments for the Indian masses and which in turn, were going to totally change India in many ways. Well, in these respects historically at the same time it is also equally true that the British imperialists, both in Britain and the Indian subcontinent, who were calling shots in their respective domains during their empire-days were in no mood in fact then to give straight away a kind of full political independence at the earliest (*but India too then suddenly, was not at all prepared for the same.*), but in spite of their imperialistically selfish 'compulsions'; during the beginning decades of the 20th century itself, they had already begun mooting some innovative ideas regarding 'Home-Rule' in India, so that agitations towards freedom-movement are ironed-out. And all these further led to many unexpected developments; and the British themselves, were overtaken by all kinds of subsequent developments.

No wonder that, the man's living-history has many ways of its continuous making and periodical path-breaking manifestations. In these significantly historic contexts specifically pertaining to India then, **Marquis of Zetland** has written that, "**That the relationship between Great Britain and India is unparalleled in history is undeniable**. How then did it arise? Not certainly, as a result of any premeditated plan—though a tendency has been observable in the controversies of recent times to attribute our presence there today in the prosecution of a deliberate exploitation and aggrandizement, Such ideas are borne out of the heat of controversy and little basis in fact (*how true indeed.*).... **That the bands of merchant adventurers, which sailed round the Cape with Charters granted by Queen Elizabeth in 1600 and by William III in 1698, were inspired by the hope of material gain there is no need to deny. But if commerce had remained the sole motive of the British in India, the history of British India would never have been written, for there would have been no British India about which to write it. It was when the activities of the traders were diverted into other channels that they became the unconscious instruments of a new and wholly unforeseen destiny**. They fought—and conquered—only because in no other way could they suppress the disorders which militated against the successful prosecution of their trade....and by the close of the seventeenth century it was already clear that the Moghul rule had reached and passed its zenith. **The breadth of vision and tolerance of Akbar had given place to the religious bigotry of Aurungzeb**.... Throughout the eighteenth century India was a prey to all those tyrannies and disorders which have been the invariable experience of eastern peoples when thrones have wilted and tottered to the dust... Such then, were the provocations which led to the imposition of the authority over the peninsular.... The fact could be emphasised, for example, that India possesses sixteen modern universities, that in one of them—the Calcutta University—she can claim the largest university in the world. It is not too much to say, indeed, that Sir Michael Sadler and his colleagues on the Commission of 1917 were staggered by the magnitude of the institution with its 58 affiliated colleges clotted over vast geographical areas and its 27,000 students. But the achievements of the British in this direction can only be properly assessed by those acquainted with the conditions under which the system of education has had to be set up....**according to the Census of 1921, 2,500,000 passed the test of literacy in *English*—imbued with culture of the Western world. For in the ranks of this minority are to be found great lawyers and eminent judges, fine scholars and brilliant scientists, capable administrators, talented writers, engineers, doctors, artists, architects and captains of modern industry. The deficiencies of the system may be readily admitted; but Indians themselves should be cautious in their denunciation of it, for it is an outstanding feature in a number of factors which have gone to make the political developments of today possible. Without unifying influence of British administration—a dynamic force of extraordinary power in view of the many and startling diversities of the Indian continent in which I have had occasion to refer—no political Constitution, such as that which is now in sight, would have been conceivable. And with all its drawbacks from an educational point of view, the imposition and wide diffusion of the English tongue has been a unifying influence of remarkable effect.**" (Zetland, 1934-35, pp. 14, 15, 16, 27, 28 & 29)

What the Marquis of Zetland had then referred to as the 'Constitution' pertained to the British India; but at the same time and also historically no less important, indeed remains all the more true in many historic ways the multiple contributions of English-education and thereby, brilliantly discerning exposure of the educated and talented Indians to the Western thoughts subsequently and very significantly, contributed in the historic phenomena of the so fast and momentous making of politically independent India's unique Constitution; which enabled India to transform herself in an unique and peaceful manner, into a sovereign democratic Republic in the today's world. And obviously therefore, in the present contexts of our discussions in hand and therein at this precise point; can any travesty of any nature be detected in the above indeed momentously far-reaching historic perceptions, pertaining to the then British India of the bygone imperial-era? And in their historically momentous continuity, **in these far-reaching multiple-aspects of the above enunciated perceptive thoughts; historic seeds of the Indian Renaissance, as well as the multi-dimensional Indian freedom-movement got sown and sprouted, and thereby in turn the subsequent reincarnation of India into united sovereign democratic Republic too got fructified and thus multiply changing India and her subsequent living-history forever**.

In these specific contexts and in a recent historic continuity, as time passed after the historic 1947 all these fruitfully and in a very integrative manner, did also significantly begin to contribute in transforming India into a fast developing big economy and a modern industrial-cum-military power, among the international comity of nations. Interestingly in the above contexts, the thoughtful perceptions of the **Marquis of Zetland** and **Nani A. Palkhiwala** too as since already quoted earlier in this regard, do contain many significant similarities.

Now in these specific perspectives and in our continuing discussions we may notice a couple of highly significant as well as also, far-reaching history-making developments taking place for the future of India and her millions of peoples. In retrospect, we may also now historically find them to be no small matters. Hence now, while specifically focusing on them in India's precise historic contexts we are able to specify these developments as—the advent and meaningfully rapid fruition of the **'English-education' in India**; and so leading to the **substantial awareness of, and purposefully thought-provoking connectivity with the Western-thoughts** and thereby these in turn, **giving the effectively enduring as well as highly futuristic births to the multiple-thoughts concerning the intellectual awakening towards the aspects of constitutionalism and also, after India's independence the peoples' representatives-based phenomena for the constitution-making too, and the dawn of parliamentary democracy** as well.

We may recall that little earlier in some contexts, we have since noted that in the continuously expanding man's history-making canvas; one thing leads to another and hence many things in their own history-making ways, lead to many unforeseen kinds of developments in the lives of different Nation-States. And interestingly here, we do find them getting substantiated historically in the then British India as the above precisely spelled-out developments together, initially triggered by the British with their typically long-term imperialistic-vision in India and then subsequently in fact led to the aspects of Indian Renaissance, multiple sprouting of the India's freedom-movement and thereby after independence, India fruitfully transforming herself into an united Democratic Republic in today's modern world. Hence, this also undeniably remains a historic reality that in their mutually influencing totality, the above mentioned developments during the first half of 20th century in the British India and, then spilling-over into the initial years of now the united sovereign democratic India and, further in continuity therein; irrespective of the motives involved from time to time, the unprecedentedly history-making multiple contributions of the British then in India, cannot be minimized in their historic perspectives; whatever may remain any one's perceptions on either side of the 'divide' in this regard. It has therefore been rightly said that the man's living-history since continuously in the making, spares none and also favours none whatsoever.

While keeping in our minds, so many initiatives of different kinds and in different contexts, since having historically taken by the British imperialists in India from time to time during their colonial-rule and out of these, we may at this precise point, like to look into now the historic aspects of the **'Home-Rule'** and the then visit of the **'Simon Commission'** to British-India. Historically here, one important point since coming on the surface remains that the **British ideas regarding the aspects of Home-Rule during the very beginning of the 20th century itself, first got sprouted for, exclusively concerning their European neighbour Ireland**. Hence in fact, these ideas subsequently for the British-India were going to be considered much later. We may recall that earlier in their relevant contexts, we have since also discussed some aspects of the highly turmoil-ridden saga of the historic sway of much earlier prolonged aspects of the hegemonic and exploitative aspects of the medieval-era British imperialism. Ireland is not so vast like the Indian subcontinent in far-off Asia and in addition also, she happens to be the Britain's next-door neighbour in Europe, and hence all these too gave the British some distinct strategic advantages. Well, all these in a context; are altogether different matters, and since they also have already been discussed earlier.

The above mentioned ideas concerning the aspects of the historic proposal for the "Home-Rule' in the Ireland then were in fact intensely debated in the British Parliament during the initial decades of the 20th century itself. The historic records of these debates in the Westminster in London then do also show the high quality of these historic parliamentary debates and, in addition also admittedly, very admirable qualitative standards of the British parliamentary democracy, which indeed remains a highly inspiring beacon-light for the other nations in this regard and hence in many respects, much more so now, for the modern sovereign democratic India and her Parliament too. The magnificently unique and highly valuable historic records in this regard, are indeed very vast since it covers almost the 500 years of the continuing British parliamentary democracy. However, **Sir John Simon**'s book titled *Comments and Criticisms*, 1930 (a historic collection of some of his speeches and, edited by **D. Rowland Evans**) on the aspects of the then 'Home-Rule' in the Ireland it is editorially recorded that, "There were two General Elections in 1910. The first of them was fought on **Mr. Lloyd George**'s budget, and the second on the proposal to limit the powers of the House of Lords (*equivalent to the 'Rajya Sabha' in India*) in the manner which was afterwards embodied in the Parliament Act, 1911. The King's speech in February, 1912, announced the intention of the Government to introduce a Home Rule Bill and to the Address **Mr. F.E. Smith** (now Lord Birkenhead) on February 19 moved an Amendment to the effect that it would be improper to proceed with a Home Rule Bill before revising the composition of the Second Chamber.... In reply to him, **Sir John Simon**, who had become Solicitor General in **Mr. Asquith**'s Government in October 1910, delivered the speech reproduced below.... (*And for our relevant purposes at this precise point, only some of the important extracts from this speech are quoted below*): *And accordingly in the pages 85, 97, 98, 100 and 101; Sir John Simon was on record to have said that*.... 'I thought that until the last half-dozen years no constitutional authority that the House of Lords had any right to insist on an appeal to the people. The party opposite is supposed to be in special charge of the Constitution. I should have thought that a Constitution which was representative in its nature, and which depended essentially upon the periodic choice of the constituencies, would have met the case (*interesting, isn't?*).... The only terms upon which the Government can be turned out is if the Irish Members vote for you. I thought that high constitutional doctrine demanded that in these matters the Irish vote did not count. I thought that your full-blooded Unionist was a man who said, 'A Irishman shall have no Assembly in his country where he shall be represented. This is the only place to which he shall come, but when he comes here he does not count.' (*Well, now historically in an* *interesting context here; don't we smell some strong odours of the multiple-facets of since the highly fixated-imperialism, on the both sides of political-divide of the then British Parliament; especially when Britannia was imperially ruling the oceanic-waves, almost all over the world.)*.... We shall bear with great composure the taunts hon. Gentlemen opposite level against us in this regard for two reasons—first, Because it is only one more proof how utterly they misunderstood the real mind and temper of their own countrymen; and, secondly, because we believe that the golden moment which was spoken of by **Mr. Gladstone** twenty years ago has returned, and it will be our consolation and satisfaction, in spite of the taunts of hon. Gentlemen opposite, at length to effect by a tardy measure of justice a real reconciliation between the Irish and the British people." (*Well, quite interesting and geopolitically visionary indeed, and at the same time, some genuine sincerity too is reflected in above quoted words and the intended meaning therein; but in retrospect, we do also find in the above quoted lines from Sir Simon's speech; so apparent kinds of the unchallenged British imperialistic-riders as well, so that the flag of hegemonic supremacy of the British-imperialism over others, could always be maintained high and flying. And hence interestingly, almost identically similar kinds of the British imperialistic designs were waiting to be subsequently offered to the then British India. No wonder that the man's living-history since ages does show that, indeed the multiple-aspects of imperialism have always had many kinds of the differently situations-based facets too.*).

Further regarding the historicity of the fairly known **'Simon Commission', since also known as 'The Indian Statutory Commission'** on page 143 of the above referred book the Editor has mentioned that, "The Government of India Act 1919, provided for the appointment, after an interval of ten years, of a Commission to inquire into the working of the Act and to make recommendations as to any changes in the Constitution of British India.... After the return of the Commission from India and while it was considering its Report, an announcement was made by the Viceroy, **Lord Irwin**, that in the judgment of His Majesty's Government 'It is implicit in the Declaration of 1917 that the natural issue of India's constitutional progress, as there contemplated, is the attainment of Dominion status.' This declaration led to debates in both Houses, and in the Commons the Chairman of the Commission (*i.e. Sir John Simon*) made the following speech (on November 7, 1929)" *and here some important extracts of this speech, from pp. 145, 147, 148 and 149 of the same book; are....* "The Statutory Commission, of course under the terms of an Act of Parliament, has a particular function which nobody seeks, which nobody in authority ever has sought either to enlarge or to diminish. Let me just point out this further, so that there shall be no misunderstanding. When the Commission wrote a little time back its letter making the announcement—which I am very glad to feel has met with universal approval both in Britain and India—that we contemplated in our Report dealing with the extremely difficult question of the relations between the Indian States (*i.e. native princely states*) and British India, and went on to suggest that after we had reported the Government might think it well—as they do—to appoint a conference including representatives from both the States and the British India.... 'The Commission is in no sense an instrument of the Government of India or of the British Government, but it enters upon a duty laid upon it by the King Emperor as a completely independent and unfettered body.'... Every competent person understands that, and that will make it abundantly plain to the House that it was the only proper course for the Commission to take, as we did on September 24th, to explain to the Secretary of State (*i.e. for India*) that, whatever the Government might contemplate, it would be far better if the Commission were not associated with it."

We also find some perceptive comments of Marquis of Zetland on the 'Simon Commission' where he has written that, "But while they have thus shared with their British colleagues control of the administration of India, they have, of course, remained equally with them responsible for their acts not to the Indian peoples (***Naturally under any kind of the hegemonic imperialism, the native peoples of an imperially colonised nation, perhaps just don't count at all.***), but to Parliament in this country....a development which began a good deal later that I have been speaking, but which since the year 1861, has been proceeding side by side with it. I mean the development under which Indians were invited to assist in framing legislation. Under the Councils Act of 1861 a small numbers of members, including one or two Indian public men, who added to the Executive Council of the Governor-General for the purpose of making laws, and the body thus constituted was known as the Legislative council.... Incidentally let me remind you that an additional argument in favour of the creation of a federal assembly, upon which the Commission lay stress, that it open the door to the inclusion of the Indian States, which after all, cover something like two-fifths of the total area of India and embrace not far short of a quarter of the population, in an All-India system. For if provinces differing as widely as the provinces of British India can enter a federal system without jeopardizing their own internal autonomy, why should it not be possible for other self-contained units such as the Native States to do so too? Whether they will do so must, of course, remain a matter for their own decision....**You cannot build-up a working constitution on catch-phrases, however, attractive. Such phrases as 'self-determination', 'provincial-autonomy', dominion status', and so on, give expression to abstract ideas. Such ideas may be admirable as indicating the goal at which you desire to aim; but they are useless if you try to use them as bricks in the erection of the edifice which you desire to build. To pave the way to the goal at which you aim, you must employ not abstract ideas, but concrete and practicable proposals**. It is because I believe that the recommendations contained in the Report are of this character that I regard it as an outstanding contribution towards the solution of the problem with which the peoples of India and the peoples of this country are jointly faced." (Zetland, 1934-35, pp. 33, 34, 46, 48 & 49)

Specifically in the above contexts presently in hand in our on-going discussions, the above quoted comments and also reflectively analytical perceptions of the Marquis of Zetland and for that matter, even the speeches of Sir John Simon in their own distinctly different ways do even now contextually have some valuable messages for the modern united democratic India if at all, presently under the totally changed scenario and therefore, with the desired objectivity and also in their appropriate perspectives; both now the independent Indian peoples and their leaders may care to gainfully look into them and pick-up some relevant ideas, for 'honing' them to improve the quality, tone and tenor of their own parliamentary debates in New Delhi. Are these even now tall-orders for the Indians or, are they not constitutionally indeed, as per the quite legitimate expectations from the Indian political-class? At this precise point now in sovereign democratic India, the Indians may also need to again contextually look in the highly perceptive thoughts of Nani A. Palkhiwala since already quoted earlier and as we may recall from his book titled, 'We the Nation'. In fact, we may also say that these too in a context are the useful segments of indeed, the timelessly valuable as well as the far-reaching relevance of the man's living-history for the man himself. And here comes obviously, some of the **proactively history-making domain of the multiple-aspects of 'essential strategic-culture and thoughts'; in any given scenario and the timeline, for any particular Nation-State**.

It is already a well-known history that since the early decades of the 18th century itself the British had already begun making deep geostrategic and politico-military advances and spreading their imperial tentacles in various ways, into the Indian heartlands. And thereby finally after the momentous Battle of Plassey in 1757, in which the British under the leadership of **Robert Clive** came out victorious over the **Indian Nawab Sirajuddawla,** with his shrewd machinations and even without much fighting; and consequently perhaps the historically destined establishment of the British imperial-rule over the Indian subcontinent, was then fast coming to its multiple-fruition. And exactly after one hundred years, and i.e. in 1857, due to differently triggered rebellious upheaval, India was brought under the direct imperial-rule of the British Crown, as one of its numerous colonies in many parts of the world and thus in the process, India finally became the then historic **'Jewel in the British Imperial Crown'**. And hence at the same time, during the 17th century onwards, we also notice the **advent of unprecedented Industrial-Revolution** in Europe in which again perhaps Britain was in the lead to begin with, and its multiple-fallouts then gave new impetus to the advancement and consolidation of the multiply imperialistic gains of the then leading European seafaring Imperial-Powers and thereby in turn the further making of the man's living-history too, turned out to be highly chequered. **In these historic perspectives, we also find that one thing leads to another and then, many things lead to many other different things and thereby in turn, to the many kinds of consequent developments and events as well**. Historically, the highly eventful 18th-19th centuries and then the first half of the 20th century powerfully belonged to Europe and the USA in many decisive ways. And thereby owing to the fast fructifying Industrial-Revolution the multi-dimensional and also accelerated advancements of the sciences and technologies, triggered many new kinds of developments as well as rivalries among nations in the world and hence, even more intensely in the Europe itself. Consequent to all these we notice that the first half of 20th century brought to the man, two highly devastating World Wars and these in turn, had their own momentously different as well as new kinds of far-reaching fallouts, for the man and his subsequent living-history. We would since be discussing these aspects contextually in more details, in the next forthcoming seventh chapter.

And related to the above, we have since already noted that on 15th August 1947, India got her independence from the British imperial-rule. This and little earlier during the end of Second World War in 1945 itself, atomic weapon was used against Japan; and all these upheavals together led to many new challenges and tragedies too, both for India and Britain in their own respective ways. India was partitioned and a glimpse of the new challenges facing India, have already been reflected in the above quoted first address to nation, since given by the first President of the new Republic of India in the year 1952. **What all have indeed been achieved by India since the 1950s and also, how far she has succeeded in her 'tryst with destiny' in this regard; would be critically delved upon later**. But besides India, many developments and events since having taken place during the decade of 1940s, also triggered many new kinds of tragedies and challenges for the British nation as well. It was a queer situation in many ways for Britain too, as in spite of achieving the final victory over Hitler's Nazi Germany; Britain found herself crippled in many ways. And then in addition, she also had to face the very serious loss of her 'Jewel in the British Imperial Crown' too, which resulted in a great blow to her economy, geopolitical-power and also, to her once the unprecedented imperial glory in the world.

Professor Peter Clarke, a well-known author on British history and, once Professor of the Modern British History at Cambridge has very precisely written that, "Here was the British Empire decked out to celebrate its greatest victory, despite everything. It was a thought that had already become inescapable... At midnight on 14-15 August 1947, a date which Jawaharlal Nehru saluted as marking 'a tryst with destiny', the two new states of India and Pakistan celebrated their independence. It was the end of the Raj, conventionally regarded as 'the jewel in the crown'. The title of Emperor of India was held by King George VI in succession from his great-grandmother,

Queen Victoria, revived from the days of the Mughal Empire to crown Britain's now self-consciously imperialist mission. On 15 August 1947, the title lapsed after seventy years of vice-regal pomp. The last Viceroy departed Delhi (though Lord Mountbatten continued for a time as Governor General of the new Dominion of India). **'As long as we rule in India, we are the greatest power in the world,' one of his most pompous predecessors, Lord Curzon, had once proclaimed. 'If we lose it we shall drop straight to a third rate power,'....** By 1947 such prophecies came home to roost. Though subsequently reinvented simply as the Commonwealth of Nations, the British Empire was now in the hands of the liquidators. Churchill's thousand-year Reich had barely outlasted Hitler's. It all happened so quickly, so quickly... It was less than five years since Churchill's 'end of the beginning'. On that occasion he had stood up in the Mansion House, flushed with the news of the 8th Army's victory at El Alamein, fully determined to make most of it, free for the moment to ignore the tiresome sermons about India that his friend Roosevelt had been preaching in their correspondence. Churchill had thus defied his critics but not defied history... It was just short of three years since the Quebec conference of September 1944. **Roosevelt and Churchill** had met on that occasion, apparently as equals and with a warmth never subsequently recaptured, both anticipating victory with an easy optimism that was quickly overtaken by events... It was two and half years since Yalta. Attending this meeting of the Big Three in February 1945 as British Prime Minister, Churchill sat alongside Stalin and Roosevelt as they made plans to divide and rule the post-war world. Many promises had meanwhile withered on their Crimean vine. Indeed, within weeks of Churchill's departure from office in July 1945, victory turned out to be more costly than even he had imagined." (*In retrospect now, we do indeed find this highly tragic scenario; as quite a nightmare then, for the British nation.*). (Clarke, 2000, pp. xviii & xix)

All these in totality were indeed multiply great blows to the British people and their economic as well as, geopolitical psyche as a Nation-State. Great Britain, once many imperially majestic things to the all kinds of so many peoples across the world in many ways; indeed stood pathetically devastated, in spite of the true splendour of victory in the World War II and that too, being as the most important member of the Allies after the USA. It was an unprecedentedly queer situation for the British nation, in the post-war world. **From the seat of a crowning-glory of being the imperial ruler of indeed a globally intercontinental British-Empire, where the 'Sun once never used to set'; amazingly she got reduced to some kind of a pathetic 'basket-case' for the USA across the Atlantic, and even where she once imperially ruled. But in spite of all these sudden downfalls within few years, she had not become irrecoverable and also, had not lost her traditional as well as multiply well-placed and again always vibrant: the multiple-aspects of 'essential strategic-culture and thoughts'**. Eventually, these aspects within a few decades were destined to turn the corner, for Great Britain in many ways, as substantially recovered and also subsequently a middle-ranking nation being closely linked to her European neighbours in today's geopolitically-cum-economically, the totally changed scenario in the world. These are no mean achievements of the British peoples. But at the same time, immediately after the Second World War in a way the so many seriously daunting crisis-situations, historically indeed also being no mean-affairs; and in which again the British nation was fairly able to successfully sail-through, towards altogether different kinds of the victory and peace.

Hence, specifically pertaining to the extreme seriousness of the above mentioned crisis-years during the last phase of the World War II, and precisely for Britain in a perspective **Professor Peter Clarke** has said that, "Roosevelt's tactical manoeuvring was characteristic. He sought to counter any public perception that he was treating Britain too well not only by murmuring to Morgenthau that the **Lend-Lease** must be cut back, but also by loudly talking up its benefits to the USA. **The opportunity was the publication on 25 November of his own regular Report to Congress on Lend-Lease and a Government White Paper in London**. The story was geared towards highlighting **Reverse Lend-Lease**. BRITISH AID TO ALLIES: Mr. Roosevelt's tribute was *The Times'* headline, 'From the day our first soldiers in 1942,' the President's report told Congress, 'one-third of all the supplies and equipment currently required by United States troops in the British Isles has been provided under reverse Lend-Lease.' The report showed that British mutual aid to all countries totalled 605 billion Pounds, of which the USA received 605 million Pounds, while British mutual aid to the USSR totalled 270 million Pounds. This was a time when the pound sterling was worth $ 4.03, so the value reverse Lend-Lease to the USA had been about $ 2.5 billion, with the clock still running. In fact, although the books obviously did not balance in dollar terms, in proportion to the relative size of the two economies, the contributions were closely comparable... Such perspectives had hardly been common on the other side of the Atlantic. Gallop poll evidence from 1943 provides instructively contrasting answers to the questions: which of the Allies had so far made the biggest contribution towards winning the war? The Russians were named by 50 percent of the British respondents, putting themselves next at 42 percent, and the Americans at 3 percent. In American sample, 55 percent said the United States, 32 percent the Soviet Union and only 9 percent the Great Britain. This is admittedly only a statistical snapshot, taken before D-Day shifted the focus. But it would be naïve to think that rooted attitudes were easily changed. We are virtually doing all the fighting in the Pacific and have made virtually all the advance in France,' the *Chicago Tribune* had claimed on 2 August 1944. 'This is an American-made victory and the peace must be an American peace.' And part of any peace, as over 70 percent of Americans continued to tell the opinion polls, should be repayment of Lend-Lease.... The circulation of the President's report in 1944, much as it cheered the British, was inevitably limited; likewise its impact on American opinion. Nevertheless, the facts that it presented directly challenged the idea that Lend-Lease was simply an unreciprocated drain on American benevolence. It showed that 31 percent of all supplies and equipment used by American forces in Europe up to 30 June 1944 had been supplied by the British, without payment. The report claimed that it would have taken 1000 ships to send these supplies across the Atlantic; instead, the shipping space was used for direct military purposes. The results were exactly in line with the policy's original rationale: that Lend-Lease enabled the most effective prosecution of the war against Hitler, whether by the United States or others; that it brought the invasion of Europe forward by months; that it gave the Allied armies their striking power, which had now taken them to the borders of Germany. And reverse Lend-Lease was its obverse: the face that made it unnecessary to launch a thousand ships." (*These are so true and, historically precise as well. So far as the aspects of Lend-Lease as since mentioned above, and some additional post-war years' loan as advanced by the USA to UK were concerned; all these were subsequently settled by means of financial settlements through their mutual discussions as well as, payments in installments by the UK to USA. We would further be discussing these aspects in some more details in the next chapter,*). (Clarke, 2007, pp. 103 & 104)

Now coming to the historically indeed far-reaching and, path-breaking also towards subsequent making of the Indian living-history during the post-1857 era and onward in continuity, to now the present times and somewhat reassuringly further in the future as well we do find that the **aspects of English education as since brought by the British**, did play substantially multiple-roles towards shaping the future of India; and hence contextually, **the advent of Indian Renaissance** too, remains a part of it. This remains an irrefutable fact of history and hence, especially more so in today's high-tech-cum-super-fast modern-era. **And therefore, the essential aspects of geopolitical**

imperatives as well as, the paramount compulsions of national interests: demand from the Indian peoples and their community-cum-political leaders, irrespective of their individual and sectarian perceptions or leanings; to view and understand these historic realities in their proper perspectives and also, desired strategic positivity and with socio-political prudent consistency; towards India' further well-being and multiple advancements for the centuries to come. These perceptive aspects and expectations now are in no way 'tall-orders' for the vast citizenry of the united and sovereign democratic India, during the present 21st century as well as also, for the further numerous centuries gradually to come one after another in future, naturally over the unending 'canvas of time'; but these are befittingly indeed, as per the India's magnificent civilisational legacy and its enchanting richness; since spanning over thousands of years and in addition also, these still being so vibrant and futuristically enduring and hence, so much multiply reassuring too. Man's living-history since always in the making in this world and so at the same time, it demands from man to take the numerous historic developments and events and their many kinds of fallouts too regardless to accept and critically understand them in their desired contextual perspectives, in ones' paramount interests, during the present times and for the future as well. In these precise contexts presently in hand these are the crux of the multiply visionary prudence and hence, do also remain part of the **multiple-aspects of 'essential strategic-culture and thoughts'**; about which we have since discussed about a number of times earlier in different perspectives, to substantiate their indeed timeless and paramount significance in this world.

India and her multitude of populace have unfortunately been rather consistently found wanting in the paramount domain of evolving, cultivating, developing and imbibing the multiple-aspects of 'essential strategic-culture and thoughts' and hence consequently these have indeed been a great and tragic negation too of an otherwise ancient and magnificent Indian civilisation.

In the above multiple contexts, **Jawaharlal Nehru** has very thoughtfully written that, "Our study of history has shown us that life is often very cruel and callous, To get excited over it, or merely to blame people is foolish, and does not help. It is much more sensible to try to understand the causes of poverty and misery and exploitation, and then try to remove them. If we fail to do so, and fall back in the march of events, we are bound to suffer. India fell back in this way. She became a bit of a fossil; her society was crystallized in old tradition, her social system lost its energy and life and began to stagnate. It is not surprising that India suffered. The British happened to be the great agents to make her suffer. If they had not been there, perhaps some other people might have acted in the same way... One great benefit the English did confer on India. The very impact of her new and vigorous life shook up India and brought about a feeling of political unity and nationality. Perhaps such a shock, painful as it was, was needed to rejuvenate our ancient country and people. English education, intended to produce clerks, also put Indian in touch with current Western thought. A new class began to arise, the English-educated class, small in numbers but cut off from the masses, but still destined to take the lead in the in the new nationalist movements. (*This too was indeed, another manifestation of the Indian renaissance.*). This class, at first, was full of admiration of England and the English ideas of liberty. Just then people in England were talking a great deal about liberty and democracy. All this was rather vague, and in India England was ruling despotically for her benefit. But it was hoped, rather optimistically, that England would confer freedom on India at the right time (*quite a wishful thinking indeed.*). The impact of Western ideas on India had its effect.... The English-educated class grew slowly in the cities, and at the same time a new middle class arose consisting of professional people—that is, lawyers and doctors and the like, and merchants and traders. There had been of course, a middle class in the past, but this was largely crushed by the early British policy. The new *bourgeoisie* or middle class was a direct outcome of British rule; in a sense they were the hangers-on of this rule." (*Well, it is natural that different authors have given their different kinds of perceptive interpretations not only of these historic developments and events; but almost on all other subject-matters, in their writings from time to time*). (Nehru, 1934, pp. 505, 506, 507 & 508)

In these historic contexts, **two things did play decisive roles in their own distinct ways: one was of course the multiple impacts and also historic fallouts of English-education in the British India and; the second one being the multiply glorious and the unique magnificence of since the still cherished and very much surviving legacies of the multiple-aspects of the very ancient Indian civilisation**. Hence, in the British India then, in the last 18th, 19th and during the first decades of the 20th century, these great imperialistic games at times even with somewhat positive intentions, which the imperial British then had launched principally for their own purposes, **the ready socio-cultural as well as socio-political 'canvas' needed for historic manifestations of the Indian renaissance,** though being quite age-old and ancient, but it was very much there and breathing and thereby, this ancient 'canvas' since being ready and hence, needed to be brought back to life for use in this regard. And needless to say that's what precisely had happened then, during those momentous times in India and which the British who were the harbingers of English education in India too, had never intended these subsequently fast developments and hence never had imagined that to begin with, they would soon themselves be overtaken by these events. **In fact in India then, these highly momentous and the far-reaching historic developments had brought awakening among the peoples and thereby connectivity with the contemporary Western thoughts, became instrumental for the manifestations of Indian renaissance, which then began yearning for reforms and these in turn, triggered many kinds of socio-political revolutions too.** Thus in the end, culminating into the Indian political independence in 1947, wherein the crippling blows of the Nazi Germany in Europe, and of Japan in the Pacific to the British imperial-might during the Second World War were also decisive too, towards the independence of India from the British colonial-rule. Contextually this is in fact, a classic case of the enormous complexities as well as, the interrelated-cum-mutually-influencing multi-dimensional making of the man's living-history, in certain specific geographical territories and also, as per the certain timeline specific far-reaching emerging historic situations over there.

Specifically in the above multiple contexts **Professor S.R. Sharma** with his precise thoughts, has written that, "The Seven Years' War (*between England and France*) definitely marked the ascendency of England. **Though England lost the American colonies (U.S.A.) after this, she was more than compensated for that loss by her acquisition of India**. The work begun at Arcot, Plassey, Buxar, in the days of Clive in the eighteenth century was completed in the nineteenth by Wellesley and Dalhousie. The final overthrow of the Marathas (1818) who had succeeded the sovereignty of the Mughals was not less significant than the overthrow of Napoleon (1815) only three years earlier: both marked a new era—one in India and the other in Europe.... Here we must not lose sight of happenings in England and Europe at the same time. It was an epoch of reforms and revolutions, economic, political and social. Both nationalism and democracy (the two great moulding forces of nineteenth century Europe) derived a new impulse and significance from the Industrial Revolution. The economic changes in agriculture, industry, and commerce—confirmed and extended the scope of democracy as well as nationalism... **But if imperialism invaded India, neither could the wave of liberalism be damned with the countries of its origin. Steamships,**

railways and telegraphs were not calculated to keep the world divided into oases and deserts. British Liberalism was bound to leaven the conquests of British conservatism. This was the significance of the simultaneous extinction in India of Indian feudalism and the English East India Company's rule. (*Well agreed, but the moot point remains that feudalism has not at all vanished from India; and sadly its numerous kinds of still lingering tentacles, are even now both socio-culturally as well as socio-psychologically, getting manifested and also keep surfacing in many ways in India.*).... We can touch only on a few phases of the **Indian awakening in the nineteenth century. It was significant that the year of the Great Rising also witnessed the foundation of the three modern Universities of Bombay, Madras and Calcutta. But it would be a mistake to suppose that the Indian Renaissance has been the product of Western education alone. It has been the resultant of several forces acting at the same time. The Rising of 1857 was more a social revolt than a mere mutiny of the army or even a political rebellion. Its repression was necessary not only for the security of British rule, but also for the creation of a New India. It was an event of epoch-making for India and Asia as the fall of the Bastille was for France and Europe. The Rousseau of the Indian Revolution was Raja Ram Mohan Roy (1774-1833), the founder of the Brahmo Samaj.** He was followed by an army of great reformers like **Devendra Nath Tagore (Rabindranath's father)** and **Keshab Chandra Sen** in Bengal, and **Swami Dayananda Saraswati** (1824-83), the founder of the Arya Samaj in the Punjab and **Mahadev Govind Ranade** (1842-1900), the founder of the Prarthna Samaj in Maharashtra. Much useful work in the national uplift was also done by **Swami Vivekananda**, the apostle of a reformed faith, who carried the message of Awakened India to Europe and America (1895-97). Similar work was also done by **Sir Saiyyad Ahmad Khan** (1817-07) to put new life into the paralysed Muslim community. He founded the Muhammadan Anglo-Oriental College (1875) which later developed into the Aligarh Muslim University.... Under such auspices was born the Indian National Congress in 1885, the one organ that in course of time was to be the champion of Renaissant India—the instrument of the Indian *Risorgimento*. Its aims were enunciated as follows:

> 'The objects of the **Indian National Congress** are the attainment by the people of a system of government similar to that enjoyed by the self-governing members of the British Empire, and a participation by them in the rights and responsibilities of the Empire on equal terms with those members. These objects are to be achieved by constitutional means by bringing about a steady reform of the existing system of administration and by promoting national unity, fostering public spirit, and developing and organising the intellectual, moral, economic and industrial resources of the country." (Sharma, 1948, pp. 395-398)

In retrospect we do find that, the above mentioned original aims and objectives in fact, of the Indian National Congress since established in 1885 when the British hegemonic imperialism was at its zenith in the subcontinent, were quite craftily worded document; geopolitically with very shrewd intentions to ensure the more smooth and continued survival of the British imperial-rule in the whole of India of those times, by even rather somewhat encouragingly allowing the English educated native Indians to actively participate in the administration on a selective basis and, being always in line with their larger geostrategic and politico-military thought-processes, for the maintenance of British Empire. **Well in any case historically a beginning was made in India then, by the British imperialists, with an intention to involve the native Indians as well. And it is a different matter that in course of time, the same Congress grew more and more uncompromisingly nationalistic in both thoughts and actions and thereby, getting transformed as a great challenge to the British imperial-rule in India.** In fact, instead of controlling and piloting the subsequent events and developments, interestingly the British were overtaken by their increasing momentum then, and the rest is history. It was historically, a queer situation for both the British as well as Indians and hence both were being swept away in overwhelming manners, by the then fast waves of the multiple-making of living-history itself in India. **It is also equally true that the present-day Congress Political Party of 21st century independent India ironically, has now nothing left in common with the original Congress of those times**. During all these prolonged years meanwhile, lot of water have already flown down the numerous rivers in Indian peninsula especially after 1947, and hence, there have been many deeds and misdeeds on the part of Congressmen of various shades and characters, which have since ensured this pathetic decline.

So far as the **multiple-aspects of Indian Renaissance** as being discussed now, are concerned; this had even developed or manifested itself into many kinds of socio-religious-cum-spiritual dimensions as well, on the larger civilisational canvas of India. Their many kinds of the historic as well as even highly illuminating glimpses may be found in the writings and discourses of **Swami Vivekananda, Swami Dayananda Saraswati and other similarly illustrious personalities,** then appearing and also momentously making-history on to the Indian scene of those times.

Specifically in the above multiple-contexts **Sri Aurobindo**, who made to the ICS but did not care to even join, stands as a class of his own superb intellect, scholarship, spiritual attainment through marvels of Yoga, and his proactive nationalistic-cum-spiritual thoughts and writings. On the many aspects of renaissance and spiritualism his booklet has mentioned that, "the example of **Mahadev Govind Ranade** presents itself in my mind as the very type of this peculiar action so necessary to a period of large and complex formation. If a foreigner were to ask us what this Mahratta economist, reformer, patriot precisely did that we give him so high a place in our memory we should find it a little difficult to answer. We should have to point to those activities of a mass of men in which his soul and thoughts were present as a formless former of things to the great figures of present-day Indian life who received the breath of his life.... Very different was the manner of working Dayananda. Here was one who did not infuse himself informally into the indeterminate soul of things, but stamped his figure indelibly as in bronze on men and things. Here was one whose formal works are the very children of his of his spiritual body, children fair and robust and full of vitality, the image of their creator. Here was one who knew definitely and clearly the work he was sent to do, choose his materials determines his condition, with a sovereign clairvoyance of the spirit and executed his conception with the puissant mastery of the born worker.... Dayananda accepted the Veda as his rock of firm foundation he took it for his guiding view of life, his rule of inner existence and his inspiration for external work, but he regarded as even more, the word of Eternal Truth on which man's knowledge of God and his relations with the Divine Being and with his fellows can be rightly and securely founded.... Neither Mr. Tilak nor his speeches really require any presentation or foreword. His speeches are, like the featureless Brahman self-luminous. Straightforward, lucid, never turning aside from the point which they mean to hammer in or wrapping it up in ornamental verbiage, they read like a series of self-evident proposition.... The indomitable will and that unwavering devotion have been the whole meaning of Mr. Tilak's life; they are the reasons of his immense hold on the people. For he does not owe his pre-eminent to any of the causes which have usually made for political leading in India, wealth and great social position, professional success, recognition of Government, a power of fervid oratory or fluent speech; for he had none of these things to help him. He owes it to himself alone and to the things his life has meant and because he has meant it with his whole mind and his whole soul. He has kept back nothing for himself for other aims, but he has given all to his country.... Mr. Tilak was the first political leader to break through the routine of its somewhat academic methods, to bridge the gulf between the present and the past and to restore continuity to the political life of the nation. He developed a language and a spirit and he used methods which Indianised the movement and brought into it the masses.... There are many who, lamenting the bygone glories of this great and ancient nation, speak as if the Rishis of old, the inspired creators of thought and civilisation, were a miracle of our heroic age, not to be repeated among degenerate men and in our distressful present. This is error and thrice an error. Ours is the eternal land, the eternal people, the eternal religion, whose strength, greatness, holiness may be overclouded but never, for a moment, and utterly cease.... The Rishi is different from saint. His life may not have been distinguished by superior holiness or his character by an ideal beauty. He is not great by what he was himself but by what he has expressed. A great and vivifying message had to be given to a nation or to humanity, and God has chosen this mouth on which to shape the words of the message. A momentous vision had to be revealed; and it is his eyes which the Almighty first unseals, The message which he has received, the vision which has been vouchsafed on him, he declares to the world that all the strength that is in him, and in one supreme moment of inspiration expresses it in words which merely to be uttered to stir men's inmost natures, clarify their minds, seize their hearts and impel them to things which would have been impossible to them in their ordinary moments. Those words are the mantra which he was born to reveal and of mantra he is the seer....Yet it is the Bankim of these later works and not the Bankim of the great creative masterpieces who will rank among the Makers of modern India. The earlier Bankim was only a poet stylist—the later Bankim was a seer and nation-builder.... As he had divined the linguistic need of his country's future, so he divined also its political needs. He, first of our great publicists, understood the hollowness and inutility of the method of political agitation which prevailed in his time and exposed it with merciless satire....**he had a positive vision of what was needed for the salvation of the country. He saw that the force from above must be met by a mightier reacting force from below—the strength of repression by an insurgent national strength**." (Aurobindo, 2001, pp. 10, 11, 17, 31, 32, 35, 53 & 54)

And while delving little deeper into these thoughtfully very precise comments of **Sri Aurobindo** on **Sri Bankim Chandra**, we do learn that the historic literary writings as well as the soul stirring songs of Bankim, were indeed able to momentously transform the multiple aspects of the then fast manifesting and also enlarging—**the Indian Renaissance**, into the far-reaching and politically pro-active-cum-consistent struggles for the India's independence from the British colonial-rule. From then on in many ways, **these Indian Renaissance-based relentless struggles for freedom** historically began gathering more and more intensity and, also expanding momentum as well. Along with many other and intellectually highly accomplished personalities preceding him, **Gandhiji** with his so many distinguished, educated, and dedicated followers; finally came onto the then so much ferment-soaked Indian political scene, to indeed historically provide a much needed non-violent leadership, for the final attainment of India's political freedom from the British rule. **But all of them too, were no less educated and influenced by the marvels of contemporary Western thoughts as well as, the socio-political brilliance of the British concepts of human-values, democracy and the multiple-aspects of their constitutionalism. And in these historically momentous dimensions the English-education since initiated by the imperial British with largely an altogether different imperial motive, had indeed played multiply far-reaching roles.** Hence while objectively deliberating upon them the contextual need remains that these momentous and historic aspects may always be viewed in their desired and fair perspectives, totally devoid of any subjective emotionalism.

While objectively viewing these overwhelming historic developments and events then taking place in India, as serious challenges to the British colonial-rule we also historically need to accept in the larger perspectives as well as, with the requisite fair-mindedness in this regard, that the British in India then, were some kinds of the 'sophisticated' imperialists in many ways. While it is indeed historically true that the British in India then, were principally interested to serve their unique British-Empire but at the same time while doing so directly and indirectly, they did provide many new kinds of valuable assets as well as inputs for the subsequent future of India then, and even after attaining independence from the British imperial-rule. We may recall at this precise point that, a historic glimpse of the same has already been given in the earlier quoted paragraphs from a book of **Nani A. Palkhiwala**, and in addition at the same time in the various earlier contextually quoted paragraphs, from the books of few British authors as well. But historically it remains always true and, since also deliberated upon by us earlier in different contexts, and also in the earlier chapters as well and thereby, highlighting the facts that since ancient of times, the multiple-aspects of the imperialism and colonial-empires too, always have had their overwhelmingly fixated and exploitative imperial vested-interests.

Hence, so far as the aspects of securing ones' freedom from the tentacles of the alien imperialism since powerfully imposed, have never been any easy matters, for the subjugated peoples in their own lands. But if the concerned imperial-power, though being very powerful but, adequately civilised, constitutionalism-oriented and at the same time, geopolitically even somewhat more sophisticated then the goings may not be so extremely suffering. Well, it is understandable here that these perceptions may be termed as the mere matters of conjecture and hence debatable; but while somewhat critically reviewing all the much more ruthless imperialism of other imperial powers in different parts of the world—such as that of Japan, Tsarist Russia, and the later period USSR and, even certain other European imperial-powers and so on and then, we may opine that India, which had to be colonised in any case then, by any of the rival European seafaring imperial-powers; she was indeed lucky and remained relatively better-off, to have perforce landed under the British imperial-rule. **But it is also true and since also substantiated by the man's living-history that, no imperial-power ever voluntarily relinquish its hold over the imperial-colonies in alien lands and hence it would have been too much to expect then from the imperial Britain, to voluntarily grant freedom just on India's asking.** But historically it was indeed a matter of great credit to both the U.K. as well as for India that the subsequent parting itself became quite nostalgic and in many ways, heart-warming as well with great possibilities and geopolitical promises too, for the close bilateral friendship and mutually beneficial permanent relationship for future. So far as the historic aspects of the Indian political freedom are concerned, India had to relentlessly struggle for it in many ways and then rather in a way, to deservingly secure it for her peoples.

In the above precise contexts, India's one of the well-known Journalists, and author, **Shri M.J. Akbar,** while delivering the tenth **Major General Samir Sinha Memorial Lecture** on 21 February 2012, in the USI of India, New Delhi, very perceptively and indeed interestingly with great historic perspectives, had said that, "**Britain created an international empire after it became a major force in India with victory in the battles of Plassey (1757) and Buxar (1765). The later was more significant, and I cannot quite comprehend why its victor, Major Munro, is not given at least as much credit as Clive succeeded in winning the battle for space in history books because he was corrupt enough to buy a place in the House of Lords**... After 1765, the power of Indian wealth and human resource, as much the steel of Sepoy Army, helped the British to conquer the substantial